T0344880

POLARIMETRIC SCATTERING AND SAR INFORMATION RETRIEVAL

POLARIMETRIC SCATTERING AND SAR INFORMATION RETRIEVAL

Ya-Qiu Jin
Fudan University, P. R. China

Feng Xu
Intelligent Automation Inc., USA

This edition first published 2013
© 2013 John Wiley & Sons Singapore Pte.Ltd.

Registered office
John Wiley & Sons Singapore Pte. Ltd., 1 Fusionopolis Walk, #07-01 Solaris South Tower, Singapore 138628

For details of our global editorial offices, for customer services and for information about how to apply for permission to reuse the copyright material in this book please see our website at www.wiley.com.

Library of Congress Cataloging-in-Publication Data

Jin, Ya-Qiu.
  Polarimetric scattering and SAR information retrieval / Ya-Qiu Jin and Feng Xu.
      pages cm.
  Includes bibliographical references and index.
  ISBN 978-1-118-18813-2 (cloth)
  1. Synthetic aperture radar. 2. Electromagnetic waves–Scattering. I. Title.
  TK6592.S95J56 2013
  621.3848′5–dc23   2012038931

ISBN: 978-1-118-18813-2

Set in 10/12 pt Times by Thomson Digital, Noida, India
Printed and bound in Singapore by Markono Print Media Pte Ltd

Contents

Preface

Polarimetric imagery of synthetic aperture radar (POL-SAR) technology has been one of the most important advances in spaceborne microwave remote sensing during recent decades. Remote sensing data and images have resulted from complex interactions of electromagnetic (EM) waves with natural media, that is, terrain surface, atmosphere, ocean, and so on. The great amount of remote sensing data/images presents rich sources for quantitative description and monitoring of our Earth's environments. However, information retrieval from POL-SAR remote sensing images requires a full understanding of the polarimetric scattering mechanism and SAR imaging principles. The study of EM wave scattering for remote sensing applications, especially for POL-SAR technology, has become an active and interdisciplinary area.

This book presents an account of some of the progress made in research on both the theoretical and numerical aspects of POL-SAR information retrieval.

We begin in Chapter 1 with some basics of polarimetric scattering, for example, the concepts of the scattering matrix, the polarization vector, the four Stokes parameters, and the Mueller matrix solutions for volumetric and surface scattering.

In Chapter 2, the vector radiative transfer (VRT) theory of natural media is presented. The VRT takes account of multiple scattering, emission, and propagation of random scattering media. Its analytical solution provides insights into the underlying mechanism of EM wave–terrain surface interaction. The components of VRT, that is, the scattering, absorption, and extinction coefficients, and the phase matrix of non-spherical scatterers, are introduced. The first-order Mueller matrix solution of VRT for a vegetation canopy model is derived. Polarization indices, eigen-analysis, and entropy are presented. The statistics of multi-look SAR images and the covariance matrix are also investigated.

Chapter 3 first introduces some fundamentals for POL-SAR technology. Based on POL-SAR imaging technology, a new approach to SAR imaging simulation for comprehensive scenarios is developed. It is called MPA (mapping and projection algorithm). In the MPA, scattering contributions from scatterers are cumulatively summed on the slant range plane (the mapping plane), while their extinction and shadowing effects are cumulatively multiplied on the ground range plane (the projection plane). Some parameterized models of random non-spherical scattering media, for example, buildings, the vegetation canopy, rough and tilted surfaces, and so on, are incorporated into the MPA simulation scheme. It finally constructs simulated POL-SAR images of a comprehensive terrain scene. The MPA presents a complete SAR raw data simulation chain, based on scattering models of natural and artificial media and a time-domain raw-data generator.

Chapter 4 deals with the simulation of bistatic POL-SAR images as well as bistatic polarimetry and signal processing. Bistatic SAR (BISAR) with separated transmitter and receiver flying on different platforms has become of great interest. Most BISAR research is focused on hardware engineering realization and signal processing algorithms, with a little on landbased bistatic experiments or theoretical modeling of bistatic scattering. The BI-MPA is developed as a fast and efficient tool for monostatic and bistatic imaging simulations. It involves the physical scattering process of multiple terrain objects. Based on simulated BISAR images, bistatic polarimetry and the unified bistatic polarization bases are discussed. Finally, raw data processing of stripmap BISAR is explored.

Chapter 5 first presents some basic concepts of radar polarimetry and target decomposition, and then develops deorientation theory for surface classification of POL-SAR images. The novel "deorientation" concept is introduced to reduce the influence of randomly fluctuating orientation and reveal the generic characteristics of the targets for better classification. The parameterized model of a random non-spherical scattering medium and its Mueller matrix solution for deorientation validation are developed. A new set of parameters of polarimetric scattering targets is defined to describe target orientation, orderly or random, and scattering surface classification.

Chapter 6 presents some inversion applications using POL-SAR images. It can be found that, as the surface is tilted, the orientation angle ψ at the maximum of co-pol or cross-pol signature can shift from $\psi = 0$. This shift can be applied to convert the surface slopes. Making use of the Mueller matrix solution and morphological processing, the digital elevation mapping of the Earth's surface is inverted from single-pass polarimetric SAR image data. Using the deorientation theory of Chapter 5, the imaging positions of first-, second-, and third-order scattering of a bridge body can be located. Using the thinning, clustering algorithm, and Hough transform, the lines indicating different orders of scattering are used to invert the bridge height.

Chapter 7 deals with automatic reconstruction of building objects from multi-aspect SAR images. This approach is called as AMAR (automatic multi-aspect reconstruction). Reconstruction of three-dimensional objects from SAR images has become a key issue for target information retrieval. It has been noted that scattering from man-made objects produces bright spots at submeter resolution, or presents strip/block images at meter resolution. The linear profile of a building object is regarded as the most prominent characteristic. The POL-CFAR detector, Hough transform, probabilistic description of the detected façade images, and the maximum-likelihood estimation of building objects from multi-aspect observed façade images are discussed. Eventually, in association with a hybrid priority rule of inversion reliability, an automatic algorithm is designed to match multi-aspect façade images and reconstruct the building objects.

Chapter 8 discusses the Faraday rotation (FR) on a POL-SAR image at UHF/VHF bands. As the polarized electromagnetic wave propagates through the Earth's ionosphere, the polarization vectors are rotated as a result of the anisotropic ionosphere and the action of the geomagnetic field. This rotation is called the FR effect. The FR may decrease the difference between co-polarized backscattering, enhance cross-polarized echoes, and mix different polarized terms. Since the FR is proportional to the square power of the wavelength, it has an especially serious impact on SAR observations operating at a frequency lower than L band. For example, the FR angle at P band can reach dozens of degrees. In this chapter, the Mueller matrix with FR is derived, and recovered from FR. The algorithm of phase unwrapping is also studied.

Chapter 9 presents an algorithm for change detection. Multi-temporal SAR observations provide fast and practicable technical means for surveying and assessing terrain surface changes. How to detect and automatically analyze information on change in terrain surfaces is a key issue in remote sensing. In this chapter, two-threshold expectation maximum (EM) and Markov random field (MRF) algorithms (2EM-MRF) are developed to detect the change direction of backscattering enhanced, reduced, and unchanged regimes from the SAR difference image. As examples, this approach is applied to change detection of urban areas and earthquake regions.

Chapter 10 presents the temporal Mueller matrix solution for polarimetric scattering and radiative transfer as a pulse wave incident upon an inhomogeneous random medium. The Mueller matrix solution for seven scattering mechanisms describes temporal polarimetric scattering and propagation through inhomogeneous scattering media, and, as examples, is applied to image simulation of pulse echoes from the lunar regolith layer, and for monitoring debris flows and landslides.

Chapter 11 introduces a novel numerical approach, that is, the bidirectional analytic ray tracing (BART) method. The basic idea is to launch ray tubes from both the source and observation, trace among the facets, and then record the illumination areas of rays shooting on facets and edges. For each pair of forward and backward rays illuminating the same facet/edge, a scattering path from source to observation is constructed by linking the forward and backward rays, and the corresponding scattering contributions are added up to give the total scattering. This fast computation is applied to the simulation of composite scattering from an electrically large target over a randomly rough surface.

Finally, Chapter 12 deals with a numerical train of simulated polarized scattering–imaging–reconstruction for a complex-shaped electrically large object above a dielectric rough surface. Target detection and reconstruction have been of great interest in both civilian and defense applications. The step-frequency (SF) radar technique is employed to 3D SAR imaging using the downward-looking spotlight mode to alleviate the sheltering and layover effect. The BART of Chapter 11 is employed to calculate the scattering matrix of a 3D target over a rough surface. Some simulation examples of 3D imaging and reconstruction of perfectly electrically conducting targets, that is, a square frustum, and a tank-like target over a rough surface, as well as discussions on the accuracy and computational efficiency, are presented.

This book is based on research progress in our laboratories over the past decade. It is assumed that readers are familiar with material normally covered in undergraduate courses on electromagnetics and some basis of radar remote sensing. However, in each chapter, some fundamentals of these topics are given whenever needed.

This book is intended as a textbook for a graduate course on advanced topics in electromagnetic scattering and SAR remote sensing, and is also a useful reference for researchers working on electromagnetic scattering and POL-SAR microwave remote sensing with various applications. New advances on POL-SAR, such as POL-INSAR, tomography SAR, inverse SAR, counter-SAR, and so on, remain for further study.

This work was partly supported by several projects of the National Science Foundation of China.

Ya-Qiu Jin and Feng Xu
Shanghai and Rockville
October 2012

1

Basics of Polarimetric Scattering

1.1 Polarized Electromagnetic Wave

1.1.1 Jones Vector and Scattering Matrix

Radar transmits an electromagnetic (EM) wave and receives the scattered wave from the target. The polarization of an EM wave is defined as the time–space variation of the electric field vector in the plane perpendicular to its propagation direction. As depicted in Figure 1.1, let us first define a coordinate system $(\hat{v}, \hat{h}, \hat{k})$, where $\hat{k}$ denotes the propagation direction, and $\hat{v}, \hat{h}$ denote the directions of vertical and horizontal linear polarizations, respectively. (In this book, unless otherwise specified, vectors are denoted as **bold** letters, for example, **E**, and matrices are **bold** letters with a bar, for example, $\bar{\mathbf{S}}$. A letter with a hat (e.g., $\hat{n}$) denotes a unit vector.) A completely polarized monochromatic EM wave can be represented by two orthogonal components along $\hat{v}, \hat{h}$ (Kong, 2005)

$$
\begin{aligned}
\mathbf{E}(z,t) &= E_{i0}\hat{e}_i e^{ik_i \cdot r} = \hat{v}E_v + \hat{h}E_h \\
E_v &= E_{0v}\cos(kz - \omega t + \varphi_v), \quad E_h = E_{0h}\cos(kz - \omega t + \varphi_h)
\end{aligned}
\tag{1.1}
$$

where E_v, E_h are the corresponding vertical and horizontal components of the electric field, k is the wavenumber and ω is the angular frequency. In addition, we have the following relationships, which can be readily derived:

$$
\begin{aligned}
\mathbf{k} &= k(\hat{x}\sin\theta\cos\phi + \hat{y}\sin\theta\sin\phi + \hat{z}\cos\phi) = k\hat{k} \\
\hat{h} &= \frac{(\hat{z}\times\hat{k})}{|\hat{z}\times\hat{k}|} = -\hat{x}\sin\phi + \hat{y}\cos\phi, \\
\hat{v} &= \hat{h}\times\hat{k} = \hat{x}\cos\theta\cos\phi + \hat{y}\cos\theta\sin\phi - \hat{z}\sin\phi
\end{aligned}
\tag{1.2}
$$

In spherical coordinates, where the source is located at the origin, we have $\hat{v} = \hat{\theta}$, $\hat{h} = \hat{\phi}$, $\hat{k} = \hat{r}$.

Imagine when an EM wave passes a certain position in space; the tip of the electric field vector will form a certain trajectory. The shape of such a trajectory is determined by the

Polarimetric Scattering and SAR Information Retrieval, First Edition. Ya-Qiu Jin and Feng Xu.
© 2013 John Wiley & Sons Singapore Pte. Ltd. Published 2013 by John Wiley & Sons Singapore Pte. Ltd.

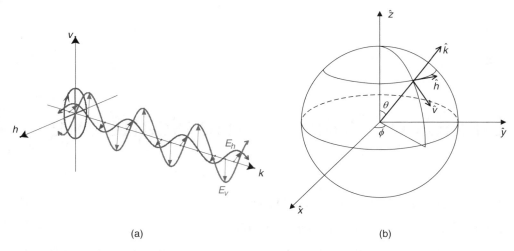

(a)   (b)

Figure 1.1 (a) Polarized EM wave and polarization vector. (b) Coordinate system of the polarization vector

amplitude ratio and phase difference between the two orthogonal components, that is, the polarization ratio,

$$|\rho| = E_{0h}/E_{0v}, \quad \varphi = \varphi_h - \varphi_v \tag{1.3}$$

The polarization of an EM wave is named after the shape of such a trajectory, that is, linear polarization (when $\varphi = m\pi$, $m = 0, 1, 2, \ldots$), circular polarization (when $\varphi = m\pi + \pi/2$, $m = 0, 1, 2, \ldots$, and $E_{0v} = E_{0h}$) or elliptical polarization (in general). The trajectories of

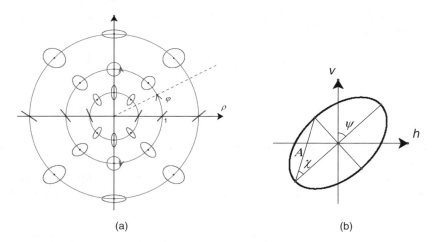

(a)   (b)

Figure 1.2 (a) Plane of polarization ratio. (b) Parameters of elliptical polarization

different polarizations are depicted accordingly in the $(|\rho|, \varphi)$ plane as in Figure 1.2a. In fact, any polarization trajectory can be expressed using the following ellipse equation:

$$\left(\frac{E_v}{E_{0v}}\right)^2 - 2\frac{E_v E_h}{E_{0v} E_{0h}}\cos\varphi + \left(\frac{E_h}{E_{0h}}\right)^2 = \sin\varphi \tag{1.4}$$

There are three geometric parameters of such an elliptical trajectory: amplitude A, elliptical angle χ, and orientation ψ, as illustrated in Figure 1.2b, that is

$$A = \sqrt{E_{0v}^2 + E_{0h}^2},$$

$$|\sin 2\chi| = 2\frac{E_{0v} E_{0h}}{E_{0v}^2 + E_{0h}^2}|\sin\varphi|, \quad \tan 2\psi = 2\frac{E_{0v} E_{0h}}{E_{0v}^2 - E_{0h}^2}\cos\varphi \tag{1.5}$$

where the sign of χ indicates the rotation direction, that is, counter-clockwise $(+)$ or clockwise $(-)$, the elliptical angle χ determines the shape, and ψ indicates orientation.

We can write Equation 1.1 in the form of a complex vector,

$$\mathbf{E} = \begin{bmatrix} E_v \\ E_h \end{bmatrix} = \begin{bmatrix} E_{0v} e^{i\varphi_v} \\ E_{0h} e^{i\varphi_h} \end{bmatrix} \tag{1.6}$$

which is often referred to as the Jones vector. It can be found from Equation 1.3 that the polarization ratio is in fact the ratio of these two elements

$$\rho = |\rho| e^{i\varphi} = \frac{E_{0h} e^{i\varphi_h}}{E_{0v} e^{i\varphi_v}} \tag{1.7}$$

The Jones vector can also be expressed using the elliptical parameters,

$$\mathbf{E} = A e^{i\varphi_0} \begin{bmatrix} \cos\psi & -\sin\psi \\ -\sin\psi & \cos\psi \end{bmatrix} \begin{bmatrix} \cos\chi \\ i\sin\chi \end{bmatrix} \tag{1.8}$$

where φ_0 is the absolute phase. It can be seen from Equation 1.8 that transformation of the Jones vector would correspondingly modify polarization characteristics. This is important in radar polarimetry, as will be discussed later in Chapter 5.

A completely polarized wave incident upon a point target or deterministic simple target would produce a completely polarized scattered wave, which is often called coherent scattering. As a polarized wave $\mathbf{E}_i = \hat{e}_i E_{i0} e^{i\mathbf{k}_i \cdot \mathbf{r}}$ incident on a scatterer of volume v_0 and dielectric constant $\varepsilon_{sn} (= \varepsilon_s/\varepsilon_0)$, the wave equation can be written as (Jin, 1994)

$$\nabla \times \nabla \times \mathbf{E}(\mathbf{r}) + k^2 \mathbf{E}(\mathbf{r}) = \begin{cases} -i\omega\varepsilon_0(\varepsilon_{sn} - 1), & \mathbf{r} \in v_0 \\ 0, & \mathbf{r} \notin v_0 \end{cases} \tag{1.9}$$

where the time-harmonic factor is $e^{-i\omega t}$ $(i = -j = \sqrt{-1})$ and r is the distance from the scattering point to the observation point. Solving Equation 1.9 under the far-field approximation yields the scattering $\mathbf{E}_s$ wave

$$
\begin{aligned}
\mathbf{E}_s(r) &= -\frac{e^{ikr}}{r}k^2 \int_{v_0} \hat{k}_s \times [\hat{k}_s \times \mathbf{E}(\mathbf{r}')][\varepsilon_{sn}(\mathbf{r}') - 1]e^{-i\mathbf{k}_s \cdot \mathbf{r}'} d\mathbf{r}' \\
&= \frac{e^{ikr}}{r}k^2 \int_{v_0} (\hat{v}_s \hat{v}_s + \hat{h}_s \hat{h}_s)[\varepsilon_{sn}(\mathbf{r}') - 1]\mathbf{E}(\mathbf{r}')e^{-i\mathbf{k}_s \cdot \mathbf{r}'} d\mathbf{r}' \qquad (1.10) \\
&\equiv \frac{e^{ikr}}{r}\mathbf{\bar{S}} \cdot \mathbf{E}_i = \frac{e^{ikr}}{r}\begin{bmatrix} S_{vv} & S_{vh} \\ S_{hv} & S_{hh} \end{bmatrix} \cdot \mathbf{E}_i
\end{aligned}
$$

Here, the scattering field is now expressed by a 2×2 (dimensional) complex matrix, that is, the scattering matrix, $\mathbf{\bar{S}}$. Each element of the scattering matrix S_{pq}, $p, q = v, h$, represents the complex scattering coefficient under q-polarization incidence and p-polarization scattering.

For a single scattering, its scattering matrix has to be obtained by first calculating the internal field $\mathbf{E}(\mathbf{r}')$ in Equation 1.10 and then performing the integration. Explicit solutions exist for many canonical problems. For example, a small scatterer with its size a far smaller than the wavelength, that is, $ka \ll 1$, can be solved by Rayleigh approximation. A non-spherical particle such as an ellipsoid, a needle- or a disk-like particle with its smallest dimension d far smaller than the wavelength, that is, $kd(\varepsilon_{sn} - 1) \ll 1$, can be solved using the Rayleigh–Gans approximation (Ishimaru, 1978, 1991). A scatterer of perfectly spherical geometry can be solved using Mie theory. Explicit expressions for $\mathbf{\bar{S}}$ also exist for symmetric geometries such as cylinders, spheroids, and so on. Analytic calculations of the scattering matrix for these examples under the Rayleigh–Gans approximation are given later in this chapter. For a more complete derivation, readers are referred to the literature focusing on EM theory (e.g., Jin, 1994).

For a large scattering body with complicated geometry, we have to resort to numerical methods such as the method of momentss (MoM), finite element method (FEM), and finite difference time difference (FDTD), or high-frequency methods such as geometrical optics and physical optics. Chapter 11 will introduce a high-frequency method for calculation of scattering from very complex targets in natural environments.

In practice, fully polarimetric radar measures the entire scattering matrix $\mathbf{\bar{S}}$, while single-polarization radar (single-pol) or dual-polarization radar (dual-pol) can only measure one or two elements of the scattering matrix. Conventionally, co-pol refers to scattering with the same polarization for incidence and scattering, while cross-pol refers to different polarization for incidence and scattering.

Another important concept is the scattering coefficient or radar cross-section (RCS). It is defined as the power ratio of scattering versus incidence as if the scattered field were uniform in all directions:

$$
\sigma_{pq} = 4\pi |S_{pq}|^2 \quad (\mathrm{m}^2) \qquad (1.11)
$$

The RCS is more useful for single-polarization radar where phase information is not critical.

1.1.2 Stokes Vector and Mueller Matrix

As opposed to coherent scattering for a single deterministic target, incoherent scattering is from random or distributed targets, for example, random discrete particles, random media, and randomly rough surfaces. Incoherent scattering is studied from a statistical perspective. That is based on the second-order moment of the random field $\langle \mathbf{E}_s \mathbf{E}_s^* \rangle$, which is in the unit of power.

Incoherent scattering produces a partially polarized wave, which is usually described using the Stokes vector $\mathbf{I}$, which consists of four Stokes parameters as follows:

$$
\begin{aligned}
I &= \frac{1}{\eta} \left\langle |E_v|^2 + |E_h|^2 \right\rangle = \frac{1}{\eta} \left\langle E_{0v}^2 + E_{0h}^2 \right\rangle \\
Q &= \frac{1}{\eta} \left\langle |E_v|^2 - |E_h|^2 \right\rangle = \frac{1}{\eta} \left\langle E_{0v}^2 - E_{0h}^2 \right\rangle \\
U &= \frac{1}{\eta} 2 \operatorname{Re} \langle E_v E_h^* \rangle = \frac{1}{\eta} 2 \langle E_{0v} E_{0h} \cos \varphi \rangle \\
V &= \frac{1}{\eta} 2 \operatorname{Im} \langle E_v E_h^* \rangle = \frac{1}{\eta} 2 \langle E_{0v} E_{0h} \sin \varphi \rangle
\end{aligned}
\tag{1.12}
$$

where $*$ denotes conjugate, η denotes wave impedance, and $\langle \cdot \rangle$ denotes the ensemble average. Stokes law gives

$$
I^2 \geq Q^2 + U^2 + V^2
\tag{1.13}
$$

where the equality only holds for a completely polarized wave. For a completely polarized wave, only three of the Stokes parameters in Equation 1.12 are independent. This corresponds to the above-mentioned three geometric parameters of elliptical polarization:

$$
\mathbf{I} = \begin{bmatrix} I \\ I \cos 2\psi \cos 2\chi \\ I \sin 2\psi \cos 2\chi \\ I \sin 2\chi \end{bmatrix}
\tag{1.14}
$$

With given intensity I, any complete polarization can be mapped into a point on a sphere in 3D (Q, U, V) space. The sphere with radius I is referred to as the Poincaré sphere. Partial polarizations will be distributed inside the Poincaré sphere. Interestingly, two angles defining a 3D vector pointing to any point on the Poincaré sphere are equal to the parameters $2\chi, 2\psi$ of the elliptical polarization on that point. Thus, we can draw the distribution of polarizations on the Poincaré sphere as in Figure 1.3. It can be directly mapped to the polarization ratio plane in Figure 1.2.

Another commonly used form of Stokes parameters are defined as

$$
I_v = \frac{1}{\eta} \left\langle |E_v|^2 \right\rangle, \quad I_h = \frac{1}{\eta} \left\langle |E_h|^2 \right\rangle, \quad U = \frac{1}{\eta} 2 \operatorname{Re} \langle E_v E_h^* \rangle, \quad V = \frac{1}{\eta} 2 \operatorname{Im} \langle E_v E_h^* \rangle
\tag{1.15}
$$

These two definitions of Stokes parameters can be transformed into one another.

Incoherent scattering of random or distributed targets has a depolarization effect, which renders the scattered wave to be a partially polarized wave. Using Stokes vectors $\mathbf{I}_i, \mathbf{I}_s$ to

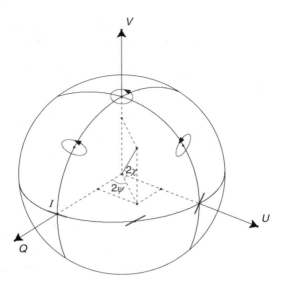

Figure 1.3 Poincaré sphere

represent the incident and scattered wave, respectively, the incoherent scattering is then
represented using a 4×4 real Mueller matrix

$$\mathbf{I}_s = \overline{\mathbf{M}} \cdot \mathbf{I}_i \tag{1.16}$$

The Mueller matrix is in fact the second-order moment of the scattering matrix. From Equations 1.10 and 1.15, we have

$$\overline{\mathbf{M}} = \begin{bmatrix} \langle |S_{vv}|^2 \rangle & \langle |S_{vh}|^2 \rangle & \mathrm{Re}\langle S_{vv}S_{vh}^* \rangle & -\mathrm{Im}\langle S_{vv}S_{vh}^* \rangle \\ \langle |S_{hv}|^2 \rangle & \langle |S_{vv}|^2 \rangle & \mathrm{Re}\langle S_{hv}S_{hh}^* \rangle & -\mathrm{Im}\langle S_{hv}S_{vh}^* \rangle \\ 2\,\mathrm{Re}\langle S_{vv}S_{hv}^* \rangle & 2\,\mathrm{Re}\langle S_{vh}S_{hh}^* \rangle & \mathrm{Re}\langle S_{vv}S_{hh}^* + S_{vh}S_{hv}^* \rangle & -\mathrm{Im}\langle S_{vv}S_{hh}^* - S_{vh}S_{hv}^* \rangle \\ 2\,\mathrm{Im}\langle S_{vv}S_{hv}^* \rangle & 2\,\mathrm{Im}\langle S_{vh}S_{hh}^* \rangle & \mathrm{Im}\langle S_{vv}S_{hh}^* + S_{vh}S_{hv}^* \rangle & \mathrm{Re}\langle S_{vv}S_{hh}^* - S_{vh}S_{hv}^* \rangle \end{bmatrix} \tag{1.17}$$

Some useful conclusions about the Mueller matrix are given here. For coherent scattering, only seven of the 16 elements of the Mueller matrix are independent, which corresponds to seven degrees of freedom in the scattering matrix (the ensemble average eliminates the absolute phase). For most natural targets, co-polarization and cross-polarization are not correlated, which makes the eight elements in the top-right and bottom-left quarters approach zero. For the remaining eight elements, the top-left quartet reflect the average power of the two polarizations and the bottom-right quartet reflect the correlation between different polarizations.

From the Mueller matrix, we can also derive polarization signatures of the target, which are defined as follows:

co-polarization signature

$$\sigma_c(\chi, \psi) = 4\pi \cdot \frac{1}{2}\mathbf{I}_s^{\mathrm{T}} \cdot \mathbf{I}_i(\chi, \psi) \tag{1.18}$$

cross-polarization signature

$$\sigma_x(\chi, \psi) = 4\pi \cdot \frac{1}{2} \left[\mathbf{I}_s^{\mathrm{T}} \cdot \mathbf{I}_s - \mathbf{I}_s^{\mathrm{T}} \cdot \mathbf{I}_i(\chi, \psi) \right] \tag{1.19}$$

where the superscript T denotes transpose. Note that polarization signatures $\sigma_c(\chi, \psi)$, $\sigma_d(\chi, \psi)$ are in fact the co-polarized (co-pol) and cross-polarized (cross-pol) scattering coefficients under elliptical polarization incidence of the parameters χ, ψ. Substituting $\chi = 0°, \psi = 0°$ or $\chi = 0°, \psi = 90°$ into Equation 1.19, it gives the horizontally or vertically co-pol and cross-pol scattering coefficients.

In addition, we have the polarization degree defined as

$$m_s = \frac{Q_s^2 + U_s^2 + V_s^2}{I_s^2} \tag{1.20}$$

and the co-pol/cross-pol phase difference as

$$\phi_{vh} = \tan^{-1} \frac{M_{43} - M_{34}}{M_{33} + M_{44}} \tag{1.21}$$

and M_{ij} denotes the i, j element of the Mueller matrix $\overline{\mathbf{M}}$. Clearly, we have the inequality $m_s \leq 1$, where $m_s = 1$ when the scattered Stokes intensity $\mathbf{I}_s$ is completely polarized. The quantity $[Q_s^2 + U_s^2 + V_s^2]^{1/2}$ denotes a point inside the Poincaré sphere, and the radius of the Poincaré sphere is equal to I.

Table 1.1 gives few examples of the scattering matrix of ideal scattering targets. The corresponding polarization signatures are given in Figure 1.4.

Table 1.1 Example scattering matrices of ideal scattering targets

Sphere, Plane, or Corner reflector	$S = \begin{bmatrix} 1 & 0 \\ 0 & -1 \end{bmatrix}$
Diplane	$S = \begin{bmatrix} 1 & 0 \\ 0 & 1 \end{bmatrix}$
Diplane (45° tilted)	$S = \begin{bmatrix} 0 & 1 \\ -1 & 0 \end{bmatrix}$
Dipole (vertically oriented)	$S = \begin{bmatrix} 1 & 0 \\ 0 & 0 \end{bmatrix}$
Dipole (45° tilted)	$S = \begin{bmatrix} 1 & 1 \\ -1 & -1 \end{bmatrix}$
Helix (right-handed)	$S = \begin{bmatrix} 1 & i \\ -i & 1 \end{bmatrix}$
Helix (left-handed)	$S = \begin{bmatrix} 1 & -i \\ i & 1 \end{bmatrix}$

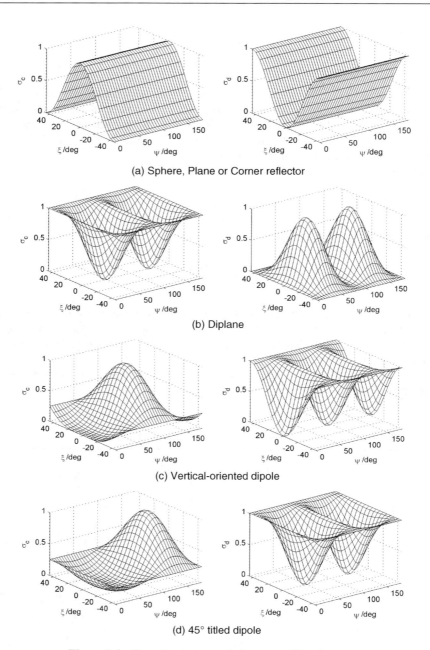

(a) Sphere, Plane or Corner reflector

(b) Diplane

(c) Vertical-oriented dipole

(d) 45° titled dipole

Figure 1.4   Co-pol and cross-pol signatures of ideal targets

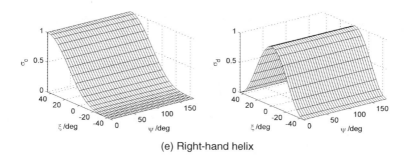

(e) Right-hand helix

Figure 1.4 (*Continued*)

1.2 Volumetric Scattering

1.2.1 Small Particle under Rayleigh–Gans Approximation

Small-particle scattering is common in natural environments, and thus is very useful for remote sensing studies. For modeling scattering of the vegetation canopy, we mainly use the Rayleigh–Gans approximation of non-spherical particles including spheroids, disk-like and needle-like particles (Ishimaru, 1978; Jin, 1994).

Under the Rayleigh approximation, the term in Equation 1.10 becomes $\exp(-i\mathbf{k}_s \cdot \mathbf{r}') \approx 1$, that is, it ignores the coherent effect of different elements inside the particle. The internal field $\mathbf{E}(\mathbf{r}')$ can be obtained using electrostatic principles. Hence, the scattering matrix can be calculated.

For non-spherical particles whose smallest dimension meets the condition $kd(\varepsilon_{sn} - 1) \ll 1$, we can still use the Rayleigh–Gans approximation, where the internal field is approximated as $\mathbf{E}(\mathbf{r}') \approx \hat{e}_i E_0 \exp(i\mathbf{k}_i \cdot \mathbf{r}')$. However, the term $\exp[i(\mathbf{k}_i - \mathbf{k}_s) \cdot \mathbf{r}']$ is kept in order to account for the coherent effect of different elements, that is

$$\mathbf{E}_s(r) = \frac{e^{ikr}}{r} \frac{k^2}{4\pi} (\hat{v}_s \hat{v}_s + \hat{h}_s \hat{h}_s) \cdot \bar{\mathbf{A}}_b \cdot \hat{e}_i E_0 \int_{v_0} [\varepsilon_{sn}(\mathbf{r}') - 1] e^{i(\mathbf{k}_i - \mathbf{k}_s) \cdot \mathbf{r}'} d\mathbf{r}' \qquad (1.22)$$

Notice that the above integral is in fact the Fourier transform of $\varepsilon_{sn} - 1$ in the direction $\mathbf{k}_i - \mathbf{k}_s$. If $\varepsilon_{sn} - 1$ is limited to within a small region, the scattering is smooth along the wave-number domain $\mathbf{k}_i - \mathbf{k}_s$, that is, it is a smooth angular distribution, which is close to Rayleigh scattering. On the other hand, the scattering will focus on a narrow forward direction, with a complicated angular distribution. In this sense, it resembles the relationship between time-domain and frequency-domain signals.

Given a small ellipsoid of semi-axes a, b, c and volume $v_0 = (4\pi/3)abc$, the term $\bar{\mathbf{A}}_b$ in Equation 1.22 can be obtained as

$$\bar{\mathbf{A}}_b = \sum_{n=1}^{3} \frac{\hat{x}_n \hat{x}_n}{1 + (\varepsilon_s/\varepsilon_0 - 1)g_n}, \quad \hat{x}_1 = \hat{x}_b, \ \hat{x}_2 = \hat{y}_b, \ \hat{x}_3 = \hat{z}_b \qquad (1.23)$$

where $\hat{x}_b, \hat{y}_b, \hat{z}_b$ denotes the local coordinates of the ellipsoid, g_1 is expressed as

$$g_1 = \frac{abc}{2} \int_0^\infty ds \frac{1}{(s+a^2)R_s}, \quad R_s = [(s+a^2)(s+b^2)(s+c^2)]^{1/2} \tag{1.24}$$

and g_2, g_3 can be expressed in the same way as g_1 with a replaced by b, c, respectively. It can be found that $g_1 + g_2 + g_3 = 1$.

The scattering matrix is thus derived as

$$S_{pq}(\theta, \phi; \theta', \phi') = \hat{p} \cdot \overline{\mathbf{T}} \cdot \hat{q}', \quad p, q = v, h$$
$$\overline{\mathbf{T}} = \sum_{n=1}^{3} T_n \hat{x}_n \hat{x}_n, \quad T_n = \frac{3}{2k} t_n (1 + it_n), \quad t_n = \frac{2}{9} k^3 abc \frac{\varepsilon_s/\varepsilon_0 - 1}{1 + (\varepsilon_s/\varepsilon_0 - 1)g_n} \tag{1.25}$$

Note that the imaginary part of T_n is included so that the scattering matrix S_{pq} can be accurate enough when later used to calculate the extinction coefficient under the optics theorem in Chapter 2.

Assuming that the ellipsoid is rotationally symmetric, that is, $a = b$, which is the case for most particles in the natural environment, then the scattering matrix can be expanded to

$$S_{vv}(\theta_s, \phi_s; \theta_i, \phi_i) = \frac{3}{2k} \left[T_1(\hat{\theta}_s \cdot \hat{\theta}_i) + (T_0 - T_1)(\hat{\theta}_s \cdot \hat{z}_b)(\hat{z}_b \cdot \hat{\theta}_i) \right]$$
$$S_{vh}(\theta_s, \phi_s; \theta_i, \phi_i) = \frac{3}{2k} \left[T_1(\hat{\theta}_s \cdot \hat{\phi}_i) + (T_0 - T_1)(\hat{\theta}_s \cdot \hat{z}_b)(\hat{z}_b \cdot \hat{\phi}_i) \right]$$
$$S_{hv}(\theta_s, \phi_s; \theta_i, \phi_i) = \frac{3}{2k} \left[T_1(\hat{\phi}_s \cdot \hat{\theta}_i) + (T_0 - T_1)(\hat{\phi}_s \cdot \hat{z}_b)(\hat{z}_b \cdot \hat{\theta}_i) \right] \tag{1.26}$$
$$S_{hh}(\theta_s, \phi_s; \theta_i, \phi_i) = \frac{3}{2k} \left[T_1(\hat{\phi}_s \cdot \hat{\phi}_i) + (T_0 - T_1)(\hat{\phi}_s \cdot \hat{z}_b)(\hat{z}_b \cdot \hat{\phi}_i) \right]$$

Note that the orientation of the particle's symmetry axis is represented by $\hat{z}_b$.

Calculating the integral of Equation 1.24, we have

$$g_1 = \frac{1}{2}(1 - g_3)$$

$$g_3 = \begin{cases} \dfrac{a^2}{a^2 - c^2} \left[1 - \dfrac{c}{\sqrt{a^2 - c^2}} \tan^{-1} \dfrac{\sqrt{a^2 - c^2}}{c} \right] & \text{if } a = b > c \\[4mm] \dfrac{a^2}{a^2 - c^2} \left[1 + \dfrac{c}{2\sqrt{c^2 - a^2}} \ln \dfrac{c - \sqrt{c^2 - a^2}}{c + \sqrt{c^2 - a^2}} \right] & \text{if } a = b < c \end{cases} \tag{1.27}$$

where the two cases correspond to prolate and oblate ellipsoids, respectively.

For a disk-like particle, with radius a, height $2h$, and volume $v_0 = 2\pi a^2 h$, and given $h \ll a$, the integral of Equation 1.22 can be derived as

$$\int_{v_0} d\mathbf{r}' \exp[-i(\mathbf{k}_i - \mathbf{k}_s) \cdot \mathbf{r}'] = \frac{2J_1\left[ka\sqrt{\sin^2\theta_s + \sin^2\theta_i - 2\sin\theta_s\sin\theta_i\cos(\phi_s - \phi_i)}\right]}{ka\sqrt{\sin^2\theta_s + \sin^2\theta_i - 2\sin\theta_s\sin\theta_i\cos(\phi_s - \phi_i)}} \quad (1.28)$$

$$g_3 = \frac{(a/h)^2}{(a/h)^2 - 1}\left[1 - \frac{1}{\sqrt{(a/h)^2 - 1}}\sin^{-1}\left(\frac{\sqrt{(a/h)^2 - 1}}{a/h}\right)\right] \quad (1.29)$$

where $J_1(\cdot)$ denotes the first-order Bessel function of the first kind.

For a needle-like particle with radius a, length $2h$, and volume $v_0 = 2\pi a^2 h$, and given $h \gg a$, the integral in Equation 1.22 can be derived as

$$\int_{v_0} d\mathbf{r}' \exp[-i(\mathbf{k}_i - \mathbf{k}_s) \cdot \mathbf{r}'] = \frac{\sin[kh(\cos\theta_i + \cos\theta_s)]}{kh(\cos\theta_i + \cos\theta_s)} \quad (1.30)$$

$$g_3 = -\frac{(a/h)^2}{1 - (a/h)^2}\left[1 + \frac{1}{2\sqrt{1 - (a/h)^2}}\log\left(\frac{1 - \sqrt{1 - (a/h)^2}}{1 + \sqrt{1 - (a/h)^2}}\right)\right] \quad (1.31)$$

Under extreme conditions, $h \ll a$ and $h \gg a$, Equations 1.29 and 1.31 are consistent with Equation 1.27.

These expressions are critical in computing the scattering matrix, phase matrix, and extinction coefficient, and for modeling the scattering behavior of random particles in the vegetation canopy.

1.2.2 Slim Cylinder

A finite-length slim cylinder is often used to model the trunk, branches or twigs of trees. Its scattering can be solved as in Equation 1.10

$$\mathbf{E}_s(\mathbf{r}) = \frac{e^{ikr}}{r}k^2 \int_{v_0}(\hat{v}_s\hat{v}_s + \hat{h}_s\hat{h}_s)[\varepsilon_{sn}(\mathbf{r}') - 1]\mathbf{E}(\mathbf{r}')e^{-i\mathbf{k}_s\cdot\mathbf{r}'}d\mathbf{r}' \quad (1.32)$$

where the internal field term $\mathbf{E}(\mathbf{r}')$ can be approximated as if it was an infinitely long cylinder, for which the explicit solution can be found. The internal field of an infinitely long cylinder can be found, for example, in Wait (1986) and Karam and Fung (1988). Substituting $\mathbf{E}(\mathbf{r}')$ into Equation 1.32 and completing the integral, it yields the explicit expression of the scattering field $\mathbf{E}_s$ and hence the scattering matrix $\overline{\mathbf{S}}$ (Karam and Fung, 1988).

Given the cylinder parameters of length $2L$ and radius a, the dielectric constant ε_{sn} ($= \varepsilon_s/\varepsilon_0$), and the fact that its axis is aligned with $\hat{z}$, the scattering matrix $\overline{\mathbf{S}}$ for a cylinder can be derived as

$$S_{vv}(\theta_s, \theta_i, \phi_s - \phi_i) = k_0^2 L(\varepsilon_{sn} - 1)\mu_{si}\{(e_{n0}B_0 \cos\theta_s \cos\theta_i - Z_0 \sin\theta_s)$$

$$+ 2\sum_{n=1}^{\infty}[(e_{nv}B_n \cos\theta_i + i\eta h_{nv}A_n)\cos\theta_s - e_{nv}Z_n \sin\theta_s]\cos[n(\phi_s - \phi_i)]\} \qquad (1.33)$$

$$S_{hh}(\theta_s, \theta_i, \phi_s - \phi_i) = k_0^2 L(\varepsilon_{sn} - 1)\mu_{si}\left\{B_0\eta h_{0h} + 2\sum_{n=1}^{\infty}[(\eta h_{nh}B_n - ie_{nh}A_n \cos\theta_i)\cos[n(\phi_s - \phi_i)]\right\}$$

$$(1.34)$$

$$S_{vh}(\theta_s, \theta_i, \phi_s - \phi_i) = -2ik_0^2 L(\varepsilon_{sn} - 1)\mu_{si}\sum_{n=1}^{\infty}[(e_{nh}B_n\cos\theta_i + i\eta h_{nh}A_n)\cos\theta_s$$

$$- e_{nh}Z_n \sin\theta_s]\sin[n(\phi_s - \phi_i)]\} \qquad (1.35)$$

$$S_{hv}(\theta_s, \theta_i, \phi_s - \phi_i) = -2ik_0^2 L(\varepsilon_{sn} - 1)\mu_{si}\sum_{n=1}^{\infty}[(\eta h_{nv}B_n - ie_{nv}A_n \cos\theta_i)\sin[n(\phi_s - \phi_i)]\} \quad (1.36)$$

where

$$A_n = \frac{k_0}{2k_{1\rho}}(Z_{n-1} - Z_{n+1}), \quad B_n = \frac{k_0}{2k_{1\rho}}(Z_{n-1} + Z_{n+1})$$

$$Z_n = \frac{a^2}{u^2 - v_s^2}[uJ_n(v_s)J_{n+1}(u) - v_sJ_n(u)J_{n+1}(v_s)]$$

$$(1.37)$$

$$u = k_{1\rho}a, \quad v_s = k_0 a \sin\theta_s, \quad v_i = k_0 a \sin\theta_i, \quad k_{1\rho} = k_0\sqrt{(\varepsilon_{sn} - \cos^2\theta_i)}$$

$$\mu_{si} = \frac{\sin[k_0L(\cos\theta_i + \cos\theta_s)]}{k_0L(\cos\theta_i + \cos\theta_s)}, \quad \eta = \sqrt{\frac{\mu_0}{\varepsilon_0}} \qquad (1.38)$$

$$e_{nv} = -\frac{i\sin\theta_i}{R_nJ_n(u)}\left[\frac{H_n^{(2)'}(v_i)}{v_iH_n^{(2)}(v_i)} - \frac{J_n'(u)}{uJ_n(u)}\right], \quad e_{nh} = -\frac{\sin\theta_i}{R_nJ_n(u)}\left[\frac{1}{v_i^2} - \frac{1}{u^2}\right] \qquad (1.39)$$

$$\eta h_{nv} = \frac{\sin\theta_i}{R_nJ_n(u)}\left[\frac{1}{v_i^2} - \frac{1}{u^2}\right]n\cos\theta_i, \quad \eta h_{nh} = -\frac{\sin\theta_i}{R_nJ_n(u)}\left[\frac{H_n^{(2)'}(v_i)}{v_iH_n^{(2)}(v_i)} - \frac{\varepsilon_{sn}J_n'(u)}{uJ_n(u)}\right] \qquad (1.40)$$

and

$$R_n = \frac{1}{2}\pi v_i^2 H_n^{(2)}(v_i)\left\{\left[\frac{H_n^{(2)'}(v_i)}{v_iH_n^{(2)}(v_i)} - \frac{J_n'(u)}{uJ_n(u)}\right] \times \left[\frac{H_n^{(2)'}(v_i)}{v_iH_n^{(2)}(v_i)} - \frac{\varepsilon_{sn}J_n'(u)}{uJ_n(u)}\right] - \left(\frac{1}{u^2} - \frac{1}{v_i^2}\right)^2 n^2 \cos^2\theta_i\right\}$$

$$(1.41)$$

Note that $J_n(u)$ and $J_n'(u)$ stand for the nth-order Bessel function and its derivative, respectively; and $H_n^{(2)}(v_i)$ and $H_n^{(2)'}(v_i)$ are the nth-order Hankel function of the second kind and its derivative, respectively.

It can be seen from Equations 1.35 and 1.36 that there is no cross-polarized backscattering, that is, $S_{vh} = S_{hv} = 0$ under the condition $\phi_s = \pi + \phi_i$.

Interestingly, it can be found that, if $k_0 L \gg 1$ and $\cos \theta_i + \cos \theta_s \neq 0$, then $\mu_{si} = 0$ in Equation 1.38; if $\cos \theta_i + \cos \theta_s = 0$, then $\mu_{si} = 1$, which means that the scattered wave propagates along the cone surface forming angle θ_i with respect to axis $\hat{z}$.

In the general case when the cylinder orientation $\hat{z}_\ell$ differs from the global coordinate $\hat{z}$, then the above expressions should be defined in its local coordinates $(\hat{x}_\ell, \hat{y}_\ell, \hat{z}_\ell)$. The scattering field in Equation 1.32 and the scattering matrix in Equations 1.33–1.36 have to be further converted to the global coordinate system via coordinate conversions, which will be introduced in subsection 1.3.7.

1.3   Surface Scattering

1.3.1   Plane Surface

Surface scattering is the most common scattering in the natural environment. The most basic model would be an infinite perfectly flat surface. Its scattering will concentrate on the specular direction, that is, reflection, which is directly solved as the Fresnel reflection coefficient.

As shown in Figure 1.5, a plane surface has a reflection matrix and a transmission matrix written as

$$\overline{\mathbf{R}} = \begin{bmatrix} R_v & 0 \\ 0 & R_h \end{bmatrix}, \quad \overline{\mathbf{T}} = \begin{bmatrix} T_v & 0 \\ 0 & T_h \end{bmatrix} = \begin{bmatrix} 1 + R_v & 0 \\ 0 & 1 + R_h \end{bmatrix} \tag{1.42}$$

where R_v, R_h denote the Fresnel coefficients of vertical and horizontal polarization

$$R_v = \frac{\varepsilon_r \cos \theta - \sqrt{\varepsilon_r - \sin^2 \theta}}{\varepsilon_r \cos \theta + \sqrt{\varepsilon_r - \sin^2 \theta}}, \quad R_h = \frac{\cos \theta - \sqrt{\varepsilon_r - \sin^2 \theta}}{\cos \theta + \sqrt{\varepsilon_r - \sin^2 \theta}} \tag{1.43}$$

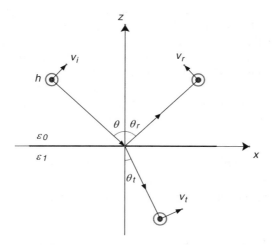

Figure 1.5   Reflection and transmission of a plane surface

where θ is the incident angle, and $\varepsilon_r = \varepsilon_1/\varepsilon_0$ is the relative dielectric constant. The reflection angle and transmission angle are related to the incident angle according to Snell's law:

$$\theta_r = \theta, \quad \sqrt{\varepsilon_1}\sin\theta_t = \sqrt{\varepsilon_0}\sin\theta \tag{1.44}$$

For incoherent scattering, the Mueller matrix of plane surface reflection and transmission can be easily derived from Equation 1.17 as

$$\mathbf{R}_{4\times4} = \begin{bmatrix} |R_v|^2 & 0 & 0 & 0 \\ 0 & |R_h|^2 & 0 & 0 \\ 0 & 0 & \mathrm{Re}(R_v R_h^*) & -\mathrm{Im}(R_v R_h^*) \\ 0 & 0 & \mathrm{Im}(R_v R_h^*) & \mathrm{Re}(R_v R_h^*) \end{bmatrix} \tag{1.45}$$

$$\mathbf{T}_{4\times4} = \varepsilon_r \cdot \begin{bmatrix} |T_v|^2 & 0 & 0 & 0 \\ 0 & |T_h|^2 & 0 & 0 \\ 0 & 0 & \dfrac{\cos\theta_t}{\cos\theta}\mathrm{Re}(T_v T_h^*) & -\dfrac{\cos\theta_t}{\cos\theta}\mathrm{Im}(T_v T_h^*) \\ 0 & 0 & \dfrac{\cos\theta_t}{\cos\theta}\mathrm{Im}(T_v T_h^*) & \dfrac{\cos\theta_t}{\cos\theta}\mathrm{Re}(T_v T_h^*) \end{bmatrix} \tag{1.46}$$

Note that the transmission Mueller matrix, Equation 1.46, contains a factor of ε_r because the Stokes vector includes wave impedance and the transmitted wave and the incident wave are in different media. Since the Mueller matrix is a real matrix, only the real part of ε_r is used here. Also note that the same letters are used to denote both scattering and Mueller matrix, simply because they are in fact the same reflection or transmission but in different forms.

1.3.2 Rough Surface

In natural environments, most surfaces are randomly rough surfaces, for example, the ground surface, sea surface, and building facets. A randomly rough surface is often described by a random height profile, that is, $z = \xi(x)$ for a 2D rough surface, and $z = \xi(x,y)$ for a 3D rough surface. When calculating rough-surface scattering, the most important entity is the correlation function of the height profile and its Fourier transform, that is, the spectral density function (or spectral function).

A Gaussian rough surface is the most basic type of rough surface. It follows the Gaussian distribution function:

$$p(\xi(x,y)) = \frac{1}{\sqrt{2\pi}\delta}\exp\left[-\xi^2/(2\delta^2)\right] \tag{1.47}$$

Let $\mathbf{r}_{1\perp} \equiv (x_1, y_1)$, the correlation function between $\mathbf{r}_{1\perp}$ and $\mathbf{r}_{2\perp}$, be

$$\delta^2 C(\mathbf{r}_{1\perp}, \mathbf{r}_{2\perp}) = \langle \xi_1(\mathbf{r}_{1\perp})\xi_2(\mathbf{r}_{2\perp}) \rangle = \delta^2 \exp\left(-\frac{|\mathbf{r}_{1\perp} - \mathbf{r}_{2\perp}|^2}{\ell^2}\right) \tag{1.48}$$

where the correlation length ℓ in both x and y directions is simply assumed to be the same (it can be different as ℓ_x, ℓ_y). Then, the joint probability density function can be derived as

$$p(\xi(x,y),\xi(x_1,y_1)) = \frac{1}{2\pi\delta^2\sqrt{1-C^2}}\exp\left[-\frac{\xi^2 - 2C\xi\xi_1 + \xi_1^2}{2\delta^2(1-C^2)}\right] \qquad (1.49)$$

$$p\left(\alpha\left[\equiv\frac{\partial}{\partial x}\xi(x,y)\right],\beta\left[\equiv\frac{\partial}{\partial y}\xi(x,y)\right]\right) = \frac{1}{2\pi\delta^2|C''(0)|}\exp\left[-\frac{\alpha^2 + \beta^2}{2\delta^2|C''(0)|}\right] \qquad (1.50)$$

where $\delta^2|C''(0)|$ is the mean-square surface slope

$$\rho_s^2 = \delta^2|C''(0)| = \frac{2\delta^2}{\ell^2} \qquad (1.51)$$

In general, a surface is considered of higher roughness if δ is larger, ℓ is smaller or ρ_s is larger.

The spectral function of a Gaussian surface can be obtained as

$$W(\boldsymbol{\beta}_\perp) = \frac{\delta^2}{(2\pi)^2}\int\limits_{-\infty}^{\infty}\int\limits_{-\infty}^{\infty} d\mathbf{r}_\perp e^{i\boldsymbol{\beta}_\perp\cdot\mathbf{r}_\perp}C(\mathbf{r}_\perp) = \frac{1}{4\pi}\delta^2\ell^2\exp\left[\frac{1}{4}\beta_\perp^2\ell^2\right] \qquad (1.52)$$

For different types of rough surfaces, the same scattering model can be applied but with different spectral functions. The most commonly used spectral functions include the Gaussian function, exponential function, and the Pierson–Moskowitz empirical spectrum, which is often used to model the wind-driven sea surface (Fung and Lee, 1982; Huang and Jin, 1995).

The scattering model of a rough surface is illustrated in Figure 1.6. We define the random height profile function as $z = \xi(x,y)$. If a polarized wave $\overline{\mathbf{E}}_i$ is incident in the direction $(\pi - \theta_i, \phi_i)$, the scattered wave $\overline{\mathbf{E}}_s$ in the direction (θ_s, ϕ_s) can be written as

$$\mathbf{E}_s(\mathbf{r}) = \frac{e^{ikr}}{r}\overline{\mathbf{S}}(\theta_s, \phi_s; \theta_i, \phi_i)\cdot\mathbf{E}_i \qquad (1.53)$$

Following Huygens' principle (Kong, 2005), Equation 1.53 can be derived as integration of the induced electric and magnetic fields on the rough surface,

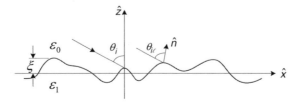

Figure 1.6　Scattering model of a rough surface

$$E_s(\mathbf{r}) = \int_A d\boldsymbol{\rho}' \{i\omega\mu_0\bar{G}(\mathbf{r},\mathbf{r}') \cdot [\hat{n} \times \mathbf{H}(\mathbf{r}')] + \nabla \times \bar{G}(\mathbf{r},\mathbf{r}') \cdot [\hat{n} \times \mathbf{E}(\mathbf{r}')]\} \qquad (1.54)$$

where $\bar{G}(\mathbf{r},\mathbf{r}')$ denotes the dyadic Green function, $\mathbf{E}(\mathbf{r}')$, $\mathbf{H}(\mathbf{r}')$ are the induced electric and magnetic fields at position $\mathbf{r}'(\in S)$, and $\hat{n}$ denotes the local normal vector at position $\mathbf{r}'$, that is

$$\hat{n} = \frac{-\alpha\hat{x} - \beta\hat{y} + \hat{z}}{\sqrt{1 + \alpha^2 + \beta^2}}, \quad \alpha = \frac{\partial}{\partial x}\xi(x,y), \quad \beta = \frac{\partial}{\partial y}\xi(x,y) \qquad (1.55)$$

Under the far-field approximation, this yields

$$E_s(\mathbf{r}) = \frac{ike^{ikr}}{4\pi r}(\bar{\mathbf{I}} - \hat{k}_s\hat{k}_s) \cdot \int_A d\boldsymbol{\rho}' \{\hat{k}_s \times [\hat{n} \times \mathbf{E}(\mathbf{r}')] + \eta_0[\hat{n} \times \mathbf{H}(\mathbf{r}')]\}e^{-i\mathbf{k}_s\cdot\mathbf{r}'} \qquad (1.56)$$

Clearly, $E_s(\mathbf{r})$ and $\mathbf{E}(\mathbf{r}')$, $\mathbf{H}(\mathbf{r}')$ are random. However, the incoherent scattering, that is, the ensemble average of the scattering wave, $\langle|\mathbf{E}_s|^2\rangle$, is of interest to us, which is further related to the correlation function and spectral function of the rough surface.

Approximate analytical solutions to the problem of rough-surface scattering can be mainly categorized into the Kirchhoff approximation (KA) approach, which is valid for large-scale roughness, the small-perturbation approximation (SPA), suitable for small-scale roughness, the two-scale approximation (TSA) approach, which combines the former two, and the integral equation method (IEM), which are introduced in detail in the following subsections.

1.3.3 Kirchhoff Approximation

If the rough surface fluctuates on a large scale, meaning that its average radius of curvature is far larger than its wavelength, then any portion of the surface can be locally seen as a flat facet and the induced current can be approximated as the equivalent tangential field. The validity condition of KA can be summarized as

$$k_1\ell > 6, \quad k_1\delta > \sqrt{10}/(\cos\theta_i - \cos\theta_s), \quad R_c > \lambda \qquad (1.57)$$

where

$$R_c = \delta\sqrt{\frac{2}{\pi}\frac{\partial^4 C(\xi)}{\partial\xi^4}\Bigg|_{\xi=0}}$$

denotes the mean radius of curvature, the wavenumber $k_1 = (2\pi/\lambda)\sqrt{\varepsilon_1}$, and ε_1 is the dielectric constant.

For a Gaussian rough surface, this translates to

$$\ell^2 > 2.76\delta\lambda \qquad (1.58)$$

Using KA, Equation 1.53 can be derived as

$$\mathbf{E}_s(r) = \frac{ike^{ikr}}{4\pi r} E_0 (\bar{\mathbf{I}} - \hat{k}_s\hat{k}_s) \int_A d\boldsymbol{\rho}' \mathbf{F}(\alpha, \beta) e^{i\mathbf{k}_d \cdot \mathbf{r}'} \tag{1.59}$$

where

$$\mathbf{F}(\alpha, \beta) = (1 + \alpha^2 + \beta^2)^{1/2} \{ -(\hat{e}_i \cdot \hat{q}_i)(\hat{n} \cdot \hat{k}_i)(1 - R_{h\ell})\hat{q}_i + (\hat{e}_i \cdot \hat{p}_i)(\hat{n} \times \hat{q}_i)(1 + R_{v\ell})$$
$$+ (\hat{e}_i \cdot \hat{q}_i)[\hat{k}_s \times (\hat{n} \times \hat{q}_i)](1 + R_{h\ell}) + (\hat{e}_i \cdot \hat{p}_i)(\hat{n} \cdot \hat{k}_i)(\hat{k}_s \times \hat{q}_i)(1 - R_{v\ell}) \} \tag{1.60}$$

$$\mathbf{k}_d \equiv \mathbf{k}_i - \mathbf{k}_s = k_{dx}\hat{x} + k_{dy}\hat{y} + k_{dz}\hat{z} \tag{1.61}$$

The local Fresnel coefficients are

$$R_{h\ell} = \frac{k\cos\theta_{i\ell} - \sqrt{k_1^2 - k^2\sin^2\theta_{i\ell}}}{k\cos\theta_{i\ell} + \sqrt{k_1^2 - k^2\sin^2\theta_{i\ell}}}, \quad R_{v\ell} = \frac{\varepsilon_1 k\cos\theta_{i\ell} - \varepsilon_0\sqrt{k_1^2 - k^2\sin^2\theta_{i\ell}}}{\varepsilon_1 k\cos\theta_{i\ell} + \varepsilon_0\sqrt{k_1^2 - k^2\sin^2\theta_{i\ell}}} \tag{1.62}$$

where k, k_1 denote the wavenumber in the upper and lower half-spaces, respectively. The local incident angle $\theta_{i\ell}$ is determined by $\cos\theta_{i\ell} = -\hat{n} \cdot \hat{k}_i$.

The integration of $\mathbf{F}(\alpha, \beta)$ can be calculated using the stationary phase method (Tsang et al., 1985; Jin, 1994), that is

$$\mathbf{E}_s(\mathbf{r}) = \frac{ike^{ikr}}{4\pi r} E_0 (\bar{\mathbf{I}} - \hat{k}_s\hat{k}_s) \cdot \mathbf{F}(\alpha_0, \beta_0) I \tag{1.63}$$

$$I = \int_A d\boldsymbol{\rho}' \exp(i\mathbf{k}_d \cdot \mathbf{r}'), \quad \mathbf{k}_d \equiv \mathbf{k}_i - \mathbf{k}_s \tag{1.64}$$

where α_0, β_0 denote the stationary phase points, and I in Equation 1.64 is a random term related to the rough surface.

The second-order moment of scattering fields consists of two components: coherent part $\mathbf{E}_m(\mathbf{r})$ (mean-square) and incoherent part $\tilde{\mathbf{E}}(\mathbf{r})$ (variance):

$$\langle |\mathbf{E}_s(\mathbf{r})|^2 \rangle = |\mathbf{E}_m(\mathbf{r})|^2 + \langle |\tilde{\mathbf{E}}(\mathbf{r})|^2 \rangle \tag{1.65}$$

$$|\mathbf{E}_m(\mathbf{r})|^2 = \frac{k^2|E_0|^2}{16\pi^2 r^2} \left[|\hat{v}_s \cdot \mathbf{F}(\alpha_0, \beta_0)|^2 + |\hat{h}_s \cdot \mathbf{F}(\alpha_0, \beta_0)|^2 \right] \cdot |\langle I \rangle|^2 \tag{1.66}$$

$$\langle |\tilde{\mathbf{E}}(\mathbf{r})|^2 \rangle = \frac{k^2|E_0|^2}{16\pi^2 r^2} \left[|\hat{v}_s \cdot \mathbf{F}(\alpha_0, \beta_0)|^2 + |\hat{h}_s \cdot \mathbf{F}(\alpha_0, \beta_0)|^2 \right] \cdot (\langle II^* \rangle - |\langle I \rangle|^2) \tag{1.67}$$

For a Gaussian rough surface, the random term is calculated as

$$|\langle I \rangle|^2 = \int dx \int dy (2L_x - |x|)(2L_y - |y|) \cdot \exp(-k_{dz}^2\delta^2) = 4\pi^2 A \exp(-k_{dz}^2\delta^2)\delta(k_{dx})\delta(k_{dy}) \tag{1.68}$$

where $2L_x \cdot 2L_y = A$, and $\delta(k_{dx})$, $\delta(k_{dy})$ are the Dirac delta functions.

The coherent part $|\mathbf{E}_m(\mathbf{r})|^2$, however, only exists in the specular direction

$$
\begin{aligned}
D_I &\equiv \langle II^* \rangle - |\langle I \rangle|^2 \\
&= A \int dx \int dy \{\exp[-k_{dz}^2 \delta^2 (1 - C(\mathbf{r}_\perp))] - \exp(-k_{dz}^2 \delta^2)\} \exp(ik_{dx}x + ik_{dy}y)
\end{aligned}
\tag{1.69}
$$

For a Gaussian rough surface, it can be explicitly derived as

$$
D_I = \pi A \sum_{m=1}^{\infty} \frac{(k_{dz}^2 \delta^2)^m}{m! m} \ell^2 \exp\left[-\frac{(k_{dx}^2 + k_{dy}^2)\ell^2}{4m} \right] \exp(-k_{dz}^2 \delta^2)
\tag{1.70}
$$

Finally, the bistatic scattering coefficient is written as

$$
\gamma_{pq}(\hat{k}_s, \hat{k}_i) = \frac{4\pi r^2 \langle |\mathbf{E}_s(\mathbf{r})| \rangle_p}{A \cos \theta_i |E_0|_q^2} = \gamma_{pq}^c(\hat{k}_s, \hat{k}_i) + \gamma_{pq}^{ic}(\hat{k}_s, \hat{k}_i), \quad p, q = v, h
\tag{1.71}
$$

where superscripts c, ic denote the coherent and incoherent parts, respectively, that is

$$
\gamma_{pq}^c(\hat{k}_s, \hat{k}_i) = \frac{4\pi |R_{p0}|^2}{\sin \theta_i} \exp(-4k^2 \delta^2 \cos^2 \theta_i) \delta(\theta_s - \theta_i) \delta(\phi_s - \phi_i) \delta_{pq}
\tag{1.72}
$$

$$
\gamma_{pq}^{ic}(\hat{k}_s, \hat{k}_i) = \frac{k^2}{4\pi A \cos \theta_i} |\hat{p}_s \cdot \mathbf{F}_q(\alpha_0, \beta_0)|^2 D_I
\tag{1.73}
$$

where $R_{p0}(p = v, h)$ is the Fresnel coefficient of flat surface at incident angle θ_i, and $\delta_{pp} = 1$, $\delta_{pq}(p \neq q) = 0$. Further,

$$
|\hat{p}_s \cdot \mathbf{F}_q(\alpha_0, \beta_0)|^2 = \frac{|\bar{k}_d|^4}{k^2 |\hat{k}_i \times \hat{k}_s|^4 k_{dz}^2} f_{pq}
\tag{1.74}
$$

$$
f_{vv} = |(\hat{h}_s \cdot \hat{k}_i)(\hat{h}_i \cdot \hat{k}_s)R_{h\ell} + (\hat{v}_s \cdot \hat{k}_i)(\hat{v}_i \cdot \hat{k}_s)R_{v\ell}|^2
\tag{1.75a}
$$

$$
f_{hh} = |(\hat{v}_s \cdot \hat{k}_i)(\hat{v}_i \cdot \hat{k}_s)R_{h\ell} + (\hat{h}_s \cdot \hat{k}_i)(\hat{h}_i \cdot \hat{k}_s)R_{v\ell}|^2
\tag{1.75b}
$$

$$
f_{vh} = |(\hat{v}_s \cdot \hat{k}_i)(\hat{h}_i \cdot \hat{k}_s)R_{h\ell} - (\hat{h}_s \cdot \hat{k}_i)(\hat{v}_i \cdot \hat{k}_s)R_{v\ell}|^2
\tag{1.75c}
$$

$$
f_{hv} = |(\hat{h}_s \cdot \hat{k}_i)(\hat{v}_i \cdot \hat{k}_s)R_{h\ell} - (\hat{v}_s \cdot \hat{k}_i)(\hat{h}_i \cdot \hat{k}_s)R_{v\ell}|^2
\tag{1.75d}
$$

where

$$
\hat{n} = \frac{k_{dx}/k_{dz}\hat{x} + k_{dy}/k_{dz}\hat{y} + \hat{z}}{(1 + k_{dx}^2/k_{dz}^2 + k_{dy}^2/k_{dz}^2)^{1/2}}
\tag{1.76}
$$

Given $k \to \infty$, the bistatic scattering coefficient can be simplified to

$$\gamma_{pq}(\hat{k}_s, \hat{k}_i) = \frac{\pi |\bar{k}_d|^4}{\cos \theta_i |\hat{k}_i \times \hat{k}_s|^4 k_{dz}^4} f_{pq} P\left(-\frac{k_{dx}}{k_{dz}}, -\frac{k_{dy}}{k_{dz}}\right) \tag{1.77}$$

Using Equation 1.50, the bistatic scattering coefficient of Equation 1.77 is written as

$$\gamma_{pq}(\hat{k}_s, \hat{k}_i) = \frac{|\bar{k}_d|^4}{\cos \theta_i |\hat{k}_i \times \hat{k}_s|^4 \frac{dz4}{k}} f_{pq} \frac{1}{2\delta^2 |C''(0)|} \exp\left[-\frac{k_{dx}^2 + k_{dy}^2}{2k_{dz}^2 \delta^2 |C''(0)|}\right] \tag{1.78}$$

and the backscattering coefficient is calculated as

$$\sigma_{hh}(\theta_i) = \sigma_{vv}(\theta_i) = \frac{|R(0)|^2}{\cos^4 \theta_i 2\delta^2 |C''(0)|} \exp\left(-\frac{\tan^2 \theta_i}{2\delta^2 |C''(0)|}\right) \tag{1.79}$$

$$\sigma_{hv}(\theta_i) = \sigma_{vh}(\theta_i) = 0 \tag{1.80}$$

where $R(0)$ is the Fresnel coefficient of the flat surface at zero degree incidence. Note that the term $\delta^2 |C''(0)|$ is in fact the square of the average slope. In the case of a Gaussian correlation function, we have $|C''(0)| = 2/\ell^2$. Some of the terms related to wavenumber can be expanded as

$$k_d^2 = 2k^2[1 + \cos \theta_i \cos \theta_s - \sin \theta_i \sin \theta_s \cos(\phi_s - \phi_i)] \tag{1.81}$$

$$k_{dz}^2 = k^2(\cos \theta_i + \cos \theta_s)^2 \tag{1.82}$$

$$|\hat{k}_i \times \hat{k}_s|^2 = 1 - [\cos \theta_i \cos \theta_s - \sin \theta_i \sin \theta_s \cos(\phi_s - \phi_i)]^2 \tag{1.83}$$

Likewise, the transmission coefficients under KA can be derived, for which readers are referred to Tsang, Kong, and Shin (1985, p. 89).

1.3.4 Small-Perturbation Approximation

SPA, also referred to as the small-perturbation method (SPM), is valid when the roughness is very small compared to the wavelength (Rice, 1963), that is, the root-mean-square height $\delta \ll \lambda$ and mean slope s is comparable to $k\delta$, which can be represented by the following inequalities (Chen and Fung, 1988):

$$k_1 \ell < 3, \quad k_1 \delta < 0.3, \quad s < 0.3 \tag{1.84}$$

From Huygens' principle, Equation 1.53, let us expand the incident field $\mathbf{E}_i$, the scattered field $\mathbf{E}_s$, and the surface field $\mathbf{E}(\mathbf{r}')$ on a rough surface into a perturbation series of $k\xi$. The zero-order term of the scattered field is the specular reflection of the plane surface. The first-order term is the incoherent part of the rough-surface scattering. Higher orders corresponds to multiple scatterings occurring on the rough surface. For detailed derivation of the SPA solution, readers are referred to the literature, such as Bass and Fuks (1979), Ulaby, Moore, and Fung (1982), and Tsang, Kong, and Shin (1985).

The first-order SPA bistatic scattering coefficient can be written as

$$\gamma_{pq}^{(SPA1)}(\hat{k}_s, \hat{k}_i) = 16\pi k^4 \cos^2\theta_s \cos\theta_i F_{pq} W(|\mathbf{k}_{\rho s} - \mathbf{k}_{\rho i}|) \tag{1.85}$$

where $W(|\mathbf{k}_{d\perp}|)$ is the spectral function and

$$F_{hh} = \left| \frac{k_1^2 - k^2}{(k_{zs} + k_{1zs})(k_{zi} + k_{1zi})} \right| \cos^2(\phi_s - \phi_i) \tag{1.86}$$

$$F_{vv} = \left| \frac{k_1^2 - k^2}{(k_1^2 k_{zs} + k^2 k_{1zs})(k_1^2 k_{zi} + k^2 k_{1zi})} [k_1^2 k^2 \sin\theta_s \sin\theta_i - k^2 k_{1zs} k_{1zi} \cos(\phi_s - \phi_i)] \right|^2 \tag{1.87}$$

$$F_{hv} = \left| \frac{k k_{1zi}(k_1^2 - k^2)}{(k_1^2 k_{zi} + k^2 k_{1zi})(k_{zs} + k_{1zs})} \right|^2 \sin^2(\phi_s - \phi_i) \tag{1.88}$$

$$F_{vh} = \left| \frac{k k_{1zs}(k_1^2 - k^2)}{(k_1^2 k_{zs} + k^2 k_{1zs})(k_{zi} + k_{1zi})} \right|^2 \sin^2(\phi_s - \phi_i) \tag{1.89}$$

For a Gaussian rough surface, it can be further expanded as

$$\gamma_{pq}^{(SPA1)}(\hat{k}_s, \hat{k}_i) = 4k^4 \cos^2\theta_s \cos\theta_i \delta^2 \ell^2 f_{pq} \exp\left(-\frac{1}{4}k_{d\rho}^2 \ell^2\right) \tag{1.90}$$

$$k_{d\rho}^2 = k_{dx}^2 + k_{dy}^2 = k^2[\sin^2\theta_s + \sin^2\theta_i - 2\sin\theta_s \sin\theta_i \cos(\phi_s - \phi_i)] \tag{1.91}$$

While in the backscattering direction, it simplifies to

$$F_{hh}^B = \left| \frac{k_{zi} - k_{1zi}}{k_{zi} + k_{1zi}} \right|^2 = |R_{h0}|^2 \tag{1.92}$$

$$F_{vv}^B = \left| \frac{(k_1^2 - k^2)(k_1^2 k^2 \sin^2\theta_i + k^2 k_{1zi}^2)}{(k_1^2 k_{zi} + k^2 k_{1zi})^2} \right|^2 \tag{1.93}$$

$$F_{vh}^B = F_{hv}^B = 0 \tag{1.94}$$

where the superscript B denotes backscattering. Hence, the backscattering coefficient can be written as

$$\sigma_{hh}^{(SPA1)} = 16\pi k^4 \cos^4\theta_i |R_{h0}|^2 W(2k \sin\theta_i) \tag{1.95}$$

$$\sigma_{vv}^{(SPA1)} = 16\pi k^4 \cos^4\theta_i W(2k \sin\theta_i) \left| \frac{(k_1^2 - k^2)(k_1^2 k^2 \sin^2\theta_i + k^2 k_{1zi}^2)}{(k_1^2 k_{zi} + k^2 k_{1zi})^2} \right|^2 \tag{1.96}$$

$$\sigma_{vh}^{(SPA1)} = \sigma_{hv}^{(SPA1)} = 0 \tag{1.97}$$

For a Gaussian rough surface, this yields

$$\sigma_{hh}^{(SPA1)} = 4k^4\delta^2\ell^2 \cos^4\theta_i |R_{h0}|^2 \exp(-k^2\ell^2 \sin^2\theta_i) \tag{1.98}$$

$$\sigma_{vv}^{(SPA1)} = 4k^4\delta^2\ell^2 \cos^4\theta_i \exp(-k^2\ell^2 \sin^2\theta_i) \left| \frac{(k_1^2 - k^2)(k_1^2 k^2 \sin^2\theta_i + k^2 k_{1zi}^2)}{(k_1^2 k_{zi} + k^2 k_{1zi})^2} \right|^2 \tag{1.99}$$

Note that the term $|R_{h0}|^2$ is the horizontal-polarization Fresnel reflection coefficient. Interestingly, the factor in $\sigma_{vv}^{(SPA1)}$ can also be written as

$$\left| \frac{(k_1^2 - k^2)(k_1^2 k^2 \sin^2\theta_i + k^2 k_{1zi}^2)}{(k^2 k_{zi} + k^2 k_{1zi})^2} \right|^2 = \left| R_{v0} + \frac{(k_1^2 - k^2)T_{v0}^2 \sin^2\theta_i}{2k_1^2 \cos^2\theta_i} \right|^2 \tag{1.100}$$

where R_{v0} and T_{v0} denote the vertically polarized Fresnel reflection coefficient and Fresnel transmission coefficient, respectively.

Clearly, the first-order SPA solution has no cross-polarization contribution: $\sigma_{vv}^{(SPA1)} \neq \sigma_{hh}^{(SPA1)}$, $\sigma_{vh}^{(SPA1)} = \sigma_{hv}^{(SPA1)} = 0$. However, $\sigma_{hv}^{(SPA2)}$, $\sigma_{vh}^{(SPA2)} \neq 0$ for the second- or higher-order solutions (Tsang, Kong, and Shin, 1985).

1.3.5 Two-Scale Approximation

In reality, natural surface roughness exists at different scales and therefore cannot be dealt with by either the KA or SPA approach only. The two-scale approximation (TSA) approach is a simple combination of KA and SPA (Wu and Fung, 1972; Jin, 1994). As illustrated in Figure 1.7, the basic assumption for the TSA model is that there is small roughness riding on top of a large fluctuation.

Based on the independent KA and SPA solutions, the TSA combines the two after corresponding modifications. The KA scattering coefficient has to be modified to take into account the dimming effect by small roughness. On the other hand, the SPA solution has to be modified to take into account the local slope.

The TSA backscattering coefficient consists of two parts:

$$\sigma_{pp}(\hat{k}_i) = \sigma_{pp}^{KA}(\hat{k}_i) + \langle \sigma_{pp}^{SPA}(\hat{k}_i) \rangle \tag{1.101}$$

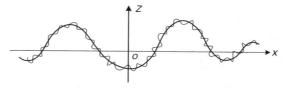

Figure 1.7 Two-scale model: small-scale roughness riding on large-scale fluctuation

The TSA bistatic coefficient is written as

$$\gamma_{pp}(\hat{k}_s, \hat{k}_i) = \gamma_{pp}^{KA}(\hat{k}_s, \hat{k}_i) + \langle \gamma_{pp}^{SPA}(\hat{k}_{s\ell}, \hat{k}_{i\ell}) \rangle \qquad (1.102)$$

The second term represents the ensemble average of the SPA solution over the distribution of local slope caused by large-scale fluctuation, that is

$$\langle \gamma_{pq}^{(SPA1)}(\hat{k}_{s\ell}, \hat{k}_{i\ell}) \rangle = \int_{-\cot \theta_2}^{\infty} d\xi_y \int_{-\cot \theta_1}^{\infty} d\xi_x \, \gamma_{pq}^{(SPA1)}(\hat{k}_{s\ell}, \hat{k}_{i\ell})$$

$$[1 + (\xi_x \cos \phi_i + \xi_y \sin \phi_i) \tan \theta_i] p(\xi_{x'}, \xi_{y'}) \qquad (1.103)$$

where $\cot \theta_1 = \cot \theta_i / \cos \phi_i$, $\cot \theta_2 = \cot \theta_i / \sin \phi_i$ is included to account for the self-shadowing effect. Let $\phi_i = 0$, then the integration ranges of $d\xi_x, d\xi_y$ become $(-\cot \theta_i, \infty), (-\infty, \infty)$, respectively. $P(\xi_{x'}, \xi_{y'})$ is the probability distribution function of local slope $(\xi_{x'}, \xi_{y'})$, which is defined as

$$\xi_{x'} \equiv \frac{\partial \xi}{\partial x'} = \xi_x \cos \phi_0 + \xi_y \sin \phi_0, \quad \xi_{y'} \equiv \frac{\partial \xi}{\partial y'} = \xi_y \cos \phi_0 - \xi_x \sin \phi_0 \qquad (1.104)$$

Then, the local coordinates $(\hat{x}_\ell, \hat{y}_\ell, \hat{z}_\ell)$ relate to global coordinates $(\hat{x}', \hat{y}', \hat{z})$ as

$$\begin{bmatrix} \hat{x}_\ell \\ \hat{y}_\ell \\ \hat{z}_\ell \end{bmatrix} = \begin{bmatrix} a_{11} & a_{12} & a_{13} \\ a_{21} & a_{22} & a_{23} \\ a_{31} & a_{32} & a_{33} \end{bmatrix} \cdot \begin{bmatrix} \hat{x}' \\ \hat{y}' \\ \hat{z} \end{bmatrix} \qquad (1.105)$$

where

$$a_{11} = \cos \theta_n \cos \phi_n, \, a_{12} = -\cos \theta_n \sin \phi_n, \, a_{13} = \sin \theta_n$$
$$a_{21} = \sin \phi_n, \, a_{22} = \cos \phi_n, \, a_{23} = 0$$
$$a_{31} = -\sin \theta_n \cos \phi_n, \, a_{32} = \sin \theta_n \sin \phi_n, \, a_{33} = \cos \theta_n$$

Furthermore, the local angles and local slopes are written as

$$\cos \theta_n = (1 + \xi_{x'}^2 + \xi_{y'}^2)^{1/2} = (1 + \xi_x^2 + \xi_y^2)^{1/2}, \, \tan \phi_n = \xi_{y'} / \xi_{x'}$$
$$\xi_{x'} = (\partial \xi / \partial \rho')(\partial \rho' / \partial x') = \tan \theta_n \cos \phi_n$$
$$\xi_{y'} = (\partial \xi / \partial \rho')(\partial \rho' / \partial y') = \tan \theta_n \sin \phi_n$$

Scattering from a randomly rough sea surface using an application of the TSA model can be found in Huang and Jin (1995).

1.3.6 Integral Equation Method

In practice, the KA is often used in circumstances of high frequency and small incident angle ($<20°$), while the SPA is suitable for cases where the frequency is low and the incident angle is large (20°–84°). The TSA combines the two types of scattering independently and thus is still limited by their respective constraints. Fung (1994) developed the integral equation method (IEM), which introduces a complementary field into the KA framework and takes into account the scattering caused by rapid fluctuations. Most of this book will employ the IEM solution to rough-surface scattering. Its validity condition is written as

$$k\delta \cdot k\ell < 1.2\sqrt{\varepsilon_1} \tag{1.106}$$

The Mueller matrix form of the IEM rough-surface scattering is written as (Fung, 1994)

$$\overline{\mathbf{M}}(\theta_s, \phi_s; \pi - \theta_i, \phi_i) = \begin{bmatrix} \langle|S_{vv}|^2\rangle & \langle|S_{vh}|^2\rangle & \mathrm{Re}\langle S_{vv}S_{vh}^*\rangle & -\mathrm{Im}\langle S_{vv}S_{vh}^*\rangle \\ \langle|S_{hv}|^2\rangle & \langle|S_{hh}|^2\rangle & \mathrm{Re}\langle S_{hv}S_{hh}^*\rangle & -\mathrm{Im}\langle S_{hv}S_{hh}^*\rangle \\ 2\mathrm{Re}\langle S_{vv}S_{hv}^*\rangle & 2\mathrm{Re}\langle S_{vh}S_{hh}^*\rangle & \mathrm{Re}\langle S_{vv}S_{hh}^*+S_{vh}S_{hv}^*\rangle & \mathrm{Im}\langle S_{vh}S_{hv}^*-S_{vv}S_{hh}^*\rangle \\ 2\mathrm{Im}\langle S_{vv}S_{hv}^*\rangle & 2\mathrm{Im}\langle S_{vh}S_{hh}^*\rangle & \mathrm{Im}\langle S_{vv}S_{hh}^*+S_{vh}S_{hv}^*\rangle & \mathrm{Re}\langle S_{vv}S_{hh}^*-S_{vh}S_{hv}^*\rangle \end{bmatrix} \tag{1.107}$$

where

$$\langle S_{pq}S_{rs}^*\rangle = \frac{k^2}{8\pi}\exp\left[-\delta^2(k_{zi}^2+k_{zs}^2)\right]\sum_{n=1}^{\infty}\delta^{2n}(I_{pq}^n I_{rs}^{n*})\frac{W^{(n)}(k_{dx}, k_{dy})}{n!}, \quad p,q,r,s=v,h \tag{1.108}$$

$$I_{pq}^n = (k_{zi}+k_{zs})^n f_{pq}\exp(-\delta^2 k_{zi}k_{zs}) + \frac{(k_{zs})^n F_{pq}(-k_{xi},-k_{yi})+(k_{zi})^n F_{pq}(-k_{xs},-k_{ys})}{2} \tag{1.109}$$

The KA coefficients are

$$f_{vv} = \frac{2R_v}{\cos\tilde{\theta}_i+\cos\theta_s}\left[\sin\tilde{\theta}_i\sin\theta_s-(1+\cos\tilde{\theta}_i\cos\theta_s)\cos(\varphi_s-\varphi_i)\right]$$

$$f_{hh} = \frac{2R_h}{\cos\tilde{\theta}_i+\cos\theta_s}\left[\sin\tilde{\theta}_i\sin\theta_s-(1+\cos\tilde{\theta}_i\cos\theta_s)\cos(\varphi_s-\varphi_i)\right] \tag{1.110}$$

$$f_{hv} = (R_h-R_v)\sin(\varphi_s-\varphi_i), \, f_{vh} = (R_v-R_h)\sin(\varphi_i-\varphi_s)$$

and the complementary field coefficients are

$$F_{vv} = -(cu_v-qt_v/\varepsilon_r)(t_vc_d+u_v\varepsilon_r c_1)+(v_v^2-ct_vu_v/q)c_2^s$$

$$F_{hh} = (cu_h-qt_h)(t_hc_d+u_hc_1)-(v_h^2-ct_hu_h/q)c_2^s$$

$$F_{hv} = (cu-qt/\varepsilon_r)(t/c_s+u\varepsilon_r/q)s_d+(u^2-ctu/q)s^2s_d$$

$$F_{vh} = (ct-qu)(u/c_s+tq)s_d+(t^2-ctu/q)s^2s_d \tag{1.111}$$

$$F_{vv}^s = -(c_su_v-q_st_v/\varepsilon_r)(t_vc_d+u_v\varepsilon_r c_1^s)+(t_v^2-c_st_vu_v/q_s)c_2^s$$

$$F_{hh}^s = (c_su_h-q_st_h)(t_hc_d+u_hc_1^s)-(t_h^2-c_st_hu_h/q_s)c_2^s$$

$$F_{hv}^s = -(c_st-q_su)(u/c_s+t/q_s)s_d-(u^2-c_stu/q_s)s_s^2s_d$$

$$F_{vh}^s = -(c_su-q_st/\varepsilon_r)(t/c_s+u\varepsilon_r/q_s)s_d-(t^2-c_stu/q_s)s_s^2s_d$$

where the numerous symbols employed are defined as

$$s = \sin\tilde{\theta}_i, \quad s_s = \sin\theta_s, \quad s_d = \sin(\varphi_s - \varphi_i)$$

$$c = \cos\tilde{\theta}_i, \quad c_s = \cos\theta_s, \quad c_d = \cos(\varphi_s - \varphi_i)$$

$$q = \sqrt{\varepsilon_r - \sin^2\tilde{\theta}_i}, \quad q_s = \sqrt{\varepsilon_r - \sin^2\theta_s}$$

$$c_1 = \frac{c_d - ss_s}{qc_s}, \quad c_1^s = \frac{c_d - ss_s}{q_s c}, \quad c_2 = s\frac{s_s - sc_d}{c_s}, \quad c_2^s = s_s\frac{s - s_s c_d}{c} \tag{1.112}$$

$$t_v = 1 + R_v, \quad t_h = 1 + R_h, \quad t = 1 + (R_v - R_h)/2$$

$$u_v = 1 - R_v, \quad u_h = 1 - R_h, \quad u = 1 - (R_v - R_h)/2$$

$$k_x = k\sin\tilde{\theta}_i\cos\varphi_i, \quad k_y = k\sin\tilde{\theta}_i\sin\varphi_i, \quad k_z = k\cos\tilde{\theta}_i$$

$$k_{sx} = k\sin\theta_s\cos\varphi_s, \quad k_{sy} = k\sin\theta_s\sin\varphi_s, \quad k_{sz} = k\cos\theta_s$$

For rough-surface transmission, the Mueller matrix $\overline{\overline{M}}(\theta_t, \phi_t; \pi - \theta_i, \phi_i)$ has exactly the same form as for scattering but with its elements replaced by (Fung, 1994)

$$\langle S_{pq}S_{rs}^* \rangle = \frac{k_1^2}{8\pi} \exp\left[-\delta^2(k_{zi}^2 + k_{zt}^2)\right] \sum_{n=1}^{\infty} \delta^{2n} (I_{pq}^n I_{rs}^{n*}) \frac{W^{(n)}(k_{1dx}, k_{1dy})}{n!}, \quad p,q,r,s = v,h \tag{1.113}$$

$$I_{pq}^n = (k_{zi} + k_{zt})^n f_{1pq} \exp(-\delta^2 k_{zi}k_{zt}) + \frac{(-k_{zt})^n F_{1pq}(-k_{xi}, -k_{yi}) + (k_{zi})^n F_{1pq}(-k_{xt}, -k_{yt})}{2} \tag{1.114}$$

$$
\begin{aligned}
f_{vv}^t &= Au_v + Bt_v, \quad f_{hh}^t = -Bt_h - Au_h \\
f_{hv}^t &= Cu + s_f t, \quad f_{vh}^t = s_f u + Ct \\
F_{vv}^t &= -(cu_v - qt_v/\varepsilon_r)(t_v c_f + u_v A_1) - (v_v^2 - ct_v u_v/q)A_2 \\
F_{hh}^t &= (cu_h - qt_h)(t_h c_f + u_h A_1)m/\varepsilon_r + (v_h^2 - ct_h u_h/q)A_2 m/\varepsilon_r
\end{aligned} \tag{1.115}
$$

$$
\begin{aligned}
F_{hv}^t &= (cu - qt/\varepsilon_r)(um/q - t/c_t)s_f - (u^2 - ctu/q)s^2 s_f/m \\
F_{vh}^t &= (ct - qu)(tm/q - t/c_t)s_f m/\varepsilon_r - (t^2 - ctu/q)s^2 s_f/m \\
F_{vv}^{tt} &= -(q_t u_v - c_t mt_v/\varepsilon_r)(t_v c_f + u_v A_1') - (c_t t_v^2/q_t - t_v u_v/m)A_2' \\
F_{hh}^{tt} &= (q_t u_h - c_t mt_h)(t_h c_f + u_h A_1')m/\varepsilon_r + (c_t t_h^2/q_t - t_h u_h/m)mA_2' \\
F_{hv}^{tt} &= (q_t t - c_t mu)(t/q_t - u/c)s_f/m + (c_t mu^2/q_t - tu)ms_t^2 s_f \\
F_{vh}^{tt} &= (q_t u - c_t mt/\varepsilon_r)(u/q_t - t/c)s_f + (c_t mt^2/q_t - tu)s_t^2 s_f
\end{aligned} \tag{1.116}
$$

and

$$s = \sin\tilde{\theta}_i, \quad s_t = \sin\tilde{\theta}_t, \quad s_f = \sin\varphi_t, \quad c = \cos\tilde{\theta}_i, \quad c_t = \cos\tilde{\theta}_t, \quad c_f = \cos\varphi_t$$

$$m = \sqrt{\varepsilon_r}, \quad q_t = \sqrt{1 - \varepsilon_r\sin^2\tilde{\theta}_t}$$

$$A = (c + sZ_x)c_f + s_f sZ_y, \quad B = c_t c_f + Z_x s_t$$

$$C = c_t s_f(c + sZ_x) + Z_y(s_t c - sc_t c_f)$$

$$Z_x = \frac{ms_t c_f - s}{mc_t - c}, \quad Z_y = \frac{ms_t s_f}{mc_t - c}$$

$$A_1 = \frac{s_t s - mc_f}{c_t mc_t}, \quad A_2 = \frac{s(s_t - sc_f/m)}{c_t}, \quad A_1' = \frac{ms_t s - c_f}{q_t c}, \quad A_2' = \frac{s_t(s - s_t c_f m)}{c} \tag{1.117}$$

$$t_v = 1 + R_v, \quad t_h = 1 + R_h, \quad t = 1 + (R_v - R_h)/2$$

$$u_v = 1 - R_v, \quad u_h = 1 - R_h, \quad u = 1 - (R_v - R_h)/2$$

$$k_x = k\sin\tilde{\theta}_i\cos\varphi_i, \quad k_y = k\sin\tilde{\theta}_i\sin\varphi_i, \quad k_z = k\cos\tilde{\theta}_i$$

$$k_{tx} = km\sin\tilde{\theta}_t\cos\varphi_t, \quad k_{ty} = km\sin\tilde{\theta}_t\sin\varphi_t, \quad k_{sz} = km\cos\tilde{\theta}_t$$

1.3.7 Tilted Surface or Oriented Object

A rough surface may be tilted as a result of topographic fluctuation. Terrain objects or build-ing surfaces may have different orientations in common global coordinates. Thus, a general method of coordinate conversion would be useful in adapting canonical scattering models for a tilted surface or oriented object.

As shown in Figure 1.8, the orientation of local coordinates $(\hat{x}', \hat{y}', \hat{z}')$ is defined by three Euler angles α, β, γ (Tsang, Kong, and Shin, 1985; Jin, 1994). Through three steps of rota-tion, $(\hat{x}', \hat{y}', \hat{z}')$ can be aligned with the global coordinates $(\hat{x}, \hat{y}, \hat{z})$: that is, first rotate along $\hat{z}'$ by angle α so that $\hat{y}'$ is located in the xoy plane ($\hat{y}'$ becomes $\hat{y}''$); then rotate along $\hat{z}'$ by angle β so that $\hat{y}''$ is overlapped with $\hat{z}$; finally, rotate along $\hat{z}$ by angle γ so that $\hat{y}''$ is overlapped with $\hat{y}$, so do the other two axes.

The coordinate conversion can be expressed in matrix form as

$$[x' \quad y' \quad z']^{\mathrm{T}} = \overline{\mathbf{T}} \cdot [x \quad y \quad z]^{\mathrm{T}}$$
$$T_{11} = \cos\gamma\cos\beta\cos\alpha - \sin\gamma\sin\alpha,$$
$$T_{12} = -\sin\gamma\cos\beta\cos\alpha - \cos\gamma\sin\alpha, \quad T_{13} = \sin\beta\cos\alpha$$
$$T_{21} = \cos\gamma\cos\beta\sin\alpha + \sin\gamma\cos\alpha, \quad\quad\quad\quad (1.118)$$
$$T_{22} = -\sin\gamma\cos\beta\sin\alpha + \cos\gamma\cos\alpha, \quad T_{23} = \sin\beta\sin\alpha$$
$$T_{31} = -\cos\gamma\sin\beta, \quad T_{32} = \sin\gamma\sin\beta, \quad T_{33} = \cos\beta$$

If the scatterer has rotational symmetry along $\hat{z}'$, without loss of generality let $\alpha = 0$ and it is simplified to

$$\overline{\mathbf{T}} = \begin{bmatrix} \cos\gamma\cos\beta & -\sin\gamma\cos\beta & \sin\beta \\ \sin\gamma & \cos\gamma & 0 \\ -\cos\gamma\sin\beta & \sin\gamma\sin\beta & \cos\beta \end{bmatrix} \quad\quad (1.119)$$

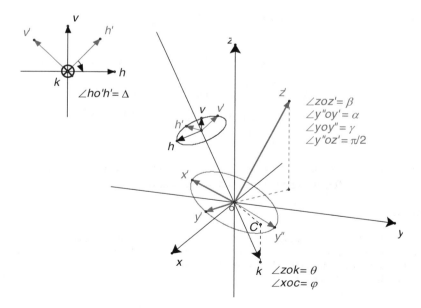

Figure 1.8 Global coordinates and local coordinates

In addition, the wave propagation vector $\hat{k}(\theta, \phi)$ defined in $(\hat{x}, \hat{y}, \hat{z})$ and the vector $\hat{k}(\theta', \phi')$ defined in $(\hat{x}', \hat{y}', \hat{z}')$ has the following conversion:

$$
\begin{aligned}
\theta' &= \cos^{-1}(-\sin\theta\cos\varphi\sin\beta\cos\gamma + \sin\theta\sin\varphi\sin\gamma\sin\beta + \cos\theta\cos\beta) \\
\cos\varphi'\sin\theta' &= \sin\theta\cos\varphi\cos\beta\cos\gamma - \sin\theta\sin\varphi\cos\beta\sin\gamma + \cos\theta\sin\beta \quad (1.120)\\
\mathrm{sgn}(\varphi') &= \mathrm{sgn}(\sin\theta\cos\varphi\sin\gamma + \sin\theta\sin\varphi\cos\gamma)
\end{aligned}
$$

Note that the polarization basis defined in two coordinate systems also differs by a certain angle Δ. Here, it is defined as the angle required to rotate $(\hat{v}', \hat{h}')$ clockwise (as seen from the $\hat{k}$ direction) to $(\hat{v}, \hat{h})$, that is

$$
\begin{cases}
\cos\Delta = \hat{h}' \cdot \hat{h} = (\hat{k} \times \dfrac{\hat{z}') \cdot (\hat{k} \times \hat{z})}{|\hat{z}' \times \hat{k}||\hat{z} \times \hat{k}|} = \dfrac{\cos\beta - \cos\theta\cos\theta'}{\sin\theta\sin\theta'} \\[4mm]
\mathrm{sgn}(\Delta) = \mathrm{sgn}[(\hat{h}' \times \hat{h}) \cdot \hat{k}] = \mathrm{sgn}[\sin(\varphi + \gamma)]
\end{cases}
\quad (1.121)
$$

Now we can conclude the coordinate conversions of the scattering matrix $\overline{S}(\theta_s, \varphi_s; \theta_i, \varphi_i)$, $\overline{S}'(\theta'_s, \varphi'_s; \theta'_i, \varphi'_i)$ and the Mueller matrix $\overline{M}(\theta_s, \varphi_s; \theta_i, \varphi_i)$, $\overline{M}'(\theta'_s, \varphi'_s; \theta'_i, \varphi'_i)$:

$$
\overline{S}(\theta_s, \varphi_s; \theta_i, \varphi_i) = \overline{U}(\Delta_s) \cdot \overline{S}'(\theta'_s, \varphi'_s; \theta'_i, \varphi'_i) \cdot \overline{U}^{-1}(\Delta_i) \quad (1.122)
$$

$$
\overline{M}(\theta_s, \varphi_s; \theta_i, \varphi_i) = \overline{U}_{4\times4}(\Delta_s) \cdot \overline{M}'(\theta'_s, \varphi'_s; \theta'_i, \varphi'_i) \cdot \overline{U}_{4\times4}^{-1}(\Delta_i) \quad (1.123)
$$

where the polarization basis rotation matrix is given as (Xu and Jin, 2006)

$$
\overline{U}(\Delta) =
\begin{bmatrix}
\cos\Delta & \sin\Delta \\
-\sin\Delta & \cos\Delta
\end{bmatrix}
\quad (1.124)
$$

$$
\overline{U}_{4\times4}(\Delta) =
\begin{bmatrix}
\cos^2\Delta & \sin^2\Delta & \dfrac{1}{2}\sin2\Delta & 0 \\[3mm]
\sin^2\Delta & \cos^2\Delta & -\dfrac{1}{2}\sin2\Delta & 0 \\[3mm]
-\sin2\Delta & \sin2\Delta & \cos2\Delta & 0 \\[2mm]
0 & 0 & 0 & 1
\end{bmatrix}
\quad (1.125)
$$

With the above equations, we can always calculate the new scattering matrix and Mueller matrix of a scatterer after being rotated to a new orientation.

References

Bass, F.G. and Fuks, I.M. (1979) *Wave Scattering from Statistically Rough Surface*, Pergamon, New York.

Chen, M.F. and Fung, A.K. (1988) A numerical study of the regions of validity of Kirchhoff and small perturbed rough surface scattering models. *Radio Science*, **24** (2), 163–170.

Fung, A.K. (1994) *Microwave Scattering and Emission Models and Their Applications*, Artech House, Norwood, MA.

Fung, A.K. and Lee, K.K. (1982) A semi-empirical sea-spectrum model for scattering coefficient estimation. *IEEE Journal of Oceanic Engineering*, **7** (4), 166–176.

Huang, X. and Jin, Y.Q. (1995) Scattering and emission from two-scale randomly rough sea surface with foam scatterers. *IEE Proceedings H – Microwaves, Antennas, and Propagation*, **142** (2), 109–114.

Ishimaru, A. (1978) *Wave Propagation and Scattering in Random Media*, Academic Press, New York.

Ishimaru, A. (1991) *Electromagnetic Wave Propagation, Radiation, and Scattering*, Prentice-Hall, Englewood Cliffs, NJ.

Jin, Y.Q. (1994) *Electromagnetic Scattering Modelling for Quantitative Remote Sensing*, World Scientific, Singapore.

Karam, M.A. and Fung, A.K. (1988) Electromagnetic scattering from a layer of finite-length randomly oriented dielectric circular cylinders. *International Journal of Remote Sensing*, **9** (6), 1109–1134.

Kong, J.A. (2005) *Electromagnetic Wave Theory*, EMW Publishing, Cambridge, MA.

Rice, S.O. (1963) *Reflection of Electromagnetic Waves by Slightly Rough Surfaces*, Interscience, New York.

Tsang, L., Kong, J.A., and Shin, B. (1985) *Theory of Microwave Remote Sensing*, John Wiley & Sons, Inc., New York.

Ulaby, F.T., Moore, R.K., and Fung, A.K. (1982) *Microwave Remote Sensing: Active and Passive*, Vol. **II**: *Radar Remote Sensing and Surface Scattering and Emission Theory*, Addison-Wesley, Reading, MA.

Wait, J.R. (1986) *Introduction to Antennas and Propagation*, Peter Peregrinus, London.

Wu, S.T. and Fung, A.K. (1972) A non-coherent model for microwave emission and backscattering from the sea surface. *Journal of Geophysical Research*, **77** (30), 5917–5929.

Xu, F. and Jin, Y.Q. (2006) Imaging simulation of polarimetric synthetic aperture radar remote sensing for comprehensive terrain scene using the mapping and projection algorithm. *IEEE Transactions on Geoscience and Remote Sensing*, **44** (11), 3219–3234.

2

Vector Radiative Transfer

Modeling of multiple scattering from layered random media is often studied in remote sensing. Radiative transfer theory is often used to describe absorption and single and multiple scattering of random media, such as rainfall, the vegetation canopy, snowpack, and so on. Vector radiative transfer theory is used to solve polarimetric scattering of complex or random media. In this chapter, we first introduce the radiative transfer equation and its components, as well as the boundary conditions, the Mueller matrix solutions, and so on, which lay the basis for the next few chapters. Some results of VRT for modeling the vegetation canopy and snowpack, as well as some numerical examples, are presented.

2.1 Radiative Transfer Equation

The VRT theory of random media is a physical–mathematical method for studying the scattering, multiple scattering, absorption, and transmission of polarized electromagnetic intensity by or through random scatterers or continuous random media. The radiative transfer (RT) theory was first studied to explain the appearance of the absorption and emission lines of stellar spectra (Chandrasekhar, 1960). Since then, RT theory has been extensively studied and applied in many areas: astrophysics, transport in neutron diffusion, heat transfer in engineering physics, atmospheric radiation, clouds and precipitation, light propagation through the atmosphere, and so on. RT theory has a simple physical meaning based on the principle of energy conservation, and takes account of absorption, multiple scattering, and emission. One of the advantages of RT theory is the capability to calculate multiple scattering numerically. Over recent decades, the RT theory of polarized electromagnetic intensity, that is, vector radiative transfer (VRT) theory, has been studied and applied to both active and passive remote sensing.

2.1.1 Specific Intensity and Stokes Vector

We define the specific intensity $I_\nu(\mathbf{r}, \hat{s})$ to express the amount of power dP flowing within a solid angle $d\Omega$ through an elementary area da in the frequency range $(\nu, \nu + d\nu)$ (see Figure 2.1):

$$dP = I_\nu(\mathbf{r}, \hat{s})\cos\theta \, da \, d\Omega \, d\nu \tag{2.1}$$

Polarimetric Scattering and SAR Information Retrieval, First Edition. Ya-Qiu Jin and Feng Xu.
© 2013 John Wiley & Sons Singapore Pte. Ltd. Published 2013 by John Wiley & Sons Singapore Pte. Ltd.

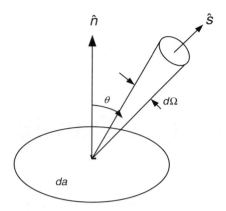

Figure 2.1   Specific intensity

where θ is the angle between the propagation direction of the specific intensity $\hat{s}$ and the outward normal of the area $\hat{n}$. Generally, $I_v(\mathbf{r}, \hat{s})$ is dependent on $\mathbf{r}$ and $\hat{s}$. If the specific intensities for all directions at a point $\mathbf{r}$ are the same, this point is known as "isotropic"; if the specific intensities for all directions and at all points in the medium are the same, then the medium is uniform and isotropic.

Integrating I_v over a small frequency interval dv, the specific intensity at the frequency v is

$$I = \int_{v-dv/2}^{v+dv/2} I_{v'} \, dv' \tag{2.2}$$

which is the definition of specific intensity.

The unit of the radial Poynting vector S_r is watts per square meter per steradian $(\text{W m}^{-2} \text{ sr}^{-1})$ and has a factor of $1/r^2$, which is different from I of Equation 2.1. The power of the spherical wave on the area $d\sigma$ is

$$dP = \mathbf{S}_r \cdot \hat{\sigma} \, d\sigma = S_r r^2 d\Omega \tag{2.3}$$

From Equations 2.1 and 2.3, we have

$$I = \frac{1}{\cos \theta \, da} r^2 S_r \tag{2.4}$$

The power of a plane wave on the plane area da is

$$dP = S_p \cos \theta \, da \tag{2.5}$$

Together with Equation 2.1, it yields

$$S_p = I \, d\Omega \tag{2.6}$$

As discussed in Chapter 1, any time-harmonic elliptically polarized electromagnetic wave with $\exp(-i\omega t)$ can be expressed as the summation of vertically and horizontally linear polarized waves, $\mathbf{E}(t) = \hat{v} E_v + \hat{h} E_h$. These two linearly polarized waves have their own amplitudes and their phases are independent. So, we have three independent variables to describe a polarized wave.

If the electromagnetic wave is perfectly monochromatic, E_v and E_h are independent of time. We can define the four Stokes parameters for a completely polarized wave as

$$I_v = \frac{1}{\eta}|E_v|^2, \quad I_h = \frac{1}{\eta}|E_h|^2, \quad U = \frac{2}{\eta}\mathrm{Re}(E_v E_h^*), \quad V = \frac{2}{\eta}\mathrm{Im}(E_v E_h^*) \tag{2.7}$$

where η is the wave impedance $\sqrt{\mu/\varepsilon}$ of the medium. All of the four Stokes parameters have the same unit as the intensity, which should be more convenient to use than the amplitude and phase with different units. We may write

$$E_v = |E_v|e^{i\phi_v}, \quad E_h = |E_h|e^{i\phi_h} \tag{2.8}$$

Substituting Equation 2.8 into Equation 2.7 yields

$$4I_v I_h = U^2 + V^2 \tag{2.9}$$

Thus, only three of the four Stokes parameters are independent.

Actually, electromagnetic waves are often quasi-monochromatic with E_v and E_h randomly fluctuating with time, that is, partially polarized waves. Thus, $|E_v|^2$, $|E_h|^2$, and $E_v E_h^*$ in the definitions of Equation 2.7 should be the averages of the stochastic process. The Stokes vector is then defined as

$$I_v = \frac{1}{\eta}\langle|E_v|^2\rangle, \quad I_h = \frac{1}{\eta}\langle|E_h|^2\rangle \tag{2.10a}$$

$$U = \frac{2}{\eta}\mathrm{Re}\langle E_v E_h^*\rangle, \quad V = \frac{2}{\eta}\mathrm{Im}\langle E_v E_h^*\rangle \tag{2.10b}$$

As shown in Equations 1.12 and 1.15 of Chapter 1, the first two Stokes parameters, I_v and I_h, sometimes are replaced by

$$I = I_v + I_h, \quad Q = I_v - I_h \tag{2.11}$$

When independent waves, for example, scattered waves from different scatterers, arrive at the observer, the total Stokes parameters are simply the sum of the Stokes parameters of each wave. For example,

$$E_v = E_{v1} + E_{v2}, \quad \langle E_{v1} E_{v2}^*\rangle = 0 \tag{2.12}$$

This gives

$$I_v = \frac{1}{\eta}|E_{v1} + E_{v2}|^2 = I_{v1} + I_{v2} + \frac{2}{\eta}\mathrm{Re}(E_{v1}E_{v2}^*) = I_{v1} + I_{v2} \tag{2.13}$$

The four Stokes parameters compose the Stokes vector – specific intensity in the VRT equation. In the infrared frequencies, it is often termed the radiance.

2.1.2   Thermal Emission and Brightness Temperature

All objects can absorb, reflect or scatter incident electromagnetic radiation, and can also emit radiation. This radiation is thermal electromagnetic emission. The emissivity e of an object is equal to its absorptivity a (Peake, 1959). The absorptivity of a body is defined as the ratio of the total thermal energy absorbed by the surface to the total thermal energy incident upon it. A black body has maximum absorptivity $a = 1$ and is a perfect emitter. The absorptivity is usually dependent upon radiation direction, frequency, and polarization.

From Planck's radiation law of quantum physics, the emission intensity of an object with emissivity e is

$$I_e = e\mu\varepsilon' \frac{h v^3}{\exp(h v/BT) - 1} \tag{2.14}$$

where the subscript e in I_e denotes thermal emission, ε' is the real part of the dielectric constant of the object, h is Planck's constant (6.634×10^{-34} J s), B is Boltzmann's constant (1.38×10^{-23} J K^{-1}), and T is the absolute temperature (K). In the Rayleigh–Jeans low-frequency limit, for example, microwave region, we can apply the approximation $h v/BT \ll 1$ to yield

$$I_e = e\frac{B\mu\varepsilon'}{\lambda^2\mu_0\varepsilon_0}T = eCT \tag{2.15}$$

where λ is the wavelength in free space, and $C = B\varepsilon'/(\lambda^2\varepsilon_0)$ if we let $\mu = \mu_0$. In microwave remote sensing, the source of thermal emission can be written as in Equation 2.15. If we consider the thermal emission of an elementary volume with unit length, the emissivity e in Equation 2.15 is then replaced by the absorption coefficient κ_a (m^{-1}). Thus, the Stokes vector of thermal emission of an elementary volume with unit length and physical temperature T is

$$\mathbf{I}_e = \overline{\overline{\boldsymbol{\kappa}}}_a \cdot C \begin{bmatrix} T \\ T \\ 0 \\ 0 \end{bmatrix} \tag{2.16}$$

where $\overline{\overline{\boldsymbol{\kappa}}}_a$ = diagonal matrix $[\kappa_{av}, \kappa_{ah}, 0, 0]$.

In passive microwave remote sensing, the thermal emissions from the target and background are the sources for radiative transfer. The brightness temperature observed by a radiometer is defined as

$$T_{Bp}(\theta, \phi) = \frac{1}{C} I_p(\theta, \phi) = e_p T \qquad (2.17)$$

where the subscript p denotes the polarization, v or h, and (θ, ϕ) is the observation direction. The brightness temperature of the object is equal to the physical temperature of the black body that can emit the same amount of radiation power as the object.

2.1.3 Vector Radiative Transfer Equation

Consider the intensity change caused by scattering and absorption through a cylindrical elementary volume with a unit cross-section and length ds embedded with n random spherical particles, as shown in Figure 2.2. From energy conservation, the intensity change can be written as a conservative form

$$dI(\mathbf{r}, \hat{s}) = -ds(\kappa_{as} + \kappa_s)nI(\mathbf{r}, \hat{s}) - ds\kappa_{ab}I(\mathbf{r}, \hat{s}) + I_e(\mathbf{r}, \hat{s})ds + n\int d\hat{\Omega}' p(\hat{s}, \hat{s}')I(\mathbf{r}, \hat{s}') \qquad (2.18)$$

Equation 2.18 means that the change dI is due to absorption $n\kappa_{as}$ and scattering $n\kappa_s$ of n particles in the volume, absorption κ_{ab} of the background medium in the volume, the thermal emission I_e from the volume, and the contribution of multiple scattering from all other directions $\hat{\Omega}'$. The function $p(\hat{s}, \hat{s}')$ coupling the incident $I(\mathbf{r}, \hat{s}')$ with $I(\mathbf{r}, \hat{s})$ is termed the phase function. Equation 2.18 is the scalar radiative transfer (RT) equation. It can be understood that the derivation of Equation 2.18 is heuristic, and independent scattering of particles is always assumed.

It is known that if the fractional volume of scatterers is higher than 0.1, the scattering medium is seen as a dense medium, and the independent scatterings assumption for dense scatterers becomes invalid. The dense-medium RT can be found in Tsang, Kong, and Shin (1984) and Jin (1994), for example.

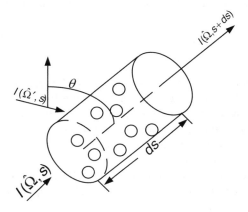

Figure 2.2   Derivation of the RT equation

Define the albedo as $\omega_0 \equiv \kappa_s/\kappa_e$ and the extinction coefficient as $\kappa_e \equiv \kappa_s + \kappa_a$ for single scattering. Then, $1 - \omega_0$ is the absorbed energy during single scattering. The phase function should be

$$\frac{1}{4\pi} \int p(\hat{s}, \hat{s}') \, d\hat{\Omega}' = \omega_0 \leq 1 \tag{2.19a}$$

So, the term for multiple scattering in Equation 2.18 is usually written as a normalized form

$$\frac{\omega_0}{4\pi} \int p_n(\hat{s}, \hat{s}') I(\mathbf{r}, \hat{s}') \, d\hat{\Omega}' \tag{2.19b}$$

The simplest case is isotropic scattering, $p(\cos \Theta) = \omega_0$, where $\Theta = \cos^{-1}(\hat{s} \cdot \hat{s}')$. For small spherical particles, it can be shown (Chandrasekhar, 1960) that

$$p(\cos \Theta) = \frac{3}{4}\omega_0(1 + \cos^2\Theta) \tag{2.20}$$

For a simple anisotropic scattering, it is usually assumed that

$$p(\cos \Theta) = \omega_0(1 + x \cos \Theta), \quad -1 \leq x \leq 1 \tag{2.21}$$

Generally, $p(\cos \Theta)$ can be expanded in terms of Legendre polynomials

$$p(\cos \Theta) = \sum_{n=0}^{\infty} \omega_n P_n(\cos \Theta) \tag{2.22}$$

where ω_n are constants.

We now extend Equation 2.18 to the case of polarized electromagnetic intensity, that is, four Stokes parameters. The vector form of the RT equation is written as

$$\frac{d\mathbf{I}(\mathbf{r}, \hat{s})}{ds} = -\overline{\overline{\kappa}}_e(\hat{s}) \cdot \mathbf{I}(\mathbf{r}, \hat{s}) + \mathbf{I}_e(\mathbf{r}, \hat{s}) + \int d\hat{s}' \overline{\overline{\mathbf{P}}}(\hat{s}, \hat{s}') \cdot \mathbf{I}(\mathbf{r}, \hat{s}') \tag{2.23}$$

where the extinction matrix $\overline{\overline{\kappa}}_e$ defines the attenuation of $\mathbf{I}(\mathbf{r}, \hat{s})$ due to scattering and absorption. The phase matrix $\overline{\overline{\mathbf{P}}}(\hat{s}, \hat{s}')$ defines angular scattering from incident direction $\hat{s}'$ to scattered direction $\hat{s}$. The physical meaning of the VRT equation is apparent: as wave $\mathbf{I}(\mathbf{r}, \hat{s})$ transfers over distance ds in direction $\hat{s}$, the differential change is the sum of three parts, that is, the attenuation defined by $\overline{\overline{\kappa}}_e$, the thermal emission $\mathbf{I}_e(\mathbf{r}, \hat{s})$, and the scattering contribution from all directions toward $\hat{s}$, which is defined by $\overline{\overline{\mathbf{P}}}(\hat{s}, \hat{s}')$.

It should be noted that all the components of the VRT equation are now in 4×4 matrix form. In the microwave region, we have microwave radiance $\mathbf{I}_e = (\kappa_{as} + \kappa_{ab})C\mathbf{T}$, where $\mathbf{T}$ is absolute temperature.

2.2 Components in Radiative Transfer Equation

2.2.1 Scattering, Absorption, and Extinction Coefficients

Following Chapter 1, when a polarized EM wave is incident upon a single scatterer, Equation 1.10 gives the scattering matrix to describe the coupling relation between the incident and scattered fields as

$$\begin{bmatrix} E_{sv} \\ E_{sh} \end{bmatrix} = \frac{e^{ikr}}{r} \begin{bmatrix} S_{vv} & S_{vh} \\ S_{hv} & S_{hh} \end{bmatrix} \cdot \begin{bmatrix} E_{vi} \\ E_{hi} \end{bmatrix} \equiv \frac{e^{ikr}}{r} \overline{\mathbf{S}} \cdot \mathbf{E}_i \tag{2.24}$$

where the elements of the scattering matrix, that is, $S_{pq}(\theta_s, \phi_s; \theta_i, \phi_i)$, $p, q = v, h$, are also referred to the scattering amplitude functions of this single scatterer.

If there are n_0 scatterers in a unit volume, the polarized electromagnetic intensity passing through the volume will lose part of this energy due to scattering. The scattering coefficients of vertical and horizontal polarizations are defined as

$$\kappa_{sv}(\theta, \phi) = n_0 \int_0^{2\pi} d\phi' \int_0^\pi d\theta' \sin\theta' [|S_{vv}(\theta', \phi'; \theta, \phi)|^2 + |S_{hv}(\theta', \phi'; \theta, \phi)|^2] \tag{2.25a}$$

$$\kappa_{sh}(\theta, \phi) = n_0 \int_0^{2\pi} d\phi' \int_0^\pi d\theta' \sin\theta' [|S_{hh}(\theta', \phi'; \theta, \phi)|^2 + |S_{vh}(\theta', \phi'; \theta, \phi)|^2] \tag{2.25b}$$

For spherical particles, clearly we have $\kappa_{sv}(\theta) = \kappa_{sh}(\theta) = \kappa_s$.

The absorption coefficient κ_a includes the absorption κ_{as} of the scatterers and the absorption κ_{ab} of the background medium. If the fractional volume of the scatterers is $f_s = n_0 v_0$ and the medium occupies $1 - f_s$, then we have

$$\kappa_{ab} = 2k''(1 - f_s) \tag{2.26}$$

where k'' is the imaginary part of wavenumber $k = \omega\sqrt{\mu_0 \varepsilon}$ of the background. The absorption coefficient of the scatterer can be obtained following the definition

$$\kappa_{as} = n_0 \frac{\int d\mathbf{r}' \frac{1}{2} \varepsilon''_{,s} \omega |\mathbf{E}(\mathbf{r}')|^2}{|E_i|^2/(2\eta)} \tag{2.27}$$

where the integration is over the scatterer volume; $\mathbf{E}(\mathbf{r}')$ is the internal field of the scatterer.

If it is difficult to solve the internal field, the optical theorem can be used to calculate the extinction coefficient as (Tsang, Kong, and Shin, 1984)

$$\kappa_{ep}(\theta, \phi) = \frac{4\pi}{k} n_0 \, \text{Im}[\hat{p} \cdot \overline{\mathbf{S}}(\theta, \phi; \theta, \phi) \cdot \hat{p}], \quad p = v, h \tag{2.28}$$

where the forward scattering matrix $S_{pq}(\theta, \phi; \theta, \phi)$ is used. Note that forward scattering means that the scattering direction is the same as the incident direction. Thus, instead of solving Equation 2.27, the absorption coefficient can now be calculated as

$$\kappa_{as} = \kappa_{ep} - \kappa_{sp}, \quad p = v, h \tag{2.29}$$

2.2.2 Extinction Matrix

However, the extinction coefficient of non-spherical scatterers is in fact a 4×4 matrix, generally non-diagonal. It cannot be calculated simply by the optical theorem. The electric field in propagation consists of the coherent (average) field and the incoherent (diffuse) field. The extinction can be essentially identified with the attenuation of coherent waves, whose phases preserve coherency in propagation. From the optical theorem, Equation 2.28, the intensity extinction is $\kappa_{ep}(\theta, \phi) = (4\pi/k)n_0 \, \text{Im}[S_{pp}(\theta, \phi; \theta, \phi)]$. Thus, as a coherent wave with vertically and horizontally polarized components E_v, E_h propagates along $\hat{s}$, its propagation equations of the coherent fields are written as (Oguchi, 1983; Ishimaru and Yeh, 1984; Tsang, Kong, and Shin, 1984)

$$\frac{d}{ds} E_v = (ik + F_{vv})E_v + F_{vh}E_h \tag{2.30a}$$

$$\frac{d}{ds} E_h = (ik + F_{hh})E_v + F_{hv}E_v \tag{2.30b}$$

where $F_{pq} = i(2\pi/k)n_0\langle S_{pq}^0 \rangle$ is obtained by the optical theorem, Equation 2.28, and denotes attenuations of the co-polarized (vv or hh) and depolarized (vh or hv) waves; and S_{pq}^0 denotes the elements of the forward scattering matrix $\overline{\mathbf{S}}^0 = \overline{\mathbf{S}}(\theta, \varphi; \theta, \varphi)$. The angular brackets, $\langle \rangle$, denote the ensemble average. Using the matrix eigenvalue technique to solve Equations 2.30a and 2.30b, we obtain two characteristic propagation wavenumbers (Ishimaru and Yeh, 1984):

$$K_1 = k + n_0 \frac{\pi}{k} \left[\langle S_{vv}^0 \rangle + \langle S_{hh}^0 \rangle + R \right] \tag{2.31a}$$

$$K_2 = k + n_0 \frac{\pi}{k} \left[\langle S_{vv}^0 \rangle + \langle S_{hh}^0 \rangle - R \right] \tag{2.31b}$$

where $R = [(\langle S_{vv}^0 \rangle - \langle S_{hh}^0 \rangle)^2 + 4\langle S_{hv}^0 \rangle\langle S_{vh}^0 \rangle]^{1/2}$, the sign of R thus being determined by the sign of the real part of $\langle S_{vv}^0 \rangle - \langle S_{hh}^0 \rangle$.

The eigenvectors of Equations 2.30a and 2.30b are $[1, b_1]$ and $[b_2, 1]$ with

$$b_1 = \frac{2\langle S_{hv}^0 \rangle}{\langle S_{vv}^0 \rangle - \langle S_{hh}^0 \rangle + R} \tag{2.32a}$$

$$b_2 = \frac{2\langle S_{vh}^0 \rangle}{-\langle S_{vv}^0 \rangle + \langle S_{hh}^0 \rangle - R} \tag{2.32b}$$

If $\langle S_{hv}^0 \rangle = \langle S_{vh}^0 \rangle = 0$ for uniformly random non-spherical particles, this leads to $b_1 = b_2 = 0$. Then, K_1 and K_2 of Equations 2.31a and 2.31b are simply reduced to vertically and horizontally polarized waves, respectively,

$$K_1 = k + 2n_0 \frac{\pi}{k} \langle S_{vv}^0 \rangle \quad \text{and} \quad K_2 = k + 2n_0 \frac{\pi}{k} \langle S_{hh}^0 \rangle \tag{2.33}$$

Making

$$\frac{1}{2\eta} E_v^* \frac{d}{ds} E_v + \frac{1}{2\eta} E_v \frac{d}{ds} E_v^*$$

and using Equations 2.30a and 2.30b, this yields

$$\frac{d}{ds} I_v = 2\operatorname{Re}(F_{vv})I_v + \operatorname{Re}(F_{vh})U + \operatorname{Im}(F_{vh})V \tag{2.34a}$$

Similarly, we have

$$\frac{d}{ds} I_h = 2\operatorname{Re}(F_{hh})I_v + \operatorname{Re}(F_{vh})U - \operatorname{Im}(F_{hv})V \tag{2.34b}$$

$$\frac{d}{ds} U = 2\operatorname{Re}(F_{hv})I_v + 2\operatorname{Re}(F_{vh})I_h + [\operatorname{Re}(F_{vv}) + \operatorname{Re}(F_{hh})]U - [\operatorname{Im}(F_{vv} - \operatorname{Im}(F_{hh})]V \tag{2.34c}$$

$$\frac{d}{ds} V = -2\operatorname{Im}(F_{hv})I_v + 2\operatorname{Im}(F_{vh})I_h + [\operatorname{Im}(F_{vv}) - \operatorname{Im}(F_{hh})]U + [\operatorname{Re}(F_{vv}) + \operatorname{Re}(F_{hh})]V \tag{2.34d}$$

Equations 2.34a–2.34d yield the propagation equation of the coherent Stokes vector

$$\frac{d}{ds} \mathbf{I}_c = -\overline{\overline{\boldsymbol{\kappa}}}_e \cdot \mathbf{I}_c \tag{2.35}$$

where the subscript c denotes the coherent wave. The extinction matrix in Equation 2.35 can be readily obtained from Equations 2.34a–2.34d,

$$\overline{\overline{\boldsymbol{\kappa}}}_e(\theta, \phi) = \frac{2\pi}{k} n_0 \begin{bmatrix} 2\operatorname{Im}\langle S_{vv}^0 \rangle & 0 & \operatorname{Im}\langle S_{vh}^0 \rangle & -\operatorname{Re}\langle S_{vh}^0 \rangle \\ 0 & 2\operatorname{Im}\langle S_{hh}^0 \rangle & \operatorname{Im}\langle S_{hv}^0 \rangle & \operatorname{Re}\langle S_{hv}^0 \rangle \\ 2\operatorname{Im}\langle S_{hv}^0 \rangle & 2\operatorname{Im}\langle S_{vh}^0 \rangle & \operatorname{Im}\langle S_{vv}^0 + S_{hh}^0 \rangle & \operatorname{Re}\langle S_{vv}^0 - S_{hh}^0 \rangle \\ 2\operatorname{Re}\langle S_{hv}^0 \rangle & -2\operatorname{Re}\langle S_{vh}^0 \rangle & \operatorname{Re}\langle S_{hh}^0 - S_{vv}^0 \rangle & \operatorname{Im}\langle S_{vv}^0 + S_{hh}^0 \rangle \end{bmatrix} \tag{2.36}$$

It can be found that $\overline{\overline{\boldsymbol{\kappa}}}_e$ would reduce to a scalar if the scatterer is a perfect sphere and to a diagonal matrix if the scatterers have a uniform distribution in a medium. Note that the scattering matrix $\overline{\mathbf{S}}$ must be accurate enough in Equation 2.28, otherwise it might cause $\kappa_e \to \kappa_a$. This is the reason why an additional imaginary term it_n^2 is included in T_n of Equation 1.25.

We seek the solution of Equation 2.35 with dependence $\exp(-\beta s)$ and obtain four eigenvalues

$$\beta_1 = 2\,\text{Im}(K_1), \quad \beta_2 = 2\,\text{Im}(K_2), \quad \beta_3 = i(K_1^* - K_2), \quad \beta_4 = i(K_2^* - K_1) \tag{2.37}$$

It may be seen that β_1 and β_2 are the extinction of two characteristic waves; and β_3 and β_4 are the difference of the phase and attenuation between two waves.

The eigenvectors of Equation 2.35 can be solved by its eigenmatrix equation

$$\overline{\mathbf{E}} \cdot \boldsymbol{\beta} = \overline{\kappa}_e \cdot \mathbf{E} \tag{2.38}$$

where $\overline{\boldsymbol{\beta}}$ is a diagonal matrix with the iith element β_i, $i = 1,2,3,4$, that is, $\overline{\boldsymbol{\beta}} = \text{diag}\exp(\beta_i * s)$. The solution of Equation 2.38 is expressed as (Tsang, Kong, and Shin, 1984)

$$\overline{\mathbf{E}}(\theta, \phi) = \begin{bmatrix} 1 & |b_2|^2 & b_2 & b_2^* \\ |b_1|^2 & 1 & b_1^* & b_1 \\ 2\,\text{Re}\,b_1 & 2\,\text{Re}\,b_2 & 1 + b_1^* b_2 & 1 + b_1 b_2^* \\ -2\,\text{Re}\,b_1 & 2\,\text{Im}\,b_2 & i(1 - b_1^* b_2) & -i(1 - b_1 b_2^*) \end{bmatrix} \tag{2.39}$$

where the first column is the first eigenvector corresponding to the first eigenvalue, and so on.

Only under the condition $\langle S_{vh}^0 \rangle = \langle S_{hv}^0 \rangle = 0$ does $b_1 = b_2 = 0$, and then $\overline{\mathbf{E}}$ becomes a constant matrix with no dependence on (θ, ϕ):

$$\overline{\mathbf{E}} = \begin{bmatrix} 1 & 0 & 0 & 0 \\ 0 & 1 & 0 & 0 \\ 0 & 0 & 1 & 1 \\ 0 & 0 & i & -i \end{bmatrix} \quad \text{and} \quad \overline{\mathbf{E}}^{-1} = \begin{bmatrix} 1 & 0 & 0 & 0 \\ 0 & 1 & 0 & 0 \\ 0 & 0 & \dfrac{1}{2} & -\dfrac{1}{2}i \\ 0 & 0 & \dfrac{1}{2} & \dfrac{1}{2}i \end{bmatrix} \tag{2.40}$$

and

$$\beta_1 = 4n_0 \frac{\pi}{k}\,\text{Im}\langle S_{vv}^0 \rangle, \quad \beta_2 = 4n_0 \frac{\pi}{k}\,\text{Im}\langle S_{hh}^0 \rangle,$$

$$\beta_3 = i2n_0 \frac{\pi}{k}\left[\langle S_{vv}^0 \rangle^* - \langle S_{hh}^0 \rangle\right], \quad \beta_4 = i2n_0 \frac{\pi}{k}\left[\langle S_{hh}^0 \rangle^* - \langle S_{vv}^0 \rangle\right]$$

Letting an electromagnetic wave propagate in direction θ through distance z in a random medium with the condition $\langle S_{vh}^0 \rangle = \langle S_{hv}^0 \rangle = 0$, the scattered Stokes vector can be written as

$$\mathbf{I}_s = \overline{\mathbf{E}} \cdot \overline{\mathbf{D}} \cdot \mathbf{E}^{-1} \cdot \mathbf{I}_0 = \overline{\mathbf{E}} \cdot \begin{bmatrix} e^{-\beta_1 z \sec \theta} & 0 & 0 & 0 \\ 0 & e^{-\beta_1 z \sec \theta} & 0 & 0 \\ 0 & 0 & e^{-\beta_3 z \sec \theta} & 0 \\ 0 & 0 & 0 & e^{-\beta_4 z \sec \theta} \end{bmatrix} \cdot \overline{\mathbf{E}}^{-1} \cdot \begin{bmatrix} I_{v0} \\ I_{h0} \\ U_0 \\ V_0 \end{bmatrix}$$

$$= \begin{bmatrix} e^{-\beta_1 z \sec \theta} I_{v0} \\ e^{-\beta_2 z \sec \theta} I_{h0} \\ \frac{1}{2} (e^{-\beta_3 z \sec \theta} + e^{-\beta_4 z \sec \theta}) U_0 \\ \frac{1}{2} (e^{-\beta_3 z \sec \theta} - e^{-\beta_4 z \sec \theta}) V_0 \end{bmatrix} \tag{2.41}$$

It can be seen that the four Stokes parameters propagating in a non-spherical particle medium have different wavenumbers. Also if $\langle S_{vh}^0 \rangle \neq \langle S_{hv}^0 \rangle$, Equation 2.39 can cause coupling among the four Stokes parameters.

Note that the eigen-expansion of the extinction matrix in the form of Equations 2.39–2.41 is useful later in solving VRT equations.

2.2.3 Phase Matrix

The phase matrix $\overline{\mathbf{P}}(\theta, \phi; \theta', \phi')$ in the VRT equation is a 4×4 real matrix describing the coupling between $\mathbf{I}(\theta', \phi', z)$ and $\mathbf{I}'(\theta, \phi, z)$. The term phase matrix is inherited from the field of astronomy; however, it is not related to the phase in the signal at all. The phase matrix is derived as

$$\overline{\mathbf{P}}(\theta, \phi; \theta', \phi') = n_0 \begin{bmatrix} \langle |S_{vv}|^2 \rangle & \langle |S_{vh}|^2 \rangle & \mathrm{Re}\langle S_{vv} S_{vh}^* \rangle & -\mathrm{Im}\langle S_{vv} S_{vh}^* \rangle \\ \langle |S_{hv}|^2 \rangle & \langle |S_{vv}|^2 \rangle & \mathrm{Re}\langle S_{hv} S_{hh}^* \rangle & -\mathrm{Im}\langle S_{hv} S_{vh}^* \rangle \\ 2\mathrm{Re}\langle S_{vv} S_{hv}^* \rangle & 2\mathrm{Re}\langle S_{vh} S_{hh}^* \rangle & \mathrm{Re}\langle S_{vv} S_{hh}^* + S_{vh} S_{hv}^* \rangle & -\mathrm{Im}\langle S_{vv} S_{hh}^* - S_{vh} S_{hv}^* \rangle \\ 2\mathrm{Im}\langle S_{vv} S_{hv}^* \rangle & 2\mathrm{Im}\langle S_{vh} S_{hh}^* \rangle & \mathrm{Im}\langle S_{vv} S_{hh}^* + S_{vh} S_{hv}^* \rangle & \mathrm{Re}\langle S_{vv} S_{hh}^* - S_{vh} S_{hv}^* \rangle \end{bmatrix}$$

$$\tag{2.42}$$

where n_0 is the number of particles per unit volume.

For random scatterers, the same type of scatterers may have random distributions of orientation, size, dielectric properties, and so on. Using the n parameters $x_1, \ldots, x_n$ to represent the stochastic properties of the scatterers, the first- and second-order moments of the scattering matrix can be written as

$$\langle S_{pq} \rangle = \int_{x_1} \int_{x_2} \cdots \int_{x_n} S_{pq} p(x_1, x_2, \ldots, x_n) dx_n \ldots dx_2 dx_1$$

$$\langle S_{pq} S_{rs}^* \rangle = \int_{x_1} \int_{x_2} \cdots \int_{x_n} S_{pq} S_{rs}^* p(x_1, x_2, \ldots, x_n) dx_n \ldots dx_2 dx_1, \quad p, q, r, s = v, h \tag{2.43}$$

where $p(x_1, \ldots, x_n)$ denotes the probability distribution function (PDF). It can be very intensive to calculate the integration numerically. However, for Rayleigh–Gans non-spherical particles with random orientation, the above integration can be explicitly derived. Some expressions of the phase matrix for non-spherical particles can be found in Appendix 2A (Jin, 1994).

As a simple case, when the particle size a is much less than the wavelength, $ka \ll 1$, the Rayleigh approximation can be used. The Rayleigh–Gans approximation (Jin, 1994) has been presented in Chapter 1.

If the particle is a Mie spherical large particle, disk, needle, finite cylinder, or other shape, calculations of the scattering matrix $\overline{\mathbf{S}}$ of Equation 2.24, the scattering, absorption, and extinction coefficients of Equations 2.25a, 2.25b and 2.37 and the phase matrix of Equation 2.42 can be found in the literature (e.g., Ishimaru, 1978; Waterman, 1965, 1971, 1973; Bohren and Huffman, 1983; Tsang, Kong, and Shin, 1985; Mishchenko, Travis, and Lacis, 2002).

2.3 Mueller Matrix Solution

2.3.1 First-Order Mueller Matrix Solution

Consider an elliptically polarized wave incident on a layer of random non-spherical particles. For example, a vegetation canopy is modeled as a layer of hybrid particles consisting of leaves modeled as non-spherical particles, and trunks and branches modeled as cylinders, as shown in Figure 2.3. Non-spherical particles have a non-uniformly random orientation described by the probability density function of the Euler angle (α, β, γ). As a first case, we consider a flat underlying surface at $z = -d$; the case of a rough-surface model will be discussed in Section 2.3.2.

The VRT equation is

$$\cos\theta \frac{d}{dz}\mathbf{I}(\theta, \phi, z) = -\overline{\mathbf{\kappa}}_e(\theta, \phi) \cdot \mathbf{I}(\theta, \phi, z) + \mathbf{S}(\theta, \phi, z) \tag{2.44a}$$

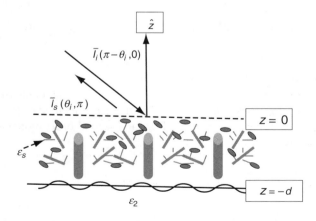

Figure 2.3 Scattering model of a layer of non-spherical particles

$$-\cos\theta \frac{d}{dz}\mathbf{I}(\pi-\theta,\phi,z) = -\overline{\kappa}_e(\theta,\phi)\cdot\mathbf{I}(\pi-\theta,\phi,z) + \mathbf{W}(\theta,\phi,z) \tag{2.44b}$$

where

$$\mathbf{S}(\theta,\phi,z) = \int_0^{2\pi} d\phi' \int_0^{\pi/2} d\theta' \sin\theta' [\overline{\mathbf{P}}(\theta,\phi;\theta',\phi')\cdot\mathbf{I}(\theta',\phi',z) + \overline{\mathbf{P}}(\theta,\phi;\pi-\theta',\phi') \\ \cdot\mathbf{I}(\pi-\theta',\phi',z)] \tag{2.45}$$

The function $\mathbf{W}$ is the same as $\mathbf{S}$ with all θ replaced by $\pi - \theta$.

The boundary conditions for transmission and reflection are written as

$$\mathbf{I}(\pi-\theta,\phi,z=0) = \mathbf{I}_0\delta(\cos\theta-\cos\theta_0)\delta(\phi-\phi_0) \tag{2.46a}$$

$$\mathbf{I}(\theta,\phi,z=-d) = \overline{\mathbf{R}}(\theta)\cdot\mathbf{I}(\pi-\theta,\phi,z=-d) \tag{2.46b}$$

where $\mathbf{I}_0$ denotes the incident wave Stokes vector with incident angles (θ_0,ϕ_0), δ denotes the Dirac delta function, $0 \le \theta < \pi/2$, and $\overline{\mathbf{R}}(\theta)$ is the reflectivity of the flat surface at $z = -d$.

Following the iterative method of the VRT equation of spherical particles, the integral VRT equations can be obtained as

$$\mathbf{I}(\pi-\theta,\phi,z) = \overline{\mathbf{E}}(\pi-\theta,\phi)\cdot\overline{\mathbf{D}}[\beta(\pi-\theta,\phi)z\sec\theta]\cdot\overline{\mathbf{E}}^{-1}(\pi-\theta,\phi)\cdot\mathbf{I}_0\delta(\cos\theta-\cos\theta_0)\delta(\phi-\phi_0) \\ +\int_z^0 dz'\left\{\overline{\mathbf{E}}(\pi-\theta,\phi)\cdot\overline{\mathbf{D}}[\beta(\pi-\theta,\phi)(z-z')\sec\theta]\cdot\overline{\mathbf{E}}^{-1}(\pi-\theta,\phi)\cdot\mathbf{W}(\theta,\phi,z)\right\} \tag{2.47a}$$

$$\mathbf{I}(\theta,\phi,z) = \overline{\mathbf{E}}(\theta,\phi)\cdot\overline{\mathbf{D}}[-\beta(\theta,\phi)(z+d)\sec\theta]\cdot\overline{\mathbf{E}}^{-1}(\theta,\phi)\cdot\overline{\mathbf{R}}(\theta)\cdot\overline{\mathbf{E}}(\pi-\theta,\phi) \\ \cdot\overline{\mathbf{D}}[-\beta(\pi-\theta,\phi)d\sec\theta]\cdot\overline{\mathbf{E}}^{-1}(\pi-\theta,\phi)\cdot\mathbf{I}_0\delta(\cos\theta-\cos\theta_0)\delta(\phi-\phi_0) \\ +\overline{\mathbf{E}}(\theta,\phi)\cdot\overline{\mathbf{D}}[-\beta(\theta,\phi)(z+d)\sec\theta]\int_{-d}^0 dz'\{\overline{\mathbf{E}}^{-1}(\theta,\phi)\cdot\overline{\mathbf{R}}(\theta)\cdot\overline{\mathbf{E}}(\pi-\theta,\phi) \\ \cdot\overline{\mathbf{D}}[-\beta(\pi-\theta,\phi)(z'+d)\sec\theta]\cdot\overline{\mathbf{E}}^{-1}(\pi-\theta,\phi)\cdot\mathbf{W}(\theta,\phi,z')\} \\ +\int_{-d}^z dz'\overline{\mathbf{E}}(\theta,\phi)\cdot\overline{\mathbf{D}}[\beta(\theta,\phi)(z'-z)\sec\theta]\cdot\overline{\mathbf{E}}^{-1}(\theta,\phi)\cdot\mathbf{S}(\theta,\phi,z') \tag{2.47b}$$

where the notation $\overline{\mathbf{D}}[\beta(\theta,\phi)z\sec\theta]$ is a 4×4 diagonal matrix with the iith element $\exp[\beta_i(\theta,\phi)z\sec\theta]$. The eigenvector matrix $\overline{\mathbf{E}}$ and eigenvalue $\beta_i, i = 1,2,3,4$, are given in Equations 2.39 and 2.37, respectively. The first term on the right-hand side (RHS) of Equations 2.47a and 2.47b is a coherent Stokes vector with $\mathbf{I}_0$ as the zeroth-order solution. Substituting these zeroth-order solutions $\mathbf{I}^{(0)}(\pi-\theta,\phi,z)$, $\mathbf{I}^{(0)}(\theta,\phi,z)$ into the functions $\mathbf{S}$ and $\mathbf{W}$ of

Equations 2.44a, 2.44b and 2.45 to obtain the first-order solution, we obtain the coupling relationship between the incident and scattered Stokes vectors by the first-order Mueller matrix as (Tsang and Kong, 1990)

$$\mathbf{I}_s^{(1)}(\theta,\phi,z=0) = \overline{\mathbf{M}}(\theta,\phi;\pi-\theta_0,\phi_0)\cdot\mathbf{I}_0 \tag{2.48}$$

The Mueller matrix is a 4×4 real matrix, which is calculated from the phase matrix of the scatterer, the reflectivity of the underlying ground surface, and the complex eigenvalues and eigenvectors of the VRT equation of the coherent Stokes vector. The elements of the first-order Mueller matrix can be explicitly derived as (Tsang, Kong, and Shin, 1984)

$$M_{mn}(\theta,\phi;\pi-\theta_0,\phi_0) = \sum_{j,i}\sec\theta\{\overline{\mathbf{E}}(\theta,\phi)\cdot\overline{\mathbf{D}}[-\beta(\theta,\phi)d\sec\theta]\cdot\overline{\mathbf{E}}^{-1}(\theta,\phi)$$

$$\cdot\overline{\mathbf{R}}(\theta)\cdot\overline{\mathbf{E}}(\pi-\theta,\phi)\}_{mj}\{\cdot\overline{\mathbf{E}}^{-1}(\pi-\theta,\phi)\cdot\overline{\mathbf{P}}(\pi-\theta,\phi;\theta_0,\phi_0)\cdot\overline{\mathbf{E}}(\theta_0,\phi_0)\}_{ji}$$

$$\times\frac{1-\exp[-\beta_j(\pi-\theta,\phi)d\sec\theta-\beta_i(\theta_0,\phi_0)d\sec\theta_0]}{\beta_j(\pi-\theta,\phi)\sec\theta+\beta_i(\theta_0,\phi_0)\sec\theta_0}\{\overline{\mathbf{E}}^{-1}(\theta_0,\phi_0)\cdot\overline{\mathbf{R}}(\theta_0)$$

$$\cdot\overline{\mathbf{E}}(\pi-\theta_0,\phi_0)\cdot\overline{\mathbf{D}}[-\beta(\pi-\theta_0,\phi_0)d\sec\theta_0]\cdot\overline{\mathbf{E}}^{-1}(\pi-\theta_0,\phi_0)\}_{in}$$

$$+\sum_{j,i}\sec\theta\{\overline{\mathbf{E}}(\theta,\phi)\cdot\overline{\mathbf{D}}[-\beta(\theta,\phi)d\sec\theta]\cdot\overline{\mathbf{E}}^{-1}(\theta,\phi)\cdot\overline{\mathbf{R}}(\theta)\cdot\overline{\mathbf{E}}(\pi-\theta,\phi)\}_{mj}$$

$$\{\overline{\mathbf{E}}^{-1}(\pi-\theta,\phi)\cdot\overline{\mathbf{P}}(\pi-\theta,\phi;\pi-\theta_0,\phi_0)\cdot\overline{\mathbf{E}}(\pi-\theta_0,\phi_0)\}_{ji}[\overline{\mathbf{E}}^{-1}(\pi-\theta_0,\phi_0)]_{in}$$

$$\times\frac{\exp[-\beta_j(\pi-\theta,\phi)d\sec\theta]-\exp[-\beta_i(\pi-\theta_0,\phi_0)d\sec\theta_0]}{\beta_i(\pi-\theta_0,\phi_0)\sec\theta_0-\beta_j(\pi-\theta,\phi)\sec\theta}$$

$$+\sec\theta\sum_{j,i}E_{mj}(\theta,\phi)\{\overline{\mathbf{E}}^{-1}(\theta,\phi)\cdot\overline{\mathbf{P}}(\theta,\phi;\theta_0,\phi_0)\cdot\overline{\mathbf{E}}(\theta_0,\phi_0)\}_{ji}$$

$$\times\frac{\exp[-\beta_j(\theta,\phi)d\sec\theta]-\exp[-\beta_i(\theta_0,\phi_0)d\sec\theta_0]}{\beta_i(\theta_0,\phi_0)\sec\theta_0-\beta_j(\theta,\phi)\sec\theta}\{\overline{\mathbf{E}}^{-1}(\theta_0,\phi_0)\}\cdot\overline{\mathbf{R}}(\theta_0)$$

$$\cdot\overline{\mathbf{E}}(\pi-\theta_0,\phi_0)\cdot\overline{\mathbf{D}}[-\beta(\pi-\theta_0,\phi_0)d\sec\theta_0]\cdot\overline{\mathbf{E}}^{-1}(\pi-\theta_0,\phi_0)\}_{in}$$

$$+\sec\theta\sum_{j,i}E_{mj}(\theta,\phi)\{\cdot\overline{\mathbf{E}}^{-1}(\theta,\phi)\cdot\overline{\mathbf{P}}(\theta,\phi;\pi-\theta_0,\phi_0)\cdot\overline{\mathbf{E}}(\pi-\theta_0,\phi_0)\}_{ji}$$

$$\times\frac{1-\exp[-\beta_j(\theta,\phi)d\sec\theta-\beta_i(\pi-\theta_0,\phi_0)d\sec\theta_0]}{\beta_i(\theta,\phi)\sec\theta+\beta_i(\pi-\theta_0,\phi_0)\sec\theta_0}[\overline{\mathbf{E}}^{-1}(\pi-\theta_0,\phi_0)]_{in}$$

$$\tag{2.49}$$

where i, j, m, n take values 1, 2, 3, 4. There are four terms in the summation on the RHS of the first-order Mueller matrix, Equation 2.49, which correspond to the scattering processes as illustrated in Figure 2.4.

Consider a backscattering problem with the transmitted wave $\mathbf{E}_0 = E_{10}\hat{e}_1 + E_{20}\hat{e}_2$ and incoming scattered wave $\mathbf{E}_s = E_{s1}\hat{e}_1 + E_{s2}(-\hat{e}_2)$, where $\hat{e}_1$, $\hat{e}_2$ are two orthogonal polarization vectors. The transmitted wave from the transmitter is directed to $\hat{e}_1 \times \hat{e}_2$, and the

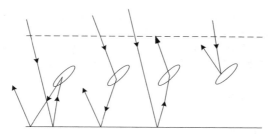

Figure 2.4 Physical meaning in the first-order Mueller matrix. © [2002] IEEE. Reprinted, with permission, from *IEEE Transactions on Geoscience and Remote Sensing*

incoming scattered wave to the receiver is directed to $\hat{e}_1 \times (-\hat{e}_2)$. The received power is proportional to

$$P_n \sim \frac{1}{2\eta}|\mathbf{E}_0 \cdot \mathbf{E}_s|^2 = \frac{1}{2\eta}|E_{01}E_{s1} - E_{02}E_{s2}|^2 = I_{v0}I_{vs} + I_{h0}I_{hs} - \frac{1}{2}U_0U_s + \frac{1}{2}V_0V_s \quad (2.50)$$

In terms of elliptical and orientation angles (χ, ψ), let the normalized and fully polarized transmitted Stokes vector be

$$\mathbf{I}_0 = \begin{bmatrix} \frac{1}{2}(1 - \cos 2\chi \cos 2\psi) \\ \frac{1}{2}(1 + \cos 2\chi \cos 2\psi) \\ -\cos 2\chi \sin 2\psi \\ \sin 2\chi \end{bmatrix} \quad (2.51)$$

Thus, the co-polarized scattering power is

$$P_n = \frac{1}{2}I_{vs}(1 - \cos 2\chi \cos 2\psi) + \frac{1}{2}I_{hs}(1 + \cos 2\chi \cos 2\psi) + \frac{1}{2}U_s\cos 2\chi \cos 2\psi + \frac{1}{2}V_s\sin 2\chi \quad (2.52)$$

where the subscript n denotes that the incident intensity is normalized. So, the co-polarized and the depolarized backscattering coefficients, respectively, are

$$\sigma_c = 4\pi \cos \theta_0 P_n \quad (2.53a)$$

$$\sigma_d = 4\pi \cos \theta_0 (I_{vs} + I_{hs} - P_n) \quad (2.53b)$$

Owing to randomness and ensemble averaging, the scattering wave may become partially polarized even for fully polarized incident waves.

2.3.2 Modeling of Vegetation Canopy over Rough Surface

Now, to take account of a rough surface in Figure 2.3, consider the VRT equations as shown as Equations 2.44a, 2.44b and 2.45. The corresponding boundary conditions are written as

$$\mathbf{I}(\pi - \theta, \phi, z = 0) = \mathbf{I}_0\delta(\cos \theta - \cos \theta_0)\delta(\phi - \phi_0)$$

$$\mathbf{I}(\theta, \phi, z = -d) = \int_0^{2\pi} d\phi' \int_0^{\pi/2} d\theta' \sin \theta' \overline{\mathbf{R}}(\theta, \phi; \theta', \phi') \cdot \mathbf{I}(\theta', \phi', z = -d) \quad (2.54)$$

It is noted that $\overline{\mathbf{R}}(\theta, \phi; \theta', \phi')$ is now the Mueller matrix of the underlying rough-surface scattering (Fung, 1994), which is different from Equations 2.46a and 2.46b.

Following the approach of Equations 2.44a, 2.44b and 2.45, the solution to Equations 2.44a, 2.44b and 2.54 should have the following form (Jin and Zhang, 2002)

$$\mathbf{I}(\theta, \phi, z) = \exp\left[-z \sec \theta \overline{\mathbf{\kappa}}_e(\theta, \phi)\right] \cdot \mathbf{A}$$
$$+ \sec \theta \exp\left[-z \sec \theta \overline{\mathbf{\kappa}}_e(\theta, \phi)\right] \int_{-d}^{z} dz' \exp\left[-z' \sec \theta \overline{\mathbf{\kappa}}_e(\theta, \phi)\right] \cdot \int_0^{2\pi} d\phi' \int_0^{\pi/2} d\theta' \sin\theta'$$
$$\cdot \left[\overline{\mathbf{P}}(\theta, \phi; \theta', \phi') \cdot \mathbf{I}(\theta', \phi', z') + \overline{\mathbf{P}}(\theta, \phi; \pi - \theta', \phi') \cdot \mathbf{I}(\pi - \theta', \phi', z')\right]$$

$$(2.55)$$

$$\mathbf{I}(\pi - \theta, \phi, z) = \exp\left[z \sec \theta \overline{\mathbf{\kappa}}_e(\pi - \theta, \phi)\right] \cdot \mathbf{B}$$
$$+ \sec \theta \exp\left[z \sec \theta \overline{\mathbf{\kappa}}_e(\pi - \theta, \phi)\right] \int_z^{0} dz' \exp\left[z' \sec \theta \overline{\mathbf{\kappa}}_e(\pi - \theta, \phi)\right] \cdot \int_0^{2\pi} d\phi' \int_0^{\pi/2} d\theta' \sin\theta'$$
$$\cdot \left[\overline{\mathbf{P}}(\pi - \theta, \phi; \theta', \phi') \cdot \mathbf{I}(\theta', \phi', z') + \overline{\mathbf{P}}(\pi - \theta, \phi; \pi - \theta', \phi') \cdot \mathbf{I}(\pi - \theta', \phi', z')\right]$$

$$(2.56)$$

where $\mathbf{A}, \mathbf{B}$ are the coefficients to be determined. Substituting Equations 2.55 and 2.56 into the boundary conditions of Equation 2.54 yields

$$\mathbf{B} = \mathbf{I}_0 \cdot \delta \tag{2.57}$$

and

$$\mathbf{A} = \exp\left[-d \sec \theta \overline{\mathbf{\kappa}}_e(\theta, \phi)\right] \overline{\mathbf{R}}(\theta, \phi; \pi - \theta_0, \phi_0) \cdot \exp\left[-d \sec \theta_0 \overline{\mathbf{\kappa}}_e(\pi - \theta_0, \phi_0)\right] \mathbf{I}_0$$
$$+ \sec \theta \exp\left[-d \sec \theta \overline{\mathbf{\kappa}}_e(\theta, \phi)\right] \int_0^{2\pi} d\phi'' \int_0^{\pi/2} d\theta'' \sin\theta'' \overline{\mathbf{R}}(\theta, \phi; \pi - \theta'', \phi'')$$
$$\cdot \exp\left[-d \sec \theta'' \overline{\mathbf{\kappa}}_e(\pi - \theta'', \phi'')\right] \cdot \int_{-d}^{0} dz' \exp\left[-z' \sec \theta'' \overline{\mathbf{\kappa}}_e(\pi - \theta'', \phi'')\right] \int_0^{2\pi} d\phi' \int_0^{\pi/2} d\theta' \sin\theta'$$
$$\cdot \left[\overline{\mathbf{P}}(\pi - \theta'', \phi''; \theta', \phi') \cdot \mathbf{I}(\theta', \phi', z') + \overline{\mathbf{P}}(\pi - \theta'', \phi''; \pi - \theta', \phi') \cdot \mathbf{I}(\pi - \theta', \phi', z')\right]$$

$$(2.58)$$

As an approximation, let us first omit the high-order scattering terms and obtain the zero-order coefficient

$$\mathbf{A}^0 = \exp\left[-d \sec \theta \overline{\mathbf{\kappa}}_e(\theta, \phi)\right] \cdot \overline{\mathbf{R}}(\theta, \phi; \pi - \theta_0, \phi_0) \cdot \exp\left[-d \sec \theta_0 \overline{\mathbf{\kappa}}_e(\pi - \theta_0, \phi_0)\right] \cdot \mathbf{I}_0 \tag{2.59}$$

and the zeroth-order solution

$$\mathbf{I}^0(\theta, \phi, z) = \exp\left[(-d - z) \sec \theta \overline{\mathbf{\kappa}}_e(\theta, \phi)\right] \overline{\mathbf{R}}(\theta, \phi; \pi - \theta_0, \phi_0) \cdot \exp\left[-d \sec \theta_0 \overline{\mathbf{\kappa}}_e(\pi - \theta_0, \phi_0)\right] \cdot \mathbf{I}_0$$
$$\mathbf{I}^0(\pi - \theta, \phi, z) = \exp\left[z \sec \theta \overline{\mathbf{\kappa}}_e(\pi - \theta, \phi)\right] \cdot \mathbf{I}_0 \cdot \delta$$

$$(2.60)$$

Substituting the zeroth-order solution of Equation 2.60 back into Equation 2.58, this gives the first-order coefficient

$$\mathbf{A}^1 = \sec\theta \exp\left[-d\sec\theta\overline{\mathbf{\kappa}}_e(\theta,\phi)\right] \int_0^{2\pi} d\phi'' \int_0^{\pi/2} d\theta'' \sin\theta'' \overline{\mathbf{R}}(\theta,\phi;\pi-\theta'',\phi'')$$
$$\cdot\exp\left[-d\sec\theta''\overline{\mathbf{\kappa}}_e(\pi-\theta'',\phi'')\right]\cdot\int_{-d}^0 dz'\exp\left[-z'\sec\theta''\overline{\mathbf{\kappa}}_e(\pi-\theta'',\phi'')\right]\int_0^{2\pi}d\phi'\int_0^{\pi/2}d\theta'\sin\theta'$$
$$\cdot\left[\overline{\mathbf{P}}(\pi-\theta'',\phi'';\theta',\phi')\cdot\mathbf{I}^0(\theta',\phi',z')+\overline{\mathbf{P}}(\pi-\theta'',\phi'';\pi-\theta',\phi')\cdot\mathbf{I}^0(\pi-\theta',\phi',z')\right]$$

$$(2.61)$$

and the first-order solution

$$\mathbf{I}^1(\theta,\phi,z) = \exp\left[-z\sec\theta\overline{\mathbf{\kappa}}_e(\theta,\phi)\right]\mathbf{A}^1 + \sec\theta\exp\left[-z\sec\theta\overline{\mathbf{\kappa}}_e(\theta,\phi)\right]$$
$$\cdot\int_{-d}^z dz'\exp\left[-z'\sec\theta\overline{\mathbf{\kappa}}_e(\theta,\phi)\right]\int_0^{2\pi}d\phi'\int_0^{\pi/2}d\theta'\sin\theta' \qquad (2.62)$$
$$\cdot\left[\overline{\mathbf{P}}(\theta,\phi;\theta',\phi')\cdot\mathbf{I}^0(\theta',\phi',z')+\overline{\mathbf{P}}(\theta,\phi;\pi-\theta',\phi')\cdot\mathbf{I}^0(\pi-\theta',\phi',z')\right]$$

Note that such iteration can be performed as many times as needed in order to obtain higher-order solutions. However, higher-order scatterings are usually trivial enough to be neglected. Thus, the scattered Stokes vector of zeroth order and first order are calculated as

$$\mathbf{I}_s = \mathbf{I}^0(\theta,\phi,z=0) + \mathbf{I}^1(\theta,\phi,z=0)$$
$$= \mathbf{A}^1 + \exp\left[-d\sec\theta\overline{\mathbf{\kappa}}_e(\theta,\phi)\right]\overline{\mathbf{R}}(\theta,\phi;\pi-\theta_0,\phi_0)\cdot\exp\left[-d\sec\theta_0\overline{\mathbf{\kappa}}_e(\pi-\theta_0,\phi_0)\right]\mathbf{I}_0$$
$$+\sec\theta\int_{-d}^0 dz'\exp\left[-z'\sec\theta\overline{\mathbf{\kappa}}_e(\theta,\phi)\right]\int_0^{2\pi}d\phi'\int_0^{\pi/2}d\theta'\sin\theta'$$
$$\cdot\left[\overline{\mathbf{P}}(\theta,\phi;\theta',\phi')\cdot\mathbf{I}^0(\theta',\phi',z')+\overline{\mathbf{P}}(\theta,\phi;\pi-\theta',\phi')\cdot\mathbf{I}^0(\pi-\theta',\phi',z')\right]$$

$$(2.63)$$

Hence, the first-order Mueller matrix solution for the model of the vegetation canopy in Figure 2.3 can be derived and consists of five terms

$$\overline{\mathbf{M}}(\theta,\phi;\pi-\theta_0,\phi_0) = \overline{\mathbf{M}}_0 + \overline{\mathbf{M}}_1 + \overline{\mathbf{M}}_2 + \overline{\mathbf{M}}_3 + \overline{\mathbf{M}}_4 \qquad (2.64)$$

Each of these five terms represents different scattering mechanisms as illustrated in Figure 2.5. Their expressions are

$$\overline{\mathbf{M}}_0 = \exp\left[-d\sec\theta\,\overline{\mathbf{\kappa}}_e(\theta,\phi)\right]\cdot\overline{\mathbf{R}}(\theta,\phi;\pi-\theta_0,\phi_0)\cdot\exp\left[-d\sec\theta_0\,\mathbf{\kappa}_e(\pi-\theta_0,\phi_0)\right] \quad (2.65)$$

$$\overline{\mathbf{M}}_1 = \sec\theta\int_{-d}^0 dz'\exp\left[z'\sec\theta\,\overline{\mathbf{\kappa}}_e(\theta,\phi)\right]\cdot\overline{\mathbf{P}}(\theta,\phi;\pi-\theta_0,\phi_0)\cdot\exp\left[z'\sec\theta_0\,\overline{\mathbf{\kappa}}_e(\pi-\theta_0,\phi_0)\right]$$

$$(2.66)$$

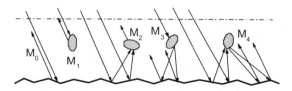

$$
\begin{aligned}
\overline{\mathbf{M}}_2 &= \sec\theta \int_{-d}^{0} dz' \exp\left[z' \sec\theta\, \overline{\mathbf{\kappa}}_e(\theta,\phi)\right] \cdot \int_{0}^{2\pi} d\phi' \int_{0}^{\pi/2} d\theta' \sin\theta' \\
&\cdot \mathbf{P}(\theta,\phi;\theta',\phi') \exp\left[(-d-z')\sec\theta'\, \overline{\mathbf{\kappa}}_e(\theta',\phi')\right] \\
&\cdot \mathbf{R}(\theta',\phi';\pi-\theta_0,\phi_0) \cdot \exp\left[-d\sec\theta_0\, \overline{\mathbf{\kappa}}_e(\pi-\theta_0,\phi_0)\right]
\end{aligned}
\tag{2.67}
$$

$$
\begin{aligned}
\overline{\mathbf{M}}_3 &= \sec\theta \exp\left[-d\sec\theta\, \overline{\mathbf{\kappa}}_e(\theta,\phi)\right] \cdot \int_{0}^{2\pi} d\phi'' \int_{0}^{\pi/2} \sin\theta'' d\theta'' \\
&\cdot \mathbf{R}(\theta,\phi;\pi-\theta'',\phi'') \int_{-d}^{0} dz' \exp\left[(-d-z')\sec\theta''\, \overline{\mathbf{\kappa}}_e(\pi-\theta'',\phi'')\right] \\
&\cdot \mathbf{P}(\pi-\theta'',\phi'';\pi-\theta_0,\phi_0) \cdot \exp\left[z'\sec\theta_0\, \overline{\mathbf{\kappa}}_e(\pi-\theta_0,\phi_0)\right]
\end{aligned}
\tag{2.68}
$$

$$
\begin{aligned}
\overline{\mathbf{M}}_4 &= \sec\theta \exp\left[-d\sec\theta\, \overline{\mathbf{\kappa}}_e(\theta,\phi)\right] \cdot \int_{0}^{2\pi} d\phi'' \int_{0}^{\pi/2} d\theta'' \sin\theta'' \\
&\cdot \mathbf{R}(\theta,\phi;\pi-\theta'',\phi'')\exp\left[(-d-z')\sec\theta''\, \overline{\mathbf{\kappa}}_e(\pi-\theta'',\phi'')\right] \\
&\cdot \int_{0}^{2\pi} d\phi' \int_{0}^{\pi/2} d\theta' \sin\theta'\, \mathbf{P}(\pi-\theta'',\phi'';\theta',\phi') \cdot \exp\left[(-d-z')\sec\theta'\, \overline{\mathbf{\kappa}}_e(\theta',\phi')\right] \\
&\cdot \mathbf{R}(\theta',\phi';\pi-\theta_0,\phi_0) \cdot \exp\left[-d\sec\theta_0\, \overline{\mathbf{\kappa}}_e(\pi-\theta_0,\phi_0)\right]
\end{aligned}
\tag{2.69}
$$

Clearly, the first term is the surface scattering directly from the underlying rough surface. The second term is direct scattering from particles. The remaining three terms are multiple scattering among particles and the rough surface, which all involve the (θ',ϕ'), (θ'',ϕ'') angular integration of incident and scattering directions of the rough surface.

Based on the discussion in Section 2.2.2, the propagator with the extinction matrix in Equations 2.65–2.69 can be rewritten as

$$
\exp\left[-d\sec\theta\, \overline{\mathbf{\kappa}}_e(\theta,\phi)\right] = \overline{\mathbf{E}}(\theta,\phi) \cdot \overline{\mathbf{D}}\left[-d\sec\theta\, \beta_{ii}(\theta,\phi)\right] \cdot \overline{\mathbf{E}}^{-1}(\theta,\phi)
\tag{2.70}
$$

Thus, the five terms of the Mueller matrix solution are written as

$$
\begin{aligned}
\overline{\mathbf{M}}_0 &= \overline{\mathbf{E}}(\theta,\phi) \cdot \overline{\mathbf{D}}\left[-d\sec\theta\, \beta_{ii}(\theta,\phi)\right] \cdot \overline{\mathbf{E}}^{-1}(\theta,\phi) \cdot \mathbf{R}(\theta,\phi;\pi-\theta_0,\phi_0) \\
&\cdot \overline{\mathbf{E}}(\pi-\theta_0,\phi_0) \cdot \overline{\mathbf{D}}\left[-d\sec\theta_0\, \beta_{ii}(\pi-\theta_0,\phi_0)\right] \cdot \overline{\mathbf{E}}^{-1}(\pi-\theta_0,\phi_0)
\end{aligned}
\tag{2.71}
$$

$$
\begin{aligned}
\overline{\mathbf{M}}_1 &= \sec\theta \int_{-d}^{0} dz'\, \overline{\mathbf{E}}(\theta,\phi) \cdot \overline{\mathbf{D}}\left[z'\sec\theta\, \beta_{ii}(\theta,\phi)\right] \cdot \overline{\mathbf{E}}^{-1}(\theta,\phi) \\
&\cdot \mathbf{P}(\theta,\phi;\pi-\theta_0,\phi_0) \cdot \overline{\mathbf{E}}(\pi-\theta_0,\phi_0) \cdot \overline{\mathbf{D}}\left[z'\sec\theta_0\, \beta_{ii}(\pi-\theta_0,\phi_0)\right] \cdot \overline{\mathbf{E}}^{-1}(\pi-\theta_0,\phi_0)
\end{aligned}
\tag{2.72}
$$

$$
\begin{aligned}
\overline{\mathbf{M}}_2 &= \sec\theta \int_{-d}^{0} dz'\, \overline{\mathbf{E}}(\theta,\phi) \cdot \overline{\mathbf{D}}\left[z'\sec\theta\, \beta_{ii}(\theta,\phi)\right] \cdot \overline{\mathbf{E}}^{-1}(\theta,\phi) \\
&\cdot \int_{0}^{2\pi} d\phi' \int_{0}^{\pi/2} d\theta' \sin\theta'\, \mathbf{P}(\theta,\phi;\theta',\phi') \cdot \overline{\mathbf{E}}(\theta',\phi') \cdot \overline{\mathbf{D}}\left[(-d-z')\sec\theta'\, \beta(\theta',\phi')\right] \cdot \overline{\mathbf{E}}^{-1}(\theta',\phi') \\
&\cdot \mathbf{R}(\theta',\phi';\pi-\theta_0,\phi_0) \cdot \overline{\mathbf{E}}(\pi-\theta_0,\phi_0) \cdot \overline{\mathbf{D}}\left[-d\sec\theta_0\, \beta_{ii}(\pi-\theta_0,\phi_0)\right] \cdot \overline{\mathbf{E}}^{-1}(\pi-\theta_0,\phi_0)
\end{aligned}
\tag{2.73}
$$

$$\overline{\mathbf{M}}_3 = \sec\theta \cdot \overline{\mathbf{E}}(\theta,\phi) \cdot \overline{\mathbf{D}}[-d\sec\theta\,\beta_{ii}(\theta,\phi)] \cdot \overline{\mathbf{E}}^{-1}(\theta,\phi)$$
$$\cdot \int_0^{2\pi} d\phi'' \int_0^{\pi/2} d\theta'' \sin\theta'' \, \overline{\mathbf{R}}(\theta,\phi;\pi-\theta'',\phi'')$$
$$\cdot \int_{-d}^0 dz' \, \overline{\mathbf{E}}(\pi-\theta'',\phi'') \cdot \overline{\mathbf{D}}[(-d-z')\sec\theta''\,\beta_{ii}(\pi-\theta'',\phi'')] \cdot \overline{\mathbf{E}}^{-1}(\pi-\theta'',\phi'')$$
$$\cdot \overline{\mathbf{P}}(\pi-\theta'',\phi'';) \cdot \overline{\mathbf{E}}(\pi-\theta_0,\phi_0) \cdot \overline{\mathbf{D}}[z'\sec\theta_0\,\beta_{ii}(\pi-\theta_0,\phi_0)] \cdot \overline{\mathbf{E}}^{-1}(\pi-\theta_0,\phi_0)$$

$$(2.74)$$

$$\overline{\mathbf{M}}_4 = \sec\theta \cdot \overline{\mathbf{E}}(\theta,\phi) \cdot \overline{\mathbf{D}}[-d\sec\theta\,\beta_{ii}(\theta,\phi)] \cdot \overline{\mathbf{E}}^{-1}(\theta,\phi)$$
$$\cdot \int_0^{2\pi} d\phi'' \int_0^{\pi/2} d\theta'' \sin\theta'' \, \overline{\mathbf{R}}(\theta,\phi;\pi-\theta'',\phi'')$$
$$\cdot \int_{-d}^0 dz' \, \overline{\mathbf{E}}(\pi-\theta'',\phi'') \cdot \overline{\mathbf{D}}[(-d-z')\sec\theta''\,\beta_{ii}(\pi-\theta'',\phi'')] \cdot \overline{\mathbf{E}}^{-1}(\pi-\theta'',\phi'')$$
$$\cdot \int_0^{2\pi} d\phi' \int_0^{\pi/2} d\theta' \sin\theta' \, \overline{\mathbf{P}}(\pi-\theta'',\phi'';\theta',\phi') \cdot \overline{\mathbf{E}}(\theta',\phi')$$
$$\cdot \overline{\mathbf{D}}[(-d-z')\sec\theta'\,\beta(\theta',\phi')] \cdot \overline{\mathbf{E}}^{-1}(\theta',\phi') \cdot \overline{\mathbf{R}}(\theta',\phi';\pi-\theta_0,\phi_0)$$
$$\cdot \overline{\mathbf{E}}(\pi-\theta_0,\phi_0) \cdot \overline{\mathbf{D}}[-d\sec\theta_0\,\beta_{ii}(\pi-\theta_0,\phi_0)] \cdot \overline{\mathbf{E}}^{-1}(\pi-\theta_0,\phi_0)$$

$$(2.75)$$

Substituting the Mueller matrix of the integral equation method (IEM) rough surface, and the phase matrix and extinction matrix of the Rayleigh–Gans non-spherical particles and cylindrical scatterers, the Mueller matrix of the vegetation scattering model in Figure 2.3 can be calculated. The integrations of (θ',ϕ'), (θ'',ϕ'') and z' can be calculated numerically using the discrete ordinate method.

Ignoring coupling among different scatterers, the overall Mueller matrix of hybrid particles can be simply calculated as the weighted sum of the Mueller matrices of each type of particle, where the weights are their fractional volumes, that is

$$\overline{\mathbf{M}}_{\text{Total}} = \frac{\sum_i f_{si}\overline{\mathbf{M}}^i}{\sum_i f_{si}}$$

$$(2.76)$$

where f_{si} and $\overline{\mathbf{M}}^i$ denote the fractional volume and the Mueller matrix of the ith type of scatterer.

2.3.3 Numerical Examples of Modeling of Vegetation Canopy

With the derived Mueller matrix solution of the VRT equations for a vegetation canopy, some numerical examples of scattering from different vegetation types are provided. The first example is a layer of branches or stems, which are modeled as slim cylinders of radius 1 cm, length 100 cm, dielectric constant $\varepsilon_s = 19.6 + i8.1$, layer depth 300 cm, fractional volume 0.0003, ground dielectric constant $\varepsilon_1 = 8 + i1$, and incident angle 30°. Consider the cylinders oriented uniformly over the range $\beta \in (0°, \beta_2°)$, $\gamma \in (0°, 360°)$, where the second Euler angle β indicates cylinder pitch and the third Euler angle γ indicates azimuthal orientation. Figure 2.6a and b shows the backscattering coefficients σ_{hh}, σ_{vv}, σ_{hv} as a function of β_2 at C and P band, respectively.

It can be seen that, as the orientation distribution gets wider, cross-polarization scattering increases and co-polarization decreases. Note that backscattering is most prominent when all

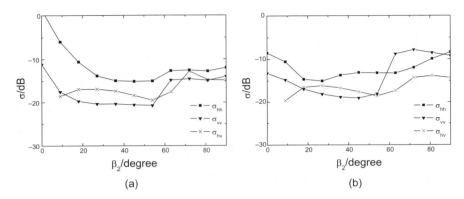

Figure 2.6 Backscattering coefficient of a layer of slim cylinders as a function of orientation β_2: (a) C band and (b) P band. © [2002] IEEE. Reprinted, with permission, from *IEEE Transactions on Geoscience and Remote Sensing*

the particles are vertically oriented ($\beta_2 = 0$), which might be due to double scattering of the cone-like reflection of the cylinder surface and the reflection of the ground surface.

Consider a layer of random needle-like particles under the Rayleigh–Gans approximation (Jin, 1994) above a flat or a rough surface. The incident angle is $30°$, while the particle radius

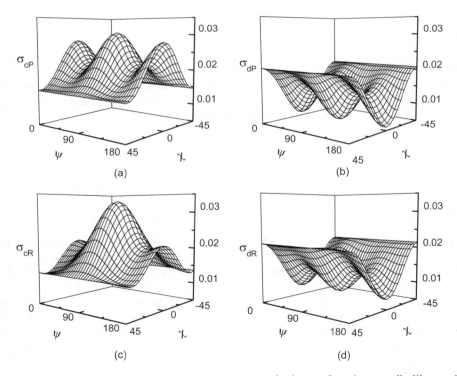

Figure 2.7 Co-polarized and cross-polarized scattering of a layer of random needle-like particles above flat and rough surfaces. © [2002] IEEE. Reprinted, with permission, from *IEEE Transactions on Geoscience and Remote Sensing*

is 0.1 cm by 0.1 cm, by 5 cm long, orientation distribution is $\beta \in [0°, 60°]$, $\gamma \in [0°, 360°]$, dielectric constant is $\varepsilon_s = 14.1 + i4.5$, depth 100 cm, fractional volume 0.003, and ground dielectric constant $\varepsilon_1 = 15 + i2$. The root-mean-square (RMS) height of the underlying rough surface is 0.5 cm and correlation length is 8 cm. Frequency is at C band, 5.3 GHz.

Figure 2.7 compares the co-polarized and depolarized backscattering coefficients σ_c, σ_d as functions of polarization angles (χ, ψ). Figure 2.7a and b are the results for an underlying flat surface, while Figure 2.7c and d are the results for an underlying rough surface. The most distinct difference between these two cases is the much stronger vertically polarized scattering in the rough-surface case. This is due to the fact that diffuse scattering from the rough surface is much weaker than the specular reflection from the flat surface, which renders weaker scattering terms, that is, $\overline{M}_2$, $\overline{M}_3$, $\overline{M}_4$, and thus makes the particle scattering term $\overline{M}_1$ more prominent, which has stronger vertically polarized scattering.

Figure 2.8 shows the scattering coefficients σ_{hh}, σ_{vv} as a function of incident angle. The fractional volume is 0.001 and depth is 50 cm. The underlying rough surface has RMS height 0.2 cm and correlation length 5 cm. The remaining parameters are the same as in Figure 2.7. It can be seen that the direct rough-surface scattering $\overline{M}_0$ decreases as the incident angle increases and so do the multiple scattering terms $\overline{M}_2$, $\overline{M}_3$, $\overline{M}_4$. The particle scattering term $\overline{M}_1$ increases as the incident angle increases. Overall, the first two terms $\overline{M}_0$ and $\overline{M}_1$ represent major scattering mechanisms. Surface scattering $\overline{M}_0$ dominates at small incidence range, while particle scattering $\overline{M}_1$ becomes dominant when the incident angle is larger than 30°. Thus, the overall scattering $\overline{M}$ (thick line in Figure 2.8) first decreases and then increases.

Figure 2.9 shows the scattering dependence on layer depth. The parameters are the same as in Figure 2.7. It can be seen that, as depth increases, the particle scattering $\overline{M}_1$ becomes stronger; the ground-surface scattering $\overline{M}_0$ decreases because of longer extinction path. Multiple scattering terms $\overline{M}_2$, $\overline{M}_3$, $\overline{M}_4$ increase initially because of more particles but decrease later due to heavier extinction. The conclusion is that surface scattering dominates when the vegetation layer is thin and the incident angle is small, and particle scattering becomes dominant otherwise.

Ground-based scatterometer observations (Wegmuller and Matzler, 1993) were taken of a wheat field at different growth stages. Figure 2.10 shows the measured σ_{hh}, σ_{vv} as functions of incident angle, where d indicates wheat height at growth stages. It can be seen that, at the beginning of growth, wheat height is only 10 cm, and scattering is mainly from the ground

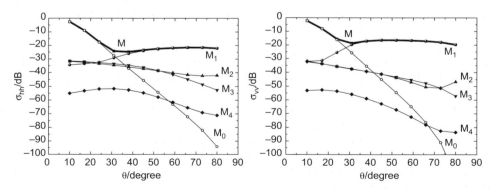

Figure 2.8 Contribution of different scattering terms as a function of incident angle. © [2002] IEEE. Reprinted, with permission, from *IEEE Transactions on Geoscience and Remote Sensing*

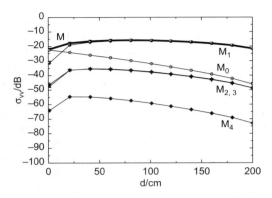

Figure 2.9 Contributions from different scattering mechanisms and their dependences on depth.
© [2002] IEEE. Reprinted, with permission, from *IEEE Transactions on Geoscience and Remote Sensing*

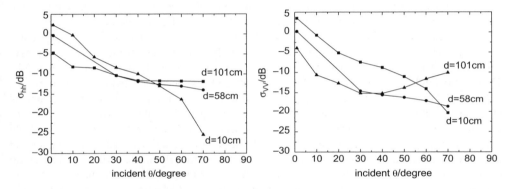

Figure 2.10 Measured scattering of a wheat field at different growth stages. © [2002] IEEE. Reprinted, with permission, from *IEEE Transactions on Geoscience and Remote Sensing*

surface and becomes reduced as the incident angle increases. At the later stage of growth, as the wheat grows to 101 cm tall, scattering from the wheat becomes dominant, which makes the overall scattering reduced at first and then increased as the incident angle increases. This finding matches well with the simulation results in Figure 2.8.

Another dataset collected by ground-based scatterometer shows measurements of a grass field at different growth stages. Table 2.1 shows the model parameters in simulations of the grass field at different heights. The simulated σ_{hh}, σ_{vv}, σ_{hv} are compared with measurements in Figure 2.11.

A validation example is compared with NASA/JPL AirSAR data collected over the boreal forest area at Prince Albert National Park, Canada. This area is covered by needle-leaf forest. Some *in situ* measured tree parameters are listed in Table 2.2 (Moghaddam and Saatchi, 1995), which are used to set the parameters in the VRT model. Trunks are modeled as vertical cylinders, branches are modeled as randomly oriented cylinders, and twigs and leaves are modeled as needle-like particles.

Figures 2.11 and 2.12 show the comparison between simulation and measurements.

Table 2.1 Model parameters[a] of grass field at different growth stages. Used with the permission of the Canadian Aeronautics and Space Institute

| Grass height/cm | Ground roughness | | Model parameters | | |
	RMS H/cm	COR L/cm	Depth/cm	Fractional volume	Orientation distribution/°
24	0.7	3.25	24	0.020	0–60
30	0.7	3.25	30	0.018	0–58
35	0.7	3.25	35	0.016	0–56
42	0.7	3.25	42	0.014	0–54

[a]Other parameters are: frequency, 4.6 GHz; incident angle, 30°; particle size, 0.1 cm × 0.1 cm × 15 cm; dielectric constant, $14.1 + i4.1$; ground soil dielectric constant, $15 + i$.

Table 2.2 Model parameter set according to *in situ* measurements[a] (Moghaddam and Saatchi, 1995). Used with the permission of the Canadian Aeronautics and Space Institute

	Radius/cm	Length/cm	Depth/cm	Orientation/°	Fractional volume	Dielectric constant
Trunk	6.5 ± 2.5	750 ± 150	1500	—	5.0×10^{-5}	$39 + i6$
Branch	0.5 ± 0.1	35 ± 9	900	70–90	3.8×10^{-4}	$39 + i6$
Twig	0.2 ± 0.04	12.5	900	0–90	2.2×10^{-4}	$39 + i6$
Leaf	0.05	1.25 ± 0.2	900	0–180	2.9×10^{-5}	$39 + i6$

[a]The number in front of the " ± " denotes the average value; the one behind the " ± " denotes the standard deviation. The ground surface is assumed to be a flat surface with dielectric constant of $7 + i1$.

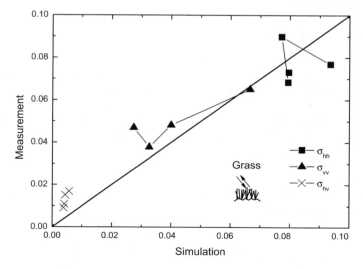

Figure 2.11 Measurement (*y* axis) versus simulation (*x* axis) of grass field scattering. © [2002] IEEE. Reprinted, with permission, from *IEEE Transactions on Geoscience and Remote Sensing*

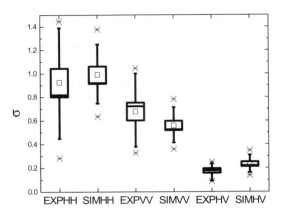

Figure 2.12 Measurement (EXP) versus simulation (SIM) of needle-leaf forest. © [2002] IEEE. Reprinted, with permission, from *IEEE Transactions on Geoscience and Remote Sensing*

2.4 Polarization Indices and Entropy

To indicate the difference between co-polarized (*vv*, *hh*) backscatterings, the polarization index is usually defined to classify the polarized scattering signature from different natural media. How one directly relates these field measurements with the concepts of entropy and eigen-analysis should be meaningful.

In this section, the relationship between entropy and co-polarized or cross-polarized back-scattering indices backscattering index is established. Thus, we can now combine the Mueller matrix solution, the entropy of the coherency matrix, and measurements of the polarization indices together and yield a quantitative description of the natural media. As examples, this theory is applied to an AirSAR image and field measurements for surface monitoring and classification (Jin and Chen, 2002).

2.4.1 Eigen-Analysis of Mueller Matrix

The Mueller matrix, as a 4×4 real matrix with complex eigenvalues and eigenvectors, is physically realizable and must satisfy the Stokes criterion together with several other restrictive conditions (Van der Mee and Hovenier, 1992). However, these restrictions do not have any direct physical interpretation in terms of the eigen-structure of the Mueller matrix. The coherency matrix $\overline{\mathbf{C}}$ has been applied to the study of polarimetric scattering of a synthetic aperture radar (SAR) image (Cloude, 1986; Jin and Cloude, 1994). Define the scattering vector as

$$\mathbf{t} = \frac{1}{2}[S_{vv} + S_{hh}, S_{vv} - S_{hh}, S_{vh} + S_{hv}, i(S_{vh} - S_{hv})]^{\mathrm{T}} \equiv \frac{1}{2}[A, B, C, iD]^{\mathrm{T}} \qquad (2.77)$$

where A, B, C, D are sequentially defined in Equation 2.77. The coherency matrix is defined as follows (Cloude, 1986; Jin and Cloude, 1994):

$$\overline{\mathbf{C}} = \langle \mathbf{t}\,\mathbf{t}^{+} \rangle = \frac{1}{4} \begin{bmatrix} \langle AA* \rangle & \langle AB* \rangle & \langle AC* \rangle & \langle AD* \rangle \\ \langle BA* \rangle & \langle BB* \rangle & \langle BC* \rangle & \langle BD* \rangle \\ \langle CA* \rangle & \langle CB* \rangle & \langle CC* \rangle & \langle CD* \rangle \\ \langle DA* \rangle & \langle DB* \rangle & \langle DC* \rangle & \langle DD* \rangle \end{bmatrix} \qquad (2.78)$$

The eigenvalue and eigenvector of the coherency matrix are defined as

$$\overline{C} = \langle \mathbf{t}\,\mathbf{t}^+ \rangle = \sum_{i=1}^{4} \lambda_i \langle \mathbf{t_i}\,\mathbf{t}_i^+ \rangle \qquad (2.79)$$

where λ_i and $\mathbf{t}_i$ are the ith eigenvalue and eigenvector, respectively. All eigenvalues are real and non-negative, and $\lambda_1 \geq \lambda_2 \geq \lambda_3 \geq \lambda_4$.

The coherency matrix is also related with the Mueller matrix as

$$M_{i+1,s+1} = \sum_t \sum_u \frac{1}{2} \text{Trace}(\overline{\sigma}_i \overline{\sigma}_t \overline{\sigma}_s \overline{\sigma}_u)[\overline{C}]_{tu} \qquad (2.80)$$

where $\overline{\sigma}_j$, $i = 0, 1, 2, 3$, are the Pauli matrices (Cloude, 1986, 2010).

Usually, the case $S_{vh} = S_{hv}$ is simply assumed. The entropy H is defined by the eigenvalues of the coherency matrix as

$$H = -\sum_{i=1}^{3} P_i \log_3 P_i \qquad (2.81)$$

where the functions P_i ($i = 1,2,3$) are expressed by the eigenvalues as

$$P_i = \lambda_i/(\lambda_1 + \lambda_2 + \lambda_3) \qquad (2.82)$$

The entropy is an important feature since it relates the randomness of the scattering media with other physical parameters, such as canopy configuration, biomass, and so on. However, to construct the eigenvalues λ_i, the functions P_i and the entropy H need full Mueller matrix measurements, as shown in Equation 2.77, and have not been directly related with the measured echo power, that is, polarimetric σ_c or σ_x.

The coherency matrix is a 4×4 positive semidefinite Hermitian and as such has real non-negative eigenvalues and complex orthogonal eigenvectors. The amplitude and difference of four eigenvalues are functionally related with the polarimetric scattering process of the terrain surface. Eigen-analysis of the coherency matrix yields better physical insight into the polarimetric scattering mechanisms than does the Mueller matrix; further, it can be employed to physically identify a Mueller matrix by virtue of the semidefinite nature of the corresponding coherency matrix.

2.4.2 Relationship between Eigenvalues, Entropy, and Polarization Indices

For the first-order solution in small albedo, cross-polarization is small, and correlations such as the terms $\langle CA* \rangle$, $\langle CB* \rangle$, $\langle DA* \rangle$, $\langle DB* \rangle$ and so on in Equation 2.78 are always very small, and may be neglected in our following derivations. Actually, for the weak assumption of azimuthally symmetric orientation, these correlations have been proved to be zero (Nighiem $et\ al.$, 1992). Applying this approximation to Equation 2.78, the eigenvalues of $\overline{C}$ have been derived as (Jin and Cloude, 1994)

$$\lambda_{1,2} = \frac{1}{8}\left[\langle|A|^2\rangle + \langle|B|^2\rangle \pm \sqrt{(\langle|A|^2\rangle - \langle|B|^2\rangle)^2 + 4\langle A^*B\rangle \langle AB^*\rangle}\right] \qquad (2.83)$$

$$\lambda_3 = \frac{1}{4}\langle|C|^2\rangle \tag{2.84}$$

It is interesting to see that, if $\langle AB*\rangle \langle A*B\rangle = \langle AA*\rangle \langle BB*\rangle$ in an ordered case, this yields $\lambda_2 = 0$.

Substituting A, B, C, and D of Equation 2.77 into Equations 2.83 and 2.84, we newly derive the eigenvalues in the power terms of $\langle|S_{vv}|^2\rangle$, $\langle|S_{hh}|^2\rangle$, $\langle|S_{hv}|^2\rangle$ as

$$\lambda_1 = \frac{1}{2}\left(\langle|S_{vv}|^2\rangle + \langle|S_{hh}|^2\rangle\right) - \lambda_2 \tag{2.85a}$$

$$\lambda_2 = \frac{1}{2}\frac{\langle|S_{vv}|^2\rangle \langle|S_{hh}|^2\rangle}{\left(\langle|S_{vv}|^2\rangle + \langle|S_{hh}|^2\rangle\right)}(1-\Delta) \tag{2.85b}$$

$$\lambda_3 = \langle|S_{hv}|^2\rangle \tag{2.85c}$$

where the configuration parameter

$$\Delta = \frac{\langle S_{vv}S_{hh}{}^*\rangle \langle S_{vv}{}^*S_{hh}\rangle}{\langle|S_{vv}|^2\rangle \langle|S_{hh}|^2\rangle} \tag{2.86}$$

varies from 0 for totally random media to 1 for ordered media, that is, non-random case.

It can be seen that the first eigenvalue λ_1 now indicates the total co-polarized vv and hh power subtracting their coherency defined by the parameter Δ of Equation 2.86, which varies from 1 for the ordered case to 0 in a totally random medium. The second eigenvalue λ_2 now takes into account the vv and hh coherent power. As Δ approaches 1, $\lambda_2 \to 0$ for ordered media, and as Δ approaches zero, λ_2 would increase to indicate the loss of the vv and hh power coherency for totally random media. The third eigenvalue λ_3 is now due to depolarization caused by the randomness of the media.

Substituting Equations 2.85a–c and 2.86 into the polarized backscattering coefficient,

$$\sigma_{pq} = 4\pi \cos\theta\langle|S_{pq}|^2\rangle \tag{2.87}$$

Thus, Equations 2.85a–c and 2.86 link the eigen-analysis with the measurements of σ_{vv}, σ_{hh}, σ_{vh}.

Substituting Equations 2.85a–c into Equation 2.82, the functions P_i, $i = 1, 2, 3$, are then derived as

$$P_1 = 1 - P_2 - P_3, \quad P_2 = X(1-\Delta)(1-\delta), \quad P_3 = \delta(1-\delta) \ll 1 \tag{2.88}$$

where

$$X \equiv \frac{\sigma_{hh}/\sigma_{vv}}{(1 + \sigma_{hh}/\sigma_{vv})^2}, \quad \delta = \frac{2\sigma_{hv}}{\sigma_{vv} + \sigma_{hh}} \tag{2.89}$$

Note that all P_i are now calculated by backscattering coefficient σ_{pq}, $p, q = v, h$, not by Equation 2.82 from the full Mueller matrix solution. Then, the entropy H of Equation 2.81 is calculated using P_i of Equations 2.88 and 2.89. It can be seen that the function P_1 is related to the total co-polarized power $\sigma_{hh} + \sigma_{vv}$; P_2 is due to the difference between σ_{hh} and σ_{vv}; and P_3 is due to depolarization σ_{hv}. They are modulated by the media configuration and randomness via the parameters Δ and δ. As the medium becomes more random or disordered, δ increases and Δ approaches zero. Vice versa, as the medium becomes more ordered, δ decreases and Δ increases. Thus, the relationship between the entropy H and backscattering measurements σ_{pq} ($pq = vv$, hh, hv) is established.

Define the co-polarization co-polarization index and cross-polarization indices cross-polarization index, respectively, as

$$CPI \equiv 10 \log_{10}(\sigma_{hh}/\sigma_{vv}) \quad (dB)$$
$$XPI \equiv \delta \tag{2.90}$$

It can be seen that the entropy H is directly related with the backscattering indices CPI and XPI via Equations 2.88–2.90.

2.4.3 Demonstration with AirSAR Imagery

Figure 2.13 presents the theoretical relationship between the entropy H and index CPI with variations of the parameters Δ and δ. The entropy H is calculated by Equations 2.88–2.90, that is, via $\sigma_{hh}, \sigma_{vv}, \sigma_{hv}$ and variations of Δ and δ. It can be seen that, as the medium becomes more random (δ increases) and CPI becomes smaller, H will be increased. Vice versa, as the medium becomes more ordered and CPI becomes larger, H will be decreased. The bold line in Figure 2.13 is for the case of $\Delta = 0$ and $\delta = 0$. More randomness will shift the H–CPI curve upwards; vice versa, a more ordered or less random medium will move the H–CPI curve downwards. The distance of the H–CPI line from this bold line indicates how random the medium might be.

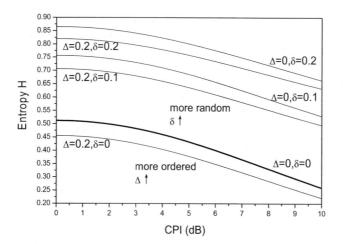

Figure 2.13 Entropy H versus index CPI with variations of Δ and δ. © [2002] IEEE. Reprinted, with permission, from *IEEE Transactions on Geoscience and Remote Sensing*

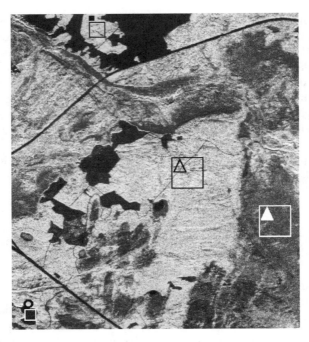

Figure 2.14 An AirSAR image of total power σ_{vv}, σ_{hh}, σ_{vh} at L band. © [2002] IEEE. Reprinted, with permission, from *IEEE Transactions on Geoscience and Remote Sensing*

Figure 2.14 presents an AirSAR image of the total power of σ_{vv}, σ_{hh}, and σ_{vh} at L band over a Jack pine (*Pinus banksiana*) area of Canada (near the Prince Albert National Park, about 103.33°W, 53.9°N). The image data from some typical areas such as the lake, an island surface, sparse trees, and thick forest (see the frames in the figure) are chosen, and their H, *CPI*, and *XPI* are compared.

To demonstrate the H–*CPI* relationship from the AirSAR imagery, Figure 2.15 shows discrete points of the H–*CPI* from four regions of Figure 2.14. First, H is rigorously calculated by polarimetric data of the Mueller matrix, that is, the eigenvalues of coherency matrix $\overline{\mathbf{C}}$ of Equation 2.84. Using the AirSAR data σ_{vv}, σ_{hh}, σ_{vh}, and the averaged δ in the respective region, the lines are calculated by Equations 2.88–2.90 with appropriate Δ to match the center of discrete data. All δ and Δ values are listed in the figure caption. The bold line (4) is for the case $\Delta = 0$ and $\delta = 0$. The lines to match discrete data indicate how far the data in the region are from the case $\Delta = 0$, $\delta = 0$, and how random the media are. It can be seen that the H–*CPI* can be well applied to surface classification.

Figure 2.16 gives a 3D graph of H versus two indices (*CPI* and *XPI*) of Equation 2.90. It can be seen that medium randomness yields high H, decreases *CPI*, and increases *XPI*. More information for surface classification can be inferred from this 3D graph. The discrete points in Figure 2.16 are the averaged data chosen from four regions of Figure 2.15, respectively.

It is certainly well known that contributions to the polarimetric echoes from the terrain canopy are made by volumetric scattering of random non-spherical particles and rough-surface scattering. Any configuration change of volumetric or surface scattering conditions

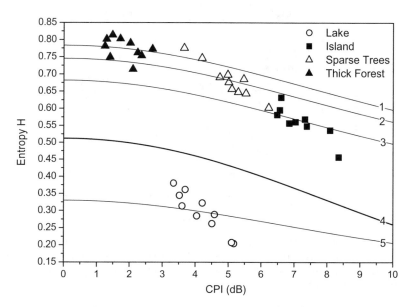

Figure 2.15 Entropy H and index CPI for surface classification from AirSAR data. Line 1: $\Delta = 0.257$, $\delta = 0.176$. Line 2: $\Delta = 0.318$, $\delta = 0.156$. Line 3: $\Delta = 0.365$, $\delta = 0.118$. Line 4: $\Delta = 0, \delta = 0$. Line 5: $\Delta = 0.731$, $\delta = 0.0265$. © [2002] IEEE. Reprinted, with permission, from *IEEE Transactions on Geoscience and Remote Sensing*

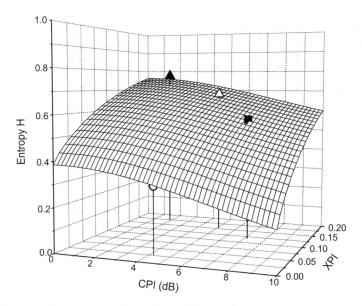

Figure 2.16 Entropy H versus two indices (CPI and XPI). The curved surface is for $\Delta = 0.3$. © [2002] IEEE. Reprinted, with permission, from *IEEE Transactions on Geoscience and Remote Sensing*

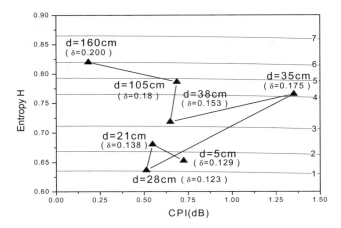

Figure 2.17 Entropy H–CPI from the corn field measurements. Line 1: $\Delta = 0.5$, $\delta = 0.12$. Line 2: $\Delta = 0.45$, $\delta = 0.13$. Line 3: $\Delta = 0.4$, $\delta = 0.15$. Line 4: $\Delta = 0.4$, $\delta = 0.15$. Line 5: $\Delta = 0.3$, $\delta = 0.2$. Line 6: $\Delta = 0.2$, $\delta = 0.2$. Line 7: $\Delta = 0$, $\delta = 0.2$. © [2002] IEEE. Reprinted, with permission, from *IEEE Transactions on Geoscience and Remote Sensing*

can cause a variation in the H–(CPI, XPI) graphs. To see an example during a period of crop growth, Figures 2.17 and 2.18 present H–CPI and H–(CPI, XPI) from corn field measurements of σ_{vv}, σ_{hh}, σ_{vh}, and δ, even though no fully polarimetric data are available in this case. The measurements were made by the scatterometer at 4.6 GHz and at an observation angle of 40° during corn growth from 18 May to 11 July 1988 (Matzler and Wegmuller, 1993). The layer thickness of the corn field is d (cm).

At the beginning of corn growth, for example, $d < 30$ cm, scattering is governed by the underlying rough surface. The CPI index might also be small due to scattering of the very roughly cultivated surface. As the canopy grows, the rough surface might be gradually smoothed or shadowed by scatterers. More random scatterers of the canopy enhance volumetric scattering and yield higher H. Specific orientation of non-spherical scatterers might also lightly increase CPI. As the canopy becomes much thicker and more random, for example, $d > 100$ cm, a totally random medium would increase H and decrease CPI. The non-

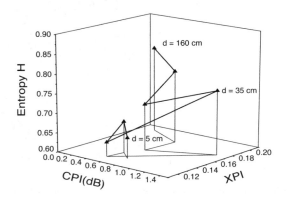

Figure 2.18 Entropy H–(CPI, XPI) from the corn field measurements. © [2002] IEEE. Reprinted, with permission, from *IEEE Transactions on Geoscience and Remote Sensing*

monotonic trend of *H–CPI* and *H–(CPI, XPI)* during corn growth, depending upon variation of volumetric and surface configurations, can be seen in Figures 2.17 and 2.18. The lines with different δ and Δ indicate characteristics of the corn growth.

The parameter δ in Figures 2.17 and 2.18 is given by the measurements of σ_{vv}, σ_{hh}, and σ_{vh} via Equations 2.87–2.90. As $\delta_{\max} = 0.25$ is reached, the medium is assumed to be totally random and $\Delta = 0$ is defined. Thus, it empirically yields $\Delta = 1 - \delta/0.25$ for the calculations in Figures 2.17 and 2.18.

Numerical simulation of polarimetric scattering from a layer of non-spherical particles over a rough surface (Jin, Zhang, and Chang, 1999) can also be applied to verification of the functional dependence of the *H* upon *CPI* and *XPI*.

2.5 Statistics of Stokes Parameters

Full understanding of the statistics of polarimetric SAR data has become of great importance for image speckle reduction, data verification, target identification, and surface classification (Sheen and Johnston, 1992; Lee, Grunes and De Grandi, 1999a). As the area illuminated by the radar contains many uniformly distributed random scatterers, the multi-look covariance matrix can be assumed to have complex Wishart distribution (Lee *et al.*, 1994). The statistics of the scattered intensity, amplitude ratio, and phase term from multi-look polarimetric SAR data have been studied by Lee *et al.* (1999b). Frery and Müller (1997) presented a clutter model for the amplitude statistics from an extremely heterogeneous background, for example, a city. Barakat (1987) first described the temporal statistics of the Stokes parameters of partially polarized light, which is applicable to single-look SAR data. Touzi and Lopes (1996) provided a numerically tractable extension to Barakat's work to include the effective phase difference by deriving the probability density functions (PDFs) of the multi-look Stokes vector as a function of coherent electric fields and complex correlation coefficient.

The aim of this section is to analytically formulate the statistics of the four Stokes parameters for multi-look SAR imagery, and to validate SAR multi-channel measurements over different terrain surfaces (Jin and Zhang, 2002).

The PDF of the first Stokes parameter, *I*, is found to be that given by Touzi and Lopes (1996). However, the PDF of the second Stokes parameter, *Q*, is much simpler than the series summation of Touzi and Lopes (1996). Under some reasonable approximations, validated by real SAR data, the PDFs of the third and fourth Stokes parameters, *U* and *V*, can be formulated analytically to replace the integral forms of Touzi and Lopes (1996). These PDFs have explicit functional dependence upon the look number and mean values of the respective Stokes parameters. To validate these statistics, the AirSAR data at multiple P, L, C bands over some typical areas such as water, sparse trees, and thick forest were chosen for comparison. In addition, an effective number of looks might be assigned to provide an excellent match with real data.

2.5.1 Multi-Look Covariance Matrix and Complex Wishart Distribution

The Mueller representation of SAR data is usually averaged by multiple pixels in both azimuth and range directions to form multi-look SAR data and to reduce the image speckle. The statistical distribution for the multi-look Mueller formulation needs to be refined. The *n*-look covariance matrix is usually defined as (Lee *et al.*, 1994)

$$\overline{\mathbf{Z}} = \frac{1}{n}\sum_{k=1}^{n}\langle\mathbf{u}(k)\mathbf{u}(k)^{\mathrm{T}}\rangle \qquad (2.91)$$

where the vector $\mathbf{u}(k)$ is the kth one-look sample; its element S_1, S_2, and S_3 denote S_{hh}, S_{vv}, and S_{hv}, respectively; the superscript T denotes complex conjugate transpose; and the angular brackets $\langle\,\rangle$ means the ensemble average. The dimension of vector $\mathbf{u}(k)$ can be taken as 1, 2, or 3. To simplify the following derivation, let $\overline{\mathbf{A}} = n\overline{\mathbf{Z}}$. Then the matrix $\overline{\mathbf{A}}$ has a complex Wishart distribution (Goodman, 1963; Lee et $al.$, 1994)

$$P_A(\overline{\mathbf{A}}) = \frac{|\overline{\mathbf{A}}|^{n-q}\exp[-\mathrm{Tr}(\overline{\mathbf{C}}-1\overline{\mathbf{A}})]}{K(n,q)|\overline{\mathbf{C}}|^n} \tag{2.92}$$

where $K(n,q) = \pi^{q(q-1)/2}\Gamma(n)\Gamma(n-1)\cdots\Gamma(n-q+1)$, q is the dimension of the vector $\mathbf{u}$ (k), the complex covariance matrix $\overline{\mathbf{C}} = E[\mathbf{u}\,\mathbf{u}^T]$, $E[v]$ is the expected value of the random variable v, and $|\overline{\mathbf{C}}|$ is the determinant of $\overline{\mathbf{C}}$. Applying the complex Wishart distribution, the PDFs of the four multi-look Stokes parameters can be derived.

2.5.2 PDFs of the Four Stokes Parameters

The Mueller matrix of multi-look SAR data is the average of single-look Mueller matrices

$$\overline{\mathbf{M}}(n) = \frac{1}{n}\sum_{k=1}^{n}\overline{\mathbf{M}}(k) \tag{2.93}$$

where n is the look number and k indicates the kth single look.

The normalized incident Stokes vector has been given as Equation 2.51. As the incident wave is a linearly polarized ($a=h$ or v), the scattered Stokes parameters of n-look SAR imagery are written as

$$I = \frac{1}{2n}\sum_{k=1}^{n}[\langle S_{aa}(k)S_{aa}^*(k)\rangle + \langle S_{ab}(k)S_{ab}^*(k)\rangle] \tag{2.94a}$$

$$Q = \frac{\pm 1}{2n}\sum_{k=1}^{n}[\langle S_{aa}(k)S_{aa}^*(k)\rangle - \langle S_{ab}(k)S_{ab}^*(k)\rangle] \tag{2.94b}$$

$$U = \frac{1}{n}\sum_{k=1}^{n}\langle\mathrm{Re}[S_{aa}(k)S_{ab}^*(k)]\rangle \tag{2.94c}$$

$$V = -\frac{1}{n}\sum_{k=1}^{n}\langle\mathrm{Im}[S_{aa}(k)S_{ab}^*(k)]\rangle \tag{2.94d}$$

where the $\pm$ signs in Equation 2.94b, respectively, indicate $+$ for h-pol and $-$ for v-pol. Since every Stokes parameter in Equations 2.94a–294d is expressed by only two scattering amplitude functions, let $q = 2$ and

$$\overline{\mathbf{A}} = \begin{bmatrix} A_{11} & A_{12} \\ A_{21} & A_{22} \end{bmatrix}, \overline{\mathbf{C}} = \begin{bmatrix} C_{11} & \sqrt{C_{11}C_{22}}|\rho_c|e^{i\theta} \\ \sqrt{C_{11}C_{22}}|\rho_c|e^{-i\theta} & C_{22} \end{bmatrix} \tag{2.95}$$

where

$$A_{11} = \sum_{k=1}^{n} \langle S_{aa}(k)S_{aa}^*(k) \rangle, \quad A_{22} = \sum_{k=1}^{n} \langle S_{ab}(k)S_{ab}^*(k) \rangle \tag{2.96a}$$

$$A_{21}^* = A_{12} = \sum_{k=1}^{n} \langle S_{aa}(k)S_{ab}^*(k) \rangle \tag{2.96b}$$

with $C_{11} = E[A_{11}]$, $C_{22} = E[A_{22}]$, and $\rho_c = E[A_{12}]/\sqrt{C_{11}C_{22}}$ is the complex correlation coefficient for $\langle S_{aa}(k)S_{ab}^*(k) \rangle$, $a \neq b$; $E[\cdot]$ means the expected value. Comparing Equation 2.95 with Equations 2.94a–2.94d, we have

$$C_{11} = E[I] \pm E[Q] \tag{2.97a}$$

$$C_{22} = E[I] \mp E[Q] \tag{2.97b}$$

$$|\rho_c| = \sqrt{E^2[U] + E^2[V]} / \sqrt{E^2[I] - E^2[Q]} \tag{2.97c}$$

The joint PDF of R_1 and R_2 has been derived by Lee *et al.* (1994) as

$$P(R_1, R_2) = \frac{n^{n+1}(R_1R_2)^{(n-1)/2} \exp\left[-\dfrac{n(R_1C_{22} + R_2C_{11})}{C_{11}C_{22}(1 - |\rho_c|^2)}\right]}{(C_{11}C_{22})^{(n+1)/2}\Gamma(n)(1 - |\rho_c|^2)|\rho_c|^{n-1}} I_{n-1}\left(2n\sqrt{\frac{R_1R_2}{C_{11}C_{22}}}\frac{|\rho_c|}{1 - |\rho_c|^2}\right) \tag{2.98}$$

where $R_1 = A_{11}/n$, $R_2 = A_{22}/n$, and $I_v(x)$ is the modified Bessel function of the first kind and vth order.

From Equations 2.94a and 2.94b, $I = (R_1 + R_2)/2$, and $Q = \pm(R_1 - R_2)/2$, the joint PDF of I and Q is written as

$$
\begin{aligned}
P(I, Q) &= \frac{2n^{n+1}(I^2 - Q^2)^{(n-1)/2}}{(C_{11}C_{22})^{(n+1)/2}\Gamma(n)(1 - |\rho_c|^2)|\rho_c|^{n-1}} \\
&\quad \times \exp\left[-\frac{n(C_{11} + C_{22})I \mp n(C_{11} - C_{22})Q}{C_{11}C_{22}(1 - |\rho_c|^2)}\right] I_{n-1}\left(2n\sqrt{\frac{I^2 - Q^2}{C_{11}C_{22}}}\frac{|\rho_c|}{1 - |\rho_c|^2}\right)
\end{aligned}
\tag{2.99}
$$

The first Stokes parameter I is total scattered power. Using the integration identity

$$\int_{-1}^{1} (1 - x^2)^{v/2} \exp(-\alpha x)I_v(\beta\sqrt{1 - x^2}) \, dx = \sqrt{2\pi}\beta^v(\alpha^2 + \beta^2)^{-v/2 - 1/4}I_{v+1/2}\left(\sqrt{\alpha^2 + \beta^2}\right) \tag{2.100}$$

and integrating Equation 2.99 with respect to Q, we obtain the PDF of I as

$$P_I(I) = \int_{-I}^{I} P(I, Q)\, dQ = \int_{-1}^{1} IP(I, xI)\, dx$$

$$= \frac{\sqrt{\pi}(2n)^{n+1/2}I^{n-1/2}\exp\left[-\dfrac{n(C_{11}+C_{22})I}{C_{11}C_{22}(1-|\rho_c|^2)}\right]I_{n-1/2}\left[\dfrac{n\sqrt{(C_{11}-C_{22})^2+4C_{11}C_{22}|\rho_c|^2}}{C_{11}C_{22}(1-|\rho_c|^2)}I\right]}{\Gamma(n)\sqrt{C_{11}C_{22}(1-|\rho_c|^2)}[(C_{11}-C_{22})^2+4C_{11}C_{22}|\rho_c|^2]^{n/2-1/4}}$$

$$(2.101)$$

Substituting Equations 2.97a–2.97c into Equation 2.101, the PDF of I is finally obtained as

$$P_I(I) = \frac{2\sqrt{\pi}n^{n+1/2}I^{n-1/2}\exp\left[-\dfrac{2nI}{E[I](1-p^2)}\right]I_{n-1/2}\left[\dfrac{2npI}{E[I](1-p^2)}\right]}{\Gamma(n)(E[I])^{n+1/2}(1-p^2)^{1/2}p^{n-1/2}}$$

$$(2.102)$$

where

$$p = \sqrt{E^2[Q]+E^2[U]+E^2[V]}/E[I]$$

$$(2.103)$$

and is defined as the polarization degree to describe partial polarization of scattered intensity propagating through random media.

Normalizing I with respect to $E[I]$, we obtain the PDF of the normalized first Stokes parameter, $S_0 = I/E[I]$, for linear incidence as

$$P_{S_0}(S_0) = \frac{2\sqrt{\pi}n^{n+1/2}S_0^{n-1/2}\exp\left[-\dfrac{2nS_0}{1-p^2}\right]I_{n-1/2}\left[\dfrac{2np}{1-p^2}S_0\right]}{\Gamma(n)(1-p^2)^{1/2}p^{n-1/2}}$$

$$(2.104)$$

This PDF has been derived in Touzi and Lopes (1996, their equation 47).

Applying the integration identity (Gradshteyn and Ryzhik, 1980, page 721)

$$\int_{1}^{\infty} (x^2-1)^{\upsilon/2}\exp(-\alpha x)J_\upsilon(\beta\sqrt{x^2-1})\, dx = \sqrt{\frac{2}{\pi}}\beta^\upsilon(\alpha^2+\beta^2)^{-\upsilon/2-1/4}K_{\upsilon+1/2}(\sqrt{\alpha^2+\beta^2})$$

$$(2.105)$$

and integrating Equation 2.99 with respect to I, we obtain

$$
P_Q(Q) = \int_{|Q|}^{\infty} P(I,Q)\,dI = \int_{1}^{\infty} |Q| P(x|Q|,Q)\,dx
$$

$$
= \frac{(2n)^{n+1/2}|Q|^{n-1/2}\exp\left[\pm\dfrac{n(C_{11}-C_{22})Q}{C_{11}C_{22}(1-|\rho_c|^2)}\right] K_{n-1/2}\left[\dfrac{n\sqrt{(C_{11}+C_{22})^2-4C_{11}C_{22}|\rho_c|^2}}{C_{11}C_{22}(1-|\rho_c|^2)}|Q|\right]}{\sqrt{\pi}\Gamma(n)\sqrt{C_{11}C_{22}(1-|\rho_c|^2)}\,[(C_{11}+C_{22})^2-4C_{11}C_{22}|\rho_c|^2]^{n/2-1/4}}
$$

(2.106)

where $K_v(x)$ is the modified Bessel function of the second kind and v th order.

Substituting Equations 2.97a–2.97c into Equation 2.106, the PDFs of the second Stokes parameter Q and its normalized parameter, $S_1 = Q/E[I]$, can be obtained. The PDF of S_1 for linearly polarized incidence is written as

$$
P_{S_1}(S_1) = \frac{2n^{n+1/2}|S_1|^{n-1/2}\exp\left[\dfrac{2nqS_1}{1-p^2}\right] K_{n-1/2}\left[\dfrac{2n\sqrt{1-u^2-v^2}|S_1|}{1-p^2}\right]}{\sqrt{\pi}\Gamma(n)(1-p^2)^{1/2}(1-u^2-v^2)^{n/2-1/4}}
$$

(2.107)

where $q = E[Q]/E[I]$, $u = E[U]/E[I]$, and $v = E[V]/E[I]$. It is noted that Equation 2.107 for the normalized second Stokes parameter is much simpler than the series summation of Touzi and Lopes (1996, their equation 55).

Let complex A_{12} in Equation 2.95 be expressed by $A_{12R} + iA_{12I}$, then Equation 2.92 becomes

$$
P_A(A_{11},A_{22},A_{12R},A_{12I}) = \frac{2[A_{11}A_{22}-(A_{12R}^2+A_{12I}^2)]^{n-2}}{\pi\Gamma(n)\Gamma(n-1)(1-|\rho_c|^2)^n C_{11}^n C_{22}^n}
$$

$$
\times \exp\left[-\frac{A_{11}C_{22}+A_{22}C_{11}-2|\rho_c|\sqrt{C_{11}C_{22}}(A_{12R}\cos\theta+A_{12I}\sin\theta)}{C_{11}C_{22}(1-|\rho_c|^2)}\right]
$$

(2.108)

where θ is the phase of the complex correlation coefficient ρ_c.

Using the integration identity (Gradshteyn and Ryzhik, 1980, page 721)

$$
\int_{1}^{\infty} (x^2-1)^{v/2}\exp(-\alpha x)J_v(\beta\sqrt{x^2-1})\,dx = \sqrt{\frac{2}{\pi}}\beta^v(\alpha^2+\beta^2)^{-v/2-1/4}K_{v+1/2}(\sqrt{\alpha^2+\beta^2})
$$

(2.109)

and integrating Equation 2.108 with respect to A_{11} for $n > 1$, we obtain

$$
\begin{aligned}
&P(A_{22}, A_{12R}, A_{12I}) \\
&= \int_{(A_{12R}^2 + A_{12I}^2)/A_{22}}^{\infty} P_A(A_{11}, A_{22}, A_{12R}, A_{12I})\, dA_{11} \\
&= \frac{2A_{22}^{n-2}\exp\left[-\dfrac{C_{11}A_{22} + C_{22}(A_{12R}^2 + A_{12I}^2)/A_{22} - 2|\rho_c|\sqrt{C_{11}C_{22}}(A_{12R}\cos\theta + A_{12I}\sin\theta)}{C_{11}C_{22}(1 - |\rho_c|^2)} \right]}{\pi\,\Gamma(n)(1 - |\rho_c|^2)C_{11}C_{22}{}^n}
\end{aligned}
$$

$$(2.110)$$

Again, using the integration identity (Gradshteyn and Ryzhik, 1980, page 340)

$$
\int_0^{\infty} x^{\nu-1} e^{-\beta/x - \gamma x}\, dx = 2\left(\frac{\beta}{\gamma}\right)^{\nu/2} K_\nu(2\sqrt{\beta\gamma})
$$

$$(2.111)$$

and integrating Equation 2.110 over A_{22}, we obtain

$$
\begin{aligned}
P(A_{12R}, A_{12I}) &= \frac{4\exp[2|\rho_c|(A_{12R}\cos\theta + A_{12I}\sin\theta)/\sqrt{C_{11}C_{22}}(1 - |\rho_c|^2)]}{\pi\Gamma(n)(1 - |\rho_c|^2)C_{11}^{(n+1)/2}C_{22}^{(n+1)/2}} \\
&\quad \cdot (A_{12R}^2 + A_{12I}^2)^{(n-1)/2} K_{n-1}[2\sqrt{(A_{12R}^2 + A_{12I}^2)/C_{11}C_{22}(1 - |\rho_c|^2)^2}]
\end{aligned}
$$

$$(2.112)$$

Since ρ_c takes into account correlation of $\langle S_{aa}(k)S_{ab}^*(k)\rangle$, $a \neq b$, its magnitude for linearly polarized incidence is always small when compared with C_{11}, C_{22}. This yields the approximations

$$
A_{12R}, A_{12I} \ll \sqrt{C_{11}C_{22}}, \qquad \frac{2|\rho_c|A_{12I}\sin\theta}{\sqrt{C_{11}C_{22}}(1 - |\rho_c|^2)} \approx 0 \quad \text{and} \quad \frac{2|\rho_c|A_{12R}\cos\theta}{\sqrt{C_{11}C_{22}}(1 - |\rho_c|^2)} \approx 0
$$

$$(2.113)$$

These approximations can be validated by SAR image data – as an example, chosen from Figure 2.21, the forest area C. It is expected that these approximations in Equation 2.113 will be valid for natural terrain at low frequency such as P, L, and C bands. Man-made structures may provide pathological cases where they fail. From Figure 2.19, it is demonstrated that

$$
\exp[2|\rho_c|A_{12I}\sin\theta/\sqrt{C_{11}C_{22}}(1 - |\rho_c|^2)] \approx 1
$$

$$(2.114a)$$

$$
\exp[2|\rho_c|A_{12R}\cos\theta/\sqrt{C_{11}C_{22}}(1 - |\rho_c|^2)] \approx 1
$$

$$(2.114b)$$

with very small deviations.

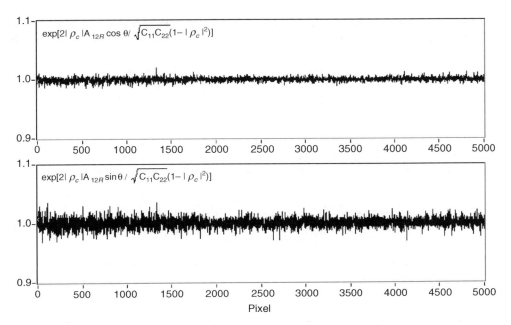

Figure 2.19 Data validation for the approximations of Equations 2.114a and b. Used with the permission of the Canadian Aeronautics and Space Institute

Substituting Equation 2.144a and the identity (Gradshteyn and Ryzhik, 1980, page 705)

$$\int_0^\infty K_\nu(\alpha\sqrt{x^2+z^2})\frac{x^{2\mu+1}}{\sqrt{(x^2+z^2)^\nu}}dx = \frac{2^\mu\Gamma(\mu+1)}{\alpha^{\mu+1}z^{\nu-\mu-1}}K_{\nu-\mu-1}(\alpha z) \tag{2.115}$$

into Equation 2.112, we obtain the integration with respect to A_{12I} as

$$P_{A_{12R}}(A_{12R}) = \frac{2|A_{12R}|^{n-1/2}\exp\left[\dfrac{2|\rho_c|\cos\theta A_{12R}}{\sqrt{C_{11}C_{22}}(1-|\rho_c|^2)}\right]K_{n-1/2}\left(\dfrac{2|A_{12R}|}{\sqrt{C_{11}C_{22}}(1-|\rho_c|^2)}\right)}{\sqrt{\pi}\,\Gamma(n)(1-|\rho_c|^2)^{1/2}C_{11}^{(2n+1)/4}C_{22}^{(2n+1)/4}} \tag{2.116}$$

Since $A_{12R} = \sum_1^n\langle\mathrm{Re}(S_{aa}(k)S_{ab}^*(k))\rangle$, the PDF of the third Stokes parameter, $U = A_{12R}/n$, is written as follows:

$$P_U(U) = \frac{2n^{n+1/2}|U|^{n-1/2}\exp\left[\dfrac{2n|\rho_c|\cos\theta}{\sqrt{C_{11}C_{22}}(1-|\rho_c|^2)}U\right]K_{n-1/2}\left(\dfrac{2n|U|}{\sqrt{C_{11}C_{22}}(1-|\rho_c|^2)}\right)}{\sqrt{\pi}\Gamma(n)(1-|\rho_c|^2)^{1/2}C_{11}^{(2n+1)/4}C_{22}^{(2n+1)/4}}$$

$$\tag{2.117}$$

Substituting Equations 2.97a–c into Equation 2.117, the PDFs of the third Stokes parameter U and its normalized parameter, $S_2 = U/E[I]$, can be obtained. The PDF of S_2 for linearly polarized incidence is

$$P_{S_2}(S_2) = \frac{2n^{n+1/2}|S_2|^{n-1/2}\exp\left(\dfrac{2n\cos\theta\sqrt{u^2+v^2}}{1-p^2}S_2\right)K_{n-1/2}\left(\dfrac{2n\sqrt{1-q^2}|S_2|}{1-p^2}\right)}{\sqrt{\pi}\,\Gamma(n)(1-p^2)^{1/2}(1-q^2)^{n/2-1/4}} \tag{2.118}$$

Similarly, employing Equation 2.114b and integrating Equation 2.112 with respect to A_{12R}, the PDFs of the fourth Stokes parameter V and its normalized parameter, $S_3 = V/E[I]$, can be obtained. The PDF of S_3 for linearly polarized incidence is written as

$$P_{S_3}(S_3) = \frac{2n^{n+1/2}|S_3|^{n-1/2}\exp\left(\dfrac{2n\sqrt{u^2+v^2}\sin\theta}{1-p^2}S_3\right)K_{n-1/2}\left(\dfrac{2n\sqrt{1-q^2}|S_3|}{1-p^2}\right)}{\sqrt{\pi}\,\Gamma(n)(1-p^2)^{1/2}(1-q^2)^{n/2-1/4}} \tag{2.119}$$

The PDFs of the normalized third and fourth Stokes parameters, Equations 2.118 and 2.119, are different from the integral forms of Touzi and Lopes (1996, their equations 64 and 65,

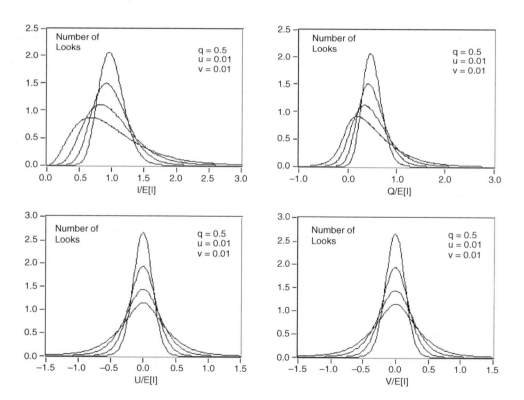

Figure 2.20 The PDFs of S_0, S_1, S_2, S_3 for different look numbers. Used with the permission of the Canadian Aeronautics and Space Institute

respectively), which cannot be analytically solved. Our approximations in Equations 2.114a and 2.114b make Equations 2.118 and 2.119 analytically formulated.

Figure 2.20 gives the PDFs of S_0, S_1, S_2, and S_3 for different look numbers n. It can be seen that increasing the look number makes the statistical distributions sharper and less skewed.

The variance of the distribution is usually defined as

$$\delta^2 = \int (x - \bar{x})^2 \mathrm{PDF}(x)\, dx \tag{2.120}$$

where $x = I, Q, U, V$, and $\bar{x}$ denotes the mean value. The skewness, as another important characteristic property of the distribution, is defined as

$$\gamma = \int (x - \bar{x})^3 \mathrm{PDF}(x)\, dx \tag{2.121}$$

The skewness characterizes the degree of asymmetry around the mean value. A positive value of skewness corresponds to a distribution with asymmetric tail extending to the right of the mean. A negative value of skewness corresponds to a distribution with asymmetric tail extending to the left.

2.5.3 Comparison with AirSAR Image Data

Three typical sites from a 16-look AirSAR image are chosen to demonstrate the statistics of the four Stokes parameters. The image is the AirSAR data acquired on 12August 1993 over the Jack pine area of Canada near the Prince Albert National Park (about W103.33°, N53.9°). The near range incident angle is 29.18°. The range and azimuthal pixel spacing is 6.66 m and 8.22 m, respectively. Part of the image (1024 × 1024 pixels) is shown in Figure 2.21, where the gray shades (red, green, and blue in the original color image) represent the total power in P, L, and C bands, respectively.

Figure 2.21 The AirSAR image. Used with the permission of the Canadian Aeronautics and Space Institute

The data histogram at three sites, marked by the rectangles A, B, and C, sequentially, are chosen to compare with the theoretical PDF statistics. Our theoretical PDFs are calculated by using the mean values of the respective Stokes parameter measurements. Site A is an area of water. At site B, the backscattering at P band is higher than that at C band. Site B is most likely a sparse growth of trees. Backscattering from a tree is usually the result of direct scattering from the tree, interaction of the tree trunk with the ground surface, and the rough land surface. The interaction of the tree trunk with the ground surface at the lower P band is usually more significant than that at the higher C band (Freeman and Durden, 1996), and it yields higher scattering at P band. As the tree canopy becomes thick, however, volumetric scattering from tree crowns is dominant and yields stronger scattering at C band than at P band, as happened at site C.

Figure 2.22a–d shows the data histogram of the four normalized Stokes parameters, S_0, S_1, S_2, S_3, at C band and on site C. The discrete points are the SAR data and the thin solid lines are the theoretical PDFs with $n = 16$. Both the lines and the discrete points are well matched. Note, however, that an effective looks number of $n = 10$, instead of $n = 16$, might give a better match, which is indicated by the bold lines in Figure 2.22a–d. This is most likely due to the assumption of the Wishart distribution for uniformly random scattering

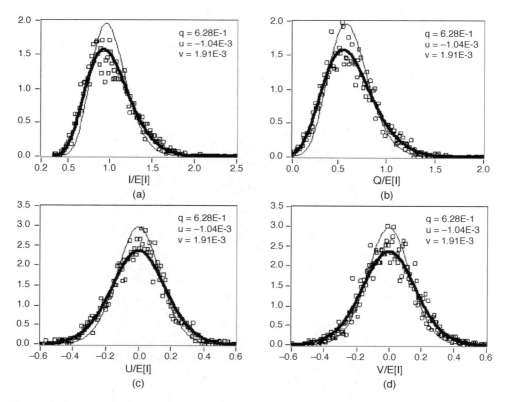

Figure 2.22 The PDFs of S_0, S_1, S_2, S_3 at C band on site C. Used with the permission of the Canadian Aeronautics and Space Institute

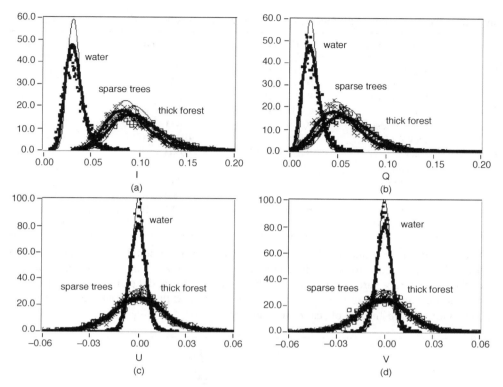

Figure 2.23 The PDFs of *I*, *Q*, *U*, *V* at C band on multiple sites A, B, and C. Used with the permission of the Canadian Aeronautics and Space Institute

media. Although the data are called 16-look data, they are actually the result of a 16-sample (pixel) integration in range and azimuth. Adjacent pixels are correlated if the data is Nyquist sampled. The PDFs are based on *N* independent samples of effective looks (corresponding to resolution cells, not pixels). Thus for this case the sampling theorem dictates that $N < 16$.

Figure 2.23a–d shows the statistics of the four Stokes parameters, *I*, *Q*, *U*, *V*, at the C band at the water site A, the sparse trees site B, and the thick forest site C. The theoretical statistics are well matched with the SAR data for all different types of surface, especially by using the bold lines of the effective look number. It sounds reasonable to assign different effective look numbers for different surface types. Also, it might be a potential application to surface classification.

Figures 2.24a–d and 2.25a–d give the PDFs of *I*, *Q*, *U*, *V* at multiple P, L, C bands at sites B and C, respectively. It is interesting to see the difference between these figures. At site B of sparse trees, stronger scattering at the lower P band than at the higher C band gives statistical patterns such as the sequence of C, L, P from the left to right in Figure 2.24a and b and from the top to bottom in Figure 2.25c and d. As a comparison, at site C of thick forest with stronger scattering from tree crowns at C band, the statistical patterns turn into the sequence of P, L, C from the left to right in Figure 2.25a and b and from the top to bottom in Figure 2.25c and d.

Table 2.3 The variance δ^2 of the four Stokes parameters on sites A, B, and C in P, L, and C bands

	Surface type	C band	L band	P band
I	water	0.119×10^{-3}	0.026×10^{-3}	0.008×10^{-3}
	sparse trees	0.618×10^{-3}	2.432×10^{-3}	1.336×10^{-2}
	thick forest	0.753×10^{-3}	0.417×10^{-3}	0.126×10^{-3}
Q	water	0.091×10^{-3}	0.016×10^{-3}	0.007×10^{-3}
	sparse trees	0.549×10^{-3}	2.431×10^{-3}	1.270×10^{-2}
	thick forest	0.675×10^{-3}	0.376×10^{-3}	0.100×10^{-3}
U	water	0.024×10^{-3}	0.003×10^{-3}	0.001×10^{-3}
	sparse trees	0.242×10^{-3}	0.592×10^{-3}	0.760×10^{-3}
	thick forest	0.284×10^{-3}	0.117×10^{-3}	0.036×10^{-3}
V	water	0.025×10^{-3}	0.003×10^{-3}	0.001×10^{-3}
	sparse trees	0.250×10^{-3}	0.631×10^{-3}	0.687×10^{-3}
	thick forest	0.265×10^{-3}	0.115×10^{-3}	0.033×10^{-3}

The variance, δ^2, of Equation 2.120 and skewness, γ, of Equation 2.121 on sites A, B, and C at multiple P, L, and C bands are listed in Tables 2.3 and 2.4. It can be seen that the area of thick forest has larger variance and is more skewed at C band, and the area of sparse trees has larger variance and is more skewed at P band. It is expected that the variance and skewness become larger as the amount of scattering increases.

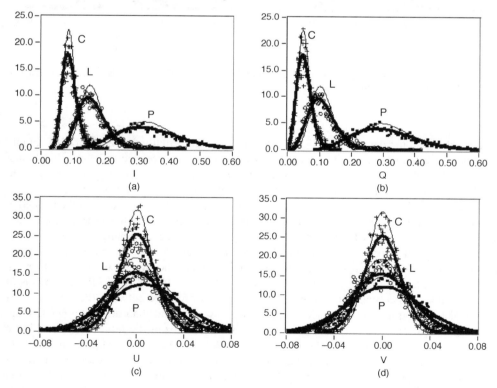

Figure 2.24 The PDFs of I, Q, U, V on site B in multiple P, L, C bands. Used with the permission of the Canadian Aeronautics and Space Institute

Table 2.4 The skewness γ of the four Stokes parameters on sites A, B, and C in P, L, and C bands

	Surface type	C band	L band	P band
I	water	0.117×10^{-5}	0.018×10^{-5}	0.005×10^{-5}
	sparse trees	0.951×10^{-5}	1.378×10^{-4}	1.524×10^{-3}
	thick forest	1.490×10^{-5}	0.743×10^{-5}	0.087×10^{-5}
Q	water	0.075×10^{-5}	0.008×10^{-5}	0.004×10^{-5}
	sparse trees	0.819×10^{-5}	1.483×10^{-4}	1.366×10^{-3}
	thick forest	1.155×10^{-5}	0.611×10^{-5}	0.080×10^{-5}
U	water	0.001×10^{-5}	0.000×10^{-5}	0.000×10^{-5}
	sparse trees	0.023×10^{-5}	0.088×10^{-5}	0.448×10^{-5}
	thick forest	0.021×10^{-5}	0.005×10^{-5}	0.000×10^{-5}
V	water	0.001×10^{-5}	0.000×10^{-5}	0.000×10^{-5}
	sparse trees	0.028×10^{-5}	0.220×10^{-5}	-0.006×10^{-5}
	thick forest	0.021×10^{-5}	0.004×10^{-5}	0.001×10^{-5}

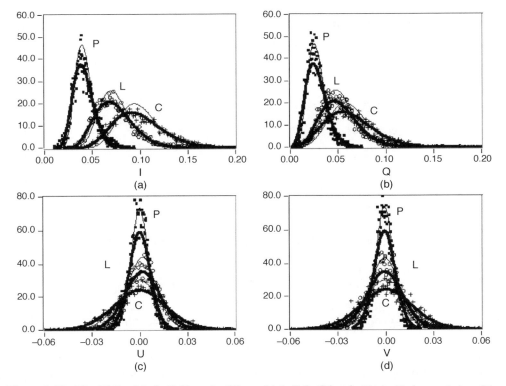

Figure 2.25 The PDFs of *I, Q, U, V* on site C in multiple P, L, C bands. Used with the permission of the Canadian Aeronautics and Space Institute

Appendix 2A: Phase Matrix of Non-Spherical Particles

Some explicit expressions of the phase matrix of non-spherical particles with random orientation under the Rayleigh–Gans approximation are provided here. First, rewrite the scattering matrix of non-spherical particles, Equation 1.26:

$$
\begin{aligned}
S_{vv} &= \frac{3}{2k}\left[T_1(\hat{\theta}_s \cdot \hat{\theta}_i) + (T_0 - T_1)(\hat{\theta}_s \cdot \hat{z}_b)(\hat{z}_b \cdot \hat{\theta}_i)\right] \\
S_{vh} &= \frac{3}{2k}\left[T_1(\hat{\theta}_s \cdot \hat{\varphi}_i) + (T_0 - T_1)(\hat{\theta}_s \cdot \hat{z}_b)(\hat{z}_b \cdot \hat{\varphi}_i)\right] \\
S_{hv} &= \frac{3}{2k}\left[T_1(\hat{\varphi}_s \cdot \hat{\theta}_i) + (T_0 - T_1)(\hat{\varphi}_s \cdot \hat{z}_b)(\hat{z}_b \cdot \hat{\theta}_i)\right] \\
S_{hh} &= \frac{3}{2k}\left[T_1(\hat{\varphi}_s \cdot \hat{\varphi}_i) + (T_0 - T_1)(\hat{\varphi}_s \cdot \hat{z}_b)(\hat{z}_b \cdot \hat{\varphi}_i)\right]
\end{aligned}
\tag{2A.1}
$$

where (θ_i, φ_i), (θ_s, φ_s) denote incident and scattered angles, respectively; and the long axis of particle $\hat{z}_b$ has orientation defined by Euler angles β, γ.

Assume a uniform distribution of Euler angles in the range $\beta \in [\beta_1, \beta_2]$ and $\gamma \in [\gamma_1, \gamma_2]$. Without loss of generality, let us rotate the whole coordinate system around axis $\hat{z}$ to make $\varphi_i \to 0$, $\varphi_s \to \varphi_s - \varphi_i$, $\gamma \to \gamma - \varphi_i$, $\gamma_1 \to \gamma_1 - \varphi_i$, and $\gamma_2 \to \gamma_2 - \varphi_i$. Then in the $(\hat{x}, \hat{y}, \hat{z})$ system, the vectors in Equation 2A.1 can be written as

$$
\begin{aligned}
\hat{z}_b &= a\hat{x} + b\hat{y} + c\hat{z} \\
\hat{\theta}_i &= r\hat{x} + s\hat{y} + t\hat{z}, \quad \hat{\varphi}_i = m\hat{x} + n\hat{y} \\
\hat{\theta}_s &= x\hat{x} + y\hat{y} + z\hat{z}, \quad \hat{\varphi}_s = p\hat{x} + q\hat{y}
\end{aligned}
\tag{2A.2}
$$

where

$$
a = -\cos\gamma \sin\beta, \quad b = \sin\gamma \sin\beta, \quad c = \cos\beta
\tag{2A.3}
$$

$$
r = \cos\theta_i \cos\phi_i = \cos\theta, \quad s = \cos\theta_i \sin\phi_i = 0, \quad t = -\sin\theta_i
\tag{2A.4}
$$

$$
x = \cos\theta_s \cos\phi_s, \quad y = \cos\theta_s \sin\phi_s, \quad z = -\sin\theta_s
\tag{2A.5}
$$

$$
p = -\sin\phi_s, \quad q = \cos\phi_s
\tag{2A.6}
$$

$$
m = -\sin\phi_i = 0, \quad n = \cos\phi_i = 1
\tag{2A.7}
$$

For the sake of succinctness, let us define intermediate variables:

$$
\begin{aligned}
A &= \hat{\theta}_s \cdot \hat{\theta}_i, \quad B = (\hat{\theta}_s \cdot \hat{z}_b)(\hat{z}_b \cdot \hat{\theta}_i), \quad C = \hat{\theta}_s \cdot \hat{\varphi}_i, \quad D = (\hat{\theta}_s \cdot \hat{z}_b)(\hat{z}_b \cdot \hat{\varphi}_i) \\
E &= \hat{\varphi}_s \cdot \hat{\theta}_i, \quad F = (\hat{\varphi}_s \cdot \hat{z}_b)(\hat{z}_b \cdot \hat{\theta}_i), \quad G = \hat{\varphi}_s \cdot \hat{\varphi}_i, \quad H = (\hat{\varphi}_s \cdot \hat{z}_b)(\hat{z}_b \cdot \hat{\varphi}_i)
\end{aligned}
\tag{2A.8}
$$

$$
\rho_1 = \frac{3}{2k}T_1, \quad \rho_2 = \frac{3}{2k}(T_0 - T_1), \quad \rho_R = 2\,\mathrm{Re}(\rho_1 \rho_2^*)
\tag{2A.9}
$$

Further assume that the only randomness is on particle orientation, that is, the remaining parameters are fixed; then ρ_1, ρ_2, ρ_R become deterministic. Hence, the second-order statistical moments of the scattering matrix can be written as

$$
\begin{bmatrix} \langle |S_{vv}|^2 \rangle \\ \langle |S_{vh}|^2 \rangle \\ \langle |S_{hv}|^2 \rangle \\ \langle |S_{hh}|^2 \rangle \end{bmatrix} = \begin{bmatrix} \langle A^2 \rangle & \langle B^2 \rangle & \langle AB \rangle \\ \langle C^2 \rangle & \langle D^2 \rangle & \langle CD \rangle \\ \langle E^2 \rangle & \langle F^2 \rangle & \langle EF \rangle \\ \langle G^2 \rangle & \langle H^2 \rangle & \langle GH \rangle \end{bmatrix} \cdot \begin{bmatrix} |\rho_1|^2 \\ |\rho_2|^2 \\ \rho_R \end{bmatrix} \tag{2A.10}
$$

$$
\begin{bmatrix} \langle S_{vv}S_{hv}^* \rangle \\ \langle S_{vh}S_{hh}^* \rangle \\ \langle S_{vv}S_{hh}^* \rangle \\ \langle S_{vv}S_{vh}^* \rangle \\ \langle S_{hv}S_{hh}^* \rangle \\ \langle S_{vh}S_{hv}^* \rangle \end{bmatrix} = \begin{bmatrix} \langle AE \rangle & \langle BF \rangle & \langle AF \rangle & \langle BE \rangle \\ \langle GC \rangle & \langle HD \rangle & \langle HC \rangle & \langle GD \rangle \\ \langle AG \rangle & \langle BH \rangle & \langle AH \rangle & \langle BG \rangle \\ \langle AC \rangle & \langle BD \rangle & \langle AD \rangle & \langle BC \rangle \\ \langle EG \rangle & \langle FH \rangle & \langle EH \rangle & \langle FG \rangle \\ \langle EC \rangle & \langle FD \rangle & \langle FC \rangle & \langle ED \rangle \end{bmatrix} \cdot \begin{bmatrix} \rho_1^2 \\ \rho_2^2 \\ \rho_1\rho_2^* \\ \rho_2\rho_1^* \end{bmatrix} \tag{2A.11}
$$

So the key is to calculate the ensemble average in Equations 2A.10 and 2A.11 and then further calculate the phase matrix. Substituting Equations 2A.3–2A.7 into Equation 2A.8 yields

$$
A = xr + zt, \quad C = y, \quad E = pr, \quad G = q \tag{2A.12}
$$

$$
\begin{aligned}
&B = xra^2 + xtac + yrab + ytbc + zrac + ztc^2, \quad D = xab + yb^2 + zbc \\
&F = pra^2 + ptac + qrab + qtbc, \quad H = pab + qb^2
\end{aligned} \tag{2A.13}
$$

Hence, Equations 2A.8 and 2A.9 can be expanded to

$$
\begin{aligned}
\langle AA \rangle =\ & x^2r^2 + 2xrzt + z^2t^2 \\
\langle BB \rangle =\ & x^2r^2\langle a^4 \rangle + z^2t^2\langle c^4 \rangle + x^2t^2\langle a^2c^2 \rangle + y^2r^2\langle a^2b^2 \rangle + y^2t^2\langle b^2c^2 \rangle \\
& + z^2r^2\langle a^2c^2 \rangle + 2x^2rt\langle a^3c \rangle + 2xr^2y\langle a^3b \rangle + 4xryt\langle a^2bc \rangle \\
& + 2xr^2z\langle a^3c \rangle + 4xrzt\langle a^2c^2 \rangle + 2xt^2y\langle abc^2 \rangle + 2xt^2z\langle ac^3 \rangle \\
& + 2y^2rt\langle ab^2c \rangle + 2yr^2z\langle a^2bc \rangle + 4yrzt\langle abc^2 \rangle + 2yt^2z\langle bc^3 \rangle + 2z^2rt\langle ac^3 \rangle \\
\langle AB \rangle =\ & x^2r^2\langle a^2 \rangle + x^2rt\langle ac \rangle + xr^2y\langle ab \rangle + xryt\langle bc \rangle + xr^2z\langle ac \rangle + xrzt\langle c^2 \rangle \\
& + ztxr\langle a^2 \rangle + zt^2x\langle ac \rangle + ztyr\langle ab \rangle + zt^2y\langle bc \rangle + z^2tr\langle ac \rangle + z^2t^2\langle c^2 \rangle \\[6pt]
\langle CC \rangle =\ & y^2 \\
\langle DD \rangle =\ & x^2\langle a^2b^2 \rangle + 2xy\langle ab^3 \rangle + 2xz\langle ab^2c \rangle + y^2\langle b^4 \rangle + 2yz\langle b^3c \rangle + z^2\langle b^2c^2 \rangle \\
\langle CD \rangle =\ & yx\langle ab \rangle + y^2\langle b^2 \rangle + yz\langle bc \rangle \\[6pt]
\langle EE \rangle =\ & p^2r^2 \\
\langle FF \rangle =\ & p^2r^2\langle a^4 \rangle + 2p^2rt\langle a^3c \rangle + 2pr^2q\langle a^3b \rangle + 4prqt\langle a^2bc \rangle \\
& + p^2t^2\langle a^2c^2 \rangle + 2pqt^2\langle abc^2 \rangle + q^2r^2\langle a^2b^2 \rangle + 2q^2rt\langle ab^2c \rangle + q^2t^2\langle b^2c^2 \rangle \\
\langle EF \rangle =\ & p^2r^2\langle a^2 \rangle + p^2rt\langle ac \rangle + pr^2q\langle ab \rangle + prqt\langle bc \rangle
\end{aligned}
$$

$$\langle GG \rangle = q^2$$
$$\langle HH \rangle = p^2 \langle a^2 b^2 \rangle + 2pq \langle ab^3 \rangle + q^2 \langle b^4 \rangle$$
$$\langle GH \rangle = qp \langle ab \rangle + q^2 \langle b^2 \rangle$$

$$\langle AE \rangle = pr^2 x + przt$$
$$\langle AF \rangle = pr^2 x \langle a^2 \rangle + przt \langle a^2 \rangle + prxt \langle ac \rangle + pt^2 z \langle ac \rangle + qr^2 x \langle ab \rangle$$
$$\qquad + qrzt \langle ab \rangle + qtxr \langle bc \rangle + qt^2 z \langle bc \rangle$$
$$\langle BE \rangle = pr^2 x \langle a^2 \rangle + prxt \langle ac \rangle + pr^2 y \langle ab \rangle + pryt \langle bc \rangle + pr^2 z \langle ac \rangle + przt \langle c^2 \rangle$$
$$\langle BF \rangle = pr^2 x \langle a^4 \rangle + 2prxt \langle a^3 c \rangle + pr^2 y \langle a^3 b \rangle + 2pryt \langle a^2 bc \rangle$$
$$\qquad + pr^2 z \langle a^3 c \rangle + 2przt \langle a^2 c^2 \rangle + pt^2 x \langle a^2 c^2 \rangle + pt^2 y \langle abc^2 \rangle$$
$$\qquad + pt^2 z \langle ac^3 \rangle + qr^2 x \langle a^3 b \rangle + 2qrxt \langle a^2 bc \rangle + qr^2 y \langle a^2 b^2 \rangle$$
$$\qquad + 2qryt \langle ab^2 c \rangle + qr^2 z \langle a^2 bc \rangle + 2qrzt \langle abc^2 \rangle + qt^2 x \langle abc^2 \rangle$$
$$\qquad + qt^2 y \langle b^2 c^2 \rangle + qt^2 z \langle bc^3 \rangle$$

$$\langle GC \rangle = qy$$
$$\langle HD \rangle = px \langle a^2 b^2 \rangle + py \langle ab^3 \rangle + pz \langle ab^2 c \rangle + qx \langle ab^3 \rangle + qy \langle b^4 \rangle + qz \langle b^3 c \rangle$$
$$\langle HC \rangle = yp \langle ab \rangle + yq \langle b^2 \rangle$$
$$\langle GD \rangle = qx \langle ab \rangle + yq \langle b^2 \rangle + qz \langle bc \rangle$$

$$\langle AG \rangle = qxr + qzt$$
$$\langle BH \rangle = pxr \langle a^3 b \rangle + pxt \langle a^2 bc \rangle + pyr \langle a^2 b^2 \rangle + pyt \langle ab^2 c \rangle$$
$$\qquad + pzr \langle a^2 bc \rangle + pzt \langle abc^2 \rangle + qxr \langle a^2 b^2 \rangle + qxt \langle ab^2 c \rangle$$
$$\qquad + qyr \langle ab^3 \rangle + qyt \langle b^3 c \rangle + qzr \langle ab^2 c \rangle + qzt \langle b^2 c^2 \rangle$$
$$\langle AH \rangle = pxr \langle ab \rangle + pzt \langle ab \rangle + qxr \langle b^2 \rangle + qzt \langle b^2 \rangle$$
$$\langle BG \rangle = qxr \langle a^2 \rangle + qxt \langle ac \rangle + qyr \langle ab \rangle + qyt \langle bc \rangle + qzr \langle ac \rangle + qzt \langle c^2 \rangle$$
$$\langle AC \rangle = yxr + yzt$$

$$\langle BD \rangle = x^2 r \langle a^3 b \rangle + 2xry \langle a^2 b^2 \rangle + 2xrz \langle a^2 bc \rangle + x^2 t \langle a^2 bc \rangle$$
$$\qquad + 2xty \langle ab^2 c \rangle + 2xtz \langle abc^2 \rangle + y^2 r \langle ab^3 \rangle + 2yrz \langle ab^2 c \rangle$$
$$\qquad + y^2 t \langle b^3 c \rangle + 2yt \langle b^2 c^2 z \rangle + z^2 r \langle abc^2 \rangle + z^2 t \langle bc^3 \rangle$$
$$\langle AD \rangle = x^2 r \langle ab \rangle + xzt \langle ab \rangle + yxr \langle b^2 \rangle + yzt \langle b^2 \rangle + zxr \langle bc \rangle + z^2 t \langle bc \rangle$$
$$\langle BC \rangle = yxr \langle a^2 \rangle + yxt \langle ac \rangle + y^2 rab + y^2 t \langle bc \rangle + yzr \langle ac \rangle + yzt \langle c^2 \rangle$$

$$\langle EG \rangle = qpr$$
$$\langle FH \rangle = p^2 r \langle a^3 b \rangle + p^2 t \langle a^2 bc \rangle + 2pqr \langle a^2 b^2 \rangle + 2pqt \langle ab^2 c \rangle$$
$$\qquad + q^2 r \langle ab^3 \rangle + q^2 t \langle b^3 c \rangle$$
$$\langle EH \rangle = p^2 r \langle ab \rangle + prq \langle b^2 \rangle$$
$$\langle FG \rangle = qpr \langle a^2 \rangle + qpt \langle ac \rangle + q^2 r \langle ab \rangle + q^2 t \langle bc \rangle$$

$$\langle EC \rangle = pry$$
$$\langle ED \rangle = pxr \langle ab \rangle + pry \langle b^2 \rangle + prz \langle bc \rangle$$
$$\langle FC \rangle = yqt \langle bc \rangle + ypr \langle a^2 \rangle + ypt \langle ac \rangle + yqr \langle ab \rangle$$
$$\langle FD \rangle = pxr \langle a^3 b \rangle + pyr \langle a^2 b^2 \rangle + pzr \langle a^2 bc \rangle + pxt \langle a^2 bc \rangle + pyt \langle ab^2 c \rangle \qquad \text{(2A.14)}$$
$$\qquad + pzt \langle abc^2 \rangle + qxr \langle a^2 b^2 \rangle + qyr \langle ab^3 \rangle + qzr \langle ab^2 c \rangle + qxt \langle ab^2 c \rangle$$
$$\qquad + qyt \langle b^3 c \rangle + qzt \langle b^2 c^2 \rangle$$

Note that the ensemble average in the above expressions can be derived by calculating the integration of terms involving a, b, c over the range $\beta \in [\beta_1, \beta_2]$, $\gamma \in [\gamma_1, \gamma_2]$, which can be readily derived as

$$\langle a^2 \rangle = f(2,0)g(0,2), \quad \langle b^2 \rangle = f(2,0)g(2,0), \quad \langle c^2 \rangle = f(0,2)$$
$$\langle a^4 \rangle = f(4,0)g(0,4), \quad \langle b^4 \rangle = f(4,0)g(4,0), \quad \langle c^4 \rangle = f(0,4)$$

$$\langle ab \rangle = -f(2,0)g(1,1), \quad \langle bc \rangle = f(1,1)g(1,0), \quad \langle ac \rangle = -f(1,1)g(0,1)$$

$$\begin{aligned}
\langle ab^3 \rangle &= -f(4,0)g(3,1), & \langle a^3 b \rangle &= -f(4,0)g(1,3) \\
\langle ac^3 \rangle &= -f(1,3)g(0,1), & \langle a^3 c \rangle &= -f(3,1)g(0,3) \\
\langle bc^3 \rangle &= f(1,3)g(1,0), & \langle b^3 c \rangle &= f(3,1)g(3,0)
\end{aligned} \qquad (2A.15)$$

$$\begin{aligned}
\langle a^2 b^2 \rangle &= f(4,0)g(2,2), & \langle a^2 c^2 \rangle &= f(2,2)g(0,2) \\
\langle b^2 c^2 \rangle &= f(2,2)g(2,0), & \langle a^2 bc \rangle &= f(3,1)g(1,2) \\
\langle b^2 ac \rangle &= -f(3,1)g(2,1), & \langle c^2 ab \rangle &= -f(2,2)g(1,1)
\end{aligned}$$

where the integral functions f and g are defined as

$$f(m,n) = \langle \sin^m \beta \cdot \cos^n \beta \rangle = \frac{1}{\cos \beta_1 - \cos \beta_2} \int_{\beta_1}^{\beta_2} \sin \beta \, d\beta \sin^m \beta \cdot \cos^n \beta$$
$$g(m,n) = \langle \sin^m \gamma \cdot \cos^n \gamma \rangle = \frac{1}{\gamma_2 - \gamma_1} \int_{\gamma_1}^{\gamma_2} d\gamma \sin^m \gamma \cdot \cos^n \gamma \qquad (2A.16)$$

Finally, the integrations in Equation 2A.15 can be derived as

$$f(2,0) = -\frac{1}{3}(\vartheta_2^2 u_2 - \vartheta_1^2 u_1 + 2u_2 - 2u_1)$$
$$f(0,2) = -\frac{1}{3}(u_2^3 - u_1^3)$$
$$f(1,1) = \frac{1}{3}(\vartheta_2^3 - \vartheta_1^3)$$

$$f(4,0) = -\frac{1}{5}(\vartheta_2^3 \vartheta_2 u_2 - \vartheta_1^3 \vartheta_1 u_1) - \frac{4}{15}(\vartheta_2^2 u_2 - \vartheta_1^2 u_1) - \frac{8}{15}(u_2 - u_1)$$
$$f(0,4) = -\frac{1}{5}(u_2^3 u_2^2 - u_1^3 u_1^2)$$
$$f(1,3) = -\frac{1}{5}(u_2^3 u_2 \vartheta_2 - u_1^3 u_1 \vartheta_1) + \frac{1}{15}(u_2^2 \vartheta_2 - u_1^2 \vartheta_1) + \frac{2}{15}(\vartheta_2 - \vartheta_1)$$

$$f(3,1) = \frac{1}{5}(\vartheta_2^3 \vartheta_2^2 - \vartheta_1^3 \vartheta_1^2)$$
$$f(2,2) = -\frac{1}{5}(u_2^3 \vartheta_2^2 - u_1^3 \vartheta_1^2) - \frac{2}{15}(u_2^3 - u_1^3)$$

$$g(1,0) = v_1 - v_2$$
$$g(0,1) = \xi_2 - \xi_1$$
$$g(1,1) = \frac{1}{2}(\xi_2^2 - \xi_1^2)$$

$$g(0,2) = \frac{1}{2}(\xi_2 v_2 - \xi_1 v_1 + \gamma_2 - \gamma_1)$$

$$g(2,0) = \frac{1}{2}(\xi_1 v_1 - \xi_2 v_2 + \gamma_2 - \gamma_1)$$

$$g(2,2) = -\frac{1}{4}(\xi_2 v_2^3 - \xi_1 v_1^3) + \frac{1}{8}(\xi_2 v_2 - \xi_1 v_1) + \frac{1}{8}(\gamma_2 - \gamma_1)$$

$$g(0,3) = \frac{1}{3}(\xi_2 v_2^2 - \xi_1 v_1^2 + 2\xi_2 - 2\xi_1)$$

$$g(3,0) = -\frac{1}{3}(\xi_2^2 v_2 - \xi_1^2 v_1 + 2v_2 - 2v_1)$$

$$g(1,2) = -\frac{1}{3}(v_2^3 - v_1^3)$$

$$g(2,1) = \frac{1}{3}(\xi_2^3 - \xi_1^3)$$

$$g(0,4) = \frac{1}{4}(\xi_2 v_2^3 - \xi_1 v_1^3) + \frac{3}{8}(\xi_2 v_2 - \xi_1 v_1) + \frac{3}{8}(\gamma_2 - \gamma_1)$$

$$g(4,0) = -\frac{1}{4}(\xi_2^3 v_2 - \xi_1^3 v_1) - \frac{3}{8}(\xi_2 v_2 - \xi_1 v_1) + \frac{3}{8}(\gamma_2 - \gamma_1)$$

$$g(3,1) = \frac{1}{4}(\xi_2 \xi_2^3 - \xi_1 \xi_1^3)$$

$$g(1,3) = -\frac{1}{4}(v_2 v_2^3 - v_1 v_1^3)$$

(2A.17)

where

$$
\begin{aligned}
\vartheta_1 &= \sin \beta_1, & \vartheta_2 &= \sin \beta_2 \\
u_1 &= \cos \beta_1, & u_2 &= \cos \beta_2 \\
\xi_1 &= \sin \gamma_1, & \xi_2 &= \sin \gamma_2 \\
v_1 &= \cos \gamma_1, & v_2 &= \cos \gamma_2
\end{aligned}
$$

(2A.18)

References

Barakat, R. (1987) Statistics of the Stokes parameters. *Journal of the Optical Society of America A*, **4** (7), 1256–1263.

Bohren, C.F. and Huffman, D.R. (1983) *Absorption and Scattering of Light by Small Particles*, John Wiley & Sons, Inc., New York.

Chandrasekhar, S. (1960) *Radiative Transfer*, Dover, New York.

Cloude, S.R. (1986) Group theory and polarization algebra. *Optik*, **75** (1), 26–36.

Cloude, S.R. (2010) *Polarisation Applications in Remote Sensing*, Oxford University Press, London.

Freeman, A. and Durden, S.L. (1996) A three-component scattering model for polarimetric SAR data. *IEEE Transactions on Geoscience and Remote Sensing*, **36** (3), 963–973.

Frery, A.C. and Müller, H.J. (1997) A model for extremely heterogeneous clutter. *IEEE Transactions on Geoscience and Remote Sensing*, **35** (3), 648–659.

Fung, A.K. (1994) *Microwave Scattering and Emission Models and Their Applications*, Artech House, Norwood, MA.

Goodman, N.R. (1963) Statistical analysis based on a certain multivariate complex Gaussian distribution (an introduction). *Annals of Mathematical Statistics*, **34** (1), 152–180.

Gradshteyn, I.S. and Ryzhik, I.M. (1980) *Table of Integrals, Series, and Products*, Academic Press, New York.

Ishimaru, A. (1978) *Wave Propagation and Scattering in Random Media*, Academic Press, New York.

Ishimaru, A. and Yeh, C.W. (1984) Matrix representations of the vector radiative transfer theory for randomly distributed nonspherical particles. *Journal of the Optical Society of America A*, **1** (4), 359–364.

Jin, Y.Q. (1994) *Electromagnetic Scattering Modelling for Quantitative Remote Sensing*, World Scientific, Singapore.

Jin, Y.Q. and Chen, F. (2002) Polarimetric scattering indexes and information entropy of the SAR imagery for surface monitoring. *IEEE Transactions on Geoscience and Remote Sensing*, **40** (11), 2502–2506.

Jin, Y.Q. and Cloude, S.R. (1994) Numerical eigenanalysis of the coherency matrix for a layer of random nonspherical scatterers. *IEEE Transactions on Geoscience and Remote Sensing*, **32** (6), 1179–1185.

Jin, Y.Q. and Zhang, N. (2002) Statistics of four Stokes parameters in multi-look polarimetric synthetic aperture radar (SAR) imagery. *Canadian Journal of Remote Sensing*, **28** (4), 610–619.

Jin, Y.Q., Zhang, N.X., and Chang, M. (1999) The Mueller and coherency matrices solution for polarimetric scattering simulation for tree canopy in SAR imaging at C band. *Journal of Quantitative Spectroscopy and Radiative Transfer*, **63** (2), 599–611.

Lee, J.S., Hoppel, K.W., Mango, S.A., and Miller, A.R. (1994) Intensity and phase statistics of multilook polarimetric and interferometric SAR imagery. *IEEE Transactions on Geoscience and Remote Sensing*, **32** (5), 1017–1027.

Lee, J.S., Grunes, M.R., and De Grandi, G. (1999a) Polarimetric SAR speckle filtering and its implication for classification. *IEEE Transactions on Geoscience and Remote Sensing*, **37** (5), 2363–2373.

Lee, J.S. *et al.* (1999b) Unsupervised classification using polarimetric decomposition and the complex Wishart classifier. *IEEE Transactions on Geoscience and Remote Sensing*, **37** (5), 2249–2257.

Matzler, C. and Wegmuller, U. (1993) *Active and Passive Microwave Signature Catalogue (2–12 GHz)*, Institute of Applied Physics, University of Berne, Berne.

Mishchenko, M.I., Travis, L.D., and Lacis, A.A. (2002) *Scattering, Absorption and Emission of Light by Small Particles*, Cambridge University Press, Cambridge.

Moghaddam, M. and Saatchi, S. (1995) Analysis of scattering mechanisms in SAR imagery over boreal forest: results from BOREAS '93. *IEEE Transactions on Geoscience and Remote Sensing*, **33** (5), 1290–1296.

Nighiem, S.V., Yueh, S.H., Kwok, R., and Li, F.K. (1992) Symmetry properties on polarimetric remote sensing. *Radio Science*, **27** (5), 693–701.

Oguchi, T. (1983) Electromagnetic wave propagation and scattering in rain and other hydrometeors. *Proceedings of the IEEE*, **71**, 1029–1078.

Peake, W.H. (1959) Interaction of electromagnetic waves with some natural surfaces. *IEEE Transactions on Antennas and Propagation*, **7** (2), 324–329.

Sheen, D.R. and Johnston, L.P. (1992) Statistical and spatial properties of forest clutter measured with polarimetric synthetic aperture radar (SAR). *IEEE Transactions on Geoscience and Remote Sensing*, **30** (3), 578–588.

Touzi, R. and Lopes, A. (1996) Statistics of the Stokes parameters and of the complex coherence parameters in one-look and multilook speckle fields. *IEEE Transactions on Geoscience and Remote Sensing*, **34** (2), 519–588.

Tsang, L. and Kong, J.A. (1990) *Polarimetric Remote Sensing*, Elsevier, New York.

Tsang, L., Kong, J.A., and Shin, R. (1984) Radiative transfer theory for active remote sensing of a layer of nonspherical particles. *Radio Science*, **19**, 629–639.

Tsang, L., Kong, J.A., and Shin, R. (1985) *Theory of Microwave Remote Sensing*, John Wiley & Sons, Inc., New York.

Van der Mee, C.V.M. and Hovenier, J.W. (1992) Structure of matrices transforming Stokes parameters. *Journal of Mathematical Physics*, **33** (10), 3574–3584.

Waterman, P.C. (1965) Matrix formulation of electromagnetic scattering. *Proceedings of the IEEE*, **53**, 805–812.

Waterman, P.C. (1971) Symmetry, unitarity, and geometry in electromagnetic scattering. *Physical Review D*, **3**, 825–839.

Waterman, P.C. (1973) Numerical solution of electromagnetic scattering problems, in *Computer Techniques for Electromagnetics* (ed. R. Mittra), Pergamon, Oxford, pp. 97–157.

Wegmuller, U. and Matzler, C. (1993) *Active and Passive Microwave Signature Catalogue on Bare Soil (2–12 GHz)*, Institute of Applied Physics, University of Berne, Berne.

3

Imaging Simulation of Polarimetric SAR: Mapping and Projection Algorithm

The technology of remote sensing and Earth observation from space has progressed in recent years. High-resolution and all-weather operational synthetic aperture radar (SAR) has now become one of the most important techonologies. Among the wide range of SAR-related remote sensing research areas, a fundamental research topic is to understand the underlying SAR imaging mechanisms and principles based on the physics of electromagnetic (EM) wave interaction with natural environments, and thereafter to develop quantitative approaches for information retrieval and target recognition via forward theoretical modeling and parametric scattering simulation.

SAR imaging simulation (Xu and Jin, 2006) involves modeling the process of SAR imaging from the virtual scene of geometrical and physical models to the formed SAR images. It can provide aids to SAR system design, image interpretation, and the development of signal processing algorithms, as well as data validation. It is extremely valuable, particularly because of the expensive and limited resources available for real airborne and spaceborne SAR experiments.

3.1 Fundamentals of SAR Imaging

3.1.1 Ranging and Pulse Compression

Pulse radar senses the target by transmitting a narrow pulse and extracting geometric and geophysical information of the target from the received response profile. In free space, the time delay of the response is related with target range as

$$2R = c\Delta t \tag{3.1}$$

with c being the speed of light. Theoretically, targets at different ranges can be differentiated given that the pulse width is narrow enough. Such capability of radar to

Polarimetric Scattering and SAR Information Retrieval, First Edition. Ya-Qiu Jin and Feng Xu.
© 2013 John Wiley & Sons Singapore Pte. Ltd. Published 2013 by John Wiley & Sons Singapore Pte. Ltd.

differentiate different targets along the range dimension is referred to as "ranging", and its metric "range resolution" is defined as the minimum range difference of two targets that can be differentiated.

In practice, a narrow pulse is not favored due to its high peak-to-average ratio (PAR). Using a narrower pulse to achieve higher range resolution would compromise average transmitted power and hence reduce the radar's operation range.

On the other hand, pulse compression is a technique to improve range resolution by expanding the signal bandwidth via modulation. The most commonly used waveform is linear frequency modulation (LFM). An LFM pulse can be written in complex form as

$$p(t) = \Pi(t/T_p) \cdot \exp(j2\pi f_c t + j\pi K t^2) \tag{3.2}$$

where $\Pi(\cdot)$ denotes a unit rectangular window function over $[-0.5, 0.5]$, T_p denotes pulse width, f_c is carrier frequency, and K is modulation slope. In the frequency domain, the energy of an LFM pulse is evenly distributed in a band of $B = KT_p$ centered at f_c (Franceschetti and Schirinzi, 1990).

The backscattered signal from a target at range R is a delayed version of the transmitted pulse $p(t - 2R/c)$ (ignoring the target's scattering coefficient). After down-conversion to baseband, its matched filter reference signal should be

$$p_r(t) = \Pi(t/T_p) \cdot \exp(-j\pi K t^2) \tag{3.3}$$

Note that, in the frequency domain, the reference signal has the same amplitude as the transmitted pulse but conjugate phase. Thus, the matched filtering is in fact compensating the phase of the received signal in the frequency domain while preserving its rectangular window amplitude. Hence, the output time-domain signal of the matched filter is approximately a sinc function (Curlander and McDonough, 1991)

$$s(t) = \frac{\sin[\pi B(t - 2R/c)]}{\pi B t(t - 2R/c)} \exp\left(-j\frac{4\pi R}{\lambda}\right) \tag{3.4}$$

where the time delay and the associated phase shift are preserved. Clearly, it becomes a narrow pulse located at the time delay carrying the Doppler phase shift. Such kinds of technique that compress a long pulse into a narrow pulse without losing range and phase information are referred to as "pulse compression". The width of the sinc function, $1/B$, determines the range resolution in space, that is

$$r_R = \frac{c}{2B} \tag{3.5}$$

Figure 3.1 shows the steps of pulse compression side by side in the time domain and frequency domain, respectively. Figure 3.2 shows an example of ranging via pulse compression. The four point targets appear as narrow pulses at the corresponding locations. Note that the right two targets are too close in space and it is difficult to differentiate between them after pulse compression.

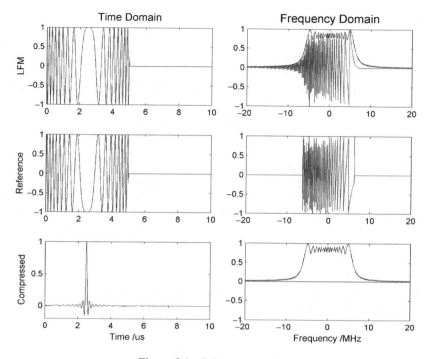

Figure 3.1   Pulse compression

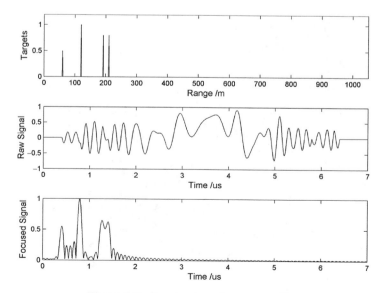

Figure 3.2   Ranging by pulse compression

3.1.2 Synthetic Aperture and Azimuth Focusing

Ranging only provides a one-dimensional range profile. Conventional real aperture radar (RAR) achieves azimuth resolution via scanning a narrow beam antenna. Thus, azimuth resolution is determined by beamwidth. However, as the target distance becomes large, it is impossible to achieve an azimuth resolution that is comparable to range resolution. On the contrary, synthetic aperture radar (SAR) takes advantage of radar movement and coherently synthesizes signals collected at different positions. It forms a large synthetic aperture that has a very narrow equivalent beamwidth and thus achieves high-resolution imaging in azimuth.

Figure 3.3 shows the stripmap imaging geometry of an airborne SAR. As the airborne platform flies along $\hat{x}$, the radar looks toward $\hat{y}$. The illumination area and position are determined by the antenna direction and pattern. As the radar flies over a distance equal to the length of the illumination area, the target is always in the radar's field of view. The radar collects echoes from the target during this period, and then obtains the target image via pulse compression in the range direction and synthetic aperture processing in the azimuth. Thus, the length of the synthetic aperture is exactly the length of the illumination area. The most classical synthetic aperture technique is Doppler modulation.

Let us assume that the radar flies along $\hat{x}$ at speed v and repeatedly sends LFM pulses at interval T_r. The pulse repetition frequency (PRF) is defined as $1/T_r$. Hence, the transmitted signal is expressed as

$$s_0(t) = \sum_n p(t - nT_r) \tag{3.6}$$

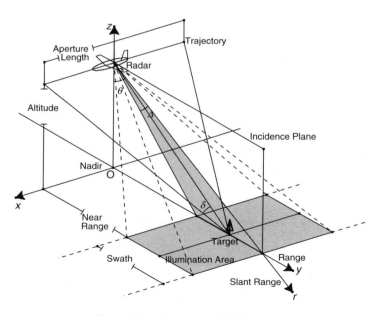

Figure 3.3   Geometry of SAR imaging

Given a point target at (x, r) with scattering coefficient $\sigma(x, r)$, its echoes, after down-conversion to baseband, can be written as

$$s(t) = \sigma \cdot \varpi(t) \cdot s_0[t - 2R(t)/c] = \sum_n \sigma \cdot \varpi(t) \cdot p[t - nT_r - 2R(t)/c] \qquad (3.7)$$

where ϖ denotes the antenna gain and we omit the parameters (x, r) for the sake of simplicity. In Equation 3.7, $R(t)$ denotes the target distance to the radar, which can be calculated as

$$R(t) = \sqrt{(vt - x)^2 + r^2} \approx r + \frac{(vt - x)^2}{2r} \qquad (3.8)$$

where the approximation is taken based on the condition $r \gg vt - x$.

Once the radar has received the response signal, pulse compression is immediately performed. Following from Equation 3.4, the signal after pulse compression is

$$s_1(t) = \sum_n \sigma \cdot \varpi(t) \cdot \delta_R[t - nT_r - 2R(t)/c] \cdot \exp\left[-j\frac{4\pi R(t)}{\lambda}\right] \qquad (3.9)$$

where $\delta_R(\cdot)$ denotes the impulse function after pulse compression, which is the sinc function in Equation 3.4. Given the fact that the radar's movement within one pulse period is negligible with respect to wavelength, we can assume that the radar "stopped" and transmitted/received one pulse and then "moved" to the next position. This approximation is also called "stop-and-go". Hence, we have

$$R(t) \approx R(n'T_r), \quad \varpi(t) \approx \varpi(n'T_r), \quad n' = \lceil t/T_r \rceil \qquad (3.10)$$

where $\lceil \cdot \rceil$ denotes rounding to the next largest integer, and $n'T_r$ is the timing when the pulses are transmitted, which is often referred to as "slow time". On the contrary, "fast time" is defined as the time within one pulse transmit/receive window, that is, $\tau = t - nT_r$.

Let us rearrange the 1D time-domain signal in Equation 3.9 into a 2D matrix on the plane of fast time versus slow time, where the nth column represents the response signal of the nth pulse. Hence, the signal in Equation 3.9 is now rewritten as

$$s_1(n, \tau) = \sigma \cdot \varpi(nT_r) \cdot \exp\left[-j\frac{4\pi R(nT_r)}{\lambda}\right] \cdot \delta_R[\tau - 2R(nT_r)/c] \qquad (3.11)$$

Figure 3.4 illustrates the receiving and rearrangement of the signal. Owing to the variation of the target-to-radar distance, the echoes from the same target will show up along a parabolic trajectory rather than a straight line. This phenomenon is known as range migration (RM). Such a trajectory in the $n - \tau$ plane can be easily derived as

$$\Delta\tau(n) = 2R_m(n)/c$$

$$R_m(n) = R(nT_r) - r = \frac{(vnT_r - x)^2}{2r} \qquad (3.12)$$

where R_m is the extent of range migration.

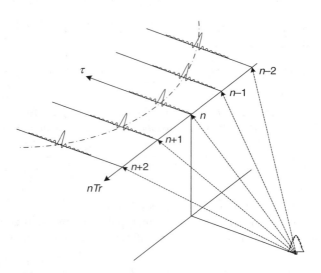

Figure 3.4 Stop-and-go model and slow time versus fast time

Letting $n_x T_r = \lceil x/v \rceil$, then the n_x th pulse is transmitted at the closest point to the target along the radar's trajectory, which is the center of the synthetic aperture. Given the synthetic aperture length L, the number of pulses transmitted within the aperture is

$$N_L = \frac{L}{vT_r} \tag{3.13}$$

Clearly, the target is only illuminated by these pulses because of the antenna pattern. This fact is reflected in the definition of the antenna pattern function, that is

$$\varpi(nT_r) = \begin{cases} \varpi^0(n - n_x, r), & n \in [\,n_x - N_L/2, \quad n_x + N_L/2\,] \\ 0, & \text{otherwise} \end{cases} \tag{3.14}$$

Note that the antenna pattern is assumed to be perfectly zero outside its field of view, and $\varpi^0(n, r)$ is the variation of the antenna gain inside its field of view. It is often assumed that the antenna gain is uniform within its field of view, that is, $\varpi^0 = 1$.

Now let us align all echoes from the same target along a straight line. The peaks of each echo now become a 1D signal, which is written as

$$s_1(n, \tau = 2R(nT_r)/c) = C \cdot \Pi\left(\frac{n - n_x}{N_L}\right) \cdot \exp\left[j\pi K_d(n - n_x)^2 T_r^2\right]$$

$$C = \sigma \cdot \delta_R(0) \cdot \varpi^0(r) \cdot \exp\left(-j\frac{4\pi r}{\lambda}\right) \tag{3.15}$$

$$K_d = -\frac{2v^2}{r\lambda}$$

where C denotes a constant coefficient, and K_d denotes the Doppler modulation rate. Comparing to Equation 3.2, we find that s_1 is in fact an LFM pulse. The same matched filtering technique can be applied to obtain azimuth compression. Its bandwidth, also known as Doppler bandwidth, is written as

$$B_d = -K_d N_L T_r = \frac{2vL}{r\lambda} \qquad (3.16)$$

The corresponding time resolution $1/B_d$ translates into space as azimuth resolution, that is

$$r_A = \frac{v}{B_d} = \frac{r\lambda}{2L} \qquad (3.17)$$

Note that azimuth resolution is inversely proportional to aperture length and proportional to range and wavelength.

3.1.3 SAR Imaging Algorithm

As seen from Figure 3.4, large range migration will lead to a single target's responses extending across several pixels in range. The Doppler processing in azimuth dimension cannot be directly applied because responses from targets at different positions cannot be aligned at the same time. In practice, similar factors such as platform motion error would also render direct Doppler processing inapplicable. Thus, various SAR imaging algorithms have been developed to achieve azimuth focusing.

The most basic algorithm in the time domain is back-projection (BP) (Soumekh, 1999), which first extracts signal samples at each response that correspond to the target at particular locations, and then coherently sums up after removing the Doppler phase shift. This algorithm works in the time domain and has very high computational complexity. Many accelerated versions of BP have been developed (e.g., Yegulalp, 1999).

The most basic frequency-domain algorithm is the range Doppler algorithm (RD algorithm) (Wu, Liu, and Jin, 1982). It first converts a 2D pulse compressed signal into the azimuth frequency domain (known as Doppler domain, as the azimuth frequency corresponds to the Doppler frequency shift) and then corrects the range migration for all targets simultaneously in the Doppler domain. After that, matched filtering can be applied directly. The RD algorithm requires interpolation when correcting range migration. Other frequency-domain algorithms include the chirp scaling algorithm (CSA) and the wavenumber domain algorithm (Raney *et al.*, 1994; Soumekh, 1999; Cumming and Wong, 2005), among others. However, only RD is introduced here, as it is regarded as the simplest algorithm for understanding the underlying SAR imaging principles. For a comprehensive description of SAR systems and imaging algorithms, readers are referred to Curlander and McDonough (1991), Soumekh (1999), and Cumming and Wong (2005).

Let us consider a 2D distributed target defined in coordinates (x, r). Without loss of generality, let the scene be centered at $x = 0$, $r = 0$. Let R_0 be the distance of the scene center to the radar trajectory. For simplicity, the following descriptions use spatial variables instead of time variables, that is

$$x = vnT_r, \quad r = c\tau/2 \qquad (3.18)$$

In the spatial domain, the range function is now defined as

$$R(x, r) = \sqrt{x^2 + (R_0 + r)^2} \approx R_0 + r + \frac{x^2}{2(R_0 + r)} \tag{3.19}$$

The received raw signal (see Figure 3.5a) can be written in the form of a convolution of target scene scattering coefficients $\sigma(x, r)$ and transmitted pulse, that is

$$
\begin{aligned}
s(x', r') &= \iint dx\, dr \cdot \sigma(x, r) \cdot \Pi_L(x' - x) \cdot \Pi_D(r' - r - \Delta R) \cdot \exp(-j\phi) \\
\Pi_L(x) &= \Pi(x/L), \quad \Pi_D(r) = \Pi(r/D), \quad D = cT_p/2 \\
\phi &= \frac{4\pi}{\lambda}[R_0 + r + \Delta R] + \frac{4\pi K}{c^2}(r' - r - \Delta R)^2 \\
\Delta R &= R(x' - x, r) - R_0 - r
\end{aligned}
\tag{3.20}
$$

where a uniform antenna pattern within the field of view is assumed. Converting the raw signal into the frequency domain, this yields

$$
\begin{aligned}
S(\xi, \eta) &= \iint dx'\, dr'\, s(x', r') \exp(-jx'\xi - jr'\eta) \\
&= \iint dx\, dr\, \sigma(x, r) \exp(-jx\xi - jr\eta) G(r, \xi, \eta)
\end{aligned}
\tag{3.21}
$$

where

$$
\begin{aligned}
G(r, \xi, \eta) &= \iint dp\, dq\, \Pi_L(p) \Pi_T(q - \Delta R) \exp(-j\phi) \exp(-jp\xi - jq\eta) \\
\phi &= \frac{4\pi}{\lambda}[R_0 + r + \Delta R] + \frac{4\pi K}{c^2}(q - \Delta R)^2 \\
\Delta R &= R(p, r) - R_0 - r
\end{aligned}
\tag{3.22}
$$

The integration of q can be easily derived as

$$G(r, \xi, \eta) = G_R(p, r, \eta) \int dp\, \Pi_L(p) \exp[-jR(p, r)4\pi/\lambda - jp\xi] \tag{3.23}$$

The first term is a range-dependent term and can be written as

$$G_R(p, r, \eta) = \frac{c\tau}{2} d\left(\frac{c\eta}{2\pi K T_p}\right) \cdot \exp\left[j\frac{c^2\eta^2}{16\pi K} + j\eta R_0 + j\eta r - j\eta R(p, r)\right] \tag{3.24}$$

where $d(\cdot)$ consists of Fresnel integration (Franceschetti and Schirinzi, 1990). Generally, it is approximately a rectangular window, that is, $d(x) \approx \Pi(x/2)$ (cf. Figure 3.1). Through pulse compression, the phase of $G_R(p, r, \eta)$ is removed and $S(\xi, \eta)$ can now be converted back to the space domain in η dimension. It yields the pulse compressed signal in the Doppler domain (see Figure 3.5b and c), that is

$$S_1(\xi, r') = \iint dx\, dr\, \sigma(x, r) \exp(-jx\xi) G(r, \xi, r') \tag{3.25}$$

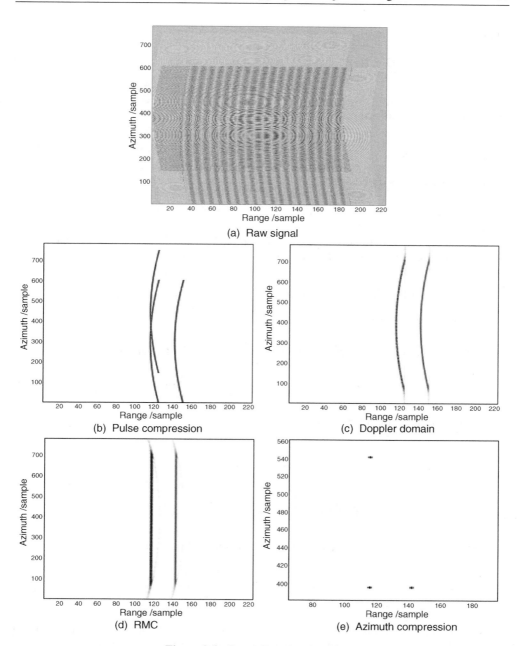

(a) Raw signal

(b) Pulse compression

(c) Doppler domain

(d) RMC

(e) Azimuth compression

Figure 3.5 Range Doppler algorithm

where

$$G(r, \xi, r') = g_R[r' - r - R_m(p, r)]$$
$$\cdot \iint dp \; \Pi_L(p) \exp[-jR(p, r)4\pi/\lambda - jp\xi] \tag{3.26}$$

where the range focusing impulse response is

$$g_R(r) \approx \text{sinc}\left(2\pi K T_p r/c\right) \tag{3.27}$$

and range migration is derived as

$$R_m(p, r) = R(p, r) - R_0 \tag{3.28}$$

Note that the function to be integrated in Equation 3.26 has a rapidly varying phase term. Thus, it can be derived by using the stationary phase approximation (Jin, 1994):

$$G(r, \xi, r') = g_R[r' - r - R_m(\tilde{p}, r)] \cdot \Pi_L(\tilde{p}) \cdot \exp[-j\phi(\tilde{p})] \cdot \sqrt{-2j\pi/\ddot{\phi}(\tilde{p})}$$
$$\phi(p) = R(p, r)4\pi/\lambda + p\xi \tag{3.29}$$

From $\dot{\phi}(\tilde{p}) = 0$, the stationary point and its second-order derivative are solved as

$$\tilde{p} = \frac{\lambda\xi}{4\pi}(R_0 + r), \quad \ddot{\phi}(\tilde{p}) = \frac{4\pi}{\lambda(R_0 + r)} \tag{3.30}$$

Substituting into Equation 3.29 yields

$$G(r, \xi, r') = g_R[r' - r - R_m(\xi)] \cdot \sqrt{-j\lambda\frac{R_0 + r}{2}}$$
$$\times \Pi_L\left[\frac{\lambda\xi}{4\pi}(R_0 + r)\right] \cdot \exp\left[-j\frac{4\pi}{\lambda}(R_0 + r)\left(1 + \xi^2 + \frac{\lambda^2\xi^2}{32\pi^2}\right)\right] \tag{3.31}$$

where the range migration can be expressed as a function of the Doppler domain

$$R_m(\xi) = r + \frac{(R_0 + r)}{2}\left(\frac{\lambda\xi}{4\pi}\right)^2 \tag{3.32}$$

Now the range migration correction (RMC) can be performed collectively in the Doppler domain, that is, a coordinate shift of $G(r, \xi, r')$ from r' to $r'_m = r' - R_m(\xi)$. The target's responses are aligned in the azimuth dimension after RMC (see Figure 3.5d). The next step is to correct the phase and amplitude modulation in the azimuth dimension, that is

$$S_2(\xi, r') = \iint dx \, dr \, \sigma(x, r) g_R(r'_m - r) \cdot \exp(-jx\xi)\Pi_L\left[\frac{\lambda\xi}{4\pi}(R_0 + r)\right] \tag{3.33}$$

Note that a non-uniform antenna pattern, if included in Equation 3.20, can now be compensated at this step. Finally, converting back to the space domain completes the RDA process and generates the final SAR image (see Figure 3.5e):

$$I(x', r'_m) = \iint dx \, dr \, \sigma(x, r) \cdot g_R(r'_m - r) \cdot g_A(x' - x) \tag{3.34}$$

where the azimuth impulse response is given as

$$g_A(x' - x) = \text{sinc}\left[\frac{2\pi L x}{\lambda(R_0 + r)}\right] \tag{3.35}$$

It can be seen that the SAR image is a convolution of scene scattering coefficients map $\sigma(x, r)$ and impulse responses $g_R \cdot g_A$. Thus, $g_R \cdot g_A$ is also referred to as the point target response (PTR) or point spread function (PSF). It simply represents the capability of imaging and focusing of any particular SAR system. Several quantitative metrics of SAR focusing capability are the peak side-lobe ratio (PSLR), integral side-lobe ratio (ISLR), and main-lobe width. In practice, window functions are often further applied in the frequency domain to reduce side-lobe level by widening the main-lobe width.

In Figure 3.5, an example of the imaging process of three point targets is shown step by step: (a) raw signal, (b) range focused signal after pulse compression, (c) converted to Doppler domain, (d) Doppler domain signal after RMC, and (e) azimuth compressed SAR image. In reality, SAR signals after down-conversion to baseband are first sampled into discrete digital signals. The Nyquist theorem dictates that the sampling frequency must be larger than the frequency bandwidth in order to avoid alias. Thus, in fast time, sampling (analog to digital conversion) f_S has to be greater than signal bandwidth B, while in slow time, the sampling frequency (that is, PRF) must be larger than the Doppler bandwidth B_d.

Clearly, the pixel size of the SAR image is determined by the sampling frequency f_S and PRF as

$$\tilde{r}_R = \frac{c}{2f_S} = \frac{B}{f_S} r_R, \quad \tilde{r}_A = \frac{v}{\text{PRF}} = \frac{B_d}{\text{PRF}} r_A \tag{3.36}$$

It can be seen that the pixel size (pixel spacing) is deemed to be smaller than the resolution. The ratio of resolution versus pixel spacing equals that of sampling frequency versus bandwidth. In practice, this ratio is usually selected from 1.0 to 2.0.

Owing to the coherent summation nature of SAR processing, a random phase radar signal will give rise to multiplicative noise in the focused SAR image, which is known as speckle. Understanding the characteristics of speckle is crucial to SAR image processing, interpretation, and simulation.

Another important concept of the SAR image is the number of looks. A multi-look SAR image is obtained by separating Doppler-domain data into multiple sub-bands, focusing independently, and then summing up incoherently. The multi-look SAR image can be seen as dividing the original aperture into N sub-apertures and each sub-aperture can be used to generate one single-look SAR image. By incoherently averaging multiple single-look SAR images, we are able to reduce speckle on the SAR image. However, this is achieved by sacrificing azimuth resolution because of the smaller aperture size for each sub-aperture.

Multi-look SAR images can also be obtained by simply averaging over the spatial domain, that is, boxcar smoothing. Single-look SAR images preserve the phase information and each pixel contains a 2×2 complex scattering matrix. Multi-look SAR images are second-order statistical moments and each pixel is a Mueller matrix or covariance/coherency matrix.

A typical procedure for designing the parameters of a SAR system and its platform include the following steps:

1. select the carrier frequency f_c and wavelength $\lambda = c/f_c$, and the required range resolution r_R;
2. determine the frequency bandwidth $B = c/2r_R$, and then the sampling frequency $f_S > B$ and pixel size $\tilde{r}_R = r_R B/f_S$;
3. given the required azimuth resolution r_A, determine the antenna azimuth beamwidth in $\theta_A = \lambda/2r_A$;
4. given platform velocity v and Doppler bandwidth $B_d = v/r_A$, determine PRF $> B_d$ and pulse repetition interval (PRI) $T_r = 1/\text{PRF}$, and then the azimuth pixel size $\tilde{r}_A = r_A B_d/\text{PRF}$;
5. given the distance from platform trajectory to scene center R_0, determine the synthetic aperture length $L = R_0 \theta_A$ and synthetic time $T_A = L/v$, as well as the Doppler modulation rate $f_{dr} = -B_d/T_A$;
6. finally select the pulse width $T_p < T_r$ and LFM modulation rate $K = B/T_p$.

This section has briefly introduced the principles and basic concepts of SAR imaging, which provides the foundation for the SAR imaging simulation to be presented in the following sections.

3.2 Mapping and Projection Algorithm

Given the sophisticated mechanisms of EM wave scattering and SAR signal collection, it is hard to tackle SAR imagery without the aid of forward simulation of scattering based on physical modeling. Most forward problems are limited in their scope to one homogeneous terrain type, for instance, vegetation canopy (Jin, 1994, 2005) or rough surface (Fung, 1994; Wang et al., 2005). Very few researchers have focused on inhomogeneously distributed targets such as forest or artificial constructions with high SAR resolution (Lin and Sarabandi, 1999; Franceschetti, Iodice, and Riccio, 2002).

Simulation of SAR imaging offers a different way to study an inhomogeneous terrain scene under SAR observation. It may be used to evaluate or predict the SAR observation, interpret the physics of scattering behavior, and extract the parameterized characteristics of terrain surface objects. The SAR image simulation approaches found in the previous literature may be classified into two categories.

The first type focuses on the statistical characteristics of SAR images (Holtzman et al., 1978; Raney and Wessels, 1988; Camporeale and Galati, 1991), and the others on the EM scattering process (Pike, 1985; Franceschetti et al., 1992; Brown, Sarabandi, and Gilgenbach, 2002). It directly employs a phantom scattering coefficient map and performs the raw signal generation and processing. A good application is simulation of spaceborne SAR image using an airborne SAR image. These simulations can only be used to study the speckle and geometric properties of SAR images rather than understanding the underlying scattering mechanism.

The second category focuses on the physics of wave interaction with complex terrain models (Brown, Sarabandi, and Gilgenbach, 2002; Franceschetti *et al.*, 1992). It includes not only the raw signal generation but also the EM scattering modeling. As this type of approach simulates the dynamic process of wave scattering and imaging, it is referred to as "imaging simulation" as opposed to the first category "image simulation".

Franceschetti *et al.* (1992), Franceschetti, Migliaccio, and Riccio (1998), and Franceschetti *et al.* (2003) developed a series of SAR raw signal simulators for terrain scenes, such as bare land surface, sea surface, and buildings. These simulators (Franceschetti *et al.*, 1992, 2003) incorporate the effect of coherent speckle in SAR images by modeling the ground scatter as stochastically distributed facets, and implements double and triple scattering between the building structure and the ground surface by using a geometrical optics–physical optics ray tracing model. Brown, Sarabandi, and Gilgenbach (2002) developed a simulation scheme for forestry scenes by adopting a Monte Carlo simulation of fractal trees (Lin and Sarabandi, 1999).

However, natural scenes are more complicated, as they include randomly and inhomogeneously distributed, penetrable or impenetrable objects, such as vegetation canopies, man-made structures, and perturbed surface topography. Particularly, volumetric scattering through penetrable scatterers, for example, timberland forests, crops, green plants, and so on, plays an essential or dominant role in SAR imagery.

In this section, we present a new approach to SAR imaging simulation for comprehensive scenarios (Xu and Jin, 2006), which takes account of scattering, attenuation, shadowing, and multiple scattering of spatially distributed volumetric and surface scatterers, as well as speckle simulation and raw data generation.

3.2.1 Mapping and Projection

As shown in Figure 3.1, as the radar advances, the incidence angle to the same target varies within an interval δ, which is determined by the synthetic aperture length and the slant range, that is, the distance between the target and the radar. Since the slant range is much larger than the aperture length, δ is small enough that the scattering contribution from the same target collected by the sensor at different times remains constant. This assumption enables us first to divide the whole scene into cross-track lines along the azimuth dimension and calculate the scattering contribution from the terrain objects lying in each line (included in the incidence plane) sequentially. It greatly reduces the complexity of the scattering problem of a distributed scene.

Figure 3.6 illustrates the mapping and projection principles of the imaged scattering of the terrain objects lying in the incidence plane. It is known that radar distinguishes targets at different ranges and obtains range resolution via the pulse compression technique. As shown in Figure 3.6, the near range point P in PA (the projection plane) is mapped to M in MA (the mapping plane) via arc PM, and the scatterings of all terrain objects distributed in the incidence plane PMA are mapped into MA via a group of arcs as well. Scattering contributions are summed into each pixel in MA to obtain the scattering coefficient map. The dashed lines linking to the radar in Figure 3.6 are the projection lines. Along each projection line, the terrain objects ahead shadow those behind. Notice that the pixels in the mapping plane MA

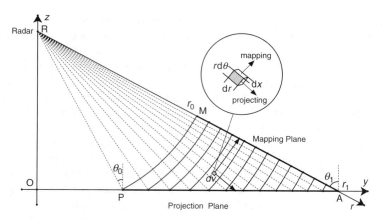

Figure 3.6 Mapping and projection. © [2006] IEEE. Reprinted, with permission, from *IEEE Transactions on Geoscience and Remote Sensing*

are equally distributed, but not those in the projection plane PA. The mapping arcs can be treated as parallel straight lines since the sensor is far away from the illumination area.

The comprehensive scenes considered in our simulation feature spatially distributed penetrable objects, for example, vegetation canopy, and some elevated or subsided objects, for example, a building, valley or target in the sky. Volumetric scattering, extinction attenuation, and shadows from penetrable scatterers on the projection line are considered. The location, size, and shape of projected shadows of elevated or subsided objects are specifically calculated.

In the polar coordinates (r, θ) of the incidence plane, the sensor is at the origin; radius r equals the slant range, and θ is the incidence angle. Given the sampling interval of each echo pulse and the elevation angle of the antenna, the imaging space is determined as $r \in [r_0, r_1]$, $\theta \in [\theta_0, \theta_1]$. Note that the half-space under plane xOy is also included in the case of subsided objects located at $z < 0$.

Since the objects ahead might project shadow on (or attenuate) those behind along the same projection line, the attenuation or shadowing attributed to the objects ahead should be counted when the scattering from the objects behind is cumulated to the mapping plane.

As shown in Figure 3.7, considering a differential element dv at position (x, r, θ) of the imaging space, its length, width, and height are dx, dr, and $r\, d\theta$, respectively. Its mapping area is, correspondingly, $dx\, dr$ in the mapping plane, and the effective projection area is $dx\, r\, d\theta$, which is the cross-section perpendicular to the incidence direction. According to the radiative transfer theory, when the incident wave arrives and penetrates through the element dv, the scattering intensity I_s (per unit area) can be written as

$$I_s(x, r, \theta) = E^+(x, r, \theta) P(x, r, \theta) E^-(x, r, \theta) I_i\, dr \tag{3.37}$$

where I_i is the incident intensity (per unit area), and E^- and E^+ are the total extinction and attenuation in the backward and forward directions, respectively. The phase function P stands for the element scattering. Letting $\kappa_e^-(x, r, \theta)$ and $\kappa_e^+(x, r, \theta)$ be the backward and forward extinction coefficients of the element dv; it causes the extinctions of

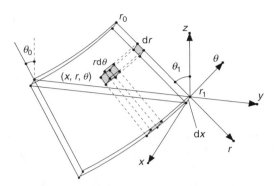

Figure 3.7 Mapping of a differential elements in space. © [2006] IEEE. Reprinted, with permission, from
IEEE Transactions on Geoscience and Remote Sensing

$\exp\left[-dr \cdot \kappa_e^-(x,r,\theta)\right]$ and $\exp\left[-dr \cdot \kappa_e^+(x,r,\theta)\right]$, respectively. Here the superscripts $+$ and $-$
denote the forward and backward directions, respectively.

Then the total extinctions, E^- and E^+, due to the elements among $r \in [r_0, r]$ along the
incident direction θ can be written as

$$E^+(x,r,\theta) = \exp\left[-\int_{r_0}^{r} dr' \, \kappa_e^+(x,r',\theta)\right]$$

$$E^-(x,r,\theta) = \exp\left[-\int_{r}^{r_0} dr' \, \kappa_e^-(x,r',\theta)\right] \tag{3.38}$$

Given the scattering intensity and the effective illumination area, the scattering power contribution of the element can be obtained as $I_s(x,r,\theta)\,dx\,r\,d\theta$. Consequently, the scattering map
$S(x,r)$ can be produced through integrating all elements along the mapping arcs as

$$S(x,r) = \int_{\theta_0}^{\theta_1} I_s(x,r,\theta) r\,dx\,d\theta$$

$$= dx\,dr \int_{\theta_0}^{\theta_1} r\,d\theta\, E^+(x,r,\theta) P(x,r,\theta) E^-(x,r,\theta) I_i(x,r,\theta) \tag{3.39}$$

Under an incidence of unit intensity, substituting Equation 3.38 into Equation 3.39 yields

$$S(x,r) = dx\,dr \int_{\theta_0}^{\theta_1} r\,d\theta \exp\left[-\int_{r_0}^{r} dr' \, \kappa_e^+(x,r',\theta)\right] \cdot P(x,r,\theta) \exp\left[-\int_{r}^{r_0} dr' \, \kappa_e^-(x,r',\theta)\right]$$

$$\tag{3.40}$$

Integrating $S(x, r)$ over the (n, i)th pixel that is, $x \in [x_n, x_{n+1})$, $s \in [s_i, s_{i+1})$, the corresponding discrete scattering map $S_{n,i}$ can be obtained as

$$
S_{n,i} = \int_{x_n}^{x_{n+1}} dx \int_{r_i}^{r_{i+1}} dr \int_{\theta_0}^{\theta_1} r\,d\theta \exp\left[-\int_{r_0}^{r} dr'\,\kappa_e^+(x, r', \theta)\right] \cdot P(x, r, \theta)\exp\left[-\int_{r}^{r_0} dr'\,\kappa_e^-(x, r', \theta)\right]
$$

$$(3.41)$$

where $x_n = nR_x$, $r_i = iR_r + r_0$, and R_x, R_r are the pixel spacing of the azimuth and range dimensions, respectively.

Owing to the complexity and inhomogeneity of the terrain objects in natural scenarios, the phase function and extinction coefficients in Equation 3.41 cannot be given explicitly. Therefore, the integration can only be done by numerical summation of all discrete elements. For simplicity, another Cartesian coordinate s is adopted instead of the polar coordinate θ as $s \equiv r\,d\theta, \hat{s} = \hat{\theta}$.

Partitioning the imaging space (x, r, s) into discrete units as $x_m = m\Delta x$, $r_p = p\Delta r$, and $s_q = q\Delta s$, Equation 3.41 can be rewritten as

$$
S_{n,i} = \sum_{m=m_n}^{m_{n+1}-1}\sum_{p=p_i}^{p_{i+1}-1}\sum_{q=0}^{Q-1}\Delta x \Delta r \Delta s \prod_{p'=0}^{p}\exp\left[-\Delta r\kappa_e^+(m, p', q)\right]
$$

$$(3.42)$$

$$
\cdot P(m, p, q)\prod_{p'=p}^{0}\exp\left[-\Delta r\kappa_e^-(m, p', q)\right]
$$

where m_n, p_i are the initial sequence numbers of the discrete units within the (n, i)th pixel. Note that the original integrals of the extinction coefficients are expressed as cumulative products of the extinctions, and the cumulative sequences are different for forward and backward extinctions due to the opposite propagation directions. This operation is for the convenience of extending this formula to the polarimetric case.

According to the vector radiative transfer (VRT) theory (Jin, 1994), the phase matrix $\overline{\mathbf{P}}$, extinction matrices $\overline{\kappa}_e^-$, $\overline{\kappa}_e^+$, and Stokes vector $\mathbf{I}$ for polarimetric SAR are adopted, instead of the scalar phase function, extinction coefficient, and intensity, respectively. Because the extinction matrices stand not only for the power attenuation but also for certain degree of depolarization, the exponential terms in Equation 3.42 must be cumulatively multiplied in accordance with the sequence of wave propagation.

As for volumetric scatterers, the volumetric scattering $\overline{\mathbf{P}}$ of each discrete unit can be expressed by the Mueller matrix $\overline{\mathbf{S}}_{\text{vol}}$ of each single scatterer or particle as (Jin, 1994)

$$
\overline{\mathbf{P}} = \frac{f_s}{v_0}\overline{\mathbf{S}}_{\text{vol}} = n_0\overline{\mathbf{S}}_{\text{vol}}
$$

$$(3.43)$$

where v_0 is the volume of each scatterer, f_s is the effective fractional volume, and n_0 denotes the number of particles included in unit volume. Correspondingly, the extinctions of independent scattering, $\overline{\kappa}_e^-$, $\overline{\kappa}_e^+$, are also proportional to n_0.

As for surface scatterers, assuming that each discrete unit includes no more than one surface scatterer, the scattering should be $\Delta x \Delta s \overline{S}_{surf}$, where $\overline{S}_{surf}$ is the Mueller matrix of the surface scatterer. Thus, the depth Δr of the discrete unit in Equation 3.42 can be omitted for surface scattering. Furthermore, a surface scatterer is considered as an impenetrable object, which totally obstructs the wave propagation. Thus, the extinction term in Equation 3.42 should be set to zero for a surface scatterer.

Summing up these two cases, Equation 3.42 is rewritten as

$$\overline{S}_{n,i} = \sum_{m=m_n}^{m_{n+1}-1} \sum_{p=p_i}^{p_{i+1}-1} \sum_{q=0}^{Q-1} \prod_{p'=0}^{p} \overline{E}_o^+(m,p',q) \overline{S}_o(m,p,q) \prod_{p'=p}^{0} \overline{E}_o^-(m,p',q) \tag{3.44}$$

$$\overline{S}_o(m,p,q) = \begin{cases} n_0 \Delta x \Delta s \Delta r \overline{S}_{vol}(m,p,q) = n_{ef} \overline{S}_{vol}(m,p,q) \\ \Delta x \Delta s \overline{S}_{surf}(m,p,q) = a_{ef} \overline{S}_{surf}(m,p,q) \end{cases} \tag{3.45}$$

$$\overline{E}_o^\pm(m,p',q) = \begin{cases} \exp\left[-\Delta r \overline{\kappa}_e^\pm(m,p',q)\right] = \exp\left[-d_{ef} \overline{\kappa}_e^\pm(m,p',q)\right] \\ 0 \end{cases} \tag{3.46}$$

where the first and second lines of $\overline{S}_o$ and $\overline{E}_o^\pm$ correspond to volumetric and surface scatterer, respectively; n_{ef} is defined as the effective number of volumetric scatterers in a volume element; a_{ef} is defined as the effective area of a surface element; and d_{ef} is defined as the effective penetration depth. The subscript o denotes scattering or extinction of a single scatterer. Equation 3.44 presents a general description of SAR imaging, which takes into account single scattering, attenuation or shadowing of all possible objects in the imaging space.

3.2.2 *Mapping and Projection Algorithm for Fast Computation*

It would be very computationally intensive to carry out computation of Equation 3.44 without any acceleration technique. Hence, a fast algorithm is further introduced to complete the computation of Equation 3.44. For consistency, it bears the same name of mapping and projection algorithm (MPA). The first step of MPA is to partition the scene and terrain objects lying in the horizontal plane xOy into multi-grids, $x_m \equiv m\Delta x$, $y_\ell \equiv \ell \Delta y$, and denote the initial grid sequence numbers as $(0,0)$. Note that each grid unit has its corresponding parts of the terrain objects overhead, for example, ground surface, vegetation, building, and so on.

As shown in Figure 3.8, two arrays $\underline{E}^\pm$ and $\underline{S}$ are deployed for temporary storage of extinction and scattering, respectively. Since the incident wave propagates along the $+y$ direction in horizontal dimension, the terrain objects never project shadow or attenuation upon those located behind them on the y axis. Thus, the attenuation or shadowing effect of each grid unit is counted as the increasing sequence of y axis and cumulatively multiplied into the array $\underline{E}^\pm$. In fact, the values stored in the array $\underline{E}^\pm$ are total attenuation or shadowing of those previously calculated grid units. Hence, the MPA calculating sequence of the terrain objects in one grid unit is as follows:

1. calculate the scattering power of the current terrain object and accumulate onto the array $\underline{S}$, where the attenuation or shadowing from other terrain objects are obtained from the array $\underline{E}^\pm$;

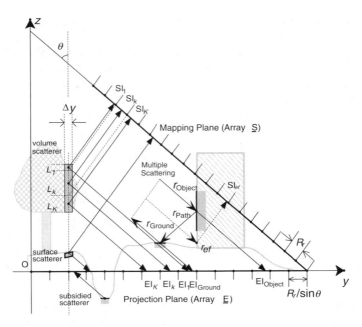

Figure 3.8 Computing procedure of MPA. © [2006] IEEE. Reprinted, with permission, from *IEEE Transactions on Geoscience and Remote Sensing*

2. calculate the attenuation or shadowing caused by the current terrain object itself, and cumulatively multiply onto the array $\mathbf{E}^{\pm}$;
3. when all grid units of the current line are counted, the scattering map is exactly produced in the array $\underline{\mathbf{S}}$.

Figure 3.8 explains the MPA computing process. In the case of a penetrable object, the part cut by a grid unit should be one or several segments of uniformly distributed particle cloud, that is, a volume scatterer. For impenetrable object, it should be an arbitrarily oriented facet, that is, a surface scatterer. As shown in Figure 3.8, the mapping plane is divided equally into discrete pixels (mapping cells). The size of the mapping cells equals the pixel spacing, that is, R_r in the range dimension and R_x in the azimuth dimension. The array $\underline{\mathbf{S}}$ is employed to store the Mueller matrices of the pixels in each line, that is, $\underline{\mathbf{S}}[i] = \bar{\mathbf{S}}_{n,i}$, which indicates the Mueller matrix of the ith pixel of the nth line. Likewise, the projection plane is also partitioned into discrete "projection cells", which are in one-to-one correspondence with the mapping cells via the mapping arcs. Hence, the projection cells are equally spaced as $R_r/\sin\theta$ in the range dimension and R_x in the azimuth dimension. In MPA, attenuation or shadowing of the scatterers lying on the jth projection line are projected onto the corresponding projection cell, which is, in fact, stored in the array element $\mathbf{E}^{\pm}[j]$.

The grid unit sizes Δx, Δy are always less than the mapping or projection cell, which are better as an integer times the grid unit, that is

$$\begin{cases} R_x = N_x \Delta x, & N_x = 1, 2, 3, \ldots \\ R_r/\sin\theta = N_y \Delta y, & N_y = 1, 2, 3, \ldots \end{cases} \tag{3.47}$$

Consider a vertical segment (length L) of volume scatterers above the current grid unit, which is larger than one cell after mapping and projection. First, the whole segment L is divided into K sub-segments in order to make each sub-segment map into one cell. Letting the length of the kth ($1 \leq k \leq K$) sub-segment be L_k, this yields $L_k = R_r/\cos\theta (k \neq 1, K)$ and $L = \sum_{k=1}^{K} L_k$. Further, let the index of the corresponding mapping cell be SI_k for the kth sub-segment, and SI_k ($1 \leq k \leq K$) must be a group of continuous sequence numbers. On the other hand, the mid-point of each sub-segment is projected onto the projection plane and the indices of the corresponding projection cells EI_k can be obtained. Note that EI_k ($1 \leq k \leq K$) might not be continuously increasing numbers.

After the mapping and projecting steps, the elements of array $\underline{\mathbf{S}}$ are updated as

$$\underline{\mathbf{S}}[\mathrm{SI}_k] := \underline{\mathbf{S}}[\mathrm{SI}_k] + \mathbf{E}^{+}[\mathrm{EI}_k] \cdot \overline{S}_o^k \cdot \mathbf{E}^{-}[\mathrm{EI}_k], \quad k = 1, \dots, K \tag{3.48}$$

where $:=$ denotes reassignment, and $\overline{S}_o^k$ is the scattering of the kth sub-segment. Assuming that volume scatterers always uniformly fill the whole sub-segment, then the effective volume in Equation 3.45 is written as $n_{ef} = \Delta x \Delta y L_k f_s / v_0$. Therefore, we have

$$\overline{S}_o^k = \Delta x \Delta y L_k f_s / v_0 \overline{\mathbf{S}}_{\mathrm{vol}} = n_{ef} \overline{\mathbf{S}}_{\mathrm{vol}} \tag{3.49}$$

where f_s is the fractional volume. Otherwise, it is necessary to estimate the total number of scatterers n_{ef} within the sub-segment, and calculate scattering from Equation 3.45.

The next step is to update the elements of the array $\mathbf{E}^{\pm}$ as

$$\underline{\mathbf{E}}^{-}[\mathrm{EI}_k] := \mathbf{E}_{o,k}^{-} \cdot \underline{\mathbf{E}}[\mathrm{EI}_k], \quad k = 1, \dots, K$$
$$\underline{\mathbf{E}}^{+}[\mathrm{EI}_k] := \underline{\mathbf{E}}^{+}[\mathrm{EI}_k] \cdot \mathbf{E}_{o,k}^{+}, \quad k = 1, \dots, K \tag{3.50}$$

where $\mathbf{E}_{o,k}^{\pm}$ is the attenuation of the kth sub-segment. The subscript o denotes attenuation or scattering of a single sub-segment. To account for the averaged attenuation, the ratio of the shadowed area of the sub-segment to the whole projection cell should be considered. The attenuation and effective depth are calculated as

$$\overline{\mathbf{E}}_{o,k}^{\pm} = \exp\left[-d_{ef}\overline{\boldsymbol{\kappa}}_e^{\pm}\right] \tag{3.51}$$

$$d_{ef} = \frac{\Delta x \Delta y L_k}{R_x R_r / \tan\theta} = \tan\theta \frac{\Delta x \Delta y L_k}{R_x R_r} \tag{3.52}$$

It can be seen from the effective depth d_{ef} of Equation 3.52 that, if we redistribute all volume scatterers of the sub-segment and make them cover the whole projection cell with the density invariant, then the layer depth of the redistributed scatterers along the projection line is seen as the effective depth. Since each projection cell contains several grid units, we can only calculate the averaged extinction of all these grid units or sub-segments. The scatterers of a single grid unit cannot cover the whole projection cell, but rather a piece of it. In order to count the effective extinction to the whole projection cell, we employ the redistribution operation to make the depth thinner and covering area wider, but preserve the density and volume invariant.

Let us now consider surface scatterers. In a similar way, larger surface scatterers, like vertical wall surfaces, cliff slopes, and so on, are segmented according to mapping cells. But, for nearly horizontal surfaces, for example, ground or roof surfaces, the facet cut by one grid unit is small enough and is projected into one projection cell, that is, $K = 1$. Hence, given the normal vector of facet $\hat{n}$, the effective area and its scattering in Equation 3.48 are calculated in two ways:

$$\overline{\mathbf{S}}_o^k = a_{ef}\overline{\mathbf{S}}_{surf} = \begin{cases} \Delta x \Delta y \overline{\mathbf{S}}_{surf} \cos\phi_2/\cos\phi_1, & K = 1 \\ \Delta x L_k \overline{\mathbf{S}}_{surf} \cos\phi_2/\sin\phi_1, & K > 1 \end{cases} \tag{3.53}$$

where ϕ_1 is the angle between $\hat{n}$ and the z axis, ϕ_2 is the angle between $\hat{n}$ and $\hat{i}$ of incidence, and L_k is the length of sub-segment. Moreover, the array $\underline{\mathbf{E}}^{\pm}$ should be set to zero over all the projection interval, that is

$$\underline{\mathbf{E}}^{\pm}[j] := 0, \qquad \mathrm{EI}_1 \leq j \leq \mathrm{EI}_K \tag{3.54}$$

Meanwhile, those subsided scatterers should be projected inversely ($z < 0$), as shown in Figure 3.8.

Cross-track lines of the whole scene are calculated sequentially. The elements of the array $\underline{\mathbf{E}}^{\pm}$ are initialized to unit matrices at the starting point of the calculation for each line, while the array $\underline{\mathbf{S}}$ should be saved before being cleared to start the next line. Note that each line should be strictly computed in the sequence of increasing y in order to precisely account for attenuation caused by the terrain objects ahead.

In the following paragraphs, we will discuss how to calculate multiple scattering mechanisms in MPA, for example, scattering between the ground and a wall, between the ground and a tree, and so on. Note that only double and triple scatterings between those terrain objects above the ground and the underlying ground surface are regarded in this work.

For the terrain objects above the ground, the corresponding ground patch of multiple scattering can be located by using the ray tracing technique. Then the length of the propagation path can be determined as well as the effective range as shown in Figure 3.8. Let the range of the counted terrain object be r_{Object}, the range of the ground patch be r_{Ground}, and the distance between them be r_{Path}. The effective range r_{ef} for double or triple scattering can be expressed as

$$\begin{aligned} r_{ef}\big|_{double} &= \frac{1}{2}\left(r_{Object} + r_{Ground} + r_{Path}\right) \\ r_{ef}\big|_{triple} &= r_{Object} + r_{Path} \end{aligned} \tag{3.55}$$

In the same way, multiple scattering of a segment of scatterers is mapped into the array $\underline{\mathbf{S}}$ by r_{ef}. The mapping interval will be first divided into sub-segments if it exceeds one cell. Since the dividing and mapping steps are the same as for single scattering, we limit our discussion on the computation of one single sub-segment, which has the corresponding index SI_{ef} of the array $\underline{\mathbf{S}}$ after mapping.

For double scattering, the propagation path consists of r_{Object}, r_{Ground}, and r_{Path}, while the attenuation or shadowing suffered through the propagation paths of r_{Object} and r_{Ground} are the same as with single scattering of the terrain object and ground patch, respectively. This means that the array $\underline{\mathbf{E}}^{\pm}$ can be used as well. However, the attenuation during the short distance from the terrain object to the ground patch, r_{Path}, is simply omitted, assuming low probability of other objects being present in between the terrain object and the ground surface. Therefore, the double scattering contribution can be written as

$$\underline{\mathbf{S}}[\mathrm{SI}_{ef}] := \underline{\mathbf{S}}[\mathrm{SI}_{ef}] + \mathbf{E}^{+}[\mathrm{EI}_{\mathrm{Ground}}] \cdot \overline{\mathbf{S}}^{2+}_{\mathrm{Ground}} \cdot \rho \overline{\mathbf{S}}^{2-}_{\mathrm{Object}} \cdot \mathbf{E}^{-}[\mathrm{EI}_{\mathrm{Object}}]$$

$$+ \mathbf{E}^{+}[\mathrm{EI}_{\mathrm{Object}}] \cdot \rho \overline{\mathbf{S}}^{2+}_{\mathrm{Object}} \cdot \overline{\mathbf{S}}^{2-}_{\mathrm{Ground}} \cdot \mathbf{E}^{-}[\mathrm{EI}_{\mathrm{Ground}}] \tag{3.56}$$

where $\mathbf{E}^{+}$ and $\mathbf{E}^{-}$ take account of the extinction during the paths; $\mathrm{EI}_{\mathrm{Ground}}$ and $\mathrm{EI}_{\mathrm{Object}}$ are the projection indices of the terrain object and the ground patch, respectively; and $\overline{\mathbf{S}}^{2\pm}_{\mathrm{Ground}}$ and $\overline{\mathbf{S}}^{2\pm}_{\mathrm{Object}}$ are the Mueller matrices of the ground patch and the terrain object along the forward and backward directions of double scattering, respectively. The coefficient $\rho = n_{ef}$ or a_{ef} is calculated similarly to Equations (3.45),(3.49) and 3.53. Note that Equation 3.56 contains two coherent scattering contributions of two opposite propagation paths.

For triple scattering (object–ground–object), the main difference is the propagation path $r_{\mathrm{Object}}-r_{\mathrm{Path}}-r_{\mathrm{Object}}$. The expression can be directly written as

$$\underline{\mathbf{S}}[\mathrm{SI}_{ef}] := \underline{\mathbf{S}}[\mathrm{SI}_{ef}] + \mathbf{E}^{+}[\mathrm{EI}_{\mathrm{Object}}] \cdot \overline{\mathbf{S}}^{2+}_{\mathrm{Object}} \cdot \overline{\mathbf{S}}^{3}_{\mathrm{Ground}} \cdot \rho \overline{\mathbf{S}}^{2-}_{\mathrm{Object}} \cdot \mathbf{E}^{-}[\mathrm{EI}_{\mathrm{Object}}] \tag{3.57}$$

where $\overline{\mathbf{S}}^{3}_{\mathrm{Ground}}$ is the Mueller matrix of the ground patch along the triple scattering path, and $\rho = n_{ef}$ or a_{ef}. It can be seen that the Mueller matrices of the terrain object are the same for both double and triple scattering.

The correct sequence to calculate each grid unit is first to count single scattering, then multiple scattering, and last its attenuation or shadowing. Note that in Figure 3.8 it only gives the case that the ground patch and the terrain object are located in the same incidence plane. However, the ground patch of multiple scattering could be located at any place around the terrain object. For simplicity, we always take the projection cell in the current incidence plane as an approximate substitute. Note that, in this approximate manipulation, only the attenuation term of the ground patch is approximated by the corresponding one in the current azimuth line. Indeed, those trees are always distributed randomly, thus the attenuation caused would not introduce intrinsic errors by this substitution. Otherwise, this problem can be tackled by expending more storage to save all the extinction matrices of the neighborhood azimuth lines that might be used in computation. Moreover, it should be noted that the focus location for multiple scattering between the terrain object and the ground patch in different azimuth lines is approximated here.

Figure 3.9 summarizes the MPA flow. In order to save processing time, calculation of the terrain object or sub-segment is skipped if the corresponding projection cell is already shadowed. Regarding the shadowing projection of the terrain objects, for example, the ground or a building, the corresponding elements of the array $\mathbf{E}^{\pm}$ should be assigned as zero or cleared. However, the elements of the array $\mathbf{E}^{\pm}$ might be useful after being cleared, particularly for counting double or triple scatterings. For instance, as shown in Figure 3.8, attenuation of the ground patch should be used for calculation of double scattering, but the element of the array $\mathbf{E}^{\pm}$ has been cleared when the shadowing of the ground patch itself is projected. Moreover, if a building is partitioned into multi-grid units, two of which belong to the same projection cell, then it would be a mistake if the shadowing projected by the first grid unit was counted to obstruct the second grid.

Set simulation parameters and load the scene;

Partition the scene into grid units;

for (next line)

{

    Initialize array $\underline{\mathbf{E}}^{\pm}$ and array $\underline{\mathbf{S}}$;
    for (next grid unit of current line) // *sequentially as increasing* y

    {

        for (next terrain object of current grid unit)

        {

            Obtain Mueller matrix and extinction matrix;
            Divide into K sub-segments according to mapping cells;
            Map and project each sub-segment to obtain SI_k and EI_k;
            Update array $\underline{\mathbf{S}}$ as in Equation (3.48); // *indices are* SI_k, EI_k
            if (contains multiple scattering)

            {

                Trace the propagation paths and calculate the effective range;
                Obtain Mueller matrices of terrain objects and ground patch for
                    multiple scattering;
                Divide into H sub-segments according to mapping cells;
                Map and project each sub-segment to obtain SI_h and EI_h;
                Update array $\underline{\mathbf{S}}$ as in Equations (3.56),(3.57); // *indices are* SI_h and EI_h

            }
            Update array $\underline{\mathbf{E}}^{\pm}$ as in Equation (3.50); // *indices are* EI_k

        }

    }
    Save array $\underline{\mathbf{S}}$ to result file;

}

Figure 3.9 Mapping and projection algorithm. © [2006] IEEE. Reprinted, with permission, from *IEEE Transactions on Geoscience and Remote Sensing*

Therefore, two flags are set up for each projection cell, that is, $\underline{\mathbf{F}}^{GS}$ and $\underline{\mathbf{F}}^{OS}$, which indicate the shadowing caused by the ground surface and other terrain objects, respectively. In the simulation process, the following rules must be fulfilled:

(a) Do not clear the array $\underline{\mathbf{E}}^{\pm}$, but rather set the flag $\underline{\mathbf{F}}^{GS}$, when the ground surface causes shadowing.
(b) Do not clear the array $\underline{\mathbf{E}}^{\pm}$, but rather set the flag $\underline{\mathbf{F}}^{OS}$, when the terrain object (except the ground surface) causes shadowing, and record the identity of the terrain object that causes the flag $\underline{\mathbf{F}}^{OS}$.
(c) Skip the calculation of the scatterers if the flag $\underline{\mathbf{F}}^{GS}$ or $\underline{\mathbf{F}}^{OS}$ of the corresponding projection cell is valid.

(d) Ignore the flag $\underline{\mathbf{F}}^{GS}$ when counting double scattering; ignore the flag $\underline{\mathbf{F}}^{OS}$ caused by itself when counting scatterings of the terrain object.

Before executing the MPA algorithm, a comprehensive terrain scene has to be constructed and described using computer data structure. It should include digital elevation mapping (DEM), the properties of the ground, and the properties and locations of the terrain objects.

3.2.3 Scattering Models for Terrain Objects

Scattering models of most common terrain objects have already been introduced in Chapters 1 and 2. In this section, we discuss how to plug those models in to the MPA simulation framework.

3.2.3.1 Vegetation Canopy

A VRT model of clouds of random non-spherical particles for the vegetation canopy is presented in Section 2.3. Leaves, small twigs, and thin stems are modeled as non-spherical dielectric particles under the Rayleigh approximation or Rayleigh–Gans approximation, while branches, trunks, and thick stems are modeled as dielectric cylinders.

As shown in Figure 3.10, a tree model is composed of crown and trunk. The tree crown is a cloud with a simple geometrical shape containing randomly oriented non-spherical particles. The tree trunk is an upright cylinder with the top covered by the crown. The ratio of the covered length is called "cover ratio". Oblate or disk-like particles and elliptic crown are adopted for broad-leaf forest, while prolate or needle-like particles and cone-like crown are adopted for needle-leaf forest. In addition, the crown shapes take a small random perturbation. Similarly, a farm field is modeled as a layer of randomly oriented non-spherical particles with perturbed layer depth.

Single scattering of tree crown and tree trunk, single scattering from rough ground surface, and interactions between the canopy and ground surface are all taken into account. However, only double scatterings with the rough ground surface in specular reflection are regarded to

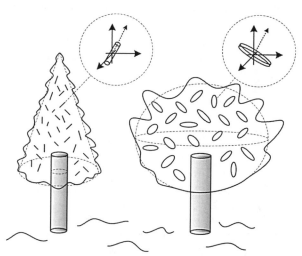

Figure 3.10 Scattering model for trees. © [2006] IEEE. Reprinted, with permission, from *IEEE Transactions on Geoscience and Remote Sensing*

make efficient computation, and those double scatterings due to angularly diffused scattering of the rough ground surface are ignored. In fact, under the assumption of moderate roughness of the ground surface, it is true that diffused scattering of the ground surface is much less than the coherent scattering (Fung, 1994) in specular direction. Here, we aim to build up the framework to simulate SAR imaging for a comprehensive scene. Simulation accuracy can be improved if a sophisticated mechanism for high-order scattering calculation is incorporated.

As the double scattering between leaves and ground is attenuated through the crown in both the paths r_{Path} and r_{Object}, we apply the attenuation of r_{Object} twice along the scattering path in order to compensate for the possible error caused by omitting the attenuation of r_{Path}. In practice, the crowns of dense forest can be treated as a parallel layer of particles. According to the MPA process, the crown part cut by each grid has its mid-point to conduct mapping and projection. Thus, the extinction depth through the path r_{Path} in Equation 3.56 equals that of the path r_{Object}. Because the extinction matrices of uniformly oriented particles for different scattering directions are the same, this approximation can be justified.

If the grid units are small enough, the segment of the canopy cut by one grid unit can be regarded as a segment fully filled with particles. Hence, its scattering and attenuation can be counted from Equations (3.49),(3.56) and (3.51), in which the Mueller matrix and the extinction matrix are calculated from the expressions given in Chapter 1. Otherwise, if the grid units are not small enough, it needs calculation of the total number of particles filling the segment. Different from the crown, the trunk is solid within a grid unit; hence its fractional volume is set to 1. The Mueller matrices and the extinction matrices of the dielectric cylinder for the trunk are also calculated from the expressions given in Chapter 1.

Since the trunk is vertically oriented, similar to the case of the wall surface of a building, the effective range r_{ef} is equivalent to the range of the intersection point of the trunk surface and the ground (Franceschetti, Iodice, and Riccio, 2002). In the same way, all the double scatterings of crown–ground should be mapped around the root.

Figure 3.11 interprets the composition of tree imaging. It can be seen that the SAR image and shadow are mapped and projected to two different sides, respectively, while double

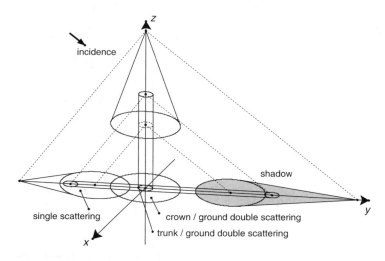

Figure 3.11 Image and shadow of a tree in SAR image. © [2006] IEEE. Reprinted, with permission, from *IEEE Transactions on Geoscience and Remote Sensing*

scattering is located around the root of the tree. This is the difference between the SAR image and an optical photograph, and should be noted particularly for morphological study of the terrain objects in high-resolution SAR imagery.

3.2.3.2 Building

Scattering of a building is seen as the surface scattering from its wall and roof surfaces and multiple interactions with the ground surface. As discussed in Chapter 1, the integral equation method (IEM) (Fung, 1994) is employed to calculate surface scattering. Note that the IEM Mueller matrix includes both incoherent and coherent scattering (Koudogbo, Combes, and Mametsa, 2004).

The building is modeled as a rectangular parallelepiped with an isosceles triangular cylinder layered upon it. The angle between the longer lateral of the parallelepiped and the x axis, that is, the orientation angle φ_o, indicates the orientation of the building. The roof angle φ_r between the tilted roof surface and the horizontal plane indicates the roof slope. Owing to the orientation of the wall surface and roof surface, the incident angle must be transformed into the local coordinates of the surface, as well as the polarization basis of wave propagation. Geometrical relationships among the wall surfaces, the roof surfaces, as well as the double and triple scatterings between the wall and ground have been discussed in Chapter 1. It should be pointed out that diffraction for building edges might be significant in high-resolution SAR imagery, which has not been studied sufficiently in the literature. For convenience, the building model is drawn in Figure 3.12, where its image and shadowing in SAR imagery are illustrated.

Figure 3.13 depicts the scattering path of wall–ground double bounce scattering. When the incident wave hits the wall at A, it reflects and shoots to the ground at B, and then scatters

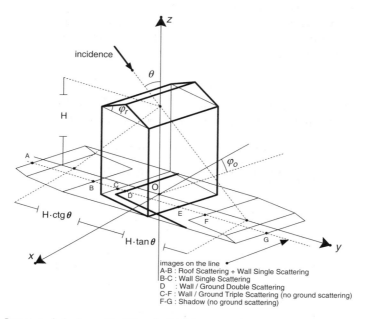

Figure 3.12 Image and shadow of building in SAR imagery. © [2006] IEEE. Reprinted, with permission, from *IEEE Transactions on Geoscience and Remote Sensing*

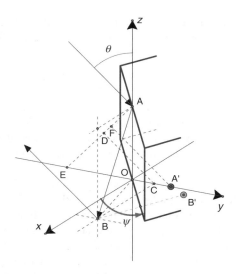

Figure 3.13 Mapping of wall–ground double bounce scattering. © [2006] IEEE. Reprinted, with permission, from *IEEE Transactions on Geoscience and Remote Sensing*

back toward the incident direction. According to SAR imaging principles, this double bounce scattering can be seen as single scattering coming from some point A', given that the range of A' equals the total range of the double scattering path.

Let us simply assume that A' is on the y axis. In the yOz plane, let line AE be perpendicular to the incident direction, that is, AE represents the wavefront. Now in the xOy plane, let line BC be perpendicular to the y axis, and line CD be perpendicular to AE, with point D on line AE. It can be shown that CD is parallel to the incident direction, meaning that it represents the path length from B back to the AE wavefront. Thus, the total path of the double bounce scattering can be calculated as

$$R_{\text{double}} = R_{A'} = R_A + |AB| + |CD| \tag{3.58}$$

where $R_{A'}$, R_A represent the two-way range from the radar to A' and A, respectively; and $|AB|$ represents the length of AB. If there is $A'F$ perpendicular to AE, then we have

$$2|A'F| = R_{A'} - R_A = |AB|+|CD| \tag{3.59}$$

From the geometry shown in Figure 3.13, it can be proved that $\angle BAO = \theta$. Thus, the position of B can be derived as

$$\begin{cases} x_B = h_A \tan\theta \cdot \cos(2\psi - \pi/2) \\ y_B = h_A \tan\theta \cdot \sin(2\psi - \pi/2) \end{cases} \tag{3.60}$$

where $h_A = |AO|$ denotes the elevation of A. Subsequently, the position of A' can be calculated as

$$y_{A'} = h_A \tan\theta \sin^2\psi \tag{3.61}$$

From the above expression, it is concluded that for any point A located on a wall of height H, that is, $h_A \in [0, H]$, its corresponding wall–ground double bounce scattering should be located inside the wall footprint at range $y_{A'} \in [0, H \tan \theta \sin^2 \psi]$.

From the mapping rule of single scattering, it is known that the single scattering from A should be mapped to point E. Thus, we conclude that single scattering from the wall itself is located outside the footprint at range $y_E \in [-H/\tan \theta, 0]$. These conclusions are critical to the 3D building reconstruction from SAR images presented in Chapter 7.

For triple scattering, its mapping point is at B's mirror point B′ with respect to the wall. However, triple and higher-order scatterings are insignificant in SAR imaging.

3.2.3.3 Ground

The normal vector of the ground facet in each grid unit is obtained from DEM data. The tangent vectors along the x and y axes on the current grid unit are used to construct the ground facet. Hence the normal vector of the current ground facet is calculated using the elevation data as

$$
\hat{n}(i_x, i_y) = \frac{\hat{t}_x \times \hat{t}_y}{\|\hat{t}_x \times \hat{t}_y\|}
$$

$$
\hat{t}_x = [2\Delta x \quad 0 \quad h(i_x + 1, i_y) - h(i_x - 1, i_y)]^{\mathrm{T}} \tag{3.62}
$$

$$
\hat{t}_y = [0 \quad 2\Delta y \quad h(i_x, i_y + 1) - h(i_x, i_y - 1)]^{\mathrm{T}}
$$

where $h(i_x, i_y)$ is the ground elevation of the grid (i_x, i_y). Consequently, ground surface scattering can be calculated in the same way as given in Section 1.3.

3.2.3.4 Other Terrain Objects

Besides the aforementioned vegetation canopy, buildings, and the ground, the natural environment might also contain rivers, roads, man-made targets, and so on. Ground-like terrain objects, for example, rivers and roads, are treated in the same way as the ground surface by using the specific parameters of roughness, dielectric constant, and elevation. Scattering from the man-made targets can be calculated by computational electromagnetics.

3.2.4 Speckle Model and Raw Data Generation

The MPA approach is based on the VRT theory, which deals with incoherent scattering power, not the coherent summation of scattering fields. However, speckle due to coherent interference is one of the critical characteristics of SAR imagery, especially for studies of SAR image filtering and optimization. Speckle simulation for mono-polarized SAR images has been studied (Holtzman *et al.*, 1978; Raney and Wessels, 1988). It is now time to extend this one-dimensional (mono-polarization) method to the multi-dimensional (fully polarimetric) case.

As the target scenario becomes sufficiently random, the resulting speckle on an SAR image resembles a Gaussian distribution (Oliver and Quegan, 2004), which is defined as completely developed speckle. The Gaussian probability distribution function (PDF) of the scattering vector, $\mathbf{k}_L = [S_{hh}, S_{hv}, S_{vh}, S_{vv}]^T$, is defined as

$$p(\mathbf{k}_L) = \frac{1}{\pi^d |\overline{\mathbf{C}}|} \exp\left(-\mathbf{k}_L^+ \cdot \overline{\mathbf{C}}^{-1} \cdot \mathbf{k}_L\right) \tag{3.63}$$

where the superscript $+$ denotes the conjugate transpose, $\overline{\mathbf{C}}$ is the covariance matrix, and d is the dimension of $\mathbf{k}_L$ (which is 4). The covariance matrix $\overline{\mathbf{C}}$ can be converted from the Mueller matrix. In this paper, the Mueller matrix from MPA fast computation is transformed to the covariance matrix $\overline{\mathbf{C}}$, and the random scattering vector of Equation 3.63 is then generated by a random number generation method (Novak and Burl, 1990) as follows:

$$\mathbf{k}_L = \sqrt{\overline{\mathbf{C}}} \cdot \mathbf{k}^0, \quad k_i^0 = I_i + jQ_i, \quad i = 1, \ldots, d \tag{3.64}$$

where I_i, Q_i are independent Gaussian random numbers with zero mean and unit variance. Given the positive semidefinite property of $\overline{\mathbf{C}}$, its lower triangular square root $\sqrt{\overline{\mathbf{C}}}$ can be obtained by Cholesky decomposition.

However, the multiplicative speckle is not suitable for strong scatterers like building parts, for which the phase information is almost constant during the integration. Moreover, the dominant scattering term, that is, wall–ground double bounce, is sensitive to the diversity in the azimuth angle θ (Schneider et al., 2005; Ferro-Famil, Reigber, and Pottier, 2005), which results in the inaccuracy of the narrow-aperture approximation. As a solution, we set an azimuth scattering diagram for the wall–ground term of building scattering, which simulates the special properties of strong scatterers through raw data generation from the previously calculated scattering map.

Nonetheless, the Gaussian distribution will become unsuitable for high-resolution work or terrain types without fully developed speckle. Therefore, it is recommended to simulate speckle with a different distribution for different terrain types or in the high-resolution case.

As we now understand, the idea behind MPA is first to analytically calculate the second-order statistics of random scattering and then to generate random coherent scattering from the second-order statistics. As opposed to directly performing a Monte Carlo coherent scattering simulation, MPA has two major advantages:

(a) It does not require deterministic models for terrain objects. Users only need to specify a statistical description about the terrain scene.
(b) It avoids computationally intensive Monte Carlo realizations by generating speckle based on a PDF model.

The scattering coefficient map has to go through the final step to be converted to an SAR image. As we know from Equation 3.34, the SAR image $s_{img}(x, r)$ is in fact a convolution of the scattering map $\sigma(x, r)$ with the PSF of the SAR system.

A simple way to generate a SAR image would be first to obtain the PSF based on the system parameters and then to perform convolution. A more sophisticated but realistic way is first to generate raw data and then to apply the SAR imaging algorithm to get the SAR image. One advantage of the latter approach would be that the generated raw data could be used to test the SAR imaging algorithm as well.

Generation of raw data from the scattering map is an inverse procedure of SAR signal processing (Franceschetti, Migliaccio, and Riccio, 1995). Suppose that the radar sends a rectangular pulse with linear frequency modulation. Then the raw signal received by the radar is expressed from the scattering map as (Wu, Liu, and Jin, 1982)

$$s_{pq}(x_n, t_i) = \delta_n + \sum_{\Delta n} \sum_{\Delta i} \varpi(\Delta n, \Delta i) \cdot S_{pq}(n + \Delta n, i + \Delta i) \cdot g(t_i - \Delta t)$$

$$g(t) = \exp(-j2\pi f_c t) \cdot p(t) \qquad (3.65)$$

$$c\Delta t/2 = \sqrt{(\Delta n \Delta x)^2 + (R_0 + \Delta i \Delta r)^2}$$

where $s_{pq}(x_n, t_i)$ is the ith sample of the nth pulse echo at the channel of pq polarization, $S_{pq}(n, i)$ denotes the n, i pixel of the scattering map, ϖ denotes the antenna pattern, $p(t)$ is the transmitted pulse, f_c is the carrier frequency, Δt is the time delay, R_0 is the distance to scene center, Δx, Δr are the pixel spacing, and δ_n is the additive noise.

Theoretically, Equation 3.65 should be in terms of integrals of a continuous scattering map, but it is, in fact, calculated by the summation of response signals from the discrete scattering map in the time domain. Nevertheless, it is rather time-consuming to run the raw data generation in the time domain. However, it does represent convenience and the possibility of carrying out a more precise simulation, which can simulate additive noise, azimuth scattering diversity of strong scatterers, and so on. If calculation efficiency is required more than accuracy, other optional effective simulation methods, such as frequency-domain generator and point spread function filtering, are preferred.

The resolution of the scattering map, namely the pixel spacing, is equal to the resolution of the sampling signal. There are three levels of resolution or spacing: (i) grid spacing, which is the element of MPA calculation; (ii) pixel spacing of scattering map or SAR image, each of which contains several grid units; and (iii) resolution of SAR image, which defines the discriminating ability. The SAR resolution is determined by the bandwidth and aperture length, while the pixel spacing is determined by the sampling rate and pulse repetition frequency (PRF). Note that the pixel spacing must be less than the SAR resolution to avoid aliasing. These two levels of resolution are determined by device specification. But the first level of grid spacing can be chosen before simulation. It is recommended that the grid spacing be chosen as one-fifth to one-tenth of the pixel spacing to satisfy the requirements of simulation accuracy within the limited cost of computation complexity.

Another technical issue to point out here involves preprocessing of the scattering map in raw signal generation. Taking a closer look at Equation 3.65, it can be seen that the scattering map is discretized on a grid with size corresponding to the pixel spacing Δx, Δr, while in MPA the scattering **S** of each unit is calculated on a grid with size equal to the resolution cell (cf. Equation 3.47). Hence, interpolation of the scattering map from the resolution cell to the pixel spacing is required before it can be used in raw signal generation.

One may wonder why we do not calculate the scattering map directly at the level of pixel spacing so that such interpolation will not be necessary. In fact, that would cause aliasing in the frequency domain. The resolution basically reflects the maximum bandwidth that the SAR system is able to handle. If we calculate the scattering map on a grid finer than the resolution, it means that we inject a wider bandwidth of data into the SAR system. The extra band of data will be wrapped around and lead to incorrect raw signals. Hence, to be on the safe side, an additional bandpass filter may be applied to the interpolated scattering map to make sure that no additional bandwidth is injected into the raw signal generation process.

Another issue is regarding discrete signal simulation. In computation simulation, the high-frequency signal is simulated at twice the single-sided bandwidth rather than twice the highest frequency. In this way, we can avoid extreme data rates while preserving all the information. However, this strategy will cause a frequency shift, which equals the remainder of the carrier frequency modulo the sampling rate, $f_{rem} = f_c \, mod \, f_s$, because the digitalization wraps the frequency band down to baseband. In order to avoid this issue, the scattering map should be pre-shifted by $-f_{rem}$ toward the opposite direction before it can be input to the raw signal generator.

3.3 Platform for SAR Simulation

A platform SARSIM (SAR simulation) has been developed for SAR simulation based on the MPA framework. It includes modules such as scene design, radar/platform design, SAR signal processor, MPA simulation, terrain scattering mode, data management, and visualization. Figure 3.14 shows the architecture of the SARSIM platform. This section gives some technical descriptions of each module and their interfaces.

3.3.1 Simulation of Individual Terrain Objects

Using NASA/JPL's AirSAR as an example, we set the radar and platform parameters as listed in Table 3.1. Two frequency bands (L, C) are considered.

The parameters of the scattering model for different terrain objects in the following simulations are listed in Table 3.2, unless specified otherwise in the text. Some parameters depend on the frequency bands.

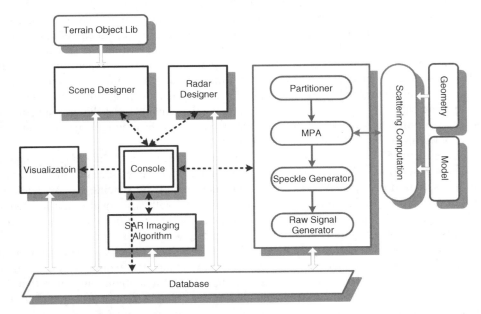

Figure 3.14   SARSIM platform implemented in a computer. © [2006] IEEE. Reprinted, with permission, from *IEEE Transactions on Geoscience and Remote Sensing*

Table 3.1 Configurations of radar and platform. © [2006] IEEE. Reprinted, with permission, from *IEEE Transactions on Geoscience and Remote Sensing*

Frequency band	L (1.26 GHz)/C (5.31 GHz)
Bandwidth	12.5 MHz/125 MHz
Sampling rate	30 MHz/300 MHz
Resolution	12 m/1.2 m (in both azimuth and slant range)
Pixel spacing	5 m/0.5 m (azimuth); 7.8 m/0.78 m (range)
Pulse width	5 μs
Pulse repeat frequency	46.3 Hz/463 Hz
Craft velocity	231.5 m s^{-1}
Altitude	8 km
Incident angle	40° at scene center

Table 3.2 Parameter settings[a] of terrain objects. © [2006] IEEE. Reprinted, with permission, from *IEEE Transactions on Geoscience and Remote Sensing*

Grid no.	2048 × 2048		Grid size	1 m × 1 m
Ground		$k\sigma = 0.30$, $k\ell = 6.0$, $\varepsilon = 5 + i0.2$ (L), $5 + i0.1$ (C)	Urban ground	$k\sigma = 0.10$, $k\ell = 0.10$, $\varepsilon = 4 + i0.1$ (L), $4 + i0.03$ (C)
Water surface		$k\sigma = 0.05$, $k\ell = 4.0$, $\varepsilon = 80 + i5$ (L), $70 + i20$ (C)	Suburban ground	$k\sigma = 0.13$, $k\ell = 4.0$, $\varepsilon = 5 + i0.2$ (L), $5 + i0.1$ (C)
Urban buildings	Wall	$k\sigma = 0.08$, $k\ell = 4.0$, $\varepsilon = 4 + i0.1$ (L), $4 + i0.03$ (C)	Suburban buildings	Wall $k\sigma = 0.10$, $k\ell = 4.0$, $\varepsilon = 4 + i0.1$ (L), $4 + i0.03$ (C)
	Roof	$k\sigma = 0.20$, $k\ell = 5.8$, $\varepsilon = 6 + i0.1$ (L), $6 + i0.03$ (C) Mean size 30 m × 20 m × 15 m, small roof angle, orderly oriented		Roof $k\sigma = 0.20$, $k\ell = 5.8$, $\varepsilon = 6 + i0.1$ (L), $6 + i0.03$ (C) Mean size 14 m × 13 m × 3.5 m, large roof angle, randomly oriented
Narrow-leaf forest		Leaves: $a = 3.5$ cm, $c = 0.1$ cm, $\varepsilon = 39.1 + i6.2$ (L), $36.1 + i9.2$ (C) Euler angles 0–60°, 0–360° $f_s = 0.005$ Crown height and width: 16 m × 8 m Trunk cover ratio 0.5 Trunk height and diameter: 3 m × 0.3 m $\varepsilon = 14.1 + i3.1$ (L), $13.1 + i4.1$ (C)	Broad-leaf forest	Leaves: $a = 0.01$ cm, $c = 2.5$ cm, $\varepsilon = 32.1 + i5.1$ (L), $29.1 + i8.1$ (C) Euler angles 0–40°, 0–360° $f_s = 0.003$ Crown height and width: 14 m × 8 m Trunk cover ratio 0.2 Trunk height and diameter: 3 m × 0.3 m $\varepsilon = 12.1 + i2.1$ (L), $11.1 + i3.1$ (C)
Farm crops		$f_s = 0.001$ to 0.005, Depth 10 cm to 90 cm, almost needle-like particles, other parameters are random		

[a] The dielectric permittivities are selected in a frequency-dependent way according to Ulaby, Moore, and Fung (1986), Franchois, Pineiro, and Lang (1998), and Matzler (1994). L and C indicate the frequency bands.

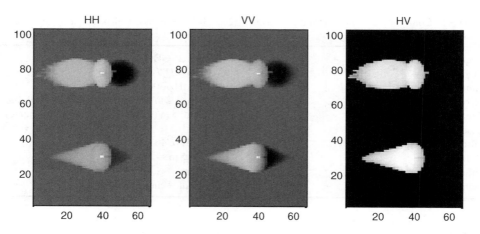

Figure 3.15 Scattering map of individual trees. Gray scale shows the normalized radar cross-section from −80 dB to 0 dB. Unless otherwise specified, horizontal axis is always range direction, while the vertical axis is azimuth direction. © [2006] IEEE. Reprinted, with permission, from *IEEE Transactions on Geoscience and Remote Sensing*

Let us look at a simple case of two different trees with, respectively, broad leaves (upper one) and narrow leaves (lower one). The grid unit size is 0.1 m × 0.1 m. The pixel spacing is 0.5 m in the azimuth dimension and 0.78 m in the range dimension. The simulated scattering map is shown in Figure 3.15.

It can be seen that the shapes and positions of the images and shadows are in accordance with the imaging mechanism and the tree models. Double scatterings mapped at the roots are especially stronger than the images of crowns and trunks on the left-hand side, while the shadows are on the right-hand side. However, in the case of a dense forest of crowded trees, double scattering on the roots is not noticeable due to the strong attenuation of the top layer of dense crowns. For a narrow-leaved tree, HH-pol (polarized) scattering is a little weaker than VV-pol, and there is also weaker attenuation of HH-pol than VV-pol. Compared to ground scattering, cross-pol (HV) of trees is much stronger than the background.

Some quantitative comparisons are given here. For a broad-leaved tree, the trunk–ground double scattering (HH) is 9.97 dB stronger than the crown scattering. The crown–ground double scattering (HH) is 3.29 dB lower than the crown scattering. The shadow makes an attenuation (HH) about 27.03 dB. For a narrow-leaved tree, the trunk–ground term is 14.69 dB higher and the crown–ground term is 1.57 dB lower than crown scattering, and shadow is 14.56 dB darker. It can be seen that double scattering becomes less prominent in dense forest due to attenuation through the top crown layer.

Another simple case is a scene of two buildings with different orientations (0° and 45°) and the same roof angle (20°). The grid unit and pixel spacing are the same as the first case. The simulated scattering map is shown in Figure 3.16.

It can be seen that the shape, position, and composition of the images and shadows are in accordance with the above analysis. Double and triple scatterings are stronger, particularly when the wall directly faces the sensor. The strip of double scattering image becomes wider and weaker if the building is not parallel to the azimuth direction, and also cross-pol scattering is enhanced due to loss of the azimuthal reflection symmetry. Scattering from the

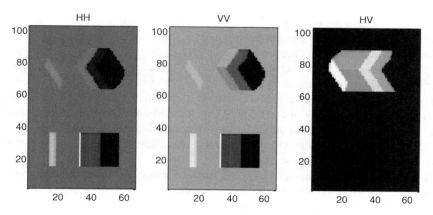

Figure 3.16 Scattering map of buildings. Gray scale shows the normalized RCS from −80 dB to 0 dB. © [2006] IEEE. Reprinted, with permission, from *IEEE Transactions on Geoscience and Remote Sensing*

left-hand side of the roof surface becomes dominant due to the roof angle. This single scattering has a different mechanism from double scattering. It is worth noting such a difference for residential buildings with tilted roofs.

Quantitative data are listed as follows. The wall–ground double scattering is at least 30 dB stronger than other single scatterings of building and about 20 dB stronger than tree scattering in Figure 3.15, which interprets the fact that urban scattering is always the strongest in real SAR images. However, in comparison with a non-oriented building, an oriented building has its scatterings at least 30 dB lower in co-pol, but 30 dB higher in cross-pol.

As mentioned above, some approximations have been made for realization of the whole simulation process, such as the narrow-aperture approximation, simplified computation of multiple scattering, incoherent modeling of the scattering of terrain objects, and speckle simulation. It is necessary to evaluate the simulation fidelity under these approximations.

First, the narrow-aperture approximation might be invalid for strong scatterers such as buildings in high-resolution imaging, particularly when the azimuth incident angle diversity is high. Assuming that the wall–ground double scattering term has a sinc profile of the scattering diagram in the azimuth direction, the narrow-aperture approximation might introduce a significant error. Though applying the scattering diagram can be helpful to diminish the error, it becomes invalid as the synthetic aperture becomes larger or the spotlight imaging mode is employed. In such cases, the MPA process should be modified to include the azimuth diversity of scattering in raw data generation.

Second, only two types of multiple scatterings are taken into account, namely the wall–ground and tree–ground interactions. It can be seen that these interactions include at least one scattering bounce as a specular reflection from a flat surface, but not all diffused or volumetric scattering. In the cases of Figures 3.15 and 3.16, those multiple scattering terms not counted indeed turn out to be much less significant. For instance, given the presumption of a smooth wall surface, double scattering with the wall scattering in non-specular direction is about 30 dB lower than that with the wall specular reflection. Double scattering contributed by leaves scattering and ground diffused scattering is about 20 dB lower than those with ground reflection. Another difficulty is due to the ground surface slope. It is difficult to trace the reflection path from the ground. Here, the ground surface is presumed flat in local scope around a single tree. Theoretically speaking, these issues can be resolved for good accuracy, but the computation efficiency might be lost.

Third, the attenuation along the propagation path r_{Path} from the terrain object to the ground is simply omitted. This path is always in a different direction, such as incidence and back-scattering, for which the attenuation effect is already calculated in the MPA procedure. This approximation might introduce artifacts into wall–ground interactions, particularly in a building area surrounded by dense trees. As mentioned above, we use the attenuation of r_{Object} as a substitute when accounting for the crown–ground interaction. In the case of Figure 3.15, this approximation makes no error for a broad-leaved tree and an acceptable error of about 0.5 dB for a narrow-leaved tree.

Finally, our simulation scheme is based on the VRT model, which incoherently cumulates the polarized scattering power and applies the polarized attenuation. Some issues relating to coherent wave propagation in a comprehensive terrain scene and the effect on imaging remain to be studied further.

3.3.2   Simulation of Comprehensive Terrain Scene

A virtual terrain scene is designed as shown in Figure 3.17. It contains different types of forests covering the hill, crops, farmland, buildings in urban and suburban regions, roads,

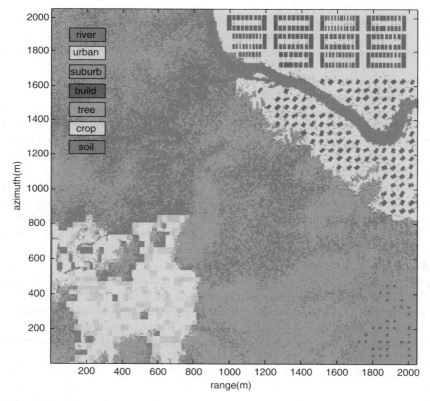

Figure 3.17   A comprehensive terrain scene designed for imaging simulation. © [2006] IEEE. Reprinted, with permission, from *IEEE Transactions on Geoscience and Remote Sensing*

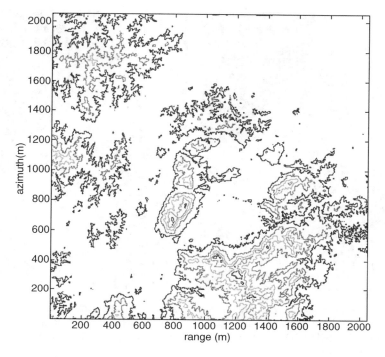

Figure 3.18 Elevation of the scene shown in Figure 3.17 (0–90 m). © [2006] IEEE. Reprinted, with permission, from *IEEE Transactions on Geoscience and Remote Sensing*

and rivers. The land surface elevation is plotted in Figure 3.18. These DEM data (Hole-Filled Seamless SRTM data V1, 2004, International Centre for Tropical Agriculture (CIAT), available from http://gisweb.ciat.cgiar.org/sig/90m_ data_tropics.htm) were collected by the SRTM mission of NASA over a region in Guangdong province, south China. In the urban area, parallel buildings are arranged as blocks with a few trees in the patio. Other buildings are randomly distributed with sparse trees around. There are dense trees randomly distributed over the hill, and the trees become denser as the elevation becomes higher. The upper area is almost broad-leaf forest, while the lower area is narrow-leaf forest. The parameters and size of each cropland are randomly selected. The grid unit size is 1.0 m × 1.0 m.

At L band with 12 m resolution (pixel spacing is 5 m), the simulated scattering map is displayed in Figure 3.19. The pseudo-color code for the left one is red = HH, green = HV, blue = VV; the Pauli decomposition for the right one is red = (HH − VV), green = (HV + VH), blue = (HH + VV). The same pseudo-color code is used throughout the rest of this section. Figure 3.20 shows the corresponding speckle-added scattering maps.

In the upper-right corner of Figure 3.19, the block images appear like parallel lines, which reveal the dominance of double scattering in the urban area. Between the buildings within the same block, the double scatterings disappear due to the obstruction of the front building. While on the other side, below the river, the buildings are oriented nearly 45° to the radar flying direction. As a result, the cross-pol scattering (green) is relatively stronger, while double scattering is reduced significantly. Moreover, the shadowing of this area is more apparent

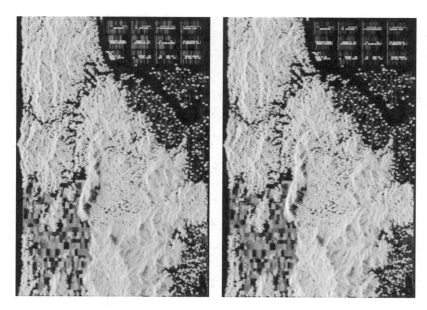

Figure 3.19 L band simulated scattering map of the comprehensive terrain scene (dB of normalized scattering power, from −50 dB to 0 dB). Left: R = HH, G = HV, B = VV. Right: R = (HH − VV), G = (HV + VH), B = (HH + VV). © [2006] IEEE. Reprinted, with permission, from *IEEE Transactions on Geoscience and Remote Sensing*

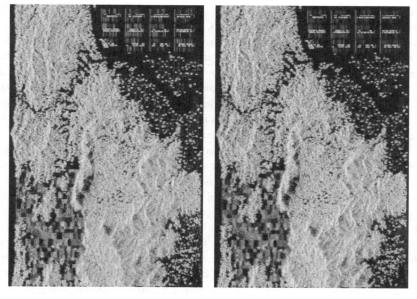

Figure 3.20 The same simulated scattering map as shown in Figure 3.19 but with speckle. © [2006] IEEE. Reprinted, with permission, from *IEEE Transactions on Geoscience and Remote Sensing*

due to the sparse distribution of the buildings. In the bottom-right rural villages, buildings are not that apparent due to their much smaller size and more random distribution.

Timberlands can be classified into two classes: broad leaves in white color, and narrow leaves in light green color due to the relatively weaker scattering of co-pol. The narrow-leaved trees produce weaker attenuation of HH-pol, and therefore stronger HH-pol scattering comes from the shadowed ground. This might be the reason why a few areas of narrow-leaved forest become a little red. In the bank zone along the river, double scatterings at tree roots are much more noticeable because of the lower tree density in this lower-elevation area.

Overlay and shadowing effects caused by mountainous topography are particularly perceptible. Since the hill or mountain is covered by dense trees, the characteristics of the overlay and shadowing are determined by the canopy covering. Besides, the bare ground surface with high slope also exhibits stronger cross-pol scattering, such as the banks along the river, bare hillsides around the farmland, and so on.

Croplands are generally uniform and dense, and appear like patches in the image. They appear in distinguishable colors and brightnesses due to the random selection of parameters such as fractional volumes, sizes, shapes, and so on.

The scattering map with coherent speckles becomes blurred and noisy. But the colors of the terrain objects remain invariant compared with the image without speckle. This indicates that the speckle simulation preserves the correlation between different polarizations.

From the scattering map with speckle in Figure 3.20, the intensity distributions of the different terrain types can be accumulated and are plotted in Figure 3.21. It appears similar to the common Rayleigh distribution (Franceschetti *et al.*, 1992). However, the major peak of the urban area is lower due to the fact that most contributions come from flat urban ground surface with weak scattering.

The raw data can be generated from the scattering map using the time-domain approach. Then we use the range Doppler algorithm to process the raw data to obtain simulated SAR images. The simulated SAR image is given in Figure 3.22, where the same pseudo-color code is applied to the left and right panels. Figure 3.23 shows the total power image where the gray scale is given on right-hand side.

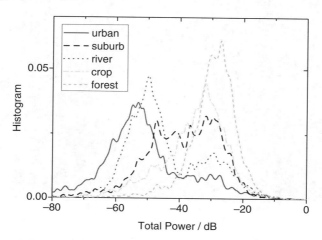

Figure 3.21 Scattering power (normalized) histogram of terrain types in simulated image in Figure 3.20.
© [2006] IEEE. Reprinted, with permission, from *IEEE Transactions on Geoscience and Remote Sensing*

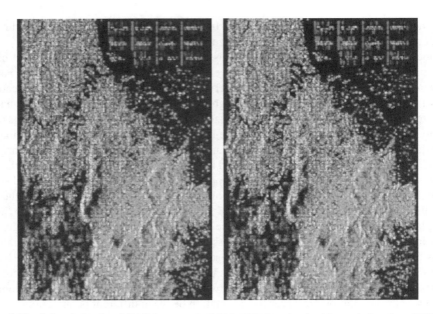

Figure 3.22 L band simulated SAR images. © [2006] IEEE. Reprinted, with permission, from *IEEE Transactions on Geoscience and Remote Sensing*

Figure 3.24 gives the simulated SAR images at C band. As a comparison with Figure 3.22 at L band, vegetation scatterings are significantly increased in the higher-frequency band, as well as attenuation being increased and wave penetrable depth being decreased. It means that trees become thinner while the shadowing becomes darker. Moreover, the profiles of the terrain objects become more distinguishable, and look more like to an optical photo.

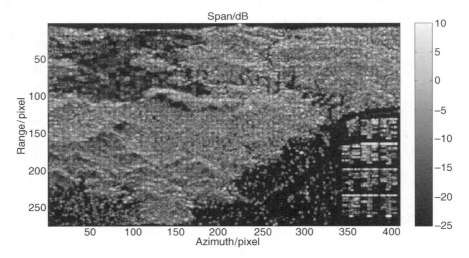

Figure 3.23 L band simulated total power SAR image. © [2006] IEEE. Reprinted, with permission, from *IEEE Transactions on Geoscience and Remote Sensing*

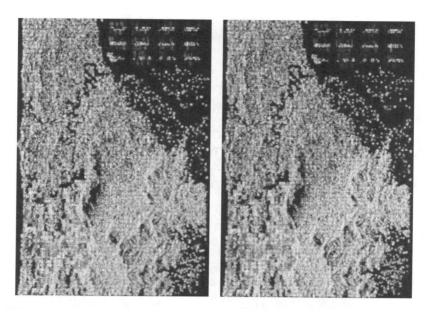

Figure 3.24   C band simulated SAR image. © [2006] IEEE. Reprinted, with permission, from *IEEE Transactions on Geoscience and Remote Sensing*

For instance, the shapes of hills, rivers, croplands, and tree shadows are more visible. In addition, the terrain objects with different properties and scattering mechanisms are more distinguishable; for example, broad-leaf forest in purple implies a larger difference between co-pol and cross-pol. However, the short penetrable depth in the higher-frequency band leads to much weaker scattering and less information directly from the underlying ground surface. Thus, different bands should be employed for imaging with different purposes.

It should be noted that scattering from urban structures decreased in accordance with the enhanced vegetation scattering. This phenomenon is partially due to the scattering model of small particles. In higher-frequency channels such as X band, the approximation for small particles in the scattering model becomes questionable, and more complicated calculations of particle scattering become necessary.

To simulate higher-resolution images (1.2 m, pixel spacing is 0.5 m), the scene of Figure 3.17 now shrinks by 10-fold to a smaller scene as shown in Figure 3.25. But the terrain objects, such as trees and buildings, preserve their original size. The grid unit size is refined to 0.1 m × 0.1 m in order to keep the precision of simulation. The rivers, urban areas, and elevation shrink proportionally. Simulated L band scattering maps, pseudo-color SAR images, and total power SAR image are given in Figures 3.26–3.28, respectively. Corresponding results for C band are given in Figures 3.29–3.31.

It can be seen that more information can be retrieved via the high-resolution SAR image. Especially, geometrical shapes and the polarimetric characteristics of the terrain objects such as trees and buildings can be clearly identified. Small-scale fluctuation and roughness of the ground surface can also be seen. Cross-pol backscattering from rough or non-flat ground becomes significant. At C band, the image and shadow become more clear and distinguishable. High-resolution SAR images carry more information, especially, about the shapes of terrain objects.

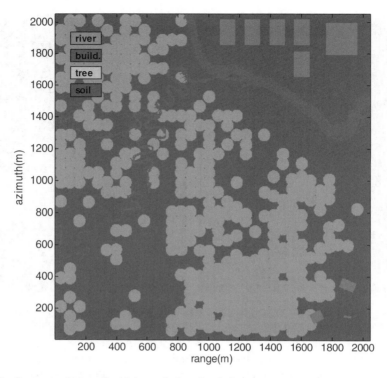

Figure 3.25 Rearranged scene for high-resolution simulation. © [2006] IEEE. Reprinted, with permission, from *IEEE Transactions on Geoscience and Remote Sensing*

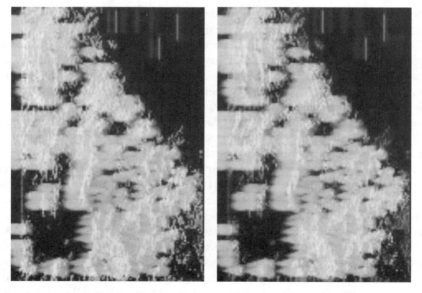

Figure 3.26 L band simulated high-resolution scattering map. © [2006] IEEE. Reprinted, with permission, from *IEEE Transactions on Geoscience and Remote Sensing*

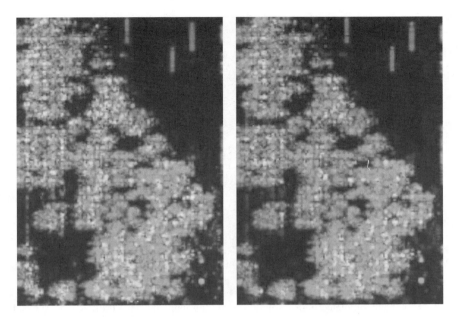

Figure 3.27 L band simulated high-resolution pseudo-color SAR images. © [2006] IEEE. Reprinted, with permission, from *IEEE Transactions on Geoscience and Remote Sensing*

Different frequencies, polarizations, and spatial resolutions should be taken into account to make the best compromise between cost and effect for terrain surface observation.

3.3.3 Extensions

The above examples of SAR simulation represent the basic and most common mode of SAR imaging. To be able to predict SAR performance for different natural environments, it may require extension to the MPA simulation framework.

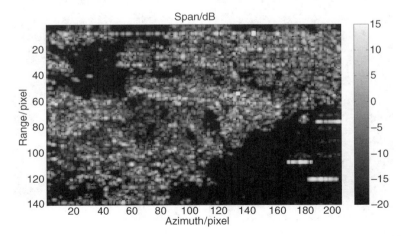

Figure 3.28 L band simulated high-resolution total power SAR image. © [2006] IEEE. Reprinted, with permission, from *IEEE Transactions on Geoscience and Remote Sensing*

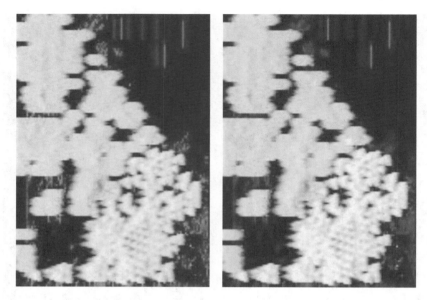

Figure 3.29 C band simulated high-resolution scattering map. © [2006] IEEE. Reprinted, with permission, from *IEEE Transactions on Geoscience and Remote Sensing*

The first possible extension is from an airborne platform to a spaceborne one. This would require calculation of the Earth's surface curvature, the relative motion between the Earth and the spacecraft, and the induced Doppler effect and geometry change.

The second extension would be from static to dynamic. SAR imaging of dynamic scenarios is important, for example, sea waves, moving targets, and so on. The additional part to be

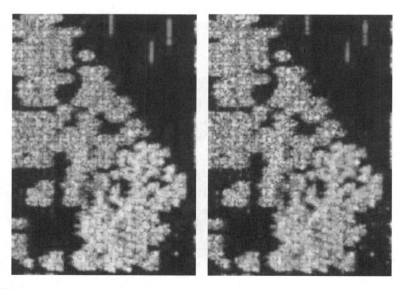

Figure 3.30 C band simulated high-resolution pseudo-color SAR images. © [2006] IEEE. Reprinted, with permission, from *IEEE Transactions on Geoscience and Remote Sensing*

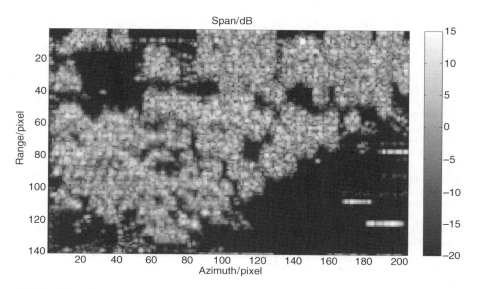

Figure 3.31 C band simulated high-resolution total power SAR image. © [2006] IEEE. Reprinted, with permission, from *IEEE Transactions on Geoscience and Remote Sensing*

considered includes the time-varying geometry of targets and environment, as well as the additional Doppler shift.

The third extension would be from the basic mode to advanced modes of SAR imaging. For example, interferometric SAR simulation would require precise phase difference calculation.

Among these extensions, the dynamic extension for sea environment simulation can be implemented by adding two steps to the current MPA chain. First, the dynamic change of the sea surface should be simulated frame by frame, and scattering and signal should be calculated for each frame at that particular time instant. Meanwhile, the Doppler shift should be accounted for based on the velocity of each scatterer. The dynamic sea surface can be modeled as a time-varying DEM, which can be generated from the Pierson–Moskowitz spectrum (Fung and Lee, 1982). In this case, there is no static scattering map. Instead, the raw signal has to be accumulated while scattering from the dynamic scene is being calculated frame by frame. Finally, the SAR image should be generated by compressing the raw signal.

References

Brown, C.G., Sarabandi, K., and Gilgenbach, M. (2002) Physics-based simulation of high-resolution polarimetric SAR images of forested areas. IEEE IGARSS'2002, Vol. **1**, pp. 466–468.

Camporeale, C. and Galati, G. (1991) Digital computer simulation of synthetic aperture systems and images. *European Transactions on Telecommunications*, **2**, 343–352.

Cumming, H. and Wong, F. (2005) *Digital Processing of Synthetic Aperture Radar Data: Algorithms and Implementation*, Artech House, Norwood, MA.

Curlander, J.C. and McDonough, R.N. (1991) *Synthetic Aperture Radar: Systems and Signal Processing*, John Wiley & Sons, Inc., New York.

Ferro-Famil, L., Reigber, A., and Pottier, E. (2005) Nonstationary natural media analysis from polarimetric SAR data using a two-dimensional time–frequency decomposition approach. *Canadian Journal of Remote Sensing*, **31** (1), 21–29.

Franceschetti, G. and Schirinzi, G. (1990) A SAR processor based on two-dimensional FFT codes. *IEEE Transactions on Aerospace Electronic Systems*, **26** (2), 356–365.

Franceschetti, G., Migliaccio, M., Riccio, D., and Schirinzi, G. (1992) SARAS: a synthetic aperture radar (SAR) raw signal simulator. *IEEE Transactions on Geoscience and Remote Sensing*, **30**, 110–123.

Franceschetti, G., Migliaccio, M., and Riccio, D. (1995) The SAR simulation: an overview. IEEE IGARSS' 1995, Vol. **3**, pp. 2283–2285.

Franceschetti, G., Migliaccio, M., and Riccio, D. (1998) On ocean SAR raw signal simulation. *IEEE Transactions on Geoscience and Remote Sensing*, **38** (1), 84–100.

Franceschetti, G., Iodice, A., and Riccio, D. (2002) A canonical problem in electromagnetic backscattering from buildings. *IEEE Transactions on Geoscience and Remote Sensing*, **40** (8), 1787–1801.

Franceschetti, G., Iodice, A., Riccio, D., and Ruello, G. (2003) SAR raw signal simulation for urban structures. *IEEE Transactions on Geoscience and Remote Sensing*, **41** (9), 1986–1994.

Franchois, A., Pineiro, Y., and Lang, R.H. (1998) Microwave permittivity measurements of two conifers. *IEEE Transactions on Geoscience and Remote Sensing*, **36** (5), 1384–1395.

Fung, A.K. (1994) *Microwave Scattering and Emission Models and Their Applications*, Artech House, Norwood, MA.

Fung, A.K. and Lee, K.K. (1982) A semi-empirical sea-spectrum model for scattering coefficient estimation. *IEEE Journal of Oceanic Engineering*, **7** (4), 166–176.

Holtzman, J.C., Frost, V.S., Abbott, J.L., and Kaupp, V.H. (1978) Radar image simulation. *IEEE Transactions on Geoscience and Remote Sensing*, **16** (1), 296–303.

Jin, Y.Q. (1994) *Electromagnetic Scattering Modeling for Quantitative Remote Sensing*, World Scientific, Singapore.

Jin, Y.Q. (2005) *Theory and Approach of Information Retrievals from Electromagnetic Scattering and Remote Sensing*, Springer, Berlin, Germany.

Koudogbo, F., Combes, P.F., and Mametsa, H.J. (2004) Numerical and experimental validations of IEM for bistatic scattering from natural and manmade rough surfaces. *Progress in Electromagnetics Research*, **46**, 203–244.

Lin, Y.C. and Sarabandi, K. (1999) A Monte Carlo coherent scattering model for forest canopies using fractal generated trees. *IEEE Transactions on Geoscience and Remote Sensing*, **37** (1), 440–451.

Matzler, C. (1994) Microwave (1–100 GHz) dielectric model of leaves. *IEEE Transactions on Geoscience and Remote Sensing*, **32** (5), 947–949.

Novak, L.M. and Burl, M.C. (1990) Optimal speckle reduction in polarimetric SAR imagery. *IEEE Transactions on Aerospace Electronics Systems*, **26** (2), 293–305.

Oliver, C. and Quegan, S. (2004) *Understanding Synthetic Aperture Radar Images*. SciTech, Raleigh, NC.

Pike, T.K. (1985) SARSIM: a synthetic aperture radar simulation model. DFVLR-Mitt. 11.

Raney, R.K. and Wessels, G.J. (1988) Spatial considerations in SAR speckle simulation. *IEEE Transactions on Geoscience and Remote Sensing*, **26** (3), 666–672.

Raney, R.K., Runge, H., Bamler, R., *et al.* (1994) Precision SAR processing using chirp scaling. *IEEE Transactions on Geoscience and Remote Sensing*, **32** (4), 786–799.

Schneider, R.Z., Papathanassiou, K., Hajnsek, I., and Moreira, A. (2005) Polarimetric interferometry over urban areas: information extraction using coherent scatterers. IEEE IGARSS'2005.

Soumekh, M. (1999) *Synthetic Aperture Radar Signal Processing with Matlab Algorithms*, John Wiley & Sons, Inc., New York.

Ulaby, F.T., Moore, R.K., and Fung, A.K. (1986) *Microwave Remote Sensing: Active and Passive*, Vol. **III**: *Volume Scattering and Emission Theory, Advanced Systems and Applications*, Artech House, Norwood, MA.

Wang, Z.L., Xu, F., Jin, Y.Q., and Ogura, H. (2005) A double Kirchhoff approximation for very rough surface scattering using the stochastic functional approach. *Radio Science*, **40**, RS4011.

Wu, C., Liu, K.Y., and Jin, M. (1982) Modeling and a correlation algorithm for spaceborne SAR signals. *IEEE Transactions on Aerospace Electronics Systems*, **AES-18** (5), 563–575.

Xu, F. and Jin, Y.Q. (2006) Imaging simulation of polarimetric SAR for a comprehensive terrain scene using the mapping and projection algorithm. *IEEE Transactions on Geoscience and Remote Sensing*, **44** (11), 3219–3234.

Yegulalp, A.F. (1999) Fast backprojection algorithm for synthetic aperture radar. Radar Conference 1999, Record of the IEEE 1999 International Radar Conference.

4

Bistatic SAR: Simulation, Processing, and Interpretation

Conventional monostatic SAR has its transmitter and receiver installed on the same platform and collects the backscattered echoes from targets within the imaged area. During the past decade, bistatic SAR (BISAR), with separated transmitter and receiver flying on different platforms, has become of great interest, because BISAR might significantly reduce the vulnerability of military systems, enhance the response due to the large angle between incidence and scattering, and present additional information from the observation, and so on (Wills, 1991; Krieger and Moreira, 2005).

Some technical problems of BISAR observation have been solved, for example, hardware technology, signal processing, platform design and realization, and so on. Since some pioneering airborne BISAR experiments were successfully performed (Yates *et al.*, 2006; Wendler *et al.*, 2003; Walterscheid *et al.*, 2006), bistatic and multistatic spaceborne SAR/InSAR programs have been put forward and are under preparation (Moccia *et al.*, 2002; Ebner *et al.*, 2004), which are laying the fundamentals for the further development of spaceborne BISAR. However, most BISAR research is focused on the engineering realization and signal processing algorithm (Loffeld *et al.*, 2004; Younis, Metzig, and Krieger, 2006), with a few on land-based bistatic experiments or theoretical modeling of bistatic scattering (McLaughlin *et al.*, 2002; Liang, Pierce, and Moghaddam, 2005; Thirion and Dahon, 2006; Franceschetti, Migliaccio, and Riccio, 1995). In order to interpret and retrieve BISAR image information, it is necessary to develop a forward scattering model and an imaging simulation approach for BISAR observation, which will be a theoretical tool for platform design, data validation, image interpretation, target classification and identification, quantitative parameter retrieval, and so on.

The model and image simulation of monostatic SAR have been studied by many authors, for example, Franceschetti, Migliaccio, and Riccio (1995), Franceschetti *et al.* (2003), Brown, Sarabandi, and Gilgenbach (2002), and Xu and Jin (2006). The mapping and projection algorithm (MPA) of Chapter 3 has provided a fast and efficient tool for monostatic imaging simulation. It involves the physical scattering process of multiple terrain objects, such as vegetation canopy, buildings, and rough ground surfaces. In this chapter, the MPA approach

Polarimetric Scattering and SAR Information Retrieval, First Edition. Ya-Qiu Jin and Feng Xu.
© 2013 John Wiley & Sons Singapore Pte. Ltd. Published 2013 by John Wiley & Sons Singapore Pte. Ltd.

is extended to the bistatic case for the BISAR simulation over complex terrain. Bistatic polarimetric characteristics of the simulated images are then studied (Xu and Jin, 2008).

This chapter is organized as follows. Section 4.1 introduces different configurations of BISAR and explains the key points of the MPA extension to the bistatic case. The scattering model of terrain objects and BISAR raw signal model for stripmap imaging are described in Section 4.2 and Section 4.6, respectively.

Simulated BISAR images of typical scenes are presented in Section 4.3, and their polarimetric behaviors are analyzed in Section 4.4. In Section 4.5, a novel unified bistatic polar basis transform for polarimetric BISAR images is proposed and the target decomposition parameters α, β, γ (see Chapter 5) are redefined to achieve better abilities for bistatic polarimetric representation. In Section 4.6, raw signal processing of stripmap BISAR is especially discussed.

4.1 Bistatic Mapping and Projection Algorithm (BI-MPA)

4.1.1 Configurations of BISAR

Since the principle of MPA in the monostatic case and the details of technical realization have been presented in Chapter 3, only the key points for BISAR application are restated here. Suppose the incident wave $\mathbf{k}$ has its y component along the $+y$ direction, that is, $k_y > 0$. Thus, the terrain objects located ahead on the y axis will not be blocked by those behind. Discrete grid units are processed sequentially as y increases, and the attenuation and shadowing effects are accumulated onto a storage array $\mathbf{E}^{\pm}$. This array is assumed to store the total attenuation or shadowing along the preceding propagation path till the grid unit in the current calculation. For each grid unit, the specific operation steps are as follows. First, count its scattering contribution and add up to another storage array $\mathbf{S}$ (indexed via mapping), where the attenuation or shadowing it should suffer is obtained from the array $\mathbf{E}^{\pm}$. Second, count its attenuation or shadowing and multiply them onto the array $\mathbf{E}^{\pm}$ (indexed via projecting). These two arrays are used for temporary storage during the computing process, and the final result produced in $\mathbf{S}$ is the scattering map for image simulation.

The complexity of MPA to the scene size N is only $o(N)$. However, the prerequisite condition for correctly performing MPA is that the calculation sequence should be coincident with the propagation of the incident wave. In bistatic circumstances, owing to the different positions and velocity vectors of the transmitter and receiver (referred to as Tx and Rx henceforth) platforms, the following issues should be taken into account.

(a) The projection and attenuation accumulation of incident and scattered waves should be performed separately, because of the split directions of incidence and scattering. Three-dimensional projection is required as well as extension of the array $\mathbf{E}^{\pm}$ from one dimension to two dimensions, if the incidence and scattering are not located in the same incidence plane.

(b) A correct sequence of calculation must be guaranteed. It should be along the incident but against the scattered direction, so as to assure that the attenuation accumulations are properly performed in both directions.

(c) In order to keep Tx and Rx antennas pointing at the same area, squint looking configurations are usually adopted, which make the position of the synthetic aperture unpredictable with respect to the imaging center at any particular moment. Additionally, the Doppler history also varies for scattering terms with different scattering paths.

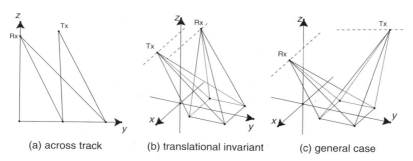

Figure 4.1 Different configurations of BISAR. © [2008] IEEE. Reprinted, with permission, from *IEEE Transactions on Geoscience and Remote Sensing*

The focusing position of each scatterer is determined by the position of its synthetic aperture and its Doppler history. Hence, it is necessary to incorporate the possible azimuth offset in mapping operation.

Generally, BISAR configurations can be categorized into three different cases, as depicted in Figure 4.1a–c, respectively.

(a) The first (across-track) case is that in which the incidence and scattering are on the same incidence plane (Figure 4.1a). We only need to make projections separately and mapping along the elliptical iso-range lines within the incidence plane.

(b) The second (translational invariant) case is where two platforms can have different positions, but with the same velocity vectors (Figure 4.1b) (Ender, 2004). In this case, the projections have to be performed in 3D space and the azimuth bias of the mapping position caused by squint looking has to be considered. However, the good property of the translational invariant case makes the mapping and projections irrelevant to the positions of the scatterers.

(c) The most general case is that two platforms have different velocity vectors (Figure 4.1c). In this case, the mapping and projections of each scatterer should be performed specifically according to the information of Tx/Rx platforms at the moment when it is illuminated. The scene size has to be restricted for each simulation time to prevent a largely varying looking angle of the sensor, which might damage the basis of valid approximation for MPA computation.

When BISAR configuration preserves the translational invariant (TI) property, its imaging process and signal processing become greatly simplified. Two typical cases of TI are the tandem and across-track configurations. The former stands for the case of two platforms flying in succession along the same trajectory (different azimuth angles of incidence and scattering), while the latter means flying along different trajectories, but on the same incidence plane (different altitude angles of incidence and scattering).

4.1.2 Three-Dimensional Projection and Mapping

Since the radars are far away from the scene, the incidence and scattering angles to the same target can be seen as invariant when the radars fly over a distance of the synthetic aperture length. Therefore, the targets always have constant scattering contributions and determined propagation paths during the imaging process. It is noted that this approximation can simplify the simulation, but might affect the precision in some cases of very large synthetic aperture.

Consider a differential volume dv in the imaging space. It has two projection lines $p_{\mathbf{v}}^-$ and $p_{\mathbf{v}}^+$ pointing to the positions of the aperture centers of Tx and Rx, respectively. Then the direct scattering contribution of dv can be written as

$$I_s(\mathbf{v}) = E^+(\mathbf{v})P(\mathbf{v})E^-(v)I_i\,dv \tag{4.1}$$

where E^-, E^+ are the total attenuations along the incidence and scattering paths, respectively; the phase function P dictates the scattering of this volume unit; and I_i, I_s are the incident and scattered Stokes vectors. The overall attenuations of incidence and scattering are calculated by integrating the corresponding extinction coefficients $\kappa_e^-(\mathbf{v})$ and $\kappa_e^+(\mathbf{v})$ along the incidence and scattering projection lines, that is,

$$E^+(\mathbf{v}) = \exp\left[-\int_{p_{\mathbf{v}}^+} d\ell'\,\kappa_e^+(\mathbf{v}')\right], \quad E^-(\mathbf{v}) = \exp\left[-\int_{p_{\mathbf{v}}^-} d\ell'\,\kappa_e^-(\mathbf{v}')\right], \quad d\ell' = \frac{dv'}{da'} \tag{4.2}$$

where da' is the averaged intersection area of dv' along the direction of wave propagation, and then $d\ell'$ should be the effective penetration depth.

Let us assume that all points that have the same mapping position as dv are distributed along an equal-range track of mapping $m_{\mathbf{v}}$ in 3D space. All scatterers lying on this track will be mapped to the same position $(a_{\mathbf{v}}, r_{\mathbf{v}})$ on the imaging plane. Thus, accumulating all scattering contributions of scatterers along this track yields the scattering power $\bar{S}(a_{\mathbf{v}}, r_{\mathbf{v}})$ of the image pixel located at $(a_{\mathbf{v}}, r_{\mathbf{v}})$, that is

$$S(a_{\mathbf{v}}, r_{\mathbf{v}}) = \int_{m_{\mathbf{v}}} I_s(\mathbf{v}) = \int_{m_{\mathbf{v}}} E^+(\mathbf{v})P(\mathbf{v})E^-(\mathbf{v})I_i\,dv \tag{4.3}$$

Under the incidence of unit power and substituting Equation 4.2 into Equation 4.3, it yields

$$S(a_{\mathbf{v}}, r_{\mathbf{v}}) = \int_{m_{\mathbf{v}}} dv\,\exp\left[-\int_{p_{\mathbf{v}}^+} dv'\,\kappa_e^+(\mathbf{v}')\right]P(\mathbf{v})\exp\left[-\int_{p_{\mathbf{v}}^-} dv'\,\kappa_e^-(\mathbf{v}')\right] \tag{4.4}$$

Note that all the above-mentioned variables should be in their vector or matrix forms when considering incidence of a polarized wave.

Using a numerical approach to compute Equation 4.4, the 3D scene is first partitioned into discrete grid units. For each grid unit, its projection lines $p_{\mathbf{v}}^\pm$ and mapping position $(a_{\mathbf{v}}, r_{\mathbf{v}})$ are first determined, attenuations are then accumulated along the lines $p_{\mathbf{v}}^\pm$, and its scattering contribution is finally added up to $S(a_{\mathbf{v}}, r_{\mathbf{v}})$. The MPA is utilized to accelerate this procedure of computation, as illustrated in Figure 4.2. Without loss of

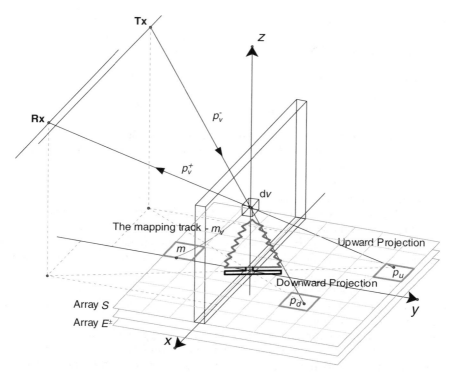

Figure 4.2 Bistatic mapping and projection. © [2008] IEEE. Reprinted, with permission, from *IEEE Transactions on Geoscience and Remote Sensing*

generality, suppose the incident/scattered waves propagate along the $+y/-y$ direction on the y axis. Then, the grid units are sequentially processed in the order of increasing y. As for each single unit, it is first projected and mapped to the corresponding cells of $\mathbf{E}^+$, $\mathbf{E}^-$ and $\mathbf{S}$, respectively. Thereafter, its scattering contribution is add to the cell of $\mathbf{S}$, where the attenuations are obtained from the cells of $\mathbf{E}^+$, $\mathbf{E}^-$. At last, attenuations caused by itself are multiplied to the cells of $\mathbf{E}^+$, $\mathbf{E}^-$. Note that both $\mathbf{E}^+$, $\mathbf{E}^-$ and $\mathbf{S}$ are 2D arrays. Each cell of $\mathbf{S}$ corresponds to a resolution pixel of the SAR image, while the sizes of the $\mathbf{E}^+$, $\mathbf{E}^-$ cells are selected so that they can be one-to-one mapped to the cells of $\mathbf{S}$.

The steps in the bistatic mapping and projection algorithm are described as follows.

1. The arrays $\mathbf{E}^+$, $\mathbf{E}^-$ are initialized as unit matrices, $\mathbf{S}$ as zero matrices. Then, visit the grid units, sequentially, along the $+y$ direction.
2. Perform 3D projections of the current grid unit along the incidence and scattering directions to the cells p_d, p_u of $\mathbf{E}^+$, $\mathbf{E}^-$, respectively.
3. Determine the mapping position of the current grid unit based on the information of its synthetic aperture and Doppler history, which corresponds to the cell m of the array $\mathbf{S}$.
4. Obtain or calculate the scattering matrix $\mathbf{S}_0$ and the upward/downward attenuation matrix $\mathbf{E}_0^\pm$ of the current grid unit.

5. Refresh the elements of $\underline{\mathbf{S}}$ as

$$\underline{\mathbf{S}}[m] := \underline{\mathbf{S}}[m] + \underline{\mathbf{E}}^+[p_u] \cdot \mathbf{S}_0 \cdot \underline{\mathbf{E}}^-[p_d] \tag{4.5}$$

Refresh the elements of $\underline{\mathbf{E}}^+$, $\underline{\mathbf{E}}^-$ as

$$\underline{\mathbf{E}}^-[p_d] := \mathbf{E}_0^- \cdot \underline{\mathbf{E}}^-[p_d] \qquad \underline{\mathbf{E}}^+[p_u] := \underline{\mathbf{E}}^+[p_u] \cdot \mathbf{E}_0^+ \tag{4.6}$$

Return to step 2, and continue to visit the next grid at the same coordinate of y or, if all grids at this coordinate have been visited, step forward to a larger coordinate of y till the whole scene is exhausted.

When realizing this algorithm, the 3D scene space is divided into 2D grids along the xOy plane, and each grid unit is actually a square column, which might contain a patch of rough ground surface, segments of vegetation canopy and/or patches of building facades. After that, the whole column of the terrain object parts (particle clouds or patches) will be processed together as a unit. In this processing, it is also required to calculate the volumetric scattering and extinction of canopy scatterers, for example, leaves and trunks, the surface scattering and shadowing of rough facets, and the effective volume, effective area, and effective penetration depth. More details of definition and calculation can be found in the monostatic SAR imaging simulation in Chapter 3.

It is noted that 3D projections are performed along average looking directions, that is, the incidence and scattering directions with respect to the aperture center. On the other hand, the mapping position, which is the same as the final focusing position in the SAR image, is determined by several factors, namely, the relative position of the synthetic aperture, the range distance, the Doppler history, and the adopted focusing method.

As for each scatterer, according to the flying and illuminating sequence of the sensors, its illuminated period (synthetic aperture) $[t_0 - \Delta t, t_0 + \Delta t]$ can be determined and its Doppler history can be written as

$$R(t) = |\mathbf{R}_R(t_0) + \mathbf{v}_R(t - t_0)| + |\mathbf{R}_T(t_0) + \mathbf{v}_T(t - t_0)| \tag{4.7}$$

where $\mathbf{R}_R$, $\mathbf{R}_T$ denote its range vectors to Rx and Tx, respectively, and $\mathbf{v}_R$, $\mathbf{v}_T$ are their velocity vectors. Its Doppler frequency is

$$f_d(t) = -\lambda^{-1} \partial R(t) / \partial t \tag{4.8}$$

where λ is the wavelength of the carrier frequency. Taking the conventional focusing method in the frequency domain as an example, given the Doppler centroid f_{dc}, the center moment t_c can be determined from

$$f_d(t_c) = f_{dc} \tag{4.9}$$

Based on the information of the positions of Tx and Rx at moment t_c, the focusing position can be determined and its coordinates in the final SAR image can be calculated as

$$i_a = [A(t_c) - A_{\min}]/b_a, \quad i_r = [R_R(t_c) + R_T(t_c) - 2R_{\min}]/2b_r \tag{4.10}$$

where A denotes the azimuth position of Rx and b_a, b_r are the azimuth and range resolutions, respectively.

Consider the case of TI configuration. Letting $\mathbf{v}_R = \mathbf{v}_T = v\hat{x}$, it is easily obtained that

$$f_d(t) = -\lambda^{-1}\left[v(\cos\vartheta_R + \cos\vartheta_T) + v^2(t - t_0)\left(\sin^2\vartheta_R R_R^{-1} + \sin^2\vartheta_T R_T^{-1}\right)\right] \qquad (4.11)$$

where $\cos\vartheta_R = x_R/R_R$, $\cos\vartheta_T = x_T/R_T$ and x_R, x_T denote the $\hat{x}$ components of $\mathbf{R}_R(t_0)$, $\mathbf{R}_T(t_0)$. Here R_R, R_T are the norms of $\mathbf{R}_R(t_0), \mathbf{R}_T(t_0)$. The angles ϑ_R, ϑ_T are just the squint angles of the radar beams with respect to the flying direction. The Doppler centroid f_{dc} is taken so as to let $t_c(\mathbf{p} = 0) = t_0$ at the reference point $\mathbf{p} = 0$. Thus, the moment t_c of each scatterer can be given as

$$t_c(\mathbf{p}) = t_0 + v^{-1}\frac{\cos\vartheta_R + \cos\vartheta_T - \cos\vartheta_R' - \cos\vartheta_T'}{\sin^2\vartheta_R' R_R'^{-1} + \sin^2\vartheta_T' R_T'^{-1}} \qquad (4.12)$$

where $\mathbf{p}$ denotes the position vector of the scatterer. The variables with primes correspond to the scatterer located at $\mathbf{p}$, for example, $\mathbf{R}_i' = \mathbf{R}_i + \mathbf{p}$, $i = R, T$.

The focusing offset is defined as the difference in the focusing position with and without taking into account the aforementioned factors. As shown in Figure 4.3, the point $\mathbf{p}$ has the same azimuth position and range as its reference point, that is, $p_x = 0$ and $R_R + R_T = R_R' + R_T'$. Apparently, this point would be mapped to the same position as its reference point according to the original mapping rules. The focusing offset can be obtained via first substituting $\mathbf{p}$ into Equation 4.12 to get t_c and then substituting t_c into Equation 4.10. As we can see, the Doppler centroid only causes a different direction of layover. In the case of $\vartheta_R = \pi - \vartheta_T$, that is, Tx and Rx have opposite squint looking angles, the constant term of Equation 4.11 turns to zero and the Doppler centroid f_{dc} becomes zero. This yields $t_c(\mathbf{p}) = t_0$ and no offset in azimuth direction under this situation. Finally, it is noted that extra Doppler offsets caused by the Earth's rotation or target motion should also be taken into account, if we consider the case of a spaceborne platform or a dynamic scene.

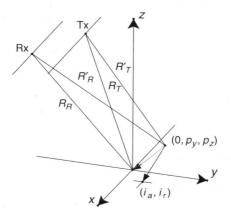

Figure 4.3 The offset of focusing position. © [2008] IEEE. Reprinted, with permission, from *IEEE Transactions on Geoscience and Remote Sensing*

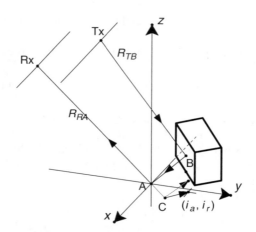

Figure 4.4 Mapping multiple scattering contributions. © [2008] IEEE. Reprinted, with permission, from *IEEE Transactions on Geoscience and Remote Sensing*

4.1.3   Multiple Scattering Terms

In comparison with the monostatic case, multiple scattering terms have different scattering paths under the bistatic configuration. Only double and triple scatterings between terrain objects and the underlying ground are considered here. Following the scattered ray from terrain object to ground, the ray tracing technique is adopted to locate the ground patch where multiple scattering occurs. An equivalent range for multiple scattering is defined as the sum of all segments of its scattering path. In addition, the positions of the scatterers that directly interact with Tx and Rx, that is, the first and last segments of its scattering path, are the main factors determining the final focusing position of multiple scattering.

As depicted in Figure 4.4, suppose that the incident wave emitted from Tx first arrives at B and then, probably after several times of scattering, arrives at A and finally is scattered back to Rx. The entire delay during this path determines its range, that is, $2R = R_{TB} + R_{BA} + R_{AR}$, where R_{BA} denotes the total length of scattering path from B to A. As viewed from Rx, it is seen equivalently as a scattering direct from a certain point C lying on the line of sight from Rx to A with the same range $2R = R_{TC} + R_{CR}$.

Select its reference point as the original mapping position of C without focusing offset. The Doppler history of multiple scattering terms can be estimated similarly, but with $\mathbf{R}_R$ replaced by $\mathbf{R}_{RA}$, and $\mathbf{R}_T$ by $\mathbf{R}_{TB}$. In the case of the TI configuration, the offset of the focusing position with respect to the reference point can be computed from Equations 4.12 and 4.10. In our experience, such offset is actually very small, for example, a few pixels, under the TI configuration.

4.2   Scattering Models and Signal Model

4.2.1   Models of Terrain Objects

The VRT model of non-spherical particles and cylinders (Jin, 1994) is employed for modeling the vegetation canopy. Leaves are modeled as Rayleigh–Gans approximated non-spherical dielectric particles. Trunks are modeled as finite-length dielectric cylinders. The shapes

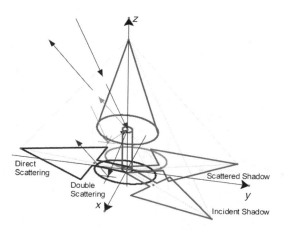

Figure 4.5 Composition of a tree in BISAR image. © [2008] IEEE. Reprinted, with permission, from *IEEE Transactions on Geoscience and Remote Sensing*

of tree crowns are assumed as oval or conical with random perturbation. It is noted that interactions between the trunks and ground surface might be overestimated in monostatic circumstances as described in Chapter 3, because the trunks were treated as absolutely upright and with no surface roughness (El-Rouby, Ulaby, and Nashashibi, 2002; Fung, 1994). In our bistatic model, the trunks are assumed to be a little tilted, for example, broad-leaf trees have tilt angles of 0–10°, and needle-leaf trees 0–5°.

Figure 4.5 explains the bistatic imaging process of a tree. One of the differences between the bistatic and monostatic cases is the shadowing area projected in both incidence and scattering directions. Additionally, owing to the split incidence and scattering directions, the double scattering terms of object–ground and ground–object are different. One reflects at the ground and diffusely scatters from the object, whereas the other scatters at the object and then reflects from the ground. These two scattering terms have different paths, which give rise to separate double scattering images. Nevertheless, it is understandable that such separate images might not be clearly observed in the real BISAR image because, as the scattering angles change, double scattering becomes weakly diffused between those two main directions.

Buildings are treated as cuboid objects with convex or flat roofs. Surface scatterings from the wall, roof, and wall–ground are major concerns in our study. All surface scatterings, from either rough or smooth surfaces, are computed by using the integral equation method (IEM) (Fung, 1994) with both incoherent and coherent parts. The composition of a building in the BISAR image is illustrated in Figure 4.6. Since the wall surfaces are usually regarded as smooth, the path of wall–ground scattering should follow reflection on the wall surface and scattering on the ground surface. Note that, owing to the bistatic configuration, the double scattering term becomes weaker than for the monostatic case, even when the wall faces straight toward the sensor.

4.2.2 Raw Signal Model for BISAR

In the VRT model, the effect of coherent superposition of radar echoes is not incorporated. Coherent speckle can be simulated via randomly generating the complex scattering matrix from the computed Mueller matrix of each pixel, in the same way as in monostatic case of Chapter 3.

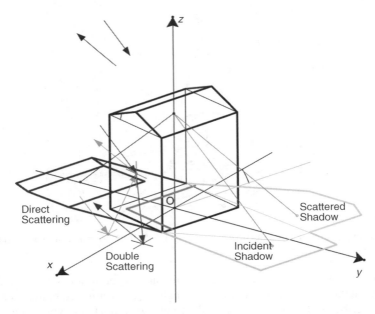

Figure 4.6 Composition of a house in BISAR image. © [2008] IEEE. Reprinted, with permission, from *IEEE Transactions on Geoscience and Remote Sensing*

The BISAR image is different from monostatic SAR images mainly in the following two aspects. One is the different scattering signature and imaging feature due to the special angular configuration. The other includes complicated signal collection, compression, image distortion, and speckle property, which are beyond the scope of this book. Hence, our raw signal simulation is limited only to the TI situation.

In the case of TI configuration, both Tx and Rx have the same velocity vectors and cooperate to perform stripmap imaging. The coordinate system adopted in following discussions is defined as follows. The origin is at the scene center. The xOy plane is the imaging plane. The platforms fly heading in the $+x$ direction. Hence, the position of Rx and Tx at a fixed moment can be denoted as (x_R, y_R, z_R) and (x_T, y_T, z_T). However, the scattering map obtained by MPA is originally defined in the grid of azimuthal position versus range distance, that is, $\gamma(x, r)$. In order to calculate the geometric relationships between the sensors and scatterers, we need to determine the corresponding position in the xOy plane of each scattering pixel (x, r). This is actually similar to the conversion of slant range to ground range in the monostatic case.

In the TI case, the expression of one-way range R in Equation 4.7 can be written in a simpler form, that is,

$$2R(x, r) = \sqrt{(x_R - x)^2 + [y_R + y(r)]^2 + z_R^2} + \sqrt{(x_T - x)^2 + [y_T + y(r)]^2 + z_T^2} \quad (4.13)$$

where (x_R, y_R, z_R) and (x_T, y_T, z_T) stand for the Rx and Tx positions at zero moment. Note that function $y(r)$ is indeed the slant-range to ground-range conversion. It is defined as the y coordinate (ground-range) of the scattering pixel $\gamma(0, r)$ (slant-range is r), which is governed by the following relationship:

$$2R(0, r) = \sqrt{x_R^2 + y_R^2 + z_R^2} + \sqrt{x_T^2 + y_T^2 + z_T^2} + 2r \tag{4.14}$$

Considering the received raw signal $s(t_n, t')$ in the form of fast and slow time $t = t' + t_n$, it can be equivalently expressed in terms of the spatial distances on range and azimuth dimensions, that is, $s(x, r)$, $r = ct'/2$, $x = vt_n$. Given the scattering map $\gamma(x, r)$ and the emitted pulse, the received raw signal $s(x, r)$ can be expressed as

$$s(x', r') = \iint dx\, dr\, \gamma(x, r)\varpi_R(x' - x)\varpi_T(x' - x)\Pi(r' - r - \Delta R)\exp(-j\phi)$$

$$\phi = \frac{4\pi}{\lambda}[R(0, 0) + r + \Delta R] + \frac{4\pi K}{c^2}(r' - r - \Delta R)^2 \tag{4.15}$$

$$\Delta R = R(x' - x, r) - R(0, 0) - r$$

where ϖ_R, ϖ_T denote the gain functions of Rx, Tx antennas, the function Π denotes a rectangular window function over $[-c\tau/4, c\tau/4]$, τ is the pulse width, and K is the chirp rate.

Consequently, the expression of the raw signal in the frequency domain can be written as

$$S(\xi, \eta) = \iint dx'\, dr'\, s(x', r')\exp(-jx'\xi - jr'\eta)$$

$$= \iint dx\, dr\, \gamma(x, r)\exp(-jx\xi - jr\eta)G(r, \xi, \eta) \tag{4.16}$$

where

$$G(r, \xi, \eta) = \iint dp\, dq\, \varpi_R(p)\varpi_T(p)\Pi(q - \Delta R)\exp(-j\phi)\exp(-jp\xi - jq\eta)$$

$$\phi = \frac{4\pi}{\lambda}[R(0, 0) + r + \Delta R] + \frac{4\pi K}{c^2}(q - \Delta R)^2 \tag{4.17}$$

$$\Delta R = R(p, r) - R(0, 0) - r$$

The integral of the range dimension can be evaluated directly, while the integral of the azimuth dimension can be approximately computed by expanding Rx/Tx Doppler terms up to second order on their own stationary phase points. It finally yields (Xu and Jin, 2009)

$$S(\xi, \eta) = G^0(\xi, \eta) \cdot \Gamma'\left(\xi, \frac{\xi\rho}{2\tan\vartheta}\right) \tag{4.18}$$

$$\Gamma'(\xi, \eta) = \iint dx\, dr\, \gamma(x, r)\sqrt{\frac{(\ell_R^0 + \ell_R^1 r)(\ell_T^0 + \ell_T^1 r)}{\ell_R^0 + \ell_T^0 + \ell_R^1 r + \ell_T^1 r}}\exp(-jx\xi - jr\eta) \tag{4.19}$$

$$G^0(\xi, \eta) = \frac{c\tau}{2} \sqrt{\frac{-4j\pi \tan \vartheta}{\xi \cos^2 \vartheta}} \cdot d\left(\frac{c\eta}{2\pi K\tau}\right) \varpi_R(\tilde{p}) \varpi_T(\tilde{p})$$

$$\cdot \exp\left[j \frac{c^2 \eta^2}{16\pi K} + j\eta R(0,0)\right]$$

$$\cdot \exp\left[-\frac{j}{2}\xi\left((\ell_R^0 + \ell_T^0)/\tan\vartheta + (x_R + x_T)\right)\right]$$

$$\cdot \exp\left\{-\frac{j}{4\tan\vartheta} \frac{\xi}{\ell_R^0 + \ell_T^0} \frac{[\cos\vartheta(x_R - x_T) - \sin\vartheta(\ell_R^0 - \ell_T^0)]^2}{\ell_R^0 + \ell_T^0}\right\} \tag{4.20}$$

where

$$\vartheta = \sin^{-1}[\xi/(4\pi/\lambda + \eta)]$$

$$\tilde{p} = \frac{\ell_T x_R + \ell_R x_T}{\ell_R + \ell_T} - \frac{2\tan\vartheta \ell_T \ell_R}{\ell_R + \ell_T}$$

$$\rho = \frac{\ell_R^1 + \ell_T^1}{2} + \frac{[\cos\vartheta(x_R - x_T) - \sin\vartheta(\ell_R^0 - \ell_T^0)]}{4(\ell_R^0 + \ell_T^0)^2}$$

$$\cdot [\cos\vartheta(\ell_R^1 + \ell_T^1)(x_R - x_T) + \sin\vartheta(\ell_R^1 \ell_R^0 - \ell_T^1 \ell_T^0 + 3\ell_R^1 \ell_T^0 - 3\ell_T^1 \ell_R^0)] \tag{4.21a}$$

$$\ell_i = \sqrt{[y_i + y(r)]^2 + z_i^2} \approx \ell_i^0 + \ell_i^1 r, \quad i = R, T$$

$$\ell_i^0 = \sqrt{y_i^2 + z_i^2}$$

$$\ell_i^1 = \frac{\sqrt{x_R^2 + y_R^2 + z_R^2}\sqrt{x_T^2 + y_T^2 + z_T^2}}{y_R\sqrt{x_T^2 + y_T^2 + z_T^2} + y_T\sqrt{x_R^2 + y_R^2 + z_R^2}} \frac{2y_i}{\sqrt{y_i^2 + z_i^2}} \tag{4.21b}$$

Note that ℓ_i has been approximated by a linear expression of r due to the fact that $r \ll \ell_i$. In Equation 4.20, $d(\cdot)$ can be computed in terms of Fresnel integrals (Franceschetti and Schirinzi, 1990) and, in general, we have

$$d\left(\frac{c\eta}{2\pi K\tau}\right) \approx \Pi_0\left(\frac{c\eta}{4\pi K\tau}\right) \tag{4.22}$$

where Π_0 is the unit rectangular window on $[-0.5, 0.5]$. It can be easily proved that, in the monostatic circumstance, Equation 4.18 will reduce to the form given in Franceschetti and Schirinzi (1990).

Note that the dependence of $\varpi_i(\tilde{p})$ on r has been omitted in Equation 4.22. In fact, $\varpi_i(\tilde{p})$ only confines the Doppler bandwidth in the azimuth dimension. Hence, varying r in $\varpi_i(\tilde{p})$ does not impact the key of compression. Subsequently, the range dependence of G^0 is extracted. Based on Equations 4.18–4.20, the raw signal can be efficiently generated from the scattering map pre-calculated by MPA. Namely, first perform a 2D fast Fourier transform (FFT) on the scattering map and then interpolate to realize the variable transform. After that, multiply the range-independent point target response G^0. And finally obtain the raw signal via 2D inverse fast Fourier transform (IFFT).

Since the approximation of second-order expansion is adopted in this derivation, it is restrained by its validity conditions (Loffeld *et al.*, 2004):

$$\left| \frac{\ell_T x_R + \ell_R x_T}{\ell_R + \ell_T} - \frac{2\tan\theta\,\ell_T\ell_R}{\ell_R + \ell_T} -x_i+\tan\theta\,\ell_i \right|^2 \ll 9|\tan\theta\,\ell_i|^2$$

$$\left| \tan\theta\,(\ell_i^0 + \ell_i^1 r) \right|^2 \ll \frac{2}{7}|\ell_i^0|^2$$

(4.23)

The first inequality comes from the second-order expansion of ϕ_i on their respective stationary phase points. It restricts the squint looking angles of Tx/Rx. The second inequality originates from the stationary phase method, which confines the size of scene with respect to the range distance. Note that the first condition greatly limits the valid range. In order to avoid this limitation, we can directly apply the stationary phase method to ϕ_i without expansion. Consequently, a more accurate iterative form is achieved as

$$G(r, \xi, \eta) = \frac{c\tau}{2} d\left(\frac{c\eta}{2\pi K\tau}\right) \exp\left[j\frac{c^2\eta^2}{16\pi K} + j\eta R(0,0) + jnr \right]$$

$$\cdot\, \varpi_R(\tilde{p}^{(n)})\varpi_T(\tilde{p}^{(n)}) \exp\left[-j\left(\frac{4\pi}{\lambda} + \eta\right) R(\tilde{p}^{(n)}, r) - j\tilde{p}^{(n)}\xi \right]$$

$$\cdot\, \sqrt{\frac{4\pi}{j\left(\frac{4\pi}{\lambda} + \eta\right)\left[\frac{1}{R_R(\tilde{p}^{(n)}, r)}\left(1 - \frac{(x_R - \tilde{p}^{(n)})^2}{R_R^2(\tilde{p}^{(n)}, r)}\right) + \frac{1}{R_T(\tilde{p}^{(n)}, r)}\left(1 - \frac{(x_T - \tilde{p}^{(n)})^2}{R_T^2(\tilde{p}^{(n)}, r)}\right) \right]}}$$

(4.24)

where the iteration is defined as

$$\tilde{p}^{(n+1)} = I(\tilde{p}^{(n)})$$

$$\tilde{p} = \frac{x_R R_T(\tilde{p}, r) + x_T R_R(\tilde{p}, r) - 2\xi\eta'^{-1} R_T(\tilde{p}, r) R_R(\tilde{p}, r)}{R_T(\tilde{p}, r) + R_R(\tilde{p}, r)} = I(\tilde{p})$$

(4.25)

The initial value of $\tilde{p}$ can be taken according to Equation 4.21a. However, Equation 4.24 has relevance to r that was not extracted. It calls for blockwise processing in the range dimension so as to keep r almost constant during each block.

In summary, the raw signal is generated according to Equation 4.18 through FFT operations, and Equation 4.24 is an option in case of the failure of Equation 4.18 caused by violation of the first condition in Equation 4.23 in a configuration with a large looking angle.

It should be noted that, before raw signal generation, a frequency-band limitation on the scattering map must be performed to prevent extra frequency components entering into the discrete simulated signal. That might cause overlapping and further blur the simulated SAR image.

The last step is to compress the raw signal into the SAR image. A number of focusing methods, for example, Natroshvili *et al.* (2006) and Ender, Walterscheid, and

Brenner (2004), can be used. As a matter of fact, compression is the inverse process of raw signal generation. In Section 4.6, the corresponding focusing method based on the raw signal model is proposed as well as an extended range Doppler method for the TI configuration.

It is noted that our approach is different from other approaches, for example, Thirion and Dahon (2006), in two aspects, namely, the scattering calculation and raw signal generation. Those other approaches require very time-consuming calculations to obtain coherent scattering within a wide frequency band at a number of different observation positions. The coherent scattering model itself probably possesses a very high level of complexity.

In this chapter, we employ the VRT statistical model and do not need to generate the exact and random terrain objects. The incoherent scattering map can be efficiently calculated in association with MPA manipulation. Then, we generate random coherent scattering from the incoherent scattering. After that, we use the frequency-domain method to generate the raw signal.

The major advantage of our approach is efficiency, and it is suitable for terrain objects with statistical scattering models. For a 1000×1000 grid scene (see Figure 4.12), it requires one hour to simulate one BISAR image and the required time will not increase too fast as the scene size increases, because of the first-order complexity of the MPA. However, the shortcoming is restricted validity in some circumstances, for example, large synthetic aperture length or imaging very large man-made targets.

4.3   Simulated BISAR Images

All the simulations are carried out under following three configurations:

(a) The monostatic configuration ("MS") with incidence angle 45°.
(b) The across-track configuration ("AT") with incidence angle (45°, 0°) and scattering angle (25°, 180°).
(c) The general translational invariant configuration ("TI") with incidence angle of (46°, 15°) and scattering angle (28°, 150°).

The parameters of the platforms and sensors are given in Table 4.1. The default parameters of the scattering models of the terrain objects are given in Table 4.2.

Table 4.1   Parameters of platforms and sensors for simulation

	Platform position (m)	Sensor specification	
(1) MS	Rx: (0, 3000, 3000)	Pulse width	5 μs
	Tx: (0, 3000, 3000)	Bandwidth	150 MHz
(2) AT	Rx: (0, 1400, 3000)	Sampling rate	180 MHz
	Tx: (0, 3000, 3000)	Range resolution	1 m
(3) TI	Rx: (−800, 1400, 3000)	Craft velocity	100 m s^{-1}
	Tx: (800, 3000, 3000)	Synthetic aperture	≈ 102 m
	Receiver (Rx), Transmitter (Tx)	PRF	120 Hz
	Carrier frequency 5.31 GHz	Azimuth resolution	1 m

Table 4.2 Parameters of scattering models[a]

Soil surface		$h = 0.57\,\text{cm}, L = 6.0\,\text{cm}, \varepsilon = 8 + 1i$
Road surface		$h = 0.10\,\text{cm}, L = 4.0\,\text{cm}, \varepsilon = 3 + 0.05i$
Water surface		$h = 0.01\,\text{cm}, L = 2.0\,\text{cm}, \varepsilon = 70 + 20i$
Building	Wall	$h = 0.08\,\text{cm}, L = 10.0\,\text{cm}, \varepsilon = 3 + 0.05i$
	Roof	$h = 0.20\,\text{cm}, L = 8.0\,\text{cm}, \varepsilon = 7 + 0.1i$
	Sizes	Length 25 m, Width 8 m, Height 8 m, Roof height 2.5 m
Needle-leaf tree	Leaf	Length 6.0 cm, Radius 0.1 cm, $\varepsilon = 19.6 + 8.7i$
		Orientation [0, 80]; [0–360]
		Fractional volume 0.001
	Crown	Cone-like, Radius 2.5–4 m, Height 10–16 m
	Trunk	Length 4 m, Radius 0.2 m, $\varepsilon = 13.1 + 4.1i$
Broad-leaf tree	Leaf	Thickness 0.02 cm, Radius 2.5 cm, $\varepsilon = 21.3 + 9.1i$
		Orientation [0, 60]; [0–360]
		Fractional volume 0.0016
	Crown	Ellipsoid-like, Radius 3–5 m, Height 6–10 m
	Trunk	Length 3 m, Radius 0.125 m, $\varepsilon = 11.1 + 3.1i$
Grass		Length 5 cm, Radius 0.1 cm, $\varepsilon = 36.1 + 9.2i$
		Orientation [0, 10]; [0–360]
		Layer depth 5 cm
		Fractional volume 0.005

[a] h and L denote the root-mean-square height and correlation length of the rough surface, respectively. Orientation [0, 80]; [0–360] means the two Euler angles of the particle are uniformly distributed among [0–80°] and [0–360°]. The grid size of the scene is 0.1 m × 0.1 m.

Let us look at a simple scene with one needle-leaf tree and one broad-leaf tree standing on flat soil ground. The simulated SAR images of this scene under the three configurations are displayed in Figure 4.7. Terrain objects appear distinctively under different configurations, as expected; for example, the shapes and positions of their images and shadows agree with our prediction. The upper half in Figure 4.7 is the broad-leaf tree, while the lower half is the

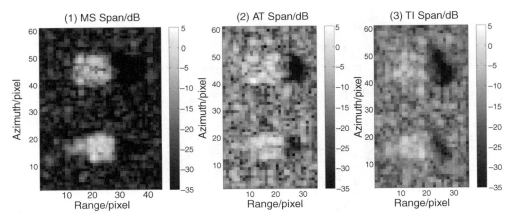

Figure 4.7 Simulated monostatic and bistatic SAR images of a two-tree scene. © [2008] IEEE. Reprinted, with permission, from *IEEE Transactions on Geoscience and Remote Sensing*

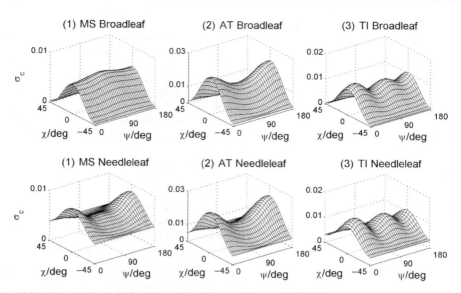

Figure 4.8   Monostatic and bistatic co-polarized signatures of different trees. © [2008] IEEE. Reprinted, with permission, from *IEEE Transactions on Geoscience and Remote Sensing*

needle-leaf tree. To study the scattering characteristics of these two trees, averaged Mueller matrices from the upper and lower halves are obtained. Co-polarized and cross-polarized signatures and polarization degrees (Jin, 1994) are presented in Figures 4.8–4.10, respectively.

In Figures 4.7 and 4.8, from monostatic case (1) to across-track (2), it can be seen that linearly co-polarized scattering from the background surface is enhanced because of the small

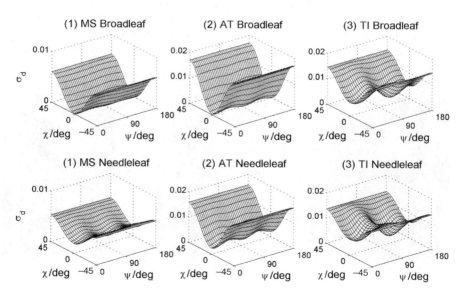

Figure 4.9   Monostatic and bistatic cross-polarized signatures of different trees. © [2008] IEEE. Reprinted, with permission, from *IEEE Transactions on Geoscience and Remote Sensing*

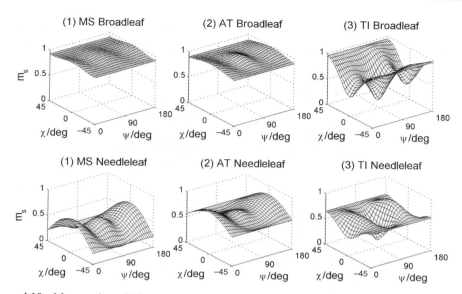

Figure 4.10 Monostatic and bistatic polarization degrees of different trees. © [2008] IEEE. Reprinted, with permission, from *IEEE Transactions on Geoscience and Remote Sensing*

observation angle. From across-track (2) to general TI (3), as the bistatic angle deviates away from the same incidence plane, the peak of the co-polarized signature also deviates away from the position of $\psi = 0°, 90°$. The depolarization effect of the needle-leaf tree is more obvious than for the broad-leaf tree, which is observed from its smaller polarization degree. The reason is due to the larger degree of asymmetry of needle-leaf than broad-leaf tree. Generally, depolarization becomes more apparent in the sequence from MS to AT and to TI cases.

The second example is a scene of two houses on a road surface with slightly different orientations, $0°$ and $15°$. The house size is a $12\,\text{m} \times 8\,\text{m} \times 10\,\text{m}$ body plus a $2.5\,\text{m}$ height convex roof. The simulated SAR images of the three configurations are shown in Figure 4.11. As we can see,

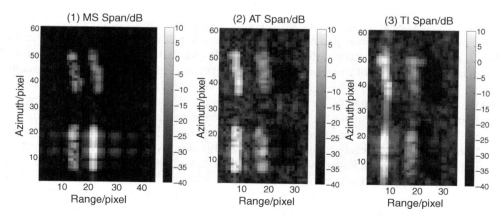

Figure 4.11 Simulated monostatic and bistatic SAR images of a two-building scene. © [2008] IEEE. Reprinted, with permission, from *IEEE Transactions on Geoscience and Remote Sensing*

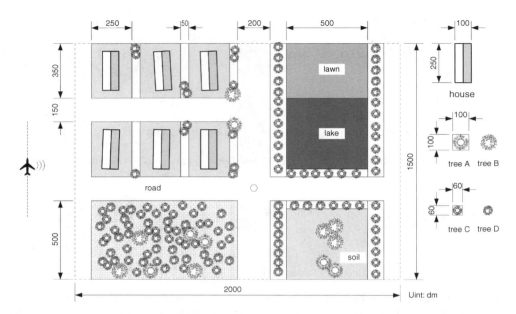

Figure 4.12 Deployment sketch of a comprehensive virtual scene used for simulation. © [2008] IEEE. Reprinted, with permission, from *IEEE Transactions on Geoscience and Remote Sensing*

the most prominent scatterings come from the nearside roof and wall–ground interaction. However, the wall–ground term is reduced significantly in the bistatic case, especially for the oriented house. These two types of scatterings are too strong so that they will cover all other scatterings. The averaged polarimetric signatures are governed by these two types of scattering.

A comprehensive virtual scene is designed for further simulation and analysis. The deployment sketch is drawn proportionally in Figure 4.12. It is a scenario of a residential district with trees, lawns, houses, roads, and a water area. The trees A, C are needle-leaf and trees B, D are broad-leaf. Their parameters are listed in Table 4.2. Note that the trees are randomly generated without overlapping.

The simulated SAR images of the three configurations are shown in Figure 4.13. To be more intuitive, the SAR images are rotated by 90° with respect to the scene, that is, the radar looks from up to bottom. The wall–ground interaction seems much stronger in the monostatic case, while the roof scattering becomes stronger in bistatic cases. The wall–ground term becomes very weak in the general TI case. On the other hand, the terrain objects with randomly distributed scatterers, for example, tree canopy, are less likely to be affected by the configuration of bistatic angles. Note that the performance of the focusing method might not be good in the general TI case and may lead to the light blurring of the image.

As an intuitive comparison, real monostatic and bistatic SAR images copied from Walterscheid *et al.* (2006) and Ender, Walterscheid, and Brenner (2004) are shown in Figure 4.14. In November 2003, the Research Institute for High-Frequency Physics and Radar Techniques performed an airborne experiment of BISAR (Walterscheid *et al.*, 2006; Ender, Walterscheid, and Brenner, 2004) using the AER-II and PAMIR systems. Several configurations of across-track mode were tested. Figure 4.14b is monostatic and Figure 4.14c is across-track with a 13° bistatic angle. No significant difference of background scattering can be found

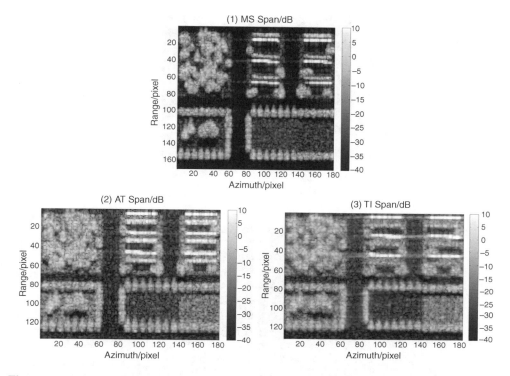

Figure 4.13 Simulated monostatic and bistatic SAR images of the scene in Figure 4.12. © [2008] IEEE. Reprinted, with permission, from *IEEE Transactions on Geoscience and Remote Sensing*

from these two images because of the close angular settings of these two configurations. Besides, these two images are focused with different methods, which might be a reason why the BISAR image seems clearer (Ender, Walterscheid, and Brenner, 2004). In the monostatic image, buildings appear darker because the wall–ground interactions are reduced for most oriented buildings. Comparing simulated and real SAR images, the features of terrain objects seem similar. Comparison of scattering intensities or polarimetric signatures has not been conducted due to the lack of real SAR data.

4.4 Polarimetric Characteristics of BISAR Image

Without multi-look processing or spatial averaging, each pixel in a polarimetric SAR image is a 2×2 scattering matrix $\bar{S}$ with four polarization components, namely, $S_{hh}, S_{hv}, S_{vh}, S_{vv}$ (under backscatter alignment convention). Using Pauli decomposition, the scattering vector is written as (Cloude and Pottier, 1996)

$$
\mathbf{k}_P = \frac{1}{\sqrt{2}} \begin{bmatrix} S_{hh} + S_{vv} \\ S_{hh} - S_{vv} \\ S_{hv} + S_{vh} \\ iS_{hv} - iS_{vh} \end{bmatrix}
\tag{4.26}
$$

(a) Optical image

(b) Monostatic SAR image   (c) Bistatic (across-track) SAR image

Figure 4.14   Real BISAR images (radar looks up–down). All adapted from Walterscheid *et al.* (2006) and Ender, Walterscheid, and Brenner (2004). © [2008] IEEE. Reprinted, with permission, from *IEEE Transactions on Geoscience and Remote Sensing*

After multi-look processing or spatial averaging, the coherency matrix is defined as (Cloude and Pottier, 1996)

$$\bar{\mathbf{T}} = \left\langle \mathbf{k}_P \cdot \mathbf{k}_P^+ \right\rangle \tag{4.27}$$

where the superscript $+$ denotes the conjugate transpose. The four eigenvalues of $\bar{\mathbf{T}}$ are used to define the entropy H, which represents the randomness of the objects. Following Cloude and Pottier (1996), the angles α, β, γ are defined as the ratios of four components of $\mathbf{k}_P$, that is,

$$\mathbf{k}_P = \|\mathbf{k}_P\| \cdot \begin{bmatrix} \cos\alpha \, e^{j\phi_1} \\ \sin\alpha \cos\beta \, e^{j\phi_2} \\ \sin\alpha \sin\beta \cos\gamma \, e^{j\phi_3} \\ \sin\alpha \sin\beta \sin\gamma \, e^{j\phi_4} \end{bmatrix} \tag{4.28}$$

In monostatic circumstances, the fourth component of $\mathbf{k}_P$, $k_{P,4}$, tends to zero and $\bar{\mathbf{T}}$ reduces to a 3×3 matrix. Moreover, α varying as $0° \rightarrow 45° \rightarrow 90°$ indicates the scattering mechanism changing from surface scattering to volumetric scattering to double scattering. The β value indicates the orientation of the object. In bistatic circumstances, Cloude (2006) also provides a new explanation of H, α.

As shown in Figure 4.15, the H, α values are computed from the simulated images. Here a 3×3 boxcar is used to perform the spatial average. In the monostatic SAR image, the H, α

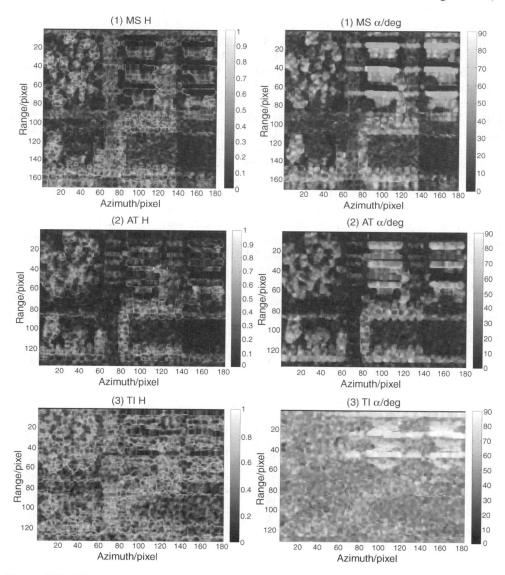

Figure 4.15 H, α maps of simulated monostatic and bistatic SAR images. © [2008] IEEE. Reprinted, with permission, from *IEEE Transactions on Geoscience and Remote Sensing*

values of different terrain surfaces agree well with the conventional interpretation. Specifically, H is high and $\alpha \rightarrow 45°$ in the forest area, while $\alpha \rightarrow 90°$ for double scattering of the buildings.

In the across-track BISAR image, basically H, α have similar appearances as in the monostatic case. This means that, under different angles of incidence and scattering but within the same incidence plane, bistatic scattering of terrain objects preserves polarimetric characteristics analogous to the monostatic case, while the major disparity is in total scattering power.

However, the behavior of the polarimetric parameters is different in the image of general TI mode. First, H is generally higher, probably due to the fact that a large bistatic angle is more likely to reveal the randomness and complexity of the object, and makes the scattering energy more uniformly distributed on different polarizations. Second, almost all terrain surfaces show $\alpha > 45°$. It is found that the scattering energy is concentrated on the fourth component of $\mathbf{k}_P$ all over the image.

It seems that α might lose its capability to represent the polarimetric characteristics under certain bistatic configurations. We may assume that the polarimetric information might be transferred to other parameters like β, γ. The β, γ of the TI case are calculated as shown in Figure 4.16. Different scattering mechanisms can be identified, somewhat more clearly in β, γ channels than in α. But, generally speaking, all α, β, γ cannot reflect polarimetric characteristics in the general TI case as clearly as in the monostatic case.

As a conclusion, conventional polarimetric parameters may largely depend on the angular settings of the sensors rather than on the inherent properties of the target. It becomes inconvenient to interpret the physical meaning of polarimetric parameters when they are severely involved with the bistatic configuration. How to extract the dependence on angular settings from polarimetric parameters is one of the fundamental issues in bistatic polarimetry.

As the coherency matrix $\bar{\mathbf{T}}$ extends to 4×4 in the bistatic case, none of the four Pauli components of $\mathbf{k}_P$ is negligible. It is inappropriate to code red–green–blue (RGB) pseudo-color images using either the first three Pauli components or the three polarizations HH, VV, and HV. Hence, in this work, the Pauli components are normalized by the total scattering power, that is,

$$P_i = |k_{P,i}|/\|\mathbf{k}_P\|, \quad i = 1, \dots, 4 \tag{4.29}$$

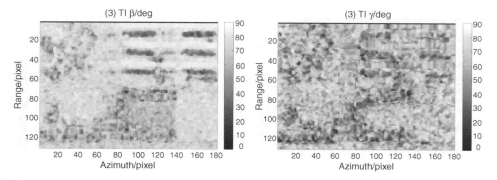

Figure 4.16 β, γ maps of simulated bistatic (TI) SAR image. © [2008] IEEE. Reprinted, with permission, from *IEEE Transactions on Geoscience and Remote Sensing*

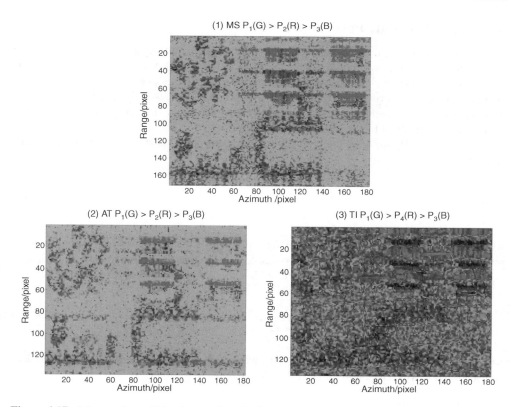

Figure 4.17 Monostatic and bistatic pseudo-color images coded by the normalized Pauli components. © [2008] IEEE. Reprinted, with permission, from *IEEE Transactions on Geoscience and Remote Sensing*

Clearly, only three of the four normalized Pauli components (NPCs) are independent. Consequently, the three largest NPCs (in the sense of averaging all over the image) are selected as the RGB channels to synthesize pseudo-color images. This kind of pseudo-color image is expected to represent the polarimetric characteristics of terrain objects in BISAR images.

The NPC color-coded images of the three configurations are shown in Figure 4.17, with the corresponding RGB channel labeled in brackets after each NPC. Both the monostatic and across-track cases have the sequence $P_1 > P_2 > P_3$, while, in the general TI case, P_4 takes the position of P_2, and the features of terrain objects also show much more diversity and randomness.

Under bistatic observation, the balance of scattering energy among the different polarization channels is simultaneously influenced by multiple independent factors, namely, the spatial orientation of the object, the different scattering mechanisms, and the angular setting of illumination and reception. This gives rise to disordered behaviors of conventional polarimetric parameters such as α, β, γ in images of the asymmetric bistatic configuration, such as the general TI case. If we redefine the polarization bases of incident and scattered waves, make them align with the corresponding bistatic geometry and thus achieve as much symmetry as possible, the bistatic polarimetric information can become more obvious and ordered, like the reciprocal and symmetric forms of coherency matrix given in Cloude (2005).

4.5 Unified Bistatic Polarization Bases

As proposed in many studies, inverse problems of bistatic scattering are usually discussed in a coordinate system determined by the bisectrix (Cloude, 2005). We believe that it is important to first transform the conventional bistatic polarization bases to a new one defined by the bisectrix, which is referred to as "unified bistatic polarization bases" because both polarization bases of the incident and scattered waves are consistently defined in parallel with the coordinates determined by the bistatic configuration.

As depicted in Figure 4.18, the conventional incident polar basis is defined in *xyz* coordinates as

$$\hat{h}_i = \frac{\hat{z} \times \hat{k}_i}{|\hat{z} \times \hat{k}_i|}, \quad \hat{v}_i = \hat{h}_i \times \hat{k}_i \tag{4.30}$$

where $\hat{k}_i$ denotes the incidence direction. The subscript i denotes incidence. The scattering polar basis has the same form but with i replaced by s, so that only the expressions with i are given in the following statements. The unified bistatic polarization bases are defined as

$$\hat{h}'_i = \frac{\hat{b} \times \hat{k}_i}{|\hat{b} \times \hat{k}_i|}, \quad \hat{v}'_i = \hat{h}'_i \times \hat{k}_i \tag{4.31}$$

where $\hat{b}$ denotes the bisectrix (Cloude, 2005), which is defined as the bisector of the incident and scattered wavevectors in the plane, that is,

$$\hat{b} = \frac{\hat{k}_s - \hat{k}_i}{|\hat{k}_s - \hat{k}_i|} \tag{4.32}$$

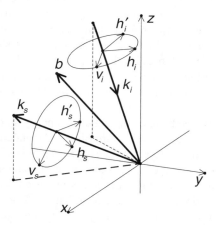

Figure 4.18 Unified bistatic polarization bases. © [2008] IEEE. Reprinted, with permission, from *IEEE Transactions on Geoscience and Remote Sensing*

The relationship between the unified bistatic polarization bases and the conventional polarization bases can be described by polarization basis rotation. Suppose the electric fields $\mathbf{E}_i$ and $\mathbf{E}'_i$ are defined in (v_i, h_i) and (v'_i, h'_i), respectively. We have

$$\mathbf{E}'_i = \bar{\mathbf{U}}_i \cdot \mathbf{E}_i \tag{4.33}$$

where $\bar{\mathbf{U}}_i$ is the rotation matrix defined as

$$\bar{\mathbf{U}}_i = \begin{bmatrix} \hat{v}_i \cdot \hat{v}'_i & \hat{h}_i \cdot \hat{v}'_i \\ \hat{v}_i \cdot \hat{h}'_i & \hat{h}_i \cdot \hat{h}'_i \end{bmatrix} \tag{4.34}$$

Obviously, the transpose $\bar{\mathbf{U}}_i^{\mathrm{T}}$ is also the inverse of $\bar{\mathbf{U}}_i$. Given the incident angle $(\pi - \theta_i, \varphi_i)$ and scattered angle (θ_s, φ_s), the elements of $\bar{\mathbf{U}}_i$ and $\bar{\mathbf{U}}_s$ can be derived as

$$
\begin{aligned}
U_{i11} = U_{i22} &= \frac{\sin^2 \theta_i \cos \theta_s + \sin \theta_i \cos \theta_i \sin \theta_s \cos (\varphi_i - \varphi_s)}{|A \sin \theta_i|} \\
U_{i21} = -U_{i12} &= \frac{\sin \theta_i \sin \theta_s \sin (\varphi_s - \varphi_i)}{|A \sin \theta_i|} \\
A^2 &= |\sin \theta_i \cos \theta_s + \cos \theta_i \sin \theta_s \cos (\varphi_i - \varphi_s)|^2 + \sin^2 \theta_s \sin^2 (\varphi_s - \varphi_i) \\
U_{s11} = U_{s22} &= \frac{\sin^2 \theta_s \cos \theta_i + \sin \theta_s \cos \theta_s \sin \theta_i \cos (\varphi_s - \varphi_i)}{|B \sin \theta_s|} \\
U_{s21} = -U_{s12} &= \frac{\sin \theta_s \sin \theta_i \sin (\varphi_s - \varphi_i)}{|B \sin \theta_s|} \\
B^2 &= |\sin \theta_s \cos \theta_i + \cos \theta_s \sin \theta_i \cos (\varphi_s - \varphi_i)|^2 + \sin^2 \theta_i \sin^2 (\varphi_i - \varphi_s)
\end{aligned}
\tag{4.35}
$$

Consequently, the relationship between the scattering matrix $\bar{\mathbf{S}}'$ under the unified bistatic polarization bases and the original scattering matrix $\bar{\mathbf{S}}$ can be expressed as

$$
\begin{aligned}
\mathbf{E}_s &= \bar{\mathbf{S}} \cdot \mathbf{E}_i, \ \mathbf{E}'_s = \bar{\mathbf{S}}' \cdot \mathbf{E}'_i \\
\mathbf{E}'_s &= \bar{\mathbf{U}}_s \cdot \mathbf{E}_s = \bar{\mathbf{U}}_s \cdot \bar{\mathbf{S}} \cdot \mathbf{E}_i = \bar{\mathbf{U}}_s \cdot \bar{\mathbf{S}} \cdot \bar{\mathbf{U}}_i^{\mathrm{T}} \cdot \mathbf{E}'_i \\
\bar{\mathbf{S}}' &= \bar{\mathbf{U}}_s \cdot \bar{\mathbf{S}} \cdot \bar{\mathbf{U}}_i^{\mathrm{T}}
\end{aligned}
\tag{4.36}
$$

This defines a transform of the scattering matrix from the conventional polarization bases to the unified bistatic polarization bases. Following Equations 4.35, it can be seen that $\bar{\mathbf{U}}_i$, $\bar{\mathbf{U}}_s$ reduce to zero in the monostatic case, which means the unified bistatic polarization bases do not work. In the across-track case, $\bar{\mathbf{U}}_i$, $\bar{\mathbf{U}}_s$ reduce to unit matrices with signs, which means that this transform only changes the sign of the scattering matrix, rather than the polarimetric parameters. Hence, only the BISAR image under the general TI configuration is regarded for the transform of unified bistatic polarization bases in the following.

The NPC color-coded image after transformation is displayed in Figure 4.19. Note that the RGB coding sequence is adapted so that its appearance agrees with the convention of human vision.

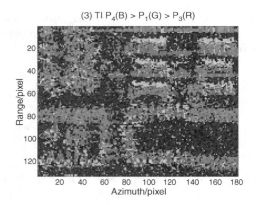

Figure 4.19 Bistatic (TI) pseudo-color images coded by the normalized Pauli components after unified bistatic polarization bases transform. © [2008] IEEE. Reprinted, with permission, from *IEEE Transactions on Geoscience and Remote Sensing*

First of all, it can be seen that P_4 occupies the first position. This is due to the fact that a certain degree of orientation persists for most of the ground surface under the observation of unified bistatic polarization bases, which leads to strong cross-polarization. So a bare ground surface appears blue with large P_4.

Second, the tree canopy appears green with large P_1, because totally random particles always maintain somewhat balanced energy among different polarizations.

Last, but not least, the red regions (large P_3) indicate double scatterings. It is known that the double scattering term gives rise to a large value of P_2 (Xu and Jin, 2005) when facing straight toward to the radar. However, the scattering energy can be transferred to P_3 if there is a certain degree of orientation (Xu and Jin, 2005). That is the reason why the double scattering that appeared darker (large P_2) in Figure 4.17 becomes red (large P_3) in Figure 4.19. This reveals a potential problem of bistatic radar polarimetry, that is, target orientation might be mixed with other information and become incapable of interpreting polarimetric parameters. Since the unified bistatic polarization bases have already fixed the reference coordinates of rotation, it becomes even more difficult to extract target orientation dependence from bistatic polarimetric scattering. This issue remains for further study.

As observed from the energy redistribution of the four Pauli components in Figure 4.19, it can be seen that α, β, γ still cannot well represent the different characteristics of terrain types. In fact, two patterns can be concluded from Figures 4.17 and 4.19, namely, the energy of single scattering is shifted between $P_1 \leftrightarrow P_4$ as the target rotates, while double scattering is shifted between $P_2 \leftrightarrow P_3$. As a result, the original definition of α (Cloude and Pottier, 1996), which preserves the roll-invariant property in the monostatic case ($P_4 = 0$), loses this virtue in the bistatic case. We modify the definition of Cloude's parameters as

$$\mathbf{k}_P = \|\mathbf{k}_P\| \cdot \begin{bmatrix} \cos\alpha \cos\gamma \, e^{j\phi_1} \\ \sin\alpha \cos\beta \, e^{j\phi_2} \\ \sin\alpha \sin\beta \, e^{j\phi_3} \\ \cos\alpha \sin\gamma \, e^{j\phi_4} \end{bmatrix} \qquad (4.37)$$

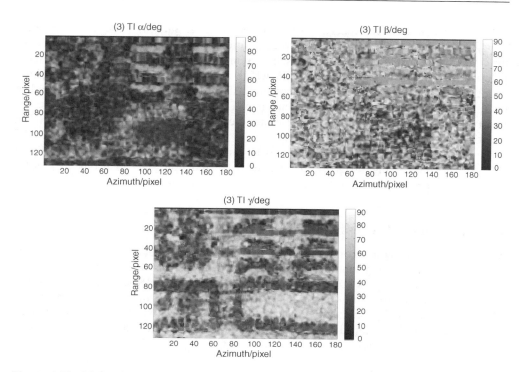

Figure 4.20 Maps of redefined α, β, γ after unified bistatic polarization bases transform. © [2008] IEEE. Reprinted, with permission, from *IEEE Transactions on Geoscience and Remote Sensing*

where the scattering vector $\bar{k}_p$ can also be the eigenvectors of the coherency matrix (Cloude and Pottier, 1996). Clearly, the new definition in Equation 4.37 is consistent with the original one in Equation 4.28 in the monostatic case. The redefined α, β, γ are calculated and plotted in Figure 4.20. Comparing with Figures 4.15 and 4.16, the redefined α, β, γ are more capable of representing the polarimetric characteristics of different scattering types. As α varies from $0° \rightarrow 45° \rightarrow 90°$, it indicates single scattering, volumetric scattering, and double scattering. The parameters β and γ indicate the orientations of double and single scattering, respectively.

Redefinition of Cloude's parameters is only a preliminary attempt of bistatic polarimetric interpretation. It remains open for further study in a systematic way and for verification using both simulation and experiments.

Through three-dimensional projection and mapping, the MPA approach for bistatic SAR imaging simulation is developed. The bistatic scattering coefficient map is calculated and produced for a complex natural scene, which contains inhomogeneously distributed penetrable objects, discontinuously protruding man-made buildings, randomly fluctuated rough ground surfaces, and so on. In the light of the explicit expression of BISAR raw signal with the range dependence extracted, the raw data can be efficiently generated from the scattering map pre-calculated by the MPA. Finally, the simulated BISAR image is obtained via BISAR focusing. Examples of complex natural scenes have been designed for BISAR imaging simulation under different bistatic configurations. The imaging characteristics of terrain objects have been investigated based on the simulated BISAR image and compared with the real BISAR image.

It is found that conventional polarimetric parameters, such as Cloude's parameters α, β, γ, might lose their ability to interpret different scattering mechanisms, because of the influence of bistatic angular settings. To circumvent this problem, the unified bistatic polarization bases transform is proposed, which aligns both polarization bases of incidence and scattering to the bistatic geometry. The parameters α, β, γ are redefined to retain the property of orientation independence. The simulated images show that the terrain types can be well identified by these modified α, β, γ after the transform of unified bistatic polarization bases.

We believe that simulated BISAR raw data can provide the samples for the study of BISAR signal processing, image evaluation, and data validation, and can also be helpful for designing the bistatic platform and sensors, image interpretation, and inversion of physical parameters.

4.6 Raw Signal Processing of Stripmap BISAR

Raw signal processing and compression are the critical steps to determine the SAR image quality. Some processing methods have been studied, such as the backprojection (BP) method in the time domain (Ding and Munson, 2002), the so-called dip move out (DMO) method introduced from the seismic community (D'Aria, Guarnieri, and Rocca, 2004), which is feasible for TI configuration, and some others (Ender, Walterscheid, and Brenner, 2004). Loffeld *et al.* developed a vectorial signal model for BISAR under the GC configuration (Loffeld *et al.*, 2004), and provided a corresponding focusing method (Natroshvili *et al.*, 2006).

To quantitatively understand the scattering mechanism and to illustrate the imaging principle are essential to interpretation of BISAR imagery and bio/geo-information retrieval. In order to realize BISAR imaging simulation of a natural scene, we need to first calculate the scattering map and then simulate the raw signal of BISAR, which is the exactly inverse process of focusing. Hence, it is necessary to first study the principle of the focusing method for raw signal simulation. However, it remains to be studied how to obtain an explicit form of the point target response with range dependence extracted for fast simulation of BISAR raw data.

In this section, our discussion is limited to the TI configuration of stripmap BISAR. The frequency-domain expression of the point target response is first derived by the stationary phase method up to the second order for the Doppler contributions of transmitter and receiver. Then, an approximate solution of the point target response is obtained with the range-dependent term extracted and the corresponding focusing method is provided. However, the validity of the approximate solution is restrictive under critical conditions. Thus, a more accurate iterative form with range dependence is also presented. Besides, the range Doppler method (RD method) is also extended to the TI case based on the derived point target response. Eventually, image simulation shows the feasibilities of the three different methods for BISAR processing as well as their advantages and disadvantages.

4.6.1 Approximate Form of the Point Target Response

The coordinates for the TI configuration are shown in Figure 4.21. The origin is the center of the scene. The imaging plane (scene plane) is the xOy plane, and $+x$ is the flying direction of the Rx/Tx. The bistatic configuration is completely determined by the positions of Rx/Tx at

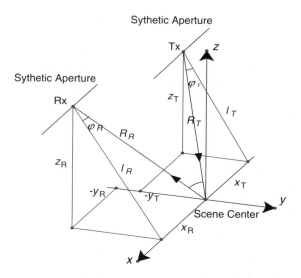

Figure 4.21   Geometry of TI configuration

their respective synthetic aperture centers, that is, (x_R, y_R, z_R), (x_T, y_T, z_T). The squint looking angles φ_R, φ_T can be written as

$$\tan \varphi_i = x_i / \ell_i^0, \quad \ell_i^0 = \sqrt{y_i^2 + z_i^2}, \quad i = R, T \tag{4.38}$$

where ℓ_R^0, ℓ_T^0 are the minimum distances to the trajectories of Rx/Tx. So the bistatic angle can be calculated similarly.

In the bistatic case, let us define a nonlinear mapping from the slant range r to the ground range y as $y(r)$. Then the one-way bistatic range R of a scatterer defined in the azimuth versus slant range grid (x, r) can be expressed as

$$2R(x, r) = \sqrt{(x_R - x)^2 + [y_R + y(r)]^2 + z_R^2} + \sqrt{(x_T - x)^2 + [y_T + y(r)]^2 + z_T^2} \tag{4.39}$$

where the slant to ground range conversion $y(r)$ is defined as the y coordinate of a scatter at $(0, r)$, that is,

$$2R(0, r) = \sqrt{x_R^2 + y_R^2 + z_R^2} + \sqrt{x_T^2 + y_T^2 + z_T^2} + 2r \tag{4.40}$$

Setting $x = 0$ in Equation 4.39 and substituting into Equation 4.40, an approximate form of $y(r)$ can be derived as

$$2r = \left(\frac{y_R}{R_R} + \frac{y_T}{R_T}\right) y + \left(\frac{R_R^2 - y_R^2}{2R_R^3} + \frac{R_T^2 - y_T^2}{2R_T^3}\right) y^2 \tag{4.41}$$

$$R_i = \sqrt{x_i^2 + y_i^2 + z_i^2}, \quad i = R, T$$

Note that both r and y are defined relative to the scene center.

After down-conversion, the raw signal $s(t)$ can be reorganized in the 2D fast time versus slow time grid. The slow time t_n is defined as the moment when Rx receives the nth pulse echo from the scene center, while the fast time $t' = t - t_n$ is defined within the receiving window of one single pulse echo. Regarding the linear relationships between the time grid (t_n, t') and the space grid (x, r), that is, $r = ct'/2$, $x = vt_n$, the point target response will be discussed in the space domain in this book.

Given the scattering map of the scene $\gamma(x, r)$, the received raw signal $s(x, r)$ can be written as

$$s(x', r') = \iint dx\, dr\, \gamma(x, r)\Omega_R(x' - x)\Omega_T(x' - x)\Pi(r' - r - \Delta R)\exp(-j\phi)$$

$$\phi = \frac{4\pi}{\lambda}[R(0,0) + r + \Delta R] + \frac{4\pi K}{c^2}(r' - r - \Delta R)^2 \tag{4.42}$$

$$\Delta R = R(x' - x, r) - R(0,0) - r$$

where Ω_R, Ω_T are the patterns of the receiving and transmitting antennas, respectively, $R(0,0) = (R_R + R_T)/2$ denotes the range of scene center, Π is a rectangular window of $[-c\tau/4, c\tau/4]$, τ is the emitted pulse width, c denotes the speed of light, and K is the chirp rate of the emitted pulse. Note that a linear frequency-modulated pulse (LFM pulse) is assumed.

The raw signal in the frequency domain can be written as

$$S(\xi, \eta) = \iint dx'\, dr'\, s(x', r')\exp(-jx'\xi - jr'\eta)$$

$$= \iint dx\, dr\, \gamma(x, r)\exp(-jx\xi - jr\eta)G(r, \xi, \eta) \tag{4.43}$$

where the point target response or point spread function is written as

$$G(r, \xi, \eta) = \iint dp\, dq\, \Omega_R(p)\Omega_T(p)\Pi(q - \Delta R)\exp(-j\phi)\exp(-jp\xi - jq\eta)$$

$$\phi = \frac{4\pi}{\lambda}[R(0,0) + r + \Delta R] + \frac{4\pi K}{c^2}(q - \Delta R)^2 \tag{4.44}$$

$$\Delta R = R(p, r) - R(0,0) - r$$

where the variable substitutions $p = x' - x$, $q = r' - r$ are applied. Similarly as in the monostatic case, the integration of q can be evaluated at first (Franceschetti and Schirinzi, 1990)

$$G(r, \xi, \eta) = \frac{c\tau}{2}d\left(\frac{c\eta}{2\pi K\tau}\right)\exp\left[j\frac{c^2\eta^2}{16\pi K} + j\eta R(0,0) + j\eta r\right]G_A(r, \xi, \eta)$$

$$G_A(r, \xi, \eta) = \int dp\, \Omega_R(p)\Omega_T(p)\exp[-j(4\pi/\lambda + \eta)R(p, r) - jp\xi] \tag{4.45}$$

where G_A denotes the azimuthal part of point target response; and $d(\cdot)$ is defined as in Franceschetti and Schirinzi (1990), and can be computed in terms of Fresnel integrals.

In most practical cases, we have

$$d\left(\frac{c\eta}{2\pi K\tau}\right) \approx \Pi_0\left(\frac{c\eta}{4\pi K\tau}\right) \tag{4.46}$$

where Π_0 denotes the unit rectangular window of $[-0.5, 0.5]$.

Following the approximation adopted in Ender, Walterscheid, and Brenner (2004), G_A is first separated into two parts contributed by the receiver and transmitter, respectively, that is,

$$G_A(r, \xi, \eta) = \iint dp\, \Omega_R(p)\Omega_T(p)\exp[-j\phi_R/2 - j\phi_T/2]$$
$$\phi_i = \eta' R_i(p, r) + \xi p, \quad i = R, T$$
$$\eta' = 4\pi/\lambda + \eta \tag{4.47}$$
$$R_i(p, r) = \sqrt{(x_i - p)^2 + [y_i + y(r)]^2 + z_i^2}$$

and then, ϕ_i, $i = R, T$, is further expanded up to second order on their own stationary phase points:

$$G_A(r, \xi, \eta) = \exp\left[-\frac{j}{2}(\phi_R(\tilde{p}_R) + \phi_T(\tilde{p}_T))\right]$$
$$\cdot \int dp\, \Omega_R(p)\Omega_T(p)\exp\left[-\frac{j}{4}\left(\ddot{\phi}_R(\tilde{p}_R)(p - \tilde{p}_R)^2 + \ddot{\phi}_T(\tilde{p}_T)(p - \tilde{p}_T)^2\right)\right] \tag{4.48}$$
$$\dot{\phi}_i(\tilde{p}_i) = \eta' \dot{R}_i(\tilde{p}_i, r) + \xi = 0$$

It is solved as

$$\tilde{p}_i = x_i - \ell_i(r)\tan\vartheta, \quad \ell_i^2(r) = [y_i + y(r)]^2 + z_i^2$$
$$\vartheta = \sin^{-1}(\xi/\eta'), \quad \ddot{\phi}_i(\tilde{p}_i) = \frac{\xi \cos^2\vartheta}{\ell_i(r)\tan\vartheta} \tag{4.49}$$

where $\tilde{p}_i$ denotes the stationary phase point, and ℓ_R, ℓ_T denote the minimum distances to the trajectories of Rx/Tx. Again, the stationary phase method (Kong, 1985) is used to evaluate the integral of Equation 4.48 and yields

$$G_A(r, \xi, \eta) = \exp\left[-\frac{j}{2}(\phi_R(\tilde{p}_R) + \phi_T(\tilde{p}_T))\right]\exp\left\{-\frac{j}{4}\frac{\ddot{\phi}_R(\tilde{p}_R)\ddot{\phi}_T(\tilde{p}_T)}{\ddot{\phi}_R(\tilde{p}_R) + \ddot{\phi}_T(\tilde{p}_T)}(\tilde{p}_R - \tilde{p}_T)^2\right\}$$
$$\cdot \Omega_R(\tilde{p})\Omega_T(\tilde{p})\sqrt{\frac{-4j\pi}{\ddot{\phi}_R(\tilde{p}_R) + \ddot{\phi}_T(\tilde{p}_T)}} \tag{4.50}$$

where the final stationary phase point is written as

$$\tilde{p} = \frac{\ddot{\phi}_R(\tilde{p}_R)\tilde{p}_R + \ddot{\phi}_T(\tilde{p}_T)\tilde{p}_T}{\ddot{\phi}_R(\tilde{p}_R) + \ddot{\phi}_T(\tilde{p}_T)} = \frac{\ell_T x_R + \ell_R x_T}{\ell_R + \ell_T} - \frac{2\tan\vartheta \ell_T \ell_R}{\ell_R + \ell_T} \tag{4.51}$$

Substituting Equation 4.49 into Equation 4.50 and making use of the linear expansion of $\ell_i(r)$, $i = R, T$,

$$\ell_i(r) \approx \ell_i^0 + \ell_i^1 r$$

$$\ell_i^0 = \sqrt{y_i^2 + z_i^2}$$

$$\ell_i^1 = \frac{\sqrt{x_R^2 + y_R^2 + z_R^2}\sqrt{x_T^2 + y_T^2 + z_T^2}}{y_R\sqrt{x_T^2 + y_T^2 + z_T^2} + y_T\sqrt{x_R^2 + y_R^2 + z_R^2}\dfrac{2y_i}{\sqrt{y_i^2 + z_i^2}}} \tag{4.52}$$

(due to $r \ll \ell_i$, this approximation of $\ell_i(r)$ is justified in most cases), it finally yields

$$G_A(r, \xi, \eta) = \Omega_R(\tilde{p})\Omega_T(\tilde{p}) \cdot \sqrt{\frac{-4j\pi \tan \vartheta}{\xi \cos^2 \vartheta}} \cdot \exp\left[-\frac{j}{2}\xi\left(\frac{(\ell_R^0 + \ell_T^0)}{\tan \vartheta} + (x_R + x_T)\right)\right]$$

$$\cdot \exp\left\{-\frac{j}{4\tan \vartheta}\frac{\xi}{\ell_R^0 + \ell_T^0}\frac{[\cos \vartheta(x_R - x_T) - \sin \vartheta(\ell_R^0 - \ell_T^0)]^2}{\ell_R^0 + \ell_T^0}\right\} \tag{4.53}$$

$$\cdot \sqrt{\frac{(\ell_R^0 + \ell_R^1 r)(\ell_T^0 + \ell_T^1 r)}{\ell_R^0 + \ell_T^0 + \ell_R^1 r + \ell_T^1 r}} \cdot \exp\left[\frac{-jr\xi\rho}{\tan \vartheta}\right]$$

where

$$\tilde{p} \approx \frac{\ell_T^0 x_R + \ell_R^0 x_T}{\ell_R^0 + \ell_T^0} - \frac{2\tan \vartheta \, \ell_T^0 \ell_R^0}{\ell_R^0 + \ell_T^0}$$

$$\rho = \frac{\ell_R^1 + \ell_T^1}{2} + \frac{\cos \vartheta(x_R - x_T) - \sin \vartheta(\ell_R^0 - \ell_T^0)}{4(\ell_R^0 + \ell_T^0)^2} \tag{4.54}$$

$$\cdot \left[\cos \vartheta(\ell_R^1 + \ell_T^1)(x_R - x_T) + \sin \vartheta(\ell_R^1 \ell_R^0 - \ell_T^1 \ell_T^0 + 3\ell_R^1 \ell_T^0 - 3\ell_T^1 \ell_R^0)\right]$$

Note that the dependence of $\Omega_i(\tilde{p})$ on r has been omitted. In fact, $\Omega_i(\tilde{p})$ only confines the Doppler bandwidth in the azimuthal dimension. Hence, varying r in $\Omega_i(\tilde{p})$ does not impact the key compression.

Reorganizing these formulas, the frequency-domain expression of the raw signal can be written as

$$S(\xi, \eta) = G^0(\xi, \eta) \cdot \Gamma'(\xi, \xi\rho/\tan \vartheta) \tag{4.55}$$

where Γ' denotes the Fourier transform of the modified scattering map γ as

$$\Gamma'(\xi, \eta) = \iint dx\, dr\, \chi(r)\gamma(x, r)\exp(-jx\xi - jr\eta)$$

$$\chi(r) = \sqrt{\frac{(\ell_R^0 + \ell_R^1 r)(\ell_T^0 + \ell_T^1 r)}{\ell_R^0 + \ell_T^0 + \ell_R^1 r + \ell_T^1 r}} \tag{4.56}$$

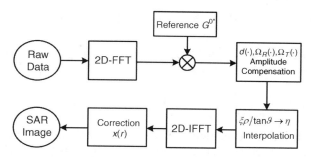

Figure 4.22   Focusing with the approximated solution of point target response

while G^0 is the point target response with range independence,

$$G^0(\xi, \eta) = \frac{c\tau}{2} \sqrt{\frac{-4j\pi\tan\vartheta}{\xi\cos^2\vartheta}} d\left(\frac{c\eta}{2\pi K\tau}\right) \Omega_R(\tilde{p})\Omega_T(\tilde{p}) \exp\left[-\frac{j}{2}\xi\left(\frac{(\ell_R^0 + \ell_T^0)}{\tan\vartheta} + (x_R + x_T)\right)\right]$$

$$\cdot \exp\left[j\frac{c^2\eta^2}{16\pi K} + j\eta R(0,0)\right] \exp\left\{-\frac{j}{4\tan\vartheta}\frac{\xi}{}\frac{[\cos\vartheta(x_R - x_T) - \sin\vartheta(\ell_R^0 - \ell_T^0)]^2}{\ell_R^0 + \ell_T^0}\right\}$$

$$(4.57)$$

Note that the window functions $d(\cdot)$, $\Omega_R(\cdot)$, $\Omega_T(\cdot)$ limit the bandwidths in the range and azimuth dimensions, as well as the values of ξ, η. Thus, it yields $|\xi|, |\eta| \ll 4\pi/\lambda$ in most cases. Under these conditions, it can be easily demonstrated that Equation 4.55 will reduce to the form of Franceschetti and Schirinzi (1990) in the monostatic case.

Based on Equations (4.55–4.57), a focusing routine is proposed, as shown in Figure 4.22. First, the raw signal is 2D FFT transformed and multiplied with the conjugate of G^0, where an amplitude compensation of $d(\cdot)$, $\Omega_R(\cdot)$, $\Omega_T(\cdot)$ is applied if necessary. Subsequently, a coordinate transform from $\xi\rho/\tan\vartheta \to \eta$ is conducted via interpolation. Eventually, the SAR image is obtained after the 2D IFFT and correction of the range-related term $\chi(r)$. This method is referred to as the approximate solution method (ASM) henceforth.

From Figure 4.22, it is found that the major computational complexity arises from the 2D FFT/IFFT steps. For $N \times M$ raw data, the complexity of ASM focusing is $O(NM\log N + NM\log M)$.

4.6.2   Validity Condition and an Iterative Solution

The validity condition of the second-order expansions on the stationary phase points is given by Ender (2004) as

$$|\tilde{p} - \tilde{p}_i|^2 \ll 9|\tilde{p}_i - x_i|^2 \ll \frac{18}{7}|\ell_i^0|^2 \tag{4.58}$$

Substituting Equations 4.49, 4.51, and 4.52 into Equation 4.58, it yields

$$\left| \frac{\ell_T x_R + \ell_R x_T}{\ell_R + \ell_T} - \frac{2 \tan \vartheta \, \ell_T \ell_R}{\ell_R + \ell_T} - x_i + \tan \vartheta \, \ell_i \right|^2 \ll 9 |\tan \vartheta \, \ell_i|^2$$

$$\left| \tan \vartheta (\ell_i^0 + \ell_i^1 r) \right|^2 \ll \frac{2}{7} |\ell_i^0|^2 \tag{4.59}$$

It is found that the second condition of Equation 4.59 originates from the last stationary phase approximation, which restricts the varying range of r, that is, the size of the scene. The first condition is due to the second-order expansion of ϕ_i. Considering the special case $\ell_R = \ell_T$, the first condition can be simplified to

$$\left| \frac{x_R}{\ell_R} - \frac{x_T}{\ell_T} \right|^2 \ll 36 |\tan \vartheta|^2 \tag{4.60}$$

which is actually restricting the squint looking angle and the bistatic angle.

Thus, the first condition makes the method invalid in the case of large angle (either squint look angle or bistatic angle). But, if the stationary phase method is directly applied to ϕ_R and ϕ_T of Equation 4.47 without expansion, this restraint can be circumvented, that is,

$$\dot{\phi}_R(\tilde{p}) + \dot{\phi}_T(\tilde{p}) = \eta' \dot{R}_R(\tilde{p}, r) + \eta' \dot{R}_T(\tilde{p}, r) + 2\xi = 0 \tag{4.61}$$

It yields

$$\tilde{p} = \frac{x_R R_T(\tilde{p}, r) + x_T R_R(\tilde{p}, r) - 2 \sin \vartheta \, R_T(\tilde{p}, r) R_R(\tilde{p}, r)}{R_T(\tilde{p}, r) + R_R(\tilde{p}, r)} = I(\tilde{p}) \tag{4.62}$$

Define an iteration $\tilde{p}^{(n+1)} = I(\tilde{p}^{(n)})$, where the initial value is taken according to Equation 4.54,

$$\tilde{p}^{(0)} = \frac{\ell_T x_R + \ell_R x_T}{\ell_R + \ell_T} - \frac{2 \tan \vartheta \, \ell_T \ell_R}{\ell_R + \ell_T} \tag{4.63}$$

It yields the iterative form of the integral

$$G_A(r, \xi, \eta) = \sqrt{\frac{-j4\pi}{4\pi/\lambda + \eta}} \frac{\Omega_R(\tilde{p}^{(n)})\Omega_T(\tilde{p}^{(n)}) \exp\left[-j(4\pi/\lambda + \eta)R(\tilde{p}^{(n)}, r) - j\tilde{p}^{(n)}\xi\right]}{\sqrt{\frac{1}{R_R(\tilde{p}^{(n)}, r)}\left(1 - \frac{(x_R - \tilde{p}^{(n)})^2}{R_R^2(\tilde{p}^{(n)}, r)}\right) + \frac{1}{R_T(\tilde{p}^{(n)}, r)}\left(1 - \frac{(x_T - \tilde{p}^{(n)})^2}{R_T^2(\tilde{p}^{(n)}, r)}\right)}} \tag{4.64}$$

Substituting Equation 4.64 into Equation 4.45, the point target response is obtained, with which the raw signal can be compressed in the frequency domain by directly applying 2D FFT/IFFT operations. Since it depends on the range r, blockwise processing in the range

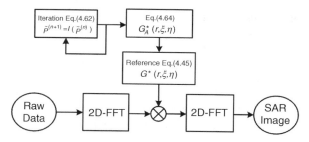

Figure 4.23   Iterative solution: focusing with the iterative solution of point target response

dimension is required to keep r nearly constant within each block. This will reduce the processing efficiency. This approach is referred to as the iterative solution method (ISM) henceforth. The flowchart of the ISM is shown in Figure 4.23. Given $N \times M$ raw data, the complexity of ISM focusing will be

$$O\left[\frac{N}{N_B} N_0 M (\log N_0 + \log M + n_I)\right] \qquad (4.65)$$

where N_B is the focusing depth, that is, the range size of one block, N_0 is the minimum length for range compression of one block, and n_I is the number of iterations.

Comparing with Ender, Walterscheid, and Brenner (2004) and Natroshvili *et al.* (2006), the approximate form in Equation 4.55 is independent of r, and can be realized quickly in the frequency domain. However, it suffers from the constraint of Equation 4.58. The iterative form in Equation 4.64 can somewhat improve the accuracy and validity in a wider range, but with a lower efficiency.

4.6.3   Extension of Range Doppler Method

The range Doppler method (RD method) is one of the classical methods for SAR signal processing. Its focusing procedure intuitively illustrates the principle of SAR imaging. The derivation of the point target response in this work is helpful to extend RD to the BISAR TI configuration. The extension needs recalculation of Doppler rate and Doppler centroid and modification of the range migration correction (RMC) coefficient.

Under the TI configuration, the Doppler history is rewritten as

$$\begin{aligned} f_d &= -\frac{1}{\lambda} \frac{2\partial R(x=vt,r)}{\partial t} = f_{dR} + f_{dT} \\ f_{di} &= -\frac{1}{\lambda} \frac{\partial R_i(vt,r)}{\partial t} = \frac{v}{\lambda} \frac{x_i - vt}{R_i(vt,r)}, \quad i = R,T \end{aligned} \qquad (4.66)$$

Expanding f_{di} up to the second order, this yields

$$f_{di} = \frac{v}{\lambda} \frac{x_i}{R_i(0,r)} - \frac{v^2 t}{\lambda R_i(0,r)} \frac{\ell_i^2(r)}{R_i^2(0,r)} \qquad (4.67)$$

Substituting Equation 4.67 into Equation 4.66 yields

$$f_d = f_{dc} + K_d t$$

$$f_{dc} = \frac{v}{\lambda}\left[\frac{x_R}{R_R(0,r)} + \frac{x_T}{R_T(0,r)}\right] \tag{4.68}$$

$$K_d = -\frac{v^2}{\lambda}\left[\frac{\ell_R^2(r)}{R_R^3(0,r)} + \frac{\ell_T^2(r)}{R_T^3(0,r)}\right]$$

where f_{dc} is the Doppler centroid and K_d is the Doppler rate.

In order to calculate the range migration in the Doppler domain, Equation 4.45 is rewritten as

$$G(r,\xi,\eta) = G_R(p,r,\eta)\int dp\, \Omega_R(p)\Omega_T(p)\exp[-jR(p,r)4\pi/\lambda - jp\xi]$$

$$G_R(p,r,\eta) = \frac{c\tau}{2}d\left(\frac{c\eta}{2\pi K\tau}\right)\exp\left[j\frac{c^2\eta^2}{16\pi K} + j\eta R(0,0) + j\eta r - j\eta R(p,r)\right] \tag{4.69}$$

Then, carry out the range compression, that is, the inverse FT of $G_R(p,r,\eta)$ on η after phase compensation,

$$G(r,\xi,r') = g_R[r' - r - R_m(p,r)]\iint dp\, \Omega_R(p)\Omega_T(p)\exp[-jR(p,r)4\pi/\lambda - jp\xi]$$

$$g_R(r) \approx \text{sinc}(2\pi K\tau r/c) \tag{4.70}$$

$$R_m(p,r) = R(p,r) - R(0,0)$$

It can be seen that the integral can be evaluated via the stationary phase method and the process is solved in the same way as Equation 4.47 with η' replaced by $4\pi/\lambda$. Thus, the range migration in the Doppler domain can be obtained as

$$R_m(\tilde{p}',r) = R(\tilde{p}',r) - R(0,0)$$

$$\tilde{p}' = \frac{\ell_T x_R + \ell_R x_T}{\ell_R + \ell_T} - \frac{2\ell_T\ell_R}{\ell_R + \ell_T}\frac{\lambda\xi}{4\pi} \tag{4.71}$$

and the RMC coefficient is

$$C_0(\xi,r) = \frac{R_m(\tilde{p}',r)}{R(0,0)} = \frac{R(\tilde{p}'(\xi),r)}{R(0,0)} - 1 \tag{4.72}$$

In monostatic case, it reduces to

$$C_0(\xi,r) = \sqrt{1 + \left(\frac{\lambda\xi}{4\pi}\right)^2} - 1 \approx \frac{1}{\sqrt{1 - (\lambda\xi/4\pi)^2}} - 1 \tag{4.73}$$

which is exactly the RMC of the monostatic RD method. Note that the derived RMC is more accurate than the conventional parabolic approximation.

Table 4.3 Parameter settings of platform and sensor

Trajectory of Rx (m)	$(-400 + 100t, 900, 3000)$	Trajectory of Tx (m)	$(800 + 100t, 3000, 3000)$
Pulse width	3 μs	Carrier frequency	5.31 GHz
Bandwidth	150 MHz	Synthetic aperture length	212 m
Sampling rate	180 MHz	Pulse repetition frequency	240 Hz
Range resolution	1 m	Azimuth resolution	0.5 m

With the Doppler rate, the centroid of Equation 4.68, and the RMC coefficient of Equation 4.72, RD processing of the TI case can be realized. The steps in the RD compression are explained in the next section. Last, but not least, it should be cautioned that a secondary range compression step is necessary if the range migration tends to become too large.

4.6.4 Simulation and Discussion

To simulate the BISAR image under the TI configuration, the parameters are chosen as shown in Table 4.3, where the trajectories of the transmitter and receiver are the coordinates referring to the scene center, that is, $x_i, y_i, z_i, \ i = R, T$, in meters, as functions of time t in seconds.

Figure 4.24 shows the results of each step of RD compression of a point target. After range compression, it shows the range history during one synthetic aperture length. Obviously, if the Doppler centroid is not zero, then the range migration is significant. After the step of RMC in the Doppler domain, the point target response is aligned along the azimuth direction and is finally focused to a point via azimuth compression. However, owing to the phase error caused by interpolation of RMC step and the omission of secondary range compression, the focusing effect in the azimuth direction is not as good as that in the range.

The point target focusing results of the approximate solution method (ASM) and the iterative solution method (ISM) are given in Figure 4.25. As a comparison, a time-domain focusing approach, the bistatic backprojection (BP) method, is also performed and its result is shown in Figure 4.25c. Their performances are evaluated by the peak side-lobe ratio (PSLR), integrated side-lobe ratio (ISLR), half-power width of main lobe (HMLW) and zero-power main-lobe width (MLW) as listed in Table 4.4. Note that BP is performed in the sense of ground range rather than slant range, which leads to a lower side-lobe level in the range dimension. As we can see, both ASM and ISM possess comparable performances but much higher efficiencies than the time-domain focusing method.

If the tapered window functions, such as Hamming window, are applied in the frequency domain, it is expected that the side-lobe level could be further suppressed and the main-lobe width would also be expanded. As shown in Table 4.4, the values in brackets are the corresponding results after applying the Hamming window in both azimuth and range dimensions. Note that they closely depend on the shape of the employed window function. It is actually a trade-off between the main-lobe width and the side-lobe level. In this case, it requires a twice expanded main lobe to suppress the side lobe downward by 20–30 dB.

Using the ASM and ISM to compress the simulated BISAR signal from the scene of multiple point targets, the results are displayed in Figure 4.26. Both methods achieve good focusing.

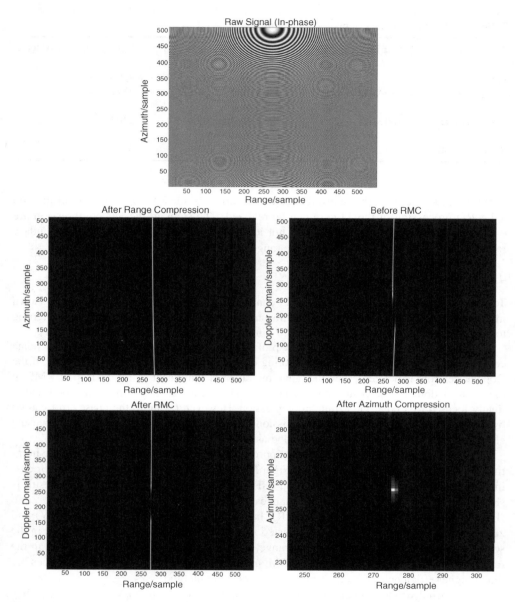

Figure 4.24 Steps of RD compression of BISAR under TI configuration (azimuth frequency offset is 86.8 Hz, frequency centroid is located at 439.5 sample). Reprinted from *International Journal of Remote Sensing*, from Taylor and Francis

However, if the bistatic angle increases, that is, if we let the azimuth biases of the receiver and transmitter be $x_R = -800$, $x_T = 2000$, the processed images are shown in Figure 4.27. It can be seen that, owing to violation of the condition of Equation 4.58, ASM leads to invalid and bad focusing. In the ISM result, large offsets of focusing positions can be observed,

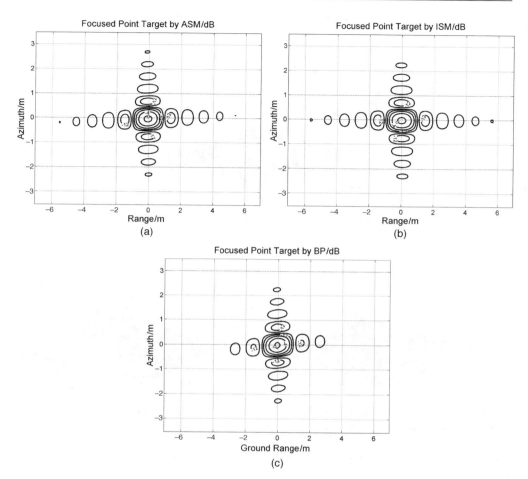

Figure 4.25 Point target responses of BISAR under TI configuration (contour lines). Reprinted from *International Journal of Remote Sensing*, from Taylor and Francis

Table 4.4 Performances of focusing[a] using approximate solution method (ASM), iterative solution method (ISM), and backprojection method (BP)

	ASM		ISM		BP	
	Range	Azimuth	Range	Azimuth	Range	Azimuth
PSLR (dB)	−13.26	−12.99	−13.20	−13.32	−15.84	−13.25
	(−39.32)	(−37.79)	(−41.78)	(−40.58)	(−36.29)	(−33.41)
ISLR (dB)	−10.41	−9.96	−9.68	−9.87	−13.54	−10.26
	(−34.01)	(−32.19)	(−26.28)	(−32.69)	(−31.65)	(−32.88)
HMLW (m)	0.90 (1.30)	0.45 (0.65)	1.00 (1.40)	0.45 (0.75)	0.70 (1.40)	0.45 (0.65)
MLW (m)	2.00 (3.70)	1.00 (1.90)	2.10 (4.10)	1.00 (1.95)	2.10 (4.10)	1.00 (2.05)

[a]PSLR, peak side-lobe ratio; ISLR, integrated side-lobe ratio; HMLW, half-power main-lobe width; MLW, zero-power main-lobe width.

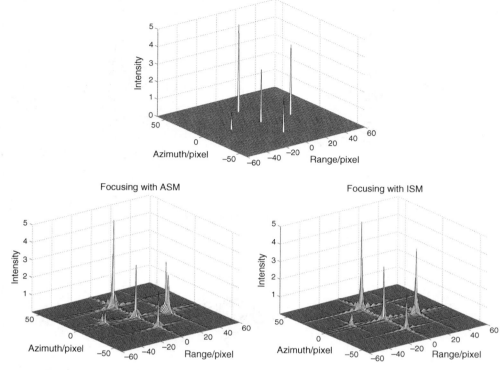

Figure 4.26   A scene of multiple point targets and its simulated images processed by ASM and ISM. Reprinted from *International Journal of Remote Sensing*, from Taylor and Francis

which is due to the intrinsic squint looking mechanism of the bistatic configuration employed in this case.

As a conclusion, the validity of ASM is constrained by the bistatic configurations, while ISM is affected by the block size of blockwise compression. A plot is given in Figure 4.28 showing

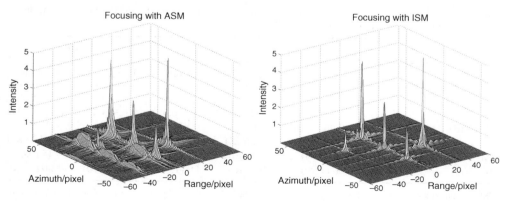

Figure 4.27   Focused images by ASM and ISM in the case of large bistatic angle. Reprinted from *International Journal of Remote Sensing*, from Taylor and Francis

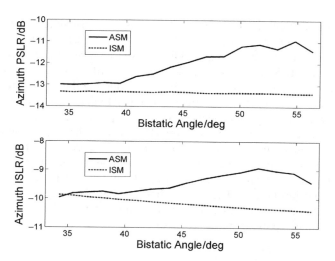

Figure 4.28 Bistatic configuration versus focusing performances. Reprinted from *International Journal of Remote Sensing*, from Taylor and Francis

the major indicator of focusing performance, that is, the azimuth PSLR and ISLR, as a function of the bistatic angle. In this case, we change the bistatic angle by linearly changing x_R from -400 to -1000 and x_T from 800 to 2600, respectively. The PSLR and ISLR are calculated from point target simulation. It can be seen that ASM tends to show poor performance as the bistatic angle exceeds $50°$, while ISM remains uninfluenced at all. Note that the bistatic angle here is only used as an indicator of different bistatic configurations. The criterion to judge whether we should choose ISM over ASM is the validity conditions given in Equation 4.57.

When we employ ISM, another issue is the selection of block size in the range dimension. As an example, a plot is given in Figure 4.29 showing the azimuth PSLR and ISLR as functions of the range offset. Range offset means the offset of the real point target position

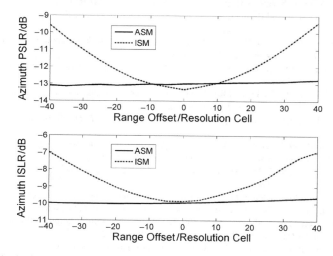

Figure 4.29 Block size versus focusing performances. Reprinted from *International Journal of Remote Sensing*, from Taylor and Francis

from the reference position. It can be seen that ASM is not sensitive to range offset, while the performance of ISM deteriorates rapidly as the range offset increases. In this case, a maximum swath width of 50 resolution cells is required for blockwise processing of ISM.

In order to realize BISAR focusing under the TI configuration, the frequency-domain expression of the point target response is derived. By adopting the approximation that Doppler terms of the receiver and transmitter are expanded up to the second order, an approximate solution (ASM) with the range dependence extracted is presented. Since the ASM condition is restrictive, a more accurate iterative solution (ISM) is also developed, which is conducted blockwisely in the range direction because of its range dependence. In terms of the expression of the point target response, the RD method is extended to the BISAR TI configuration. Simulations and comparisons prove the feasibility of these methods and explain their respective advantages and limitations. The RD method has bad focusing ability in the azimuth direction, large bistatic angle might cause ASM failure, and the ISM should be processed in the range dimension by applying block processing.

References

Brown, C.G., Sarabandi, K., and Gilgenbach, M. (2002) Physics-based simulation of high-resolution polarimetric SAR images of forested areas. Proceedings of IGARSS'02, Vol. 1, pp. 466–468.

Cloude, S.R. (2005) On the status of bistatic polarimetry theory. Proceedings of IGARSS'05, Seoul, South Korea.

Cloude, S.R. (2006) Information extraction in bistatic polarimetry. Proceedings of EUSAR'2006.

Cloude, S.R. and Pottier, E. (1996) A review of target decomposition theorems in radar polarimetry. *IEEE Transactions on Geoscience and Remote Sensing*, **34**(2), 498–518.

D'Aria, D., Guarnieri, A.M., and Rocca, F. (2004) Focusing bistatic synthetic aperture radar using dip move out. *IEEE Transactions on Geoscience and Remote Sensing*, **42**(7), 1362–1376.

Ding, Y. and Munson, D.C. (2002) A fast backprojection algorithm for bistatic SAR imaging. Proceedings of International Conference of Image Processing, Rochester, pp. 441–444.

Ebner, H., Riegger, S., Hajnsek, I. *et al.* (2004) Single-pass SAR interferometry with a Tandem TerraSAR-X configuration. Proceedings of EUSAR'04.

El-Rouby, A.E., Ulaby, F.T., and Nashashibi, A.Y. (2002) MMW scattering by rough lossy dielectric cylinders and tree trunks. *IEEE Transactions on Geoscience and Remote Sensing*, **40**(4), 871–879.

Ender, J.H.G. (2004) A step to bistatic SAR processing. Proceedings of EUSAR'04, Ulm, Germany, pp. 359–364.

Ender, J.H.G., Walterscheid, I., and Brenner, A.R. (2004) New aspects of bistatic SAR: processing and experiments. Proceedings of IGARSS'04, pp. 758–1762.

Franceschetti, G. and Schirinzi, G. (1990) A SAR processor based on two-dimensional FFT codes. *IEEE Transactions on Aerospace Electronic Systems*, **26**(2), 356–365.

Franceschetti, G., Migliaccio, M., and Riccio, D. (1995) The SAR simulation: an overview. Proceedings of IGARSS'95, Vol. 3, pp. 2283–2285.

Franceschetti, G., Iodice, A., Riccio, D., and Ruello, G. (2003) SAR raw signal simulation for urban structures. *IEEE Transactions on Geoscience and Remote Sensing*, **41**(9), 1986–1994.

Fung, A.K. (1994) *Microwave Scattering and Emission Models and Their Applications*, Artech House, Norwood, MA.

Jin, Y.Q. (1994) *Electromagnetic Scattering Modeling for Quantitative Remote Sensing*, World Scientific, Singapore.

Kong, J.A. (1985) *Electromagnetic Wave Theory*. EMW Publishing, Cambridge, MA.

Krieger, G. and Moreira, A. (2005) Multistatic SAR satellite formations: potentials and challenges. Proceedings of IGARSS'05, Vol. 4, pp. 2680–2684.

Liang, P., Pierce, L.E., and Moghaddam, M. (2005) Radiative transfer model for microwave bistatic scattering from forest canopies. *IEEE Transactions on Geoscience and Remote Sensing*, **43**(11), 2470–2483.

Loffeld, O., Nies, H., Peters, V., and Knedlik, S. (2004) Models and useful relation for bistatic SAR processing. *IEEE Transactions on Geoscience and Remote Sensing*, **42**(10), 2031–2038.

McLaughlin, D., Wu, Y., Stevens, W.G. *et al.* (2002) Fully polarimetric bistatic radar scattering behavior of forested hills. *IEEE Transactions on Antennas and Propagation*, **50**(2), 101–110.

Moccia, A., Rufino, G., D'Errico, M. *et al.* (2002) BISSAT: a bistatic SAR for Earth observation. Proceedings of IGARSS'02.

Natroshvili, K., Loffeld, O., Nies, H. *et al.* (2006) Focusing of general bistatic SAR configuration data with 2-D inverse scaled FFT. *IEEE Transactions on Geoscience and Remote Sensing*, **44**(10), 2718–2727.

Thirion, L. and Dahon, C. (2006) Modeling of the polarimetric bistatic scattering of a forested area at P-band. Proceedings of EUSAR'06.

Walterscheid, I., Brenner, A.R., Ender, J.H.G., and Loffeld, O. (2006) Bistatic SAR processing and experiments. *IEEE Transactions on Geoscience and Remote Sensing*, **44**(10), 2710–2717.

Wendler, M., Krieger, G., Horn, R. *et al.* (2003) Results of a bistatic airborne SAR experiment. Proceedings of IRS (International Radar Conference), Dresden, Germany, pp. 247–253.

Wills, N.J. (1991) *Bistatic Radar*, Artech House, Norwood, MA.

Xu, F. and Jin, Y.Q. (2005) Deorientation theory of polarimetric scattering targets and application to terrain surface classification. *IEEE Transactions on Geoscience and Remote Sensing*, **43**(10), 2351–2364.

Xu, F. and Jin, Y.Q. (2006) Imaging simulation of polarimetric synthetic aperture radar for comprehensive terrain scene using the mapping and projection algorithm. *IEEE Transactions on Geoscience and Remote Sensing*, **44** (11), 3219–3234.

Xu, F. and Jin, Y.Q. (2008) Imaging simulation of bistatic synthetic aperture radar (SAR) and its polarimetric analysis. *IEEE Transactions on Geoscience and Remote Sensing*, **46**(8), 2233–2248.

Xu, F. and Jin, Y.Q. (2009) Raw signal processing of stripmap bistatic synthetic aperture radar. *IET – Radar, Sonar and Navigation*, **3**(3), 233–244.

Yates, G., Home, A.M., Blake, A.P., and Middleton, R. (2006) Bistatic SAR image formation. *IET Proceedings on Radar, Sonar and Navigation*, **153**(3), 208–213.

Younis, M., Metzig, R., and Krieger, G. (2006) Performance prediction and verification for bistatic SAR synchronization link. Proceedings of EUSAR'06.

5

Radar Polarimetry and Deorientation Theory

Scattering modeling and SAR imaging simulation of fully polarimetric SAR have been established in the previous chapters. These forward methods will ultimately help us to understand, interpret, and retrieve the information hidden in the scattering data, response profiles, and SAR imagery. However, the fundamentals of radar polarimetry have to be introduced before we can proceed to SAR image/data and advanced applications of PolSAR remote sensing. This chapter starts with the basic concepts in radar polarimetry and target decomposition, continues with the deorientation theory, and concludes on an example application of terrain surface classification using PolSAR data.

5.1 Radar Polarimetry and Target Decomposition

5.1.1 Polarization Transformation

It is known from Chapter 1 that the polarization tip of a monochromatic single-polarization transverse electromagnetic wave follows an elliptical trajectory in the transverse plane. Its complex vectorial form can be written using elliptical polarization parameters as

$$\mathbf{E} = A e^{-j\varphi} \begin{bmatrix} \cos\psi & -\sin\psi \\ \sin\psi & \cos\psi \end{bmatrix} \begin{bmatrix} \cos\chi \\ j\sin\chi \end{bmatrix} \tag{5.1}$$

This is physically interpreted as follows: the left complex coefficient indicates absolute amplitude and phase; the middle rotation matrix transforms the right-hand side polarization vector by an angle ψ; and the right polarization vector represents an elliptical polarization that is aligned with the polarization basis. The above expression can also be written in the following form:

$$\mathbf{E} = A \begin{bmatrix} \cos\psi & -\sin\psi \\ \sin\psi & \cos\psi \end{bmatrix} \begin{bmatrix} \cos\chi & j\sin\chi \\ j\sin\chi & \cos\chi \end{bmatrix} \begin{bmatrix} e^{-j\varphi} & 0 \\ 0 & e^{j\varphi} \end{bmatrix} \begin{bmatrix} 1 \\ 0 \end{bmatrix} \tag{5.2}$$

Polarimetric Scattering and SAR Information Retrieval, First Edition. Ya-Qiu Jin and Feng Xu.
© 2013 John Wiley & Sons Singapore Pte. Ltd. Published 2013 by John Wiley & Sons Singapore Pte. Ltd.

which shows the process to transform a unit linear polarization into a general elliptical polarization via three steps.

Using matrix algebra, let us first introduce the Pauli spin matrices originating from quantum mechanics:

$$\mathbf{i} = \begin{bmatrix} 1 & 0 \\ 0 & 1 \end{bmatrix}, \quad \mathbf{j} = \begin{bmatrix} 0 & -1 \\ 1 & 0 \end{bmatrix}, \quad \mathbf{k} = \begin{bmatrix} 0 & j \\ j & 0 \end{bmatrix}, \quad \mathbf{l} = \begin{bmatrix} -j & 0 \\ 0 & j \end{bmatrix} \tag{5.3}$$

The four Pauli matrices Pauli matrix form a quaternion group via the multiplication operation, expressed as follows:

$$\begin{array}{c|cccc} \times & \mathbf{i} & \mathbf{j} & \mathbf{k} & \mathbf{l} \\ \hline \mathbf{i} & \mathbf{i} & \mathbf{j} & \mathbf{k} & \mathbf{l} \\ \mathbf{j} & \mathbf{j} & -\mathbf{i} & \mathbf{l} & -\mathbf{k} \\ \mathbf{k} & \mathbf{k} & -\mathbf{l} & -\mathbf{i} & \mathbf{j} \\ \mathbf{l} & \mathbf{l} & \mathbf{k} & -\mathbf{j} & -\mathbf{i} \end{array} \tag{5.4}$$

As will be shown later, the Pauli matrices are suitable for describing polarization vectors. According to matrix algebra (Huynen, 1970), we have the following:

$$e^{\alpha \mathbf{j}} = \cos \alpha \, \mathbf{i} + \sin \alpha \, \mathbf{j} = \begin{bmatrix} \cos \alpha & -\sin \alpha \\ \sin \alpha & \cos \alpha \end{bmatrix} = \overline{\mathbf{U}}_\psi(\alpha)$$

$$e^{\alpha \mathbf{k}} = \cos \alpha \, \mathbf{i} + \sin \alpha \, \mathbf{k} = \begin{bmatrix} \cos \alpha & j\sin \alpha \\ j\sin \alpha & \cos \alpha \end{bmatrix} = \overline{\mathbf{U}}_\chi(\alpha) \tag{5.5}$$

$$e^{\alpha \mathbf{l}} = \cos \alpha \, \mathbf{i} + \sin \alpha \, \mathbf{l} = \begin{bmatrix} e^{-j\alpha} & 0 \\ 0 & e^{j\alpha} \end{bmatrix} = \overline{\mathbf{U}}_\varphi(\alpha)$$

Thus, Equation 5.2 can be written as

$$\mathbf{E} = A\overline{\mathbf{U}}_\psi(\psi) \cdot \overline{\mathbf{U}}_\chi(\chi) \cdot \overline{\mathbf{U}}_\varphi(\varphi) \cdot \begin{bmatrix} 1 \\ 0 \end{bmatrix} = Ae^{\psi \mathbf{j}} \cdot e^{\chi \mathbf{k}} \cdot e^{\varphi \mathbf{l}} \cdot \begin{bmatrix} 1 \\ 0 \end{bmatrix} \tag{5.6}$$

Meanwhile, we have the following identities of the Pauli matrices:

$$\begin{cases} e^{(\alpha+\beta)\mathbf{j}} = e^{\alpha \mathbf{j}} \cdot e^{\beta \mathbf{j}} = e^{\beta \mathbf{j}} \cdot e^{\alpha \mathbf{j}} \\ \mathbf{k} \cdot e^{\alpha \mathbf{j}} = e^{-\alpha \mathbf{j}} \cdot \mathbf{k} \\ e^{\alpha \mathbf{j}} \cdot e^{\beta \mathbf{k}} - e^{\beta \mathbf{j}} \cdot e^{\alpha \mathbf{k}} = 2\sin \alpha \sin \beta \, \mathbf{l} \end{cases} \tag{5.7}$$

where $\mathbf{j}, \mathbf{k}, \mathbf{l}$ can be interchanged and these identities still hold. As we have noted, any polarization can be formed from a basic $[1,0]^T$ through three matrix transforms. These three matrices have the functionality to alter the corresponding parameters, as denoted in their subscripts. From the perspective of group theory, these three matrices belong to the SU(2) group

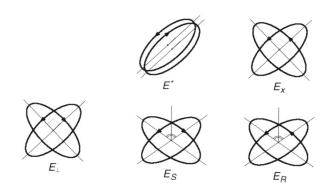

Figure 5.1 Polarization pairs

of 2×2 complex matrices (Cloude, 1986; Lopez-Martinez, Ferro-Famil, and Pottier, 2005). Although putting a matrix in the exponent makes the expressions succinct, it is not intuitive for readers. Therefore, the conventional matrix form $\overline{\mathbf{U}}_\psi(\cdot), \overline{\mathbf{U}}_\chi(\cdot), \overline{\mathbf{U}}_\varphi(\cdot)$ is employed in the rest of this chapter.

In radar polarimetry (Huynen, 1970), the polarization of an antenna is defined as the polarization of the electromagnetic wave that it radiates when used as a transmit antenna. Therefore, the antenna's polarization can be expressed in the same way as in Equation 5.1 using a polarization vector **a**. However, when an antenna is used to receive an incoming wave of different polarization **E**, it can only intercept part of the wave energy, which is measured by receiver antenna gain as

$$G = |\mathbf{a} \cdot \mathbf{E}| \tag{5.8}$$

Note that the incoming wave has opposite propagation direction, and thus its polarization vector is defined in a flipped coordinate system with respect to the antenna's system.

Given an incoming wave of polarization **E**, using antennas of different polarizations will produce different gains. As shown in Figure 5.1, several important relationships between polarizations can be summarized as follows:

(a) Co-polarization – the co-polarization of **E** is **E**.
(b) Conjugate polarization $\mathbf{E}^*$ has the same orientation but opposite rotation – the receive gain reaches a maximum $G = A^2$ when $\mathbf{a} = \mathbf{E}^*$.
(c) Cross-polarization $\mathbf{E}_x$ has the same rotation but orthogonal orientation – that is, $A_x = A$, $\varphi_{0x} = \varphi_0$, $\psi_x = \psi + \pi/2$, $\chi_x = \chi$.
(d) Orthogonal polarization $\mathbf{E}_\perp$ has the opposite rotation and orthogonal orientation – that is, $\psi_\perp = \psi + \pi/2$, $\chi_\perp = -\chi$, and the receive gain reaches zero when $\mathbf{a} = \mathbf{E}_\perp$ (note that conjugate polarization of cross-polarization is in fact orthogonal polarization).
(e) Symmetric polarization $\mathbf{E}_S$ is a flipped version of co-polarization – that is, $A_S = A$, $\varphi_{0S} = \varphi_0$, $\psi_S = -\psi$, $\chi_S = -\chi$ (symmetric polarization is a useful concept in target decomposition).
(f) Receiver polarization $\mathbf{E}_R$, is a flipped version of conjugate polarization – that is $A_R = A$, $\varphi_{0R} = -\varphi_0$, $\psi_R = -\psi$, $\chi_R = \chi$ (note that symmetric polarization of conjugate

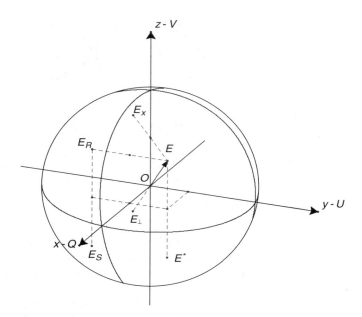

Figure 5.2 Polarization pairs on the Poincaré sphere

polarization is receiver polarization). Receiver polarization is so-called because of its physical meaning: when the receive antenna's polarization is the receiver polarization of the transmit antenna's polarization, it can intercept the maximum wave energy; this is true because of the flipped antenna coordinates between transmit and receive.

Figure 5.2 shows the geometric representation of these polarization pairs on the Poincaré sphere, which are summarized as follows: orthogonal polarizations are symmetric with respect to the origin; conjugate polarizations are reflection symmetric with respect to the xOy plane; cross-polarizations are axially symmetric with respect to the z axis; receiver polarizations are reflection symmetric with respect to the xOz plane; and symmetric polarizations are axially symmetric with respect to the x axis.

Any pair of orthogonal polarizations $(\hat{e}_x, \hat{e}_y)$ can be used to form a polarization basis, under which any polarization can be expressed in the form

$$\mathbf{E}_{(x,y)} = (\mathbf{E} \cdot \hat{e}_x)\hat{e}_x + (\mathbf{E} \cdot \hat{e}_y)\hat{e}_y = \begin{bmatrix} \mathbf{E} \cdot \hat{e}_x \\ \mathbf{E} \cdot \hat{e}_y \end{bmatrix} \tag{5.9}$$

where the subscripts denote the two bases. The most common pair of polarization bases is the linear vertical (V) and horizontal (H) polarizations. In addition, the left (L) and right (R) circular polarizations, and the rotated version of V/H are among the most used ones.

From Equation 5.6, any polarization basis can be written in the following form:

$$\begin{aligned} \hat{e}_x &= \overline{\mathbf{U}}_\psi(\psi) \cdot \overline{\mathbf{U}}_\chi(\chi) \cdot \overline{\mathbf{U}}_\varphi(\varphi) \cdot \hat{v} \\ \hat{e}_y &= \overline{\mathbf{U}}_\psi(\psi) \cdot \overline{\mathbf{U}}_\chi(\chi) \cdot \overline{\mathbf{U}}_\varphi(\varphi) \cdot \hat{h} \end{aligned} \tag{5.10}$$

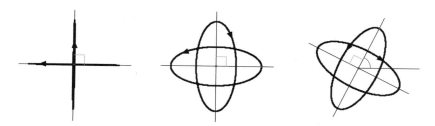

Figure 5.3   Transformation of polarization bases

Subsequently, we have

$$
\begin{aligned}
\mathbf{E}_{(x,y)} &= \left[\overline{\mathbf{U}}_\psi(\psi) \cdot \overline{\mathbf{U}}_\chi(\chi) \cdot \overline{\mathbf{U}}_\varphi(\varphi)\right]^{-1} \cdot \mathbf{E} \\
&= \overline{\mathbf{U}}_\varphi(-\varphi) \cdot \overline{\mathbf{U}}_\chi(-\chi) \cdot \overline{\mathbf{U}}_\psi(-\psi) \cdot \mathbf{E}
\end{aligned}
\tag{5.11}
$$

where $\mathbf{E}$ denotes the polarization vector under the original $(\hat{v}, \hat{h})$ basis. Clearly, it is the transformation of the polarization vector from one polarization basis to another. Figure 5.3 illustrates the process of basis transformation from linear polarization to general elliptical polarization.

As for multiple polarization or partial polarization, we have introduced the second-order statistical Stokes vector as counterpart of the polarization vector (also known as the Jones vector). In fact, the Stokes vector can also be defined from the Pauli decomposition of the second-order moment of polarization vector $\mathbf{E}$, that is

$$
2\langle \mathbf{E} \cdot \mathbf{E}^+ \rangle = I \cdot \mathbf{i} + jQ \cdot \mathbf{1} - jU \cdot \mathbf{k} + jV \cdot \mathbf{j}
\tag{5.12}
$$

where the superscript $+$ denotes conjugate transpose; the angle brackets denote the ensemble average.

In the case of completely polarized waves, the polarization vector can be one-to-one mapped to the Stokes vector. As we already know from Equation 5.6 that any polarization vector can be transformed from a unit vector via three matrix transformations, we intend to seek the similar transformation for the Stokes vector, which is found to be

$$
\mathbf{I} = A^2 \mathbf{O}_\psi(2\psi) \cdot \mathbf{O}_\chi(2\chi) \cdot \mathbf{O}_\varphi(2\varphi) \cdot \mathbf{I}_v
\tag{5.13}
$$

where $\mathbf{I}_v$ is the Stokes vector of a completely polarized wave $\mathbf{E} = [1,0]^\mathrm{T}$, while the 4×4 real matrices are defined as

$$
\mathbf{O}_\alpha(2\alpha)\big|_{n,m} = \tfrac{1}{2}\mathrm{Tr}\left[\mathbf{U}_\alpha^+(\alpha) \cdot \mathbf{p}_n \cdot U_\alpha(\alpha) \cdot \mathbf{p}_m\right], \quad \mathbf{p}_{1,2,3,4} = \mathbf{i}, j\mathbf{1}, -j\mathbf{j}, j\mathbf{k}
\tag{5.14}
$$

where the subscripts denotes the (n,m)th element, and $\mathrm{Tr}(\cdot)$ denotes trace. Therefore, the transformation matrices can be expanded as

$$\mathbf{O}_\psi(2\psi) = \begin{bmatrix} 1 & 0 & 0 & 0 \\ 0 & \cos 2\psi & -\sin 2\psi & 0 \\ 0 & \sin 2\psi & \cos 2\psi & 0 \\ 0 & 0 & 0 & 1 \end{bmatrix}$$

$$\mathbf{O}_\chi(2\chi) = \begin{bmatrix} 1 & 0 & 0 & 0 \\ 0 & \cos 2\chi & 0 & -\sin 2\chi \\ 0 & 0 & 1 & 0 \\ 0 & \sin 2\chi & 0 & \cos 2\chi \end{bmatrix} \qquad (5.15)$$

$$\mathbf{O}_\varphi(2\varphi) = \begin{bmatrix} 1 & 0 & 0 & 0 \\ 0 & 1 & 0 & 0 \\ 0 & 0 & \cos 2\varphi & -\sin 2\varphi \\ 0 & 0 & \sin 2\varphi & \cos 2\varphi \end{bmatrix}$$

Note that the absolute phase φ has disappeared after matrix multiplication in Equation 5.13, which echoes the fact that the Stokes vector has dropped absolute phase information.

Let us re-examine the Stokes vector on the Poincaré sphere as shown in Figure 5.4. It is found that these three transformation matrices represent rotation operations of the Stokes vector along three axes, respectively. Obviously, any Stokes vector can be transformed from $\hat{x}$ via two rotations.

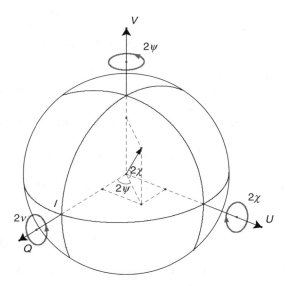

Figure 5.4 Rotation of the Stokes vector

The Stokes vector of a completely polarized wave can be represented as a vector in the 3D Poincaré space and, therefore, a similar vectorial basis transformation can be derived. However, we decided not to go further as the Stokes vector is mostly used in studies of partially polarized waves.

5.1.2 Radar Polarimetry

Polarimetric radar senses the target by transmitting a polarized wave and receiving the return signal using a polarized antenna and then deriving the signal strength and phase information. Radar polarimetry is the theory for designing the measurement method as well as interpreting the measured data of polarimetric radar.

As known from Equation 5.8, radar measurement is in fact the dot product of receive antenna polarization and returned wave polarization, while in Equation 5.9 we have shown that any polarization can be fully represented as its dot products with a pair of orthogonal polarization bases. Therefore, a pair of antennas with orthogonal polarizations has to be used in order to measure the complete polarimetric information of a returned wave. Likewise, a pair of orthogonally polarized antennas should be used to transmit a desired wave polarization.

Full polarimetry is defined as sensing a target with all possible combinations of transmit and receive polarizations. However, in practice, we only need to use one pair of orthogonally polarized antennas to obtain full polarimetric information, which will be proved in following paragraphs.

Considering the antenna polarizations $(\hat{e}_x, \hat{e}_y)$, when we use the first antenna to transmit a wave of polarization $\hat{e}_x$, the returned wave under basis $(\hat{e}_x, \hat{e}_y)$ is written as $\mathbf{E}_s = S_{xx}\hat{e}_x + S_{yx}\hat{e}_y$. Likewise, when the second antenna is used to transmit polarization $\hat{e}_y$, the returned wave is expressed as $\mathbf{E}_s = S_{xy}\hat{e}_x + S_{yy}\hat{e}_y$. Therefore, we can organize the measurements into the form of a 2×2 scattering matrix:

$$\mathbf{E}_s = \begin{bmatrix} S_{xx} & S_{xy} \\ S_{yx} & S_{yy} \end{bmatrix} \cdot \mathbf{E}_i \tag{5.16}$$

Using the polarization basis transformation in Equation 5.11, we can transform $\mathbf{E}_{(x,y)}$ into any basis $(\hat{e}'_x, \hat{e}'_y)$, that is

$$\begin{aligned} \mathbf{E}_{(x',y')} &= \overline{\mathbf{U}}_\varphi(-\varphi') \cdot \overline{\mathbf{U}}_\chi(-\chi') \cdot \overline{\mathbf{U}}_\psi(-\psi') \cdot \mathbf{E} \\ &= \overline{\mathbf{U}}_\varphi(-\varphi') \cdot \overline{\mathbf{U}}_\chi(-\chi') \cdot \overline{\mathbf{U}}_\psi(-\psi') \cdot \overline{\mathbf{U}}_\psi(\psi) \cdot \overline{\mathbf{U}}_\chi(\chi) \cdot \overline{\mathbf{U}}_\varphi(\varphi) \cdot \mathbf{E}_{(x,y)} \\ &= \overline{\mathbf{U}}_{(x',y')\leftarrow(x,y)} \cdot \mathbf{E}_{(x,y)} \end{aligned} \tag{5.17}$$

Hence, from Equation 5.16 we have

$$\begin{aligned} \mathbf{E}_{s(x',y')} &= \begin{bmatrix} S_{x'x'} & S_{x'y'} \\ S_{y'x'} & S_{y'y'} \end{bmatrix} \cdot \mathbf{E}_{i(x',y')} \\ &= \overline{\mathbf{U}}_{(x',y')\leftarrow(x,y)} \begin{bmatrix} S_{xx} & S_{xy} \\ S_{yx} & S_{yy} \end{bmatrix} \cdot \overline{\mathbf{U}}^{-1}_{(x',y')\leftarrow(x,y)} \cdot \mathbf{E}_{i(x',y')} \end{aligned} \tag{5.18}$$

which shows that full polarimetric information can be obtained from transformation of the 2×2 scattering matrix. Equation 5.18 is the basis transformation formula of scattering matrix $\mathbf{S}$.

It is now obvious that the 2×2 scattering matrix $\mathbf{S}$ bears the complete information of full polarimetry. Within one particular radar configuration, a 2×2 scattering matrix $\mathbf{S}$ fully represents a deterministic scattering target.

For a Gaussian stochastic target, it can be proved that the Mueller matrix bears the complete information of polarimetry. In fact, we showed in Section 1.1 when discussing the polarization signature surface that co-polarization and cross-polarization scattering coefficients can be derived from the Mueller matrix. Similarly, we can derive the basis transformation formulas for the Stokes vector and Mueller matrix. However, it is less useful in target decomposition and thus beyond the scope of this book.

The Mueller matrix, as a type of second-order statistical moment, has to be obtained by averaging over multiple observations. However, in SAR imaging, spatial average or multi-look processing is often used instead of repeated observations. By definition, the Mueller matrix can be expressed in terms of second-order moments of elements of the scattering matrix. Let us first vectorize the scattering matrix into a scattering vector

$$
\mathbf{k}_L = \begin{bmatrix} S_{xx} \\ S_{xy} \\ S_{yx} \\ S_{yy} \end{bmatrix}
\tag{5.19}
$$

whose covariance matrix is

$$
\overline{\mathbf{C}} = \left\langle \mathbf{k}_L \cdot \mathbf{k}_L^+ \right\rangle
\tag{5.20}
$$

Another important form of vectorization originates from Pauli decomposition, that is

$$
\mathbf{k}_P = \frac{1}{\sqrt{2}} \begin{bmatrix} S_{xx} + S_{yy} \\ S_{xx} - S_{yy} \\ S_{xy} + S_{yx} \\ -j(S_{xy} - S_{yx}) \end{bmatrix}
\tag{5.21}
$$

whose covariance matrix is often referred to as the coherency matrix, written as

$$
\overline{\mathbf{T}} = \left\langle \mathbf{k}_P \cdot \mathbf{k}_P^+ \right\rangle
\tag{5.22}
$$

Apparently, all three matrices, that is, the Mueller matrix, covariance matrix, and coherency matrix, are composed of second-order moments of scattering matrix elements, and therefore can be equivalently converted into each other.

5.1.2.1 Interconversion among Different Definitions

For historic reasons, there have been different definitions and conventions used by researchers in different areas. First of all, there is forward scattering alignment (FSA) versus backward scattering alignment (BSA). The former is often used in electromagnetic scattering and propagation studies. Therefore, the previously introduced scattering models are all defined under FSA, where the polarization basis and propagation direction construct the right-handed system $(\hat{x}, \hat{y}, \hat{k})$. However, in radar research, the returning wave is always in the opposite direction of the antenna's view. Thus, the polarization basis and antenna looking direction become a left-handed system, that is, BSA. The FSA is also referred to as the wave coordinates, while the BSA is referred to as the antenna coordinates, indicating the different references each has taken.

In the monostatic case, the only difference between the BSA and FSA is that $\hat{x}$ is flipped and thus the polarization rotation direction is also defined oppositely. The scattering matrix defined under the FSA and BSA can be simply converted as

$$\overline{\mathbf{S}}_A = \begin{bmatrix} S_{xx}^* & S_{xy}^* \\ -S_{yx}^* & -S_{yy}^* \end{bmatrix} \tag{5.23}$$

where the subscript A denotes the antenna coordinates, that is, BSA. From the above expression, it is straightforward to derive the corresponding conversion of the Mueller matrix, covariance matrix, and coherency matrix.

Another two conventions to be clarified are regarding the definition of the linear polarization basis $(\hat{v}, \hat{h})$. In many textbooks on electromagnetic theory and scattering theory (e.g., Fung, 1994; Jin, 1994; Kong, 2005; Tsang, Kong, and Shin, 1985), $(\hat{v}, \hat{h}, \hat{k})$ is defined as the wave propagation system, and this convention is followed in this book. On the contrary, much of the radar polarimetry literature (e.g., Huynen, 1970; Lopez-Martinez, Ferro-Famil, and Pottier, 2005; Cloude and Pottier, 1996; Hellmann, 2001) uses $(\hat{h}, \hat{v}, \hat{k})$. In this chapter, we use $(\hat{h}, \hat{v}, \hat{k})$ for consistency with discussions by many other authors on this subject, and indicate it simply by h/v. But generally, in all the other chapters, we adopt $(\hat{v}, \hat{h}, \hat{k})$ and indicate it by v/h. The conversion between these two is rather simple. Following Equation 5.23, the scattering matrices under these two conventions are respectively written as

$$\overline{\mathbf{S}}_{v/h} = \overline{\mathbf{S}}\big|_{x=v, y=h}, \quad \overline{\mathbf{S}}_{h/v} = \overline{\mathbf{S}}\big|_{x=h, y=v} \tag{5.24}$$

where we use subscript v/h to denote $(\hat{v}, \hat{h}, \hat{k})$ and h/v to denote $(\hat{h}, \hat{v}, \hat{k})$. Again, the corresponding conversions of the Mueller matrix, covariance matrix, and coherency matrix are straightforward.

Regarding the Stokes vector, there are two types of definitions as mentioned in Section 2.1.1, the first one, here referred to as "unified" Stokes vector, defined as I, Q; the second one, here referred to as "separated" Stokes vector, uses I_h, I_v instead. Correspondingly, we have two different versions of the Muller matrix denoted as $\mathbf{M}^{\pm}$ and $\mathbf{M}^{vh}$.

Last, but not least, the reciprocity in most backscattering scenarios is a very useful feature in target decomposition. Under backscattering reciprocity, the two cross-polarization scatterings are mutually dependent and thus the scattering matrix can be reduced to three elements and the covariance/coherency matrix reduced to 3×3. Meanwhile, the Mueller matrix is reduced to nine degrees of freedom. Therefore, all these notions have their corresponding reduced versions under backscattering reciprocity.

Table 5.1　Notations under different conventions

					h/v, ±			
	v/h				Wave coordinates (FSA)		Antenna coordinates (BSA)	
	±	vh			bistatic	monostatic	bistatic	monostatic
Scattering matrix	$\overline{\mathbf{S}}_{v/h}^{\pm}$	$\overline{\mathbf{S}}_{v/h}^{vh}$	$\overline{\mathbf{S}}_{h/v}^{vh} =$	$\overline{\mathbf{S}}$	$\overline{\mathbf{S}}^{B}$	$\overline{\mathbf{S}}_{A}$	$\overline{\mathbf{S}}_{A}^{B}$
Mueller matrix	$\overline{\mathbf{M}}_{v/h}^{\pm}$	$\overline{\mathbf{M}}_{v/h}^{vh}$	$\overline{\mathbf{M}}_{h/v}^{vh} =$	$\overline{\mathbf{M}}$	$\overline{\mathbf{M}}^{B}$	$\overline{\mathbf{M}}_{A}$	$\overline{\mathbf{M}}_{A}^{B}$
Coherency matrix	$\overline{\mathbf{T}}_{v/h}^{\pm}$	$\overline{\mathbf{T}}_{v/h}^{vh}$	$\overline{\mathbf{T}}_{h/v}^{vh} =$	$\overline{\mathbf{T}}$	$\overline{\mathbf{T}}^{B}$	$\overline{\mathbf{T}}_{A}$	$\overline{\mathbf{T}}_{A}^{B}$

All these definitions and corresponding expressions are summarized in Table 5.1, where we use superscript B to denote backscattering reciprocity. Figure 5.5 illustrates conversions among these definitions of $\mathbf{S}, \mathbf{M}, \mathbf{T}$. The dashed lines denote the boundaries of different domains, while the arrows denote conversions. The conversion equations are given in Appendix 5A.

5.1.2.2　Eigen-Analysis and Optimal Polarization

It was in the 1950s that Kennaugh (1952) first introduced the concept of optimal polarization when studying the maximized or zero reception power of an antenna. Then Huynen

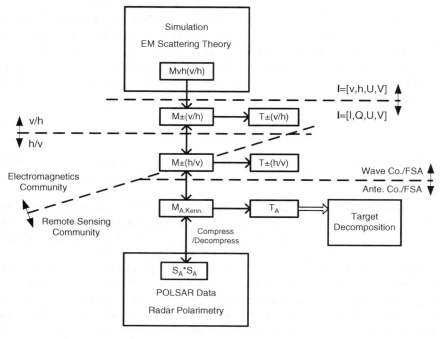

Figure 5.5　Interconversion of different conventions

(1970) further generalized it to the concept of "polarization fork" in the Poincaré sphere by using eigen-analysis of the scattering matrix and introduced the target decomposition theorem. In the 1980s, Boerner, van Zyl, and coworkers (Davidovitz and Boerner, 1986; Van Zyl, 1985) extended it to "polarization tree" and "polarization trajectory". These are just a few examples of the pioneering research that forms the fundamentals of radar polarimetry. For conciseness, we will only briefly introduce the Huynen eigen-analysis and polarization fork here.

Let us form the eigen-equation of the scattering matrix

$$\overline{\mathbf{S}} \cdot \mathbf{x} = \lambda \cdot \mathbf{x} \tag{5.25}$$

Under the backscattering reciprocity, the scattering matrix becomes a Hermitian matrix. Solving the eigen-equation yields two complex eigenvectors and eigenvalues, which can be expressed as

$$\overline{\mathbf{S}}_D = \begin{bmatrix} \lambda_1 & 0 \\ 0 & \lambda_2 \end{bmatrix} = \overline{\mathbf{U}}^{\mathrm{T}} \cdot \overline{\mathbf{S}} \cdot \overline{\mathbf{U}}, \quad \overline{\mathbf{U}} = [\mathbf{e}_1, \mathbf{e}_2] \tag{5.26}$$

where $\lambda_i, \mathbf{e}_i$, $i = 1, 2$, denote the eigenvalues and eigenvectors, respectively.

Note that Equation 5.26 resembles Equation 5.18. In fact, it can be proved (Huynen, 1970) that the eigenvector matrix $\mathbf{U}$ can be written in the form of $\mathbf{U}_{(x',y')\leftarrow(x,y)}$ as

$$\overline{\mathbf{S}}_D = \overline{\mathbf{U}}^{\mathrm{T}}(\psi, \tau, v) \cdot \overline{\mathbf{S}} \cdot \overline{\mathbf{U}}(\psi, \tau, v) \tag{5.27}$$

This implies that Equation 5.26 has the following physical meaning: under a certain polarization basis transform, any scattering matrix can be reduced to a diagonal matrix, and ψ, τ, v are the parameters of this particular basis transform. A diagonal scattering matrix means that no cross-polarization scattering exists.

Huynen further parameterized the diagonal scattering matrix as

$$\overline{\mathbf{S}}_D = m \begin{bmatrix} 1 & 0 \\ 0 & \tan^2 \gamma \end{bmatrix} e^{j2\rho} \tag{5.28}$$

Hence, a reciprocal scattering target is now parameterized into six real parameters (its original scattering matrix has six degrees of freedom). The six parameters were named by Huynen as follows:

- rotation angle ψ indicates target orientation;
- helicity angle τ indicates target asymmetry;
- hop angle v indicates multiple scattering;
- characteristic angle γ carries target intrinsic characteristics;
- absolute magnitude m is proportional to the radar cross-section;
- absolute phase ρ contains information such as target range.

The decomposition of the scattering matrix into these parameters is the core of Huynen's target decomposition theorem, upon which most later developed target decomposition theorems are more or less based.

Based on eigen-analysis, Huynen further extended Kennaugh's optimal polarization. Optimal polarizations are defined at the extrema of reception power. From Equation 5.8, radar reception power is defined as

$$G = |\mathbf{E}_R \cdot \mathbf{E}_s| = |\mathbf{E}_R^T \cdot \mathbf{E}_s| = |\mathbf{E}_R^T \cdot \overline{\mathbf{S}} \cdot \mathbf{E}_i| \qquad (5.29)$$

where $\mathbf{E}_R$ denotes the polarization of the receiver antenna. Moreover, co-polarization and cross-polarization reception power (scattering coefficient) are written as

$$\begin{cases} G_C = |\mathbf{E}_i^T \cdot \overline{\mathbf{S}} \cdot \mathbf{E}_i| \\ G_X = |\mathbf{E}_{i\perp}^T \cdot \overline{\mathbf{S}} \cdot \mathbf{E}_i| \end{cases} \qquad (5.30)$$

They are equivalent to the polarization signatures introduced in Section 1.1.2. They are functions of polarization parameters (ψ, χ). Finding the extrema of these functions yields the following characteristic polarizations:

- H point – global maximum of co-polarization (also the zero point of cross-polarization);
- V point – local maximum of co-polarization (also the zero point of cross-polarization);
- N1, N2 points – zero points of co-polarization;
- S1, S2 points – maxima of cross-polarization;
- T1, T2 points – saddle points of cross-polarization.

Now let us present these points in the Poincaré sphere, as in Figure 5.6. The following are all observed in this geometric representation. Except for N1 and N2, all the points are conjugate, that is, two ends of a diameter. HV is perpendicular to S1S2. N1 and N2 are located on

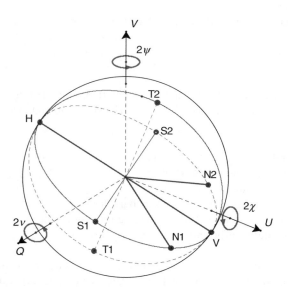

Figure 5.6   The polarization fork

the plane spanned by HV and S1S2, and they form an angle equally divided by HV. T1T2 is perpendicular to the plane spanned by HV and S1S2. It can be proved that the angle between N1 and N2 is twice the target's characteristic angle, that is, 2γ, while ψ, τ, ν are the angles that would be required to rotate HV back to the zero position along three axes, respectively.

Huynen introduced the concept of "polarization fork", assigning H as the handle, and V, N1, N2 as the tines. Boerner extended it to "polarization tree", where H is the root and all the rest are branches. Besides these, there are further concepts such as polarization trajectory (Huynen, 1970; Yang *et al.*, 2002).

5.1.2.3 Degrees of Freedom and Symmetry

Degrees of freedom (DOF) is a good indicator in radar polarimetry and target decomposition. In general, the scattering matrix has eight DOF, while the Mueller matrix and the coherency matrix of a deterministic target have seven DOF, after discarding the absolute phase by ensemble averaging. For a non-deterministic target, the Mueller matrix and coherency matrix have 16 DOF, where the Mueller matrix is a 4×4 real matrix and the coherency matrix is a 4×4 Hermitian matrix.

Under backscattering reciprocity, the scattering matrix reduces to six DOF, that is

$$\overline{\mathbf{S}}_{recip} = \begin{bmatrix} S_{hh} & S_{hv} \\ S_{hv} & S_{vv} \end{bmatrix} \tag{5.31}$$

corresponding to Huynen's six target decomposition parameters. Thus, the scattering matrix can further be vectorized as a 3D target vector:

$$\mathbf{k}_{L,recip} = \begin{bmatrix} S_{hh} \\ \sqrt{2}S_{hv} \\ S_{vv} \end{bmatrix}, \quad \mathbf{k}_{P,recip} = \frac{1}{\sqrt{2}} \begin{bmatrix} S_{hh} + S_{vv} \\ S_{hh} - S_{vv} \\ 2S_{vh} \end{bmatrix} \tag{5.32}$$

Subsequently, the covariance matrix and coherency matrix are reduced to a 3×3 Hermitian matrix with nine DOF. Meanwhile, the Mueller matrix also reduces to nine DOF though it is still a 4×4 real matrix with seven dependent equations.

Furthermore, if a backscattering target has reflection symmetry, it can be proved that its helicity angle approaches zero, that is, $\tau \to 0$. Thus, the remaining five Huynen's parameters result in five DOF in the scattering matrix. Meanwhile, it is also true that cross-polarization and co-polarization are not correlated, that is, $\langle S_{hh} \cdot S_{hv}^* \rangle = \langle S_{vv} \cdot S_{hv}^* \rangle = 0$. Thus, it reduces the Mueller matrix, covariance matrix, and coherency matrix to five DOF, which can be written in the following form:

$$\overline{\mathbf{T}}_{ref\text{-}sym} = \begin{bmatrix} a & d+je & 0 \\ d-je & b & 0 \\ 0 & 0 & c \end{bmatrix}, \quad \overline{\mathbf{C}}_{ref\text{-}sym} = \begin{bmatrix} a' & 0 & d'+je' \\ 0 & b' & 0 \\ d'-je' & 0 & c' \end{bmatrix} \tag{5.33}$$

If the target has rotation symmetry, then its rotation angle approaches zero, that is, $\psi \to 0$, and its horizontal and vertical polarizations are equivalent, which means that its characteristic angle $\gamma \to \pi/4$. Finally, we have only four Huynen's parameters left, which renders a scattering matrix with four DOF. In this case, the coherency matrix and covariance matrix have the forms:

$$\overline{\mathbf{T}}_{rot\text{-}sym} = \begin{bmatrix} a & 0 & 0 \\ 0 & b+c & j(b-c) \\ 0 & -j(b-c) & b+c \end{bmatrix} \tag{5.34a}$$

$$\overline{\mathbf{C}}_{rot\text{-}sym} = \frac{1}{2} \begin{bmatrix} a+b+c & j\sqrt{2}(b-c) & a-b-c \\ -j\sqrt{2}(b-c) & 2b+2c & j\sqrt{2}(b-c) \\ a-b-c & -j\sqrt{2}(b-c) & a+b+c \end{bmatrix} \tag{5.34b}$$

If the target has both reflection and rotation symmetry, then only three Huynen's parameters are free, while the coherency matrix and covariance matrix become

$$\overline{\mathbf{T}}_{sym} = \begin{bmatrix} a & 0 & 0 \\ 0 & b & 0 \\ 0 & 0 & b \end{bmatrix}, \quad \overline{\mathbf{C}}_{sym} = \frac{1}{2} \begin{bmatrix} a+b & 0 & a-b \\ 0 & 2b & 0 \\ a-b & 0 & a+b \end{bmatrix} \tag{5.35}$$

which means that the Mueller matrix, covariance matrix, and coherency matrix have only two DOF.

Among the above-mentioned cases, backscattering reciprocity and reflection symmetry are valid in many real applications. There are much more mathematical studies regarding polarimetry that are beyond the scope of this book, for example, studies on similarities between polarimetry and quantum mechanics (Britton, 2000) and application of group theory in polarimetry (Huynen, 1970; Cloude, 1986, 1992).

5.1.3 Target Decomposition

The fundamental task of radar polarimetry is to interpret the polarimetric scattering information regarding the target's geometric and physical properties, that is, to extract the physical parameters of the target from its scattering information, $\mathbf{S}, \mathbf{k}_P, \mathbf{k}_L, \mathbf{M}, \mathbf{C}, \mathbf{T}$. Target decomposition is the theory to solve this problem.

Target decomposition theories can be classified into two major categories: deterministic target decomposition, which focuses on the scattering matrix; and stochastic target decomposition, which focuses on the second-order statistical moment, for example, the coherency or covariance matrix.

Deterministic target decomposition includes Huynen's single-target decomposition (Huynen, 1970), Krogager's sphere–diplane–helix decomposition (Krogager, 1990), and Cameron's decomposition (Cameron and Leung, 1990; Cameron, Youssef, and Leung, 1996), to name but a few.

Stochastic target decomposition often follows two steps: first decompose into several individual deterministic targets, and then decompose each target using deterministic target decomposition. Huynen proposed the non-deterministic target decomposition

(Huynen, 1970), which decompose a partially polarized Mueller matrix into a complete polarized Mueller matrix (referred to as single-target Mueller matrix) and a noise Mueller matrix. The single-target Mueller matrix is then decomposed into Huynen's parameters.

Cloude and Pottier (1997) proposed target decomposition based on an eigen-analysis of the coherency matrix. The eigenvectors of the coherency matrix are seen as independent deterministic targets, which are weighted by its corresponding eigenvalues. Other decomposition methods include Freeman's three-component decomposition (Freeman and Durden, 1998), van Zyl's signature-analysis-based method (van Zyl, 1989), as well as multiplicative decomposition (Carrea and Wanielik, 2001; Mischenko, 1992), to name but a few. Owing to limited space, only the Cloude–Pottier decomposition is briefly introduced here.

Let us first review the important properties of the basic Pauli spin decomposition of the scattering matrix, that is

$$\overline{\mathbf{S}} = k_{P,1}\mathbf{i} + jk_{P,2}\mathbf{1} - jk_{P,3}\mathbf{k} + jk_{P,4}\mathbf{j} \tag{5.36}$$

where $k_{P,i}$ is the ith element of the Pauli target vector $\mathbf{k}_P$. The four elements of $\mathbf{k}_P$ have very useful properties. The first element $k_{P,1}$ is the weight of the scattering matrix $\mathbf{i}$, which generically represents odd-bounce surface reflections, for example, plane surface, sphere, and corner reflector. The second element $k_{P,2}$ corresponds to $j\mathbf{1}$, which is an indication of even-bounce surface reflections, for example, diplane. The third element $k_{P,3}$ corresponds to $-j\mathbf{k}$, an indication of 45°-rotated double-bounce scattering, for example, tilted diplane. The last element $k_{P,4}$ is the weight of $j\mathbf{j}$, representing cross-polarizer, which does not exist under backscattering reciprocity.

Moreover, the Pauli decomposed target vector $\mathbf{k}_P$ has a simpler polarization basis rotation transformation, that is, matrix $\mathbf{U}_\psi(\psi)$, which is written as

$$\mathbf{k}_P' = \begin{bmatrix} 1 & 0 & 0 \\ 0 & \cos\psi & \sin\psi \\ 0 & -\sin\psi & \cos\psi \end{bmatrix} \mathbf{k}_P \tag{5.37}$$

It can be seen that rotation of the polarization basis is directly reflected in the transformation between the second and third components of the Pauli vector. This is the most important property of the Pauli decomposed target vector.

In the Cloude–Pottier decomposition, the 3D Pauli vector under backscattering reciprocity is parameterized as

$$\mathbf{k}_P = \|\mathbf{k}_P\| \cdot \begin{bmatrix} \cos\alpha \cdot e^{i\phi} \\ \sin\alpha \cdot \cos\beta \cdot e^{i\sigma} \\ \sin\alpha \cdot \sin\beta \cdot e^{i\gamma} \end{bmatrix} \tag{5.38}$$

Obviously, α, β are the angles of vector $|\mathbf{k}_P|$ in its definition space, while ϕ, σ, γ are the phases of $\mathbf{k}_P$ components. It is interesting to know that via three matrix transforms $\mathbf{k}_P$ can be reduced to the unit vector $[1,0,0]^T$. This is referred to by Cloude as unitary reduction of the scattering vector (Cloude, 1995):

$$\mathbf{k}_P = \|\mathbf{k}_P\| \begin{bmatrix} \cos\alpha & \sin\alpha & 0 \\ -\sin\alpha & \cos\alpha & 0 \\ 0 & 0 & 1 \end{bmatrix} \cdot \begin{bmatrix} 1 & 0 & 0 \\ 0 & \cos\beta & \sin\beta \\ 0 & -\sin\beta & \cos\beta \end{bmatrix} \cdot \begin{bmatrix} e^{i\phi} & 0 & 0 \\ 0 & e^{i\delta} & 0 \\ 0 & 0 & e^{i\gamma} \end{bmatrix} \cdot \begin{bmatrix} 1 \\ 0 \\ 0 \end{bmatrix} \tag{5.39}$$

As one of the properties of the Pauli decomposition, the rotation matrix of β is equivalent to the polarization basis rotation transform of the orientation angle ψ. The angle α becomes roll-invariant, and is interpreted as follows: $\alpha = 0$ stands for uniform single scattering, for example, sphere; $\alpha = 45°$ stands for dipole scattering; $\alpha = 90°$ stands for diplane scattering; rough surface and other types of single scattering are between $0 \rightarrow 45°$; and double scattering between ground surface and terrain objects are between $45° \rightarrow 90°$.

As to the coherency matrix of $\mathbf{k}_P$, it is first decomposed into a sum of eigen-components

$$\langle \mathbf{k}_P \mathbf{k}_P^+ \rangle = \overline{\mathbf{T}} = \sum_i \lambda_i \mathbf{e}_i \cdot \mathbf{e}_i^+ = \sum_i \lambda_i \overline{\mathbf{T}}_i, \quad \lambda_1 \geq \lambda_2 \geq \lambda_3 \tag{5.40}$$

where $\lambda_i, \mathbf{e}_i$, $i = 1, 2, 3$, denote real eigenvalues and eigenvectors. Now by treating each eigen-component $\mathbf{T}_i$ as an independent target, the original stochastic target is in fact modeled as a probabilistic mean of three deterministic targets, where their respective probabilities are defined as

$$P_i = \lambda_i \bigg/ \sum_i \lambda_i \tag{5.41}$$

Clearly, eigenvectors $\mathbf{e}_i$ are the *de facto* scattering vectors of each independent target. Thus, each target can be parameterized according to Equation 5.38 as

$$\mathbf{e}_i = \begin{bmatrix} \cos\alpha_i \cdot e^{j\phi_i} \\ \sin\alpha_i \cdot \cos\beta_i \cdot e^{j\sigma_i} \\ \sin\alpha_i \cdot \sin\beta_i \cdot e^{j\gamma_i} \end{bmatrix} \tag{5.42}$$

Now we define an averaged α, β as

$$\bar{\alpha} = \sum_i \alpha_i P_i, \quad \bar{\beta} = \sum_i \beta_i P_i \tag{5.43}$$

and the entropy as

$$H = -\sum P_i \log_3 P_i \tag{5.44}$$

where $\bar{\alpha}$ indicates the averaged scattering mechanism of a stochastic target; and $\bar{\beta}$ indicates the averaged orientation. Concluding from Equation 5.19, when $\mathbf{T}$ is of a

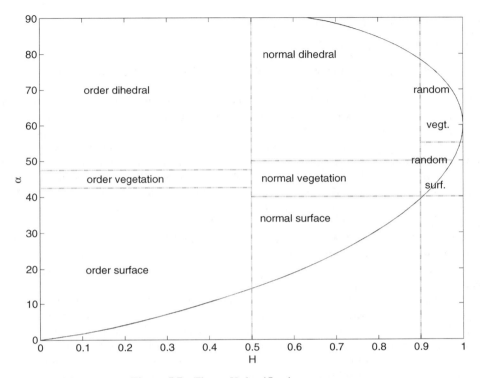

Figure 5.7   The α–H classification spectrum

deterministic target, $\lambda_1 \neq 0$, $\lambda_2 = \lambda_3 = 0$, it yields $H = 0$; when three random targets have equal probability, $\lambda_1 = \lambda_2 = \lambda_3$, it yields $H = 1$. Thus, entropy H is an indicator of randomness.

Later, Ferro-Famil, Pottier, and Lee (2000) introduced anisotropy as a complementary variable to the entropy

$$A = \frac{\lambda_1 - \lambda_2}{\lambda_1 + \lambda_2} \tag{5.45}$$

When the entropy H approaches a maximum or minimum, anisotropy does not carry additional information; while, in most circumstances, anisotropy reflects the balance between the two smaller components.

Based on these parameters, the α–H (or α–H–A) unsupervised land surface classification is introduced. As shown in Figure 5.7, $0 < \bar{\alpha} < 90°$ is divided into three sections: surface, dipole, and multiple scattering; and $0 < H < 1$ is divided into order, normal, and random. A total of eight regions are divided on the $\alpha - H$ plane. The curve on the right-hand side defines the valid boundary of possible α and H values (Cloude and Pottier, 1997).

5.2 Deorientation Theory

5.2.1 Deorientation

Radar polarimetry has become one of most advanced technologies for terrain surface observation and remote sensing, especially due to the pioneer experiments of SIR-C in 1994 and the coming launch of Radarsat-2 SAR, and other airborne flights such as AirSAR and so on. Classification of complex terrain surfaces using polarimetric SAR imagery is one of most important SAR applications.

Earlier polarimetric classification methods were based on the supervised method of coherent scattering matrix or coherent scattering data of multi-polarization channels (Kong *et al.*, 1990; Zebker *et al.*, 1991). Van Zyl (1989) developed an unsupervised method of classification based on comparison of the polarization properties of each pixel in the image with the scattering of some target classes, such as odd and even numbers of reflections and diffuse scattering.

Cameron and Leung (1990) and Cameron, Youssef, and Leung (1996) proposed a method that decomposes the maximum symmetric component and results in a classification spectrum defined on unitary polarization ratio plane. This method is mostly applicable to metal object recognition.

Cloude and Pottier (1997) presented an unsupervised method based on the entropy H and target decomposition parameters, which extracts the target decomposition parameter α and entropy H from an eigen-analysis of the coherency matrix and constructs an unsupervised classification spectrum on the $\alpha-H$ plane. Freeman and Durden (1998) suggested a three-component decomposition method of polarimetric data, which introduced a combination of surface, volume, and even-bounce scatterings. Lee and co-workers developed some classification methods combining both the target decomposition and complex Wishart classifier (Ferro-Famil, Pottier, and Lee, 2000; Lee *et al.*, 1999, 2004). These methods achieve preliminary classification sets using the $\alpha-H$, $\alpha-H-A$ methods or three- component decomposition method, and thereafter iteratively classify these preliminary sets to make final classifications.

Scattering from terrain targets (scatterers) functionally depends on the scatter orientation, shape, dielectric property, scattering mechanism, and so on. The scatter targets of complex terrain surfaces are often randomly oriented, which produces randomly fluctuating echoes. It is difficult to classify randomly oriented and randomly distributed scatter targets. Different scatters with different orientations can produce similar scattering, and, vice versa, the same scatters with random orientation can make different scattering, both of which lead to a confused classification.

In this section, a "deorientation" concept is introduced to reduce the influence of randomly fluctuating orientation and reveal the targets' generic characteristics for better terrain surface classification. Based on this deoriented classification, a detailed orientation of targets, orderly or randomly, and further classification of the scatter targets can be obtained.

5.2.1.1 Scattering Vector and its Parameterization

As a general condition, backscattering reciprocity is always assumed to be valid in this section. The target scattering vector is defined by vectorizing the scattering matrix **S** (Cloude and Pottier, 1996), one of which is Pauli vectorization and the other is lexicographical straightforward vectorization as follows:

$$\mathbf{k}_P = \frac{1}{\sqrt{2}} [S_{hh} + S_{vv} \quad S_{hh} - S_{vv} \quad 2S_x]^T \tag{5.46}$$

$$\mathbf{k}_L = [S_{hh} \quad \sqrt{2}S_x \quad S_{vv}]^T \tag{5.47}$$

where the subscripts P and L denote "Pauli" and "lexicographic", respectively, and the subscript x denotes the cross-polarization term.

The covariance matrices of the scattering vectors are

$$\overline{\mathbf{T}} = \langle \mathbf{k}_P \cdot \mathbf{k}_P^+ \rangle, \quad \overline{\mathbf{C}} = \langle \mathbf{k}_L \cdot \mathbf{k}_L^+ \rangle \tag{5.48}$$

where $\mathbf{T}$ is referred to as the coherency matrix in order to be differentiable from the covariance matrix $\mathbf{C}$. Cloude (1986) and Cloude and Pottier (1996) defined the parameterization of $\mathbf{k}_P$ in terms of α, β:

$$\mathbf{k}_P = \|\mathbf{k}_P\| \cdot \begin{bmatrix} \cos\alpha & e^{j\phi_1} \\ \sin\alpha\cos\beta & e^{j\phi_2} \\ \sin\alpha\sin\beta & e^{j\phi_3} \end{bmatrix} \tag{5.49}$$

As mentioned in the last section, α, β represent the relative scale of each complex element of $\mathbf{k}_P$ in the Pauli space, while the angles ϕ_1, ϕ_2, ϕ_3 are the phase of each element. The target scattering vector $\mathbf{k}_P$ and its parameters are depicted in the Pauli space as in the left panel of Figure 5.8.

In this book, a similar parameterization of $\mathbf{k}_L$ is defined as

$$\mathbf{k}_L = \|\mathbf{k}_L\| \cdot \begin{bmatrix} \sin c \cos a \cdot e^{j\phi_0} \\ \cos c \cdot e^{j\phi_x} \\ \sin c \sin a \cdot e^{j(\phi_0 + 2b)} \end{bmatrix} \tag{5.50}$$

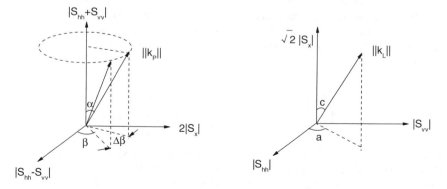

Figure 5.8 Parameterization of the scattering vector. © [2005] IEEE. Reprinted, with permission, from *IEEE Transactions on Geoscience and Remote Sensing*

From Equations 5.47 and 5.50, it yields

$$
\begin{cases}
a = \tan^{-1}(|S_{vv}|/|S_{hh}|) \\
b = \dfrac{1}{2}\arg(S_{vv}/S_{hh}) \\
c = \cos^{-1}(\sqrt{2}|S_x|/\|\bar{k}_L\|)
\end{cases}
\tag{5.51}
$$

where the operator $\arg(\cdot)$ means the phase angle of a complex number. It can be seen that the parameter a indicates the amplitude ratio of co-polarized (co-pol) scatterings S_{hh} and S_{vv}; b indicates the phase difference between S_{vv} and S_{hh}; and c indicates the cross-polarized (x-pol) scatterings. The vector $\mathbf{k}_L$ is also shown in the right panel of Figure 5.8 for comparison. As will be shown later, the straightforward vectorization of $\mathbf{k}_L$ is more intuitive and easier to process. In particular, it has the advantage of separating two co-polarization components. Thus the parameterization of $\mathbf{k}_L$ will be used in later deorientation analysis.

From Equations 5.46–5.50, the two sets of parameters a, b, c and α, β are related as

$$
\cos\alpha = \sqrt{\frac{1 + \sin 2a \cos 2b}{2 + 2/\tan^2 c}}
$$

$$
\cos\beta = \sqrt{\frac{1 - \sin 2a \cos 2b}{1 - \sin 2a \cos 2b + 2/\tan^2 c}}
\tag{5.52}
$$

When the x-pol scattering is small and negligible, it may yield $c \approx 90°$, $\beta \approx 0°$ and

$$
\cos 2\alpha = \sin 2a \cos 2b
\tag{5.53}
$$

Now, the parameter α can be determined from the parameters a, b.

5.2.1.2 Extraction of Target Orientation Information

The parameter β in the Pauli vectorization was not used in Cloude's $\alpha - H$ classification. In fact, under certain conditions, β denotes the target orientation on the plane perpendicular to line of sight (LOS) linking the observer and target (Cloude and Pottier, 1996).

Let the wave incident direction be $\hat{k}_i$, which is also the LOS. Rotating the polarization base $(\hat{h}, \hat{v})$ along the LOS, the new polarization base $(\hat{h}', \hat{v}')$ is obtained. Correspondingly, the electric field vector $\mathbf{E} = [E_h\ E_v]^T$ becomes

$$
\mathbf{E}' = \begin{bmatrix} E'_h \\ E'_v \end{bmatrix} = \overline{\mathbf{R}} \cdot \begin{bmatrix} E_h \\ E_v \end{bmatrix}^T, \quad
\overline{\mathbf{R}} = \begin{bmatrix} \cos\psi & -\sin\psi \\ \sin\psi & \cos\psi \end{bmatrix}
\tag{5.54}
$$

where ψ is the angle of rotation, and $\mathbf{R}$ is the rotation matrix, which is in fact $\mathbf{R} = \mathbf{U}_\psi$ according to the polarization basis transformation derived in the last section.

In the backscattering direction, the two bases of the incident polarization and backscattered polarization are the same (in the antenna coordinates), namely $(\hat{h}, \hat{v})$. After the transformation of basis $(\hat{h}, \hat{v})$ to $(\hat{h}', \hat{v}')$, the corresponding incident electric field vector $\mathbf{E}'_i$ and scattering electric field $\mathbf{E}'_s$ are written as $\mathbf{E}'_i = \mathbf{R} \cdot \mathbf{E}_i, \mathbf{E}'_s = \mathbf{R} \cdot \mathbf{E}_s$ that is, $\mathbf{E}_i = \mathbf{R}^T \cdot \mathbf{E}'_i$, $\mathbf{E}_s = \mathbf{R}^T \cdot \mathbf{E}'_s$. Thus,

$$\mathbf{E}'_s = \overline{\mathbf{R}} \cdot \overline{\mathbf{S}} \cdot \overline{\mathbf{R}}^T \cdot \mathbf{E}'_i = \overline{\mathbf{S}}' \cdot \mathbf{E}'_i, \quad \overline{\mathbf{S}}' = \overline{\mathbf{R}} \cdot \overline{\mathbf{S}} \cdot \overline{\mathbf{R}}^T \tag{5.55}$$

where $\mathbf{S}'$ denotes the new scattering matrix resulting from rotation of $\mathbf{S}$ by ψ. Substituting $\mathbf{R}$ into it, the elements of $\mathbf{S}'$ are expressed as

$$\begin{aligned}
S'_{hh} &= \cos^2 \psi\, S_{hh} + \cos \psi \sin \psi\, S_{hv} + \cos \psi \sin \psi\, S_{vh} + \sin^2 \psi\, S_{vv} \\
S'_{hv} &= -\cos \psi \sin \psi\, S_{hh} + \cos^2 \psi\, S_{hv} - \sin^2 \psi\, S_{vh} + \cos \psi \sin \psi\, S_{vv} \\
S'_{vh} &= -\cos \psi \sin \psi\, S_{hh} - \sin^2 \psi\, S_{hv} + \cos^2 \psi\, S_{vh} + \cos \psi \sin \psi\, S_{vv} \\
S'_{vv} &= \sin^2 \psi\, S_{hh} - \cos \psi \sin \psi\, S_{hv} - \cos \psi \sin \psi\, S_{vh} + \cos^2 \psi\, S_{vv}
\end{aligned} \tag{5.56}$$

From Equations 5.46 and 5.56, the relationship between the Pauli scattering vectors $\mathbf{k}_P, \mathbf{k}'_P$ is found to be

$$\mathbf{k}'_P = \frac{1}{\sqrt{2}} \begin{bmatrix} S'_{hh} + S'_{vv} \\ S'_{hh} - S'_{vv} \\ 2S'_{hv} \end{bmatrix} = \begin{bmatrix} 1 & 0 & 0 \\ 0 & \cos 2\psi & \sin 2\psi \\ 0 & -\sin 2\psi & \cos 2\psi \end{bmatrix} \cdot \frac{1}{\sqrt{2}} \begin{bmatrix} S_{hh} + S_{vv} \\ S_{hh} - S_{vv} \\ 2S_{hv} \end{bmatrix} = \mathbf{U} \cdot \mathbf{k}_P \tag{5.57}$$

where the transformation matrix is

$$\overline{\mathbf{U}} = \begin{bmatrix} 1 & 0 & 0 \\ 0 & \cos 2\psi & \sin 2\psi \\ 0 & -\sin 2\psi & \cos 2\psi \end{bmatrix} \tag{5.58}$$

It can be understood that the ψ rotation of the polarization basis along the LOS is equivalent to a $-\psi$ reverse rotation of the target, and vice versa. Thus, Equation 5.58 also describes the target rotation.

From Equations 5.49 and 5.57, the parameters of $\mathbf{k}_P, \mathbf{k}'_P$ are related as

$$\begin{bmatrix} \cos \alpha'\, e^{j\phi'_1} \\ \sin \alpha' \cos \beta'\, e^{j\phi'_2} \\ \sin \alpha' \sin \beta'\, e^{j\phi'_3} \end{bmatrix} = \begin{bmatrix} \cos \alpha \cdot e^{j\phi_1} \\ \sin \alpha\, [\cos 2\psi \cos \beta + e^{j(\phi_3 - \phi_2)} \sin 2\psi \sin \beta] \cdot e^{j\phi_2} \\ \sin \alpha\, [-\sin 2\psi \cos \beta + e^{j(\phi_3 - \phi_2)} \cos 2\psi \sin \beta] \cdot e^{j\phi_2} \end{bmatrix} \tag{5.59}$$

Under the condition $\phi_3 - \phi_2 \approx 0, \pi$, this becomes

$$\begin{bmatrix} \cos \alpha'\, e^{j\phi'_1} \\ \sin \alpha' \cos \beta'\, e^{j\phi'_2} \\ \sin \alpha' \sin \beta'\, e^{j\phi'_3} \end{bmatrix} = \begin{bmatrix} \cos \alpha \cdot e^{j\phi_1} \\ \sin \alpha \cos(\beta \mp 2\psi) \cdot e^{j\phi_2} \\ \sin \alpha \sin(\pm\beta - 2\psi) \cdot e^{j\phi_2} \end{bmatrix} \tag{5.60}$$

Therefore, we obtain $\alpha' = \alpha$, $\beta' = \beta - 2\psi$ or $-\beta - 2\psi$, which means that α remains invariant and β changes by 2ψ after the ψ angle rotation of the target along the LOS. In fact, the condition $\phi_3 - \phi_2 \approx 0, \pi$ is the condition of a symmetric target. This conclusion will be utilized later.

5.2.1.3   Deorientation and the Minimization of Cross-Polarization

The purpose of deorientation introduced here is to reduce or eliminate the fluctuating influence of randomly distributed target orientation on polarimetric scattering and to show the prominence of the intrinsic scatter characteristics. The deorientation operation should be designed in such a way that two identical targets with different orientations should have the same scattering after deorientation and thus can be classified as the same terrain type. In other words, we place more emphasis on the generic properties of the scatter targets, for example, type, shape, and so on, as for terrain surface classification.

The idea of deorientation is to make a certain rotation ψ_m of the target along the LOS to minimize the cross-polarization scattering (min-x-pol) of the new target scattering vector and focus on co-pol scattering terms. Note that the concept of min-x-pol is different from Huynen's optimal polarization (Huynen, 1970), because the transformation adopted here preserves the physical meaning, namely spatial rotation of the target along the LOS, rather than being a pure mathematical transform.

Here, one reason why min-x-pol is adopted is that by using min-x-pol the information can be concentrated onto the co-pol, and the corresponding parameters, for example, u, v, become more effective in describing target characteristics and make better classification.

From the third term of $\mathbf{k}'_P$, x-pol scattering $k'_{P,3}$ in Equation 5.59, min-x-pol yields

$$\min_{\psi} \frac{\left|k'_{P,3}\right|}{\left\|\overline{k}'_P\right\|} = \min_{\psi} \left| \sin \alpha \left(-\sin 2\psi \cos \beta + \cos 2\psi \sin \beta \cdot e^{j(\phi_3 - \phi_2)} \right) \cdot e^{j\phi_2} \right| \qquad (5.61)$$

Letting

$$\frac{\partial \left|k'_{P,3}\right|^2}{\partial \psi} = 0, \qquad \frac{\partial^2 \left|k'_{P,3}\right|^2}{\partial \psi^2} > 0$$

yields

$$\begin{cases} \tan 4\psi = \tan 2\beta \cos(\phi_2 - \phi_3), & \cos 2\beta \neq 0 \\ \cos 2\beta \cos 4\psi > 0 \end{cases} \qquad (5.62)$$

Assuming that $\psi_m \in [0, \pi/2]$, this gives the solution

$$\psi_m = \left[\mathrm{sgn}\{\cos(\phi_2 - \phi_3)\} \frac{2\beta - [2\beta]_\pi + [\tan^{-1}\{\tan 2\beta|\cos(\phi_2 - \phi_3)|\}]_\pi}{4} \right]_{\pi/2} \qquad (5.63)$$

where the operator $[x]_y$ means the residue of x divided by y; and sgn$\{\cdot\}$ denotes sign. Under the condition $\phi_3 - \phi_2 = 0, \pi$, Equation 5.63 becomes

$$\psi_m = \left[\text{sgn}\{\cos(\phi_2 - \phi_3)\} \cdot \frac{\beta}{2} \right]_{\pi/2} \tag{5.64}$$

This ψ_m satisfies Equation 5.60, and the min-x-pol equals zero at this time. This agrees with the condition for a symmetric target (Cameron and Leung, 1990). In other words, $\phi_3 - \phi_2 \approx \pi/2$ indicates the degree of non-symmetry or the target helicity (Cloude, Papathanassiou, and Pottier, 2001). Thus, non-helix surface scattering or volumetric scattering can always satisfy the condition $\phi_3 - \phi_2 \approx 0, \pi$.

After rotating the target along the LOS by ψ_m, $\mathbf{k}_P$ becomes the deoriented $\mathbf{k}_P^d$, that is

$$\mathbf{k}_P^d = \overline{\mathbf{U}}m \cdot \mathbf{k}_P \tag{5.65}$$

where the superscript d denotes "deoriented". The deorientation matrix $\mathbf{U}_m$ is the transform matrix in Equation 5.58 with ψ replaced by ψ_m. Similarly, we can obtain the deoriented scattering matrix $\mathbf{S}_d$ and the other form of scattering vector $\mathbf{k}_L^d$ as well. Clearly, $\cos c$ in Equation 5.50 is minimized, that is, c approaches $90°$, under which Equation 5.53 is valid.

As mentioned above, parameterization of $\mathbf{k}_L^d$ is adopted after deorientation in order to extract information from the scattering matrix more intuitively and independently. In Equation 5.52, with $\sin 2a = 0$, one of the co-pol S_{vv} or S_{hh} becomes zero and renders the phase difference between S_{vv} and S_{hh} (i.e., b) indeterminate. If $\sin c = 0$, both S_{vv} and S_{hh} become zero, resulting in both a and b being indeterminate. Thus, we define a new set of parameters u, v, w:

$$\begin{cases} u = \sin c \cos 2a \\ v = \sin c \sin 2a \cos 2b \\ w = \cos c \end{cases} \tag{5.66}$$

where w indicates the amplitude ratio of co-pol and x-pol scatterings, u indicates the amplitude ratio of co-pol scatterings in the case of little x-pol scattering ($\sin c \rightarrow 1$), and v denotes the phase difference of co-pol scatterings when both co-pols have comparable amplitudes ($\sin 2a \rightarrow 1$).

A quarter of the sphere on the u, v, w space is shown in Figure 5.9. Scattering is described within the sphere. The parameters u, v, w correspond to the axes OA, B_1B_2, OC, respectively. Since the situations $u \geq 0$ or $u < 0$ differ only in which co-pol, vv or hh, is higher, the scattering mechanism is actually the same. Hence, only the part $u \geq 0$ is plotted in this figure. It can be seen that the scattering distribution tends to point C as x-pol scattering becomes large, and tends to plane B_1B_2A as x-pol scattering becomes small. On plane B_1B_2A, the scattering distribution tends to point A as the relative difference between the co-pol scattering magnitudes becomes large, and tends to line B_1B_2 as the two co-pol scattering components become comparable. On line B_1B_2, the scattering distribution is shifted from point B_1 to B_2 as the phase difference between the two co-pol scatterings decreases.

The following is adopted to show the distribution of typical scatter targets in Figure 5.9: the helix scatterer is near point C, the dipole scatterer is near point A, the spherical scatterer

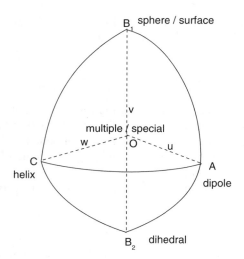

Figure 5.9 The characteristic u–v–w space. © [2005] IEEE. Reprinted, with permission, from *IEEE Transactions on Geoscience and Remote Sensing*

is near point B_1, and the dihedral scatterer is near point B_2. Note that scattering at point O has the special form

$$\overline{\mathbf{S}}_O = \begin{bmatrix} 1 & 0 \\ 0 & j \end{bmatrix} \tag{5.67}$$

Thus, we consider targets near point O as a complicated multiple scattering, a special material scatterer, or an artificial device.

Most of the simple scattering targets can be described by the parameters u, v on plane B_1B_2A. In this case, the condition of Equation 5.53 is satisfied and $v = \sin 2a \cos 2b = \cos 2\alpha$. We find that on plane B_1B_2A the parameter v is equivalent to α of Pauli parameterization.

The scattering distribution and parameter relationships are plotted in Figure 5.10, which is consistent with Cloude's explanation of α (Cloude, 1995). It can be seen from Equation 5.66

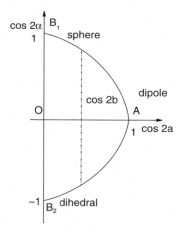

Figure 5.10 The u–v plane. © [2005] IEEE. Reprinted, with permission, from *IEEE Transactions on Geoscience and Remote Sensing*

and Figure 5.10 that the parameter v is determined by $\sin 2a$ and $\cos 2b$, while $\sin 2a$ limits the v range and $\cos 2b$ makes v vary between $-\sin 2a$ and $\sin 2a$. The set of u, v contains all of the information that α has, and can present more characteristic information about the scattering mechanisms. This demonstrates the feasibility of more precise classification by using the set of parameters u, v.

Via the deorientation method of minimizing x-pol of the target scattering vectors, different targets in the u, v, w space can be well distinguished. There exists a pair of orthogonal orientation states of min-x-pol due to the different signs of u. Define ψ as

$$\psi = \begin{cases} \psi_m & u \geq 0, \\ \psi_m + \pi/2 & u < 0, \end{cases} \quad \psi_m \in [0, \pi/2), \ \psi \in [0, \pi) \tag{5.68}$$

As $w \to 1$, the target is a helix. As $w \to 0$ and $|u| \to 1$, it is a dipole. As $w \to 0$, $|u| \to 0$, and $v \to 1$, it is a sphere. As $w \to 0$, $|u| \to 0$, and $v \to -1$, it is a dihedral. And as $w \to 0$, $|u| \to 0$, and $v \to 0$, it means multiple scattering or special target scattering. The angle ψ denotes the target orientation deviation from the min-x-pol state.

5.2.1.4 Deorientation of Non-Deterministic Targets

The Mueller matrix $\mathbf{M}$, the coherency matrix $\mathbf{T}$, and the covariance matrix $\mathbf{C}$ can be obtained from the second-order statistical moment of random target scattering. The matrix $\mathbf{M}$ is real, and both $\mathbf{T}$ and $\mathbf{C}$ are semi-positive Hermitian with positive real eigenvalues and orthogonal eigenvectors. Since $\mathbf{T}$ and $\mathbf{C}$ conserve unitary similarity transformation, they have the same eigenvalues and eigenvectors (Cloude and Pottier, 1996) and are expressed as

$$\overline{\mathbf{T}} = \sum_{i=1}^{3} \lambda_i \mathbf{k}_{P,i} \cdot \mathbf{k}_{P,i}^{+}, \quad \overline{\mathbf{C}} = \sum_{i=1}^{3} \lambda_i \mathbf{k}_{L,i} \cdot \mathbf{k}_{L,i}^{+} \tag{5.69}$$

where $\lambda_1 \geq \lambda_2 \geq \lambda_3 \geq 0$ are the eigenvalues, and $\mathbf{k}_{P,i}, \mathbf{k}_{L,i}$ are the eigenvectors, respectively. Each eigenvector is a coherent target scattering vector and can be parameterized into a set of target decomposition parameters. The entropy H is defined as (Cloude, 1986; Cloude, Papathanassiou, and Pottier, 2001; Jin and Cloude, 1994)

$$H = -\sum_{i=1}^{3} P_i \log_3 P_i, \quad P_i = \lambda_i \Big/ \sum_{i=1}^{3} \lambda_i \tag{5.70}$$

where P_i means the probability or priority of each coherent scattering vector occurring among randomly distributed targets. In the previous literature, for example, Cloude and Pottier (1997), the average parameters extracted from every eigenvector are used to describe the average characteristics of randomly distributed targets as

$$\langle y \rangle = \sum_{i=1}^{3} p_i y_i \tag{5.71}$$

where y_i denotes one of the target decomposition parameters extracted from the eigenvector $\mathbf{k}_{P,i}$.

When the first eigenvalue is significantly larger than the rest, the main properties of the targets are described by the first eigenvector (here referred to as the principal eigenvector). However, in the case of complex random targets with multiple scattering – for example, a multi-layered forest canopy with volumetric scattering of crown leaves, branches, and trunks, surface scattering of the underlying rough surface, and interactions between the canopy scatterers and the surface – the second and third eigenvalues become noticeable. Each of them indicates a kind of scattering and contains its own information about the whole layer of random targets, which might be missed by the average method adopted in the previous literature. Thus, we focus on the principal eigenvector and its parameterization.

5.2.2 Efficacy of Deorientation

In order to verify the efficacy of deorientation parameters, we use forward scattering models to study their dependences on physical parameters.

5.2.2.1 A Single Non-Spherical Particle

Consider scattering from a single non-spherical particle as shown in Figure 5.11. The local coordinates $(\hat{x}_b, \hat{y}_b, \hat{z}_b)$ of the particle are related to the principal coordinates $(\hat{x}, \hat{y}, \hat{z})$ by the Euler angles ω, ξ, η (Jin, 1994). Define the incident coordinates as $(\hat{h}, \hat{v}, -\hat{k}_i)$ and the Euler angles ω_0, ξ_0, η_0 of the local coordinates $(\hat{h}, \hat{v}, -\hat{k}_i)$ as shown in Figure 5.11. The relationships between ξ_0, η_0 and ξ, η can be derived as follows:

$$\begin{cases} \cos \xi_0 = \sin \xi \cos \eta \sin \theta + \cos \xi \cos \theta \\ \cos \eta_0 \sin \xi_0 = \cos \xi \sin \theta - \sin \xi \cos \theta \cos \eta \\ \sin \eta_0 \sin \xi_0 = -\sin \xi \sin \eta \end{cases} \quad (5.72)$$

where $\varphi_i = 0$ is assumed.

Note that, in the incident coordinates, the incident angle is always $0°$. Given the $0°$ incidence, the scattering matrix of a small spheroid under Rayleigh approximation (Jin, 1994) is

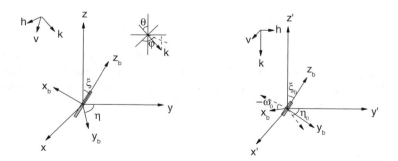

Figure 5.11 Principal coordinates and incident coordinates. © [2005] IEEE. Reprinted, with permission, from *IEEE Transactions on Geoscience and Remote Sensing*

written as

$$\begin{cases} S_{0hh} = \dfrac{vk^2}{4\pi}\left[G_a + (G_c - G_a)\sin^2\xi_0 \sin^2\eta_0\right] \\[2mm] S_{0vv} = \dfrac{vk^2}{4\pi}\left[G_a + (G_c - G_a)\sin^2\xi_0 \cos^2\eta_0\right] \\[2mm] S_{0hv} = S_{0vh} = \dfrac{vk^2}{4\pi}(G_c - G_a)\sin^2\xi_0 \sin\eta_0 \cos\eta_0 \end{cases} \tag{5.73}$$

where

$$G_a = \frac{1}{1/(\varepsilon_s - 1) + g_a}, \qquad G_c = \frac{1}{1/(\varepsilon_s - 1) + g_c} \tag{5.74}$$

and v_0 is the particle volume, k is the wavenumber, and g_a and g_c are functions related to the particle shape (the three axes of a spheroid are assumed as $\ell_a = \ell_b, \ell_c$) (Jin, 1994). One has $2g_a + g_c = 1$. Their expressions can be found in Equation 1.27. It can be seen that g_c and $\rho = \ell_c/\ell_a$ show a monotonic relationship. Hence, g_c can be used as a description of the particle shape property.

The target scattering vector $\mathbf{k}_P$ is derived as

$$\mathbf{k}_P = \frac{vk^2}{4\pi} \begin{bmatrix} 2G_a + (G_c - G_a)\sin^2\xi_0 \\[1mm] -(G_c - G_a)\cos 2\eta_0 \sin^2\xi_0 \\[1mm] (G_c - G_a)\sin 2\eta_0 \sin^2\xi_0 \end{bmatrix} \tag{5.75}$$

Further, the target decomposition parameters are derived as

$$\begin{aligned} \alpha &= \tan^{-1}\left[\frac{|G_c - G_a|\sin^2\xi_0}{\left|2G_a + (G_c - G_a)\sin^2\xi_0\right|}\right] \\[2mm] \beta &= \cos^{-1}|\cos 2\eta_0| \\[1mm] \phi_1 &= \arg\left[2G_a + (G_c - G_a)\sin^2\xi_0\right] \\[1mm] \phi_2 &= \arg\left[-(G_c - G_a)\cos 2\eta_0\right] \\[1mm] \phi_3 &= \arg\left[(G_c - G_a)\sin 2\eta_0\right] \end{aligned} \tag{5.76}$$

Clearly, $\phi_3 - \phi_2 = 0, \pi$. From Equation 5.68, the rotation angle of min-x-pol is

$$\psi_m = \begin{cases} \beta/2 & \phi_2 - \phi_3 = 0 \\ (\pi - \beta)/2 & \phi_2 - \phi_3 = \pi \end{cases} \tag{5.77}$$

and further from Equations 5.76 and 5.77, we have

$$\psi_m = \pi/2 - [\eta_0]_{\pi/2} \tag{5.78}$$

Substituting Equation 5.78 into Equation 5.65, the deoriented target scattering vector $\mathbf{k}_P^d$ is written as

$$\mathbf{k}_P^d = \frac{vk^2}{4\pi} \begin{bmatrix} 2G_a + (G_c - G_a)\sin^2 \xi_0 \\ \pm(G_c - G_a)\sin^2 \xi_0 \\ 0 \end{bmatrix} \tag{5.79}$$

where the $\pm$ sign is positive if $\eta_0 \in [0, \pi/2)$, and negative if $\eta_0 \in (\pi/2, \pi)$. Accordingly, the deoriented $\mathbf{k}_L^d$ is written as

$$\mathbf{k}_L^d = \frac{vk^2}{4\pi} \begin{bmatrix} G_a + (G_c - G_a)\sin^2 \xi_0 \\ 0 \\ G_a \end{bmatrix} \quad \text{or} \quad \mathbf{k}_L^d = \frac{vk^2}{4\pi} \begin{bmatrix} G_a \\ 0 \\ G_a + (G_c - G_a)\sin^2 \xi_0 \end{bmatrix} \tag{5.80}$$

and from Equation 5.50, the parameters are derived as

$$a = \tan^{-1}\left|\frac{G_a}{G_a + (G_c - G_a)\sin^2 \xi_0}\right| \quad \text{or} \quad \tan^{-1}\left|\frac{G_a + (G_c - G_a)\sin^2 \xi_0}{G_a}\right|$$

$$b = \pm\frac{1}{2}\arg\left[\frac{G_a}{G_a + (G_c - G_a)\sin^2 \xi_0}\right] \tag{5.81}$$

$$c = \pi/2$$

The set of parameters u, v, w, ψ are expressed as

$$\begin{cases} w = 0 \\ u = \pm\dfrac{|G_a|^2 - |G_a + (G_c - G_a)\sin^2 \xi_0|^2}{|G_a|^2 + |G_a + (G_c - G_a)\sin^2 \xi_0|^2} \\ v = 1 - \dfrac{|(G_c - G_a)\sin^2 \xi_0|^2}{|G_a|^2 + |G_a + (G_c - G_a)\sin^2 \xi_0|^2} \\ \psi = \eta_0 \end{cases} \tag{5.82}$$

Thus, we have established the relationships between the particle's parameters (e.g., g_c, ε_s) and the target decomposition parameters (e.g., u, v).

Figure 5.12 shows the relationship between $|u|$ and the orientation ξ_0 and shape g_c of a particle. As the particle shape varies from needle-like to spherical, g_c changes from $0 \rightarrow 1/3$, and $|u|$ decreases monotonically. As the particle shape varies from sphere to disk-like, g_c changes from $1/3 \rightarrow 1$ and $|u|$ increases monotonically. As the angle ξ_0 varies from $0 \rightarrow \pi/2$, $|u|$ increases monotonically in the case of a needle-like or a disk-like particle, and remains constant in the case of a spherical particle. This demonstrates that $|u|$ indicates the non-symmetry of the particle's projection along the LOS.

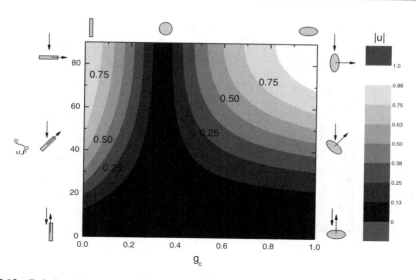

Figure 5.12 Relationship between $|u|$, ξ_0, and g_c ($\varepsilon_s = 6.5 + 2.5i$). © [2005] IEEE. Reprinted, with permission, from *IEEE Transactions on Geoscience and Remote Sensing*

Figure 5.13 shows the relationship between v and ξ_0 and g_c. It shows that v is insensitive to either the orientation or the shape of the particle.

Figure 5.14 shows the relationship between $|u|$, v, and $|\varepsilon_s|$ of the complex dielectric constant $\varepsilon_s = \varepsilon'_s + i\varepsilon''_s$ of a needle-like particle. As $|\varepsilon_s|$ increases, $|u|$ increases monotonically and v decreases in the opposite way, as shown in Equation 5.82. However, as $\tan^{-1}\left(\varepsilon''_s/\varepsilon'_s\right)$ increases, $|u|$ varies very slowly and v decreases monotonically. This means that $|\varepsilon_s|$ affects $|u|$, and $\varepsilon''_s/\varepsilon'_s$ affects v.

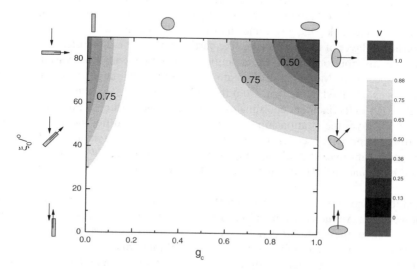

Figure 5.13 Relationship between v, ξ_0, and g_c ($\varepsilon_s = 6.5 + 2.5i$). © [2005] IEEE. Reprinted, with permission, from *IEEE Transactions on Geoscience and Remote Sensing*

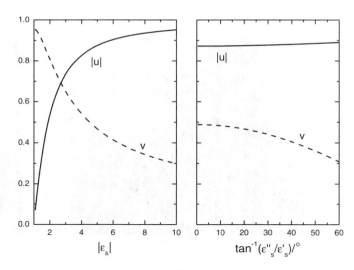

Figure 5.14 Relationship between $|u|$, v, ε_s, and $\tan^{-1}(\varepsilon_s''/\varepsilon_s') = 20°$ ($\xi_0 = \pi/2$, $g_c = 0.9$, $|\varepsilon_s| = 5$).
© [2005] IEEE. Reprinted, with permission, from *IEEE Transactions on Geoscience and Remote Sensing*

5.2.2.2 A Layer of Random Non-Spherical Particles

The coherency matrix of random non-spherical particles is calculated by

$$\overline{\mathbf{T}} = \int_{x_1}\int_{x_2}\cdots\int_{x_n} \mathbf{k}_P \cdot \mathbf{k}_P^+ \, P_n(x_n)\, dx_n \cdots P_2(x_2)\, dx_2 \, P_1(x_1)\, dx_1 \tag{5.83}$$

where $x_1, x_2, \ldots, x_n$ are the parameters of random scatter properties, and $P_1(x_1), P_2(x_2), \ldots, P_n(x_n)$ are the corresponding probability distribution functions. If the target scattering vector is $\mathbf{k}_L$, the covariance matrix $\mathbf{C}$ is then obtained.

We consider a layer of random needle-like particles whose random orientation is described by a normal distribution of the Euler angles ξ, η with means and variances $\langle\xi\rangle$, $\Delta\xi^2$, $\langle\eta\rangle$, $\Delta\eta^2$, respectively. Making a deorientation of the principal target scattering vector, a set of $\psi, |u|, v, w$ is obtained. Note that, from the Euler angles $\langle\xi\rangle$, $\langle\eta\rangle$, the corresponding Euler angles $\langle\xi_0\rangle$, $\langle\eta_0\rangle$ in the incident coordinates are obtained according to Equation 5.72.

Figure 5.15a gives the relationship between $\langle\eta_0\rangle$ and ψ on a different range $\Delta\eta$. Since ψ is derived from the principal scattering vector, it maintains a linear relationship between ψ and $\langle\eta_0\rangle$. Figure 5.15b gives the relationship between the averaged $\langle\psi\rangle$ and $\langle\eta_0\rangle$. Here $\langle\psi\rangle$ is calculated by Equation 5.71. It can be seen that the linearity between $\langle\psi\rangle$ and $\langle\eta_0\rangle$ becomes degraded.

Simulations also show that other parameters, $|u|$, v, w, and H, are insensitive to varying $\langle\eta_0\rangle$. However, as $\Delta\eta$ increases, it causes H to increase and $|u|$ to decrease. It is known that H indicates randomness and $|u|$ denotes the symmetry of targets; a larger range of $\Delta\eta$ produces overall symmetry and uniformity of the layer of random targets.

Figure 5.16a gives the relationship between $|u|$ and $\sin^2 \xi_0$, and shows the good linearity between them. Figure 5.16b shows ψ versus $\langle\xi_0\rangle$ with $\eta_0 = 45°$. It can be seen that, as $\langle\xi_0\rangle$ increases, the projection of the particle along the LOS becomes asymmetrical, and ψ

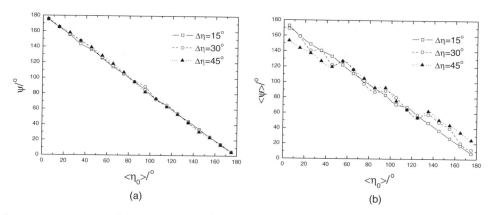

Figure 5.15 Relationship between ψ, $\langle\psi\rangle$, and $\langle\eta_0\rangle$, for various $\Delta\eta(\xi_0 = 90°$, $\Delta\xi = 15°$, $\theta_i = 40°$, $\varphi_i = 0°$, $g_c = 0.9$, $\varepsilon_s = 6.5 + 2.5i)$. © [2005] IEEE. Reprinted, with permission, from *IEEE Transactions on Geoscience and Remote Sensing*

tends to $180° - \langle\eta_0\rangle$. This indicates that there are non-orientation issues for symmetrical targets.

Let g_c be a normal distribution with mean $\langle g_c\rangle$ and variance Δg_c^2. Figure 5.17 shows $|u|$ versus g_c with different Δg_c. It shows that $|u|$ also can describe the averaged shape of a layer of random particles.

To summarize the above results, the deorientation parameters ψ, u, v, H correctly reflect the physical properties of the target. Specifically, ψ indicates the target's orientation; u indicates the target's shape; v indicates the target's dielectric properties; and H indicates the randomness of a distributed target. How to employ these parameters in terrain surface classification using SAR imagery will be the subject of the next section.

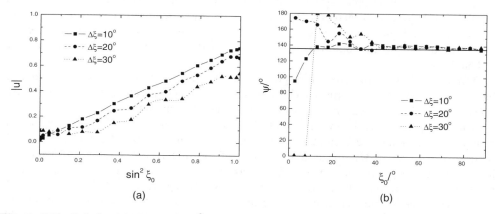

Figure 5.16 Relationship between $\sin^2\xi_0$, $|u|$, $\langle\xi_0\rangle$, and ψ, for various $\Delta\xi(\eta_0 = 45°$, $\Delta\eta = 30°$, $\theta_i = 40°$, $\varphi_i = 0°$, $g_c = 0.9$, $\varepsilon_s = 6.5 + 2.5i)$. © [2005] IEEE. Reprinted, with permission, from *IEEE Transactions on Geoscience and Remote Sensing*

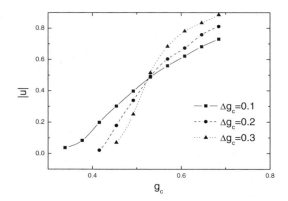

Figure 5.17 Relationship between $|u|$ and g_c, for various $\Delta g_c(\xi_0 = 90°, \Delta\xi = 15°, \eta_0 = 45°, \Delta\eta = 30°, \theta_i = 40°, \varphi_i = 0°, \varepsilon_s = 6.5 + 2.5i)$. © [2005] IEEE. Reprinted, with permission, from *IEEE Transactions on Geoscience and Remote Sensing*

5.3 Terrain Surface Classification

5.3.1 Terrain Scattering Modeling and Classification Spectrum

To better understand how the deorientation parameters can convey information regarding the terrain surface, we first use several typical terrain scattering models to study the distribution and migration of different terrain surfaces in the parameter space and then design a classification spectrum based on this knowledge.

The model of a layer of random non-spherical particles above a rough surface developed in Chapter 1 is employed here. Five typical terrain surfaces are modeled with different model parameters, as specified in Table 5.2. All cases are simulated under incidence angle 30° and frequency 1.25 GHz.

Figure 5.18 presents the distribution of some terrain surfaces on the $|u| - H$ plane. It can be seen that weak and single surface scatterings of the ground soil and oceanic surface cause small H. In the case of vegetation canopy or forests, more scattering causes increasing H, which reaches a high value for thick (multi-layering) and dense forests. As the scatterers of the dense forests becomes uniform, especially dominated by homogeneous crown leaves, H is then decreased.

Different dielectric properties of the ground soil and oceanic surface also make u identifiable on the $|u|$ axis. The uniformity and symmetry of the random scatters makes $|u|$ decrease. The types 1 and 2 of the vegetation canopy can be identified in Figure 5.18.

From Chapter 1, we know that scattering in the Mueller matrix of the vegetation canopy model includes contributions from: the underlying rough surface M_0, the scatter of trunks and leaves M_1, the double bounce of the underlying surface and trunks or leaves M_2, the triple bounce of the underlying surface and trunks or leaves M_3, and so on. Figure 5.19 presents the distribution of these scattering terms on the plane $|u| - v$. It can be seen that single scatterings are located on the area $v > 0$, double scatterings on the area $v < 0$, and high-order triple scatterings on the area around $v = 0$ because of the minimum x-pol enhanced by high-order scattering. Multi-path scattering makes large x-pol and v approaches zero.

Table 5.2 Simulation model and parameters for some typical terrain surfaces

Type	Model and parameters
Soil land	An integral equation method rough surface without the canopy (Fung, 1994), with root-mean-square height $\delta = 2 - 4$ cm, correlation length $L = 8$ cm, and dielectric constant $\varepsilon_2 = 15 + 2i$
Sea surface	Similar to the soil land, but with $\varepsilon_2 = 70 + 40i$
Vegetation type 1	A layer of non-spherical particles above a rough surface (Xu and Jin, 2004). Scattering particles: $\ell_a = 0.1$ cm, $\ell_c = 5$ cm, $\varepsilon_s = 6.5 + 2.5i$, fractional volume $f_s = 0.005$, layer depth $d = 150$ cm, Euler angle η distribution over $[0°, 30°]$ to $[0°, 90°]$, ξ distribution over $[0°, 360°]$ Underlying rough surface: $\delta = 0.3$ cm, $L = 10$ cm, and $\varepsilon_2 = 15 + 2i$
Vegetation type 2	Similar to vegetation type 1, but with $\varepsilon_s = 22.5 + 7.5i$, $\ell_a = 0.1$ cm, and $\ell_c = 3$ cm
Forest	A layer of hybrid scatterers (spheroids for leaves, and cylinders for trunks) above a rough surface (Xu and Jin, 2004). Spheroids: $\ell_a = 0.1$ cm, $\ell_c = 3$ cm, $d = 300$ cm, $\eta \in [0°, 60°]$, and $\xi \in [0°, 360°]$ Cylinders: radius $r = 15$ cm, height $h = 500$ cm, and dielectric constant $\varepsilon'_s = 10 + 2i$ Underlying rough surface: $\delta = 0.3$ cm, $L = 10$ cm, and $\varepsilon_2 = 15 + 2i$ Note that the fractional volume of the trunks is $f'_s = 0.0005$, and the fractional volume of leaves varies from $f_s = 0.0001$ to 0.005 for sparse or dense forests

Scattering contributed by different terms M_i $(i = 1, 2, 3)$ and total scattering M, as the canopy changes from sparse to dense scatterers, are shown on the $|u| - v$ plane of Figure 5.19. It can be seen that dominant scattering is gradually shifted from trunks M_2 to crown M_1. As a brief summary, H indicates the canopy randomness, u is useful for distinguishing different terrain targets, and v is helpful to take account of scattering mechanisms of different orders.

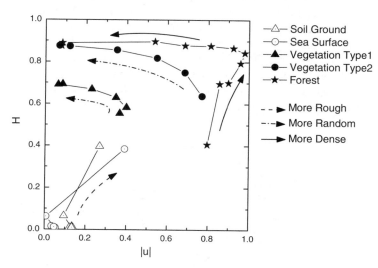

Figure 5.18 Distribution of some typical terrain surfaces on the $|u|-H$ plane. © [2005] IEEE. Reprinted, with permission, from *IEEE Transactions on Geoscience and Remote Sensing*

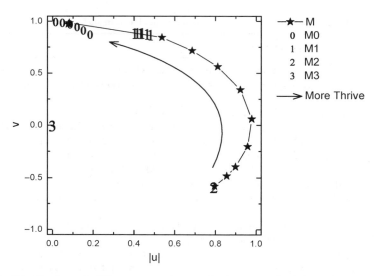

Figure 5.19 Distribution of different scattering mechanisms on the $|u|-v$ plane. © [2005] IEEE. Reprinted, with permission, from *IEEE Transactions on Geoscience and Remote Sensing*

The flowchart of the overall classification method for PolSAR data based on deorientation is shown in Figure 5.20. We classify the terrain surfaces into 19 classes by u, v, H. According to different v, we define single scattering as $(v > 0.2)$, double scattering as $(v < -0.2)$, and multiple scattering as $(-0.2 \leq v \leq 0.2)$. Further, single scattering and double scattering are divided into nine classes as shown in Figure 5.21. It is believed that triple or higher-order scatterings are not significant in most terrain surfaces at AirSAR or SIR-C frequencies. We classify the single scattering as the surface type if $H < 0.5$, as the canopy type if $0.5 \leq H \leq 0.8$, and as the forest type if $H > 0.8$. Further, each class of single scatterings is divided into three classes as $|u| < 0.3, 0.3 \leq |u| \leq 0.7$, and $|u| > 0.7$. Double scatterings are classified into simple urban area (s. urban) if $H < 0.5$, complex urban area (c. urban) if $0.5 \leq H \leq 0.8$, and forest type if $H > 0.8$, where double bounce is due to the interactions between sparse trunks and ground (s. forest). Similarly, each class of double bounce is further divided into three classes as $|u| < 0.3, 0.3 \leq |u| \leq 0.7$, and $|u| > 0.7$.

Note that the thresholds above can be tuned in a particular application. Kang *et al.* (2007) investigated a clustering method based on deorientation that in fact uses a soft boundary to

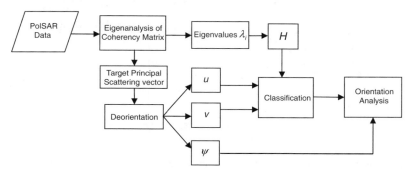

Figure 5.20 Flow-chart of classification based on deorientation. © [2005] IEEE. Reprinted, with permission, from *IEEE Transactions on Geoscience and Remote Sensing*

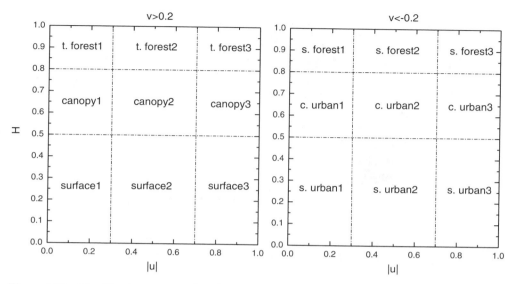

Figure 5.21 Classification spectrum based on deorientation. © [2005] IEEE. Reprinted, with permission, from *IEEE Transactions on Geoscience and Remote Sensing*

classify different terrain surfaces. It is worth mentioning that the spectrum may change in different frequency bands because terrain scattering depends on wavelength. For example, the same forest that has dominant trunk–ground double scattering in L band may become crown-scattering dominant in C band.

5.3.2 Application to SIR-C Data

As an example, the SIR-C data of NASA/JPL over China's Guangdong Huiyang district in 1994 is tested for classification application. Figure 5.22 is the image of the total power of

Figure 5.22 Total power of SIR-C data, Huiyang, L band. © [2005] IEEE. Reprinted, with permission, from *IEEE Transactions on Geoscience and Remote Sensing*

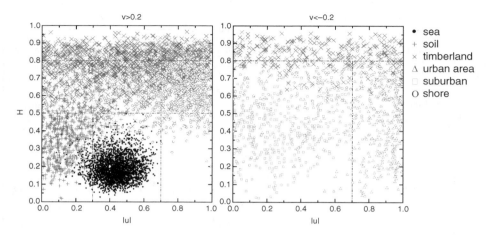

Figure 5.23 Distributions of different terrain surfaces on the u, v, H spectrum. © [2005] IEEE. Reprinted, with permission, from *IEEE Transactions on Geoscience and Remote Sensing*

SIR-C data at L band. In Figure 5.23, six regions of soil, timberland, urban area, suburban, shore, and sea are chosen according to the geographic map of this area.

The u, v, H distribution maps of these terrain surfaces are shown in Figure 5.23. It can be seen that, in relation to Figure 5.21, ocean is mainly located on the zone of surface 2, soil land on the zone of surface 1, timberland on the zone of forest, urban area on the c. urban 3 and s. urban 3 zones, suburban on the c. urban 1 and s. urban 1 zones, and shore on the zone of canopy 3. This result proves the validity and effectiveness of the u, v, H classification approach.

According to the distributions on Figure 5.23, the terrain surfaces are divided into eight classes following the decision tree of Figure 5.24: Timber, Urban 2, Urban 1, Canopy 2, Canopy 1, Surface 1, Surface 2, and Surface 3. After the u, v, H classification, a post-processing, that is, majority-rule clustering within a 5×5 window, is applied in order to make the classes distributing have continuity.

Comparing the result to geographic data, the terrain surfaces classified are well matched with the ground truth. The final classifications using u, v, H are shown in Figure 5.25. The main types of land and site names in this district are labeled in Figure 5.23 for comparison.

Comparing with the $\alpha - H$ method, Figure 5.26 gives the distributions of six regions on the $\alpha - H$ plane. It can be seen that ocean and soil land are intermixed in zone 9 and are not well separated. A dot-dashed line for possible separation is tilted on the $\alpha - H$ plane and cannot be determined by a single α. Moreover, the urban and suburban areas, and the shore and timber areas, are also more or less confused.

Following the Cloude–Pottier $\alpha - H$ classification spectrum, nine types can be classified as shown in Figure 5.27. Note that the same post-processing is applied after initial classification. Clearly, $\alpha - H$ classification has worse performance as compared to u, v, H in three aspects: confusion of sea surface and soil surface; confusion of urban and suburban; and confusion of shore and timber.

As an equal 3D parameter comparison, we further employ the anisotropy A as an additional parameter, that is, the $\alpha - H - A$ classification scheme. As an additional parameter, we first examine its classification ability in terms of those confused pairs of terrain surfaces in

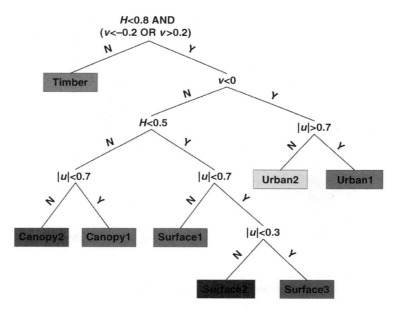

Figure 5.24 Decision tree. © [2005] IEEE. Reprinted, with permission, from *IEEE Transactions on Geoscience and Remote Sensing*

Figure 5.25 The u, v, H classification of SIR-C data over Huiyang. © [2005] IEEE. Reprinted, with permission, from *IEEE Transactions on Geoscience and Remote Sensing*

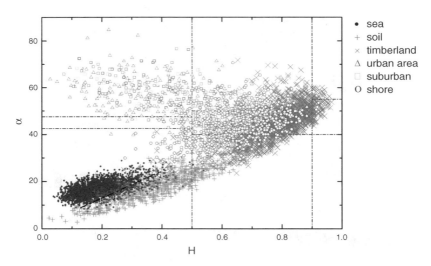

Figure 5.26 Distributions of different terrain surfaces on the $\alpha-H$ plane. © [2005] IEEE. Reprinted, with permission, from *IEEE Transactions on Geoscience and Remote Sensing*

the $\alpha-H$ plane, namely suburban and urban area, shore and timberland, and sea and soil. The normalized histogram of each pair of terrain surfaces is plotted in Figure 5.28. It can be seen that, if further divided by an extra threshold of $A > 0.5$, only the timberland and shore can be further classified, while sea and soil, and suburban and urban areas are still confused.

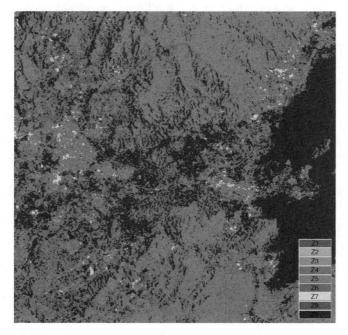

Figure 5.27 The $\alpha-H$ classification of SIR-C data over Huiyang. © [2005] IEEE. Reprinted, with permission, from *IEEE Transactions on Geoscience and Remote Sensing*

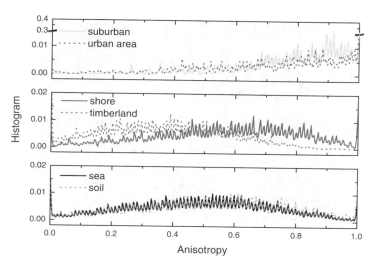

Figure 5.28 Anisotropy histogram of different terrain surfaces. © [2005] IEEE. Reprinted, with permission, from *IEEE Transactions on Geoscience and Remote Sensing*

By the condition $A > 0.5$ or $A \leq 0.5$, each class in the $\alpha-H$ plane is further divided into two classes. Thus, the final classification of 18 classes Z1-1, Z1-2, . . . , Z9-1, Z9-2 based on the $\alpha-H-A$ spectrum is shown in Figure 5.29. Note that the same 5×5 majority-rule clustering as post-processing is also applied after classification.

Figure 5.29 The $\alpha-H-A$ classification of SIR-C data over Huiyang. © [2005] IEEE. Reprinted, with permission, from *IEEE Transactions on Geoscience and Remote Sensing*

Figure 5.30   Total power of AirSAR data, boreal area of Canada. © [2005] IEEE. Reprinted, with permission, from *IEEE Transactions on Geoscience and Remote Sensing*

As a primary and simple classification method, the set of u, v, H is better than single $\alpha - H$ or $\alpha - H - A$ for classifications of certain terrain surfaces. The set of u, v enhanced the capability to distinguish different targets. Another reason for the defects in $\alpha - H$ is due to the averaged parameters over all eigenvectors of the coherency matrix and leading to tilted separation lines on the $\alpha - H$ plane. As H becomes large, the parameter α is restricted within the valid zone (Cloude and Pottier, 1997), that is, high entropy leads to compression of the dynamic range of α. It makes the separation line tilted and degrades the classification capability.

However, many excellent methods of classification have been developed that include improved methods using $\alpha - H$ or $\alpha - H - A$ as the primary step, for example, the Wishart classifier (Ferro-Famil, Pottier, and Lee, 2000; Lee *et al.*, 1999, 2004; Pottier and Lee, 1999).

5.3.3 Orientation Analysis

AirSAR data at L band of the boreal area in Canada with rich resources of vegetation and underlying flat ground have been chosen for orientation analysis. The main kind of vegetation canopy in this area is needle-leaf forest.

Figures 5.30 and 5.31 are the total power image and its deorientation classification over this area, respectively. Note that the classifications in Figure 5.31 show that some vegetation areas are classified as Urban 1 (the red area). This is probably due to the sparse vegetation with mainly double scatterings, that is, $v < 0$ and lower H. Thus, more orientation analysis is necessary to confirm the class or to resolve some of the confusion.

Figure 5.32 presents the orientation distributions of eight classes of terrain surface targets. Note that the short segment in each pixel displays the orientation angle of the parameter ψ of this pixel. Only the center pixel from each 4×4 window is displayed to avoid overlapped confusion from neighboring pixels. The following can be seen from the figure:

(a) Most areas of "Timber" show random orientation, for example, the region A. There are only a few parts with uniform orientation, for example, the region B with almost vertical orientations.

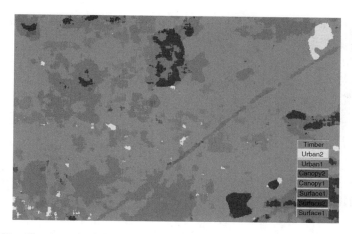

Figure 5.31 Classification of AirSAR data, boreal area of Canada. © [2005] IEEE. Reprinted, with permission, from *IEEE Transactions on Geoscience and Remote Sensing*

(b) There are almost uniform orientations of vertical or horizontal directions in the "Urban 2", for example, the region C, due to the regular features of artificial constructions.

(c) There are uniformly vertical orientations in the "Urban 1" (sparse forest of vertical trunks, instead of artificial constructions).

(d) Random orientation in the "Canopy 1" means that the vegetation canopy in this area might be disordered bush. Note that the region D should be roads inside the forest, whose randomness might be confused by the bush vegetation on the roadsides.

(e) The "Surface 2" can be divided into two subclasses by the orientation distribution, that is, uniformly horizontal oriented area, for example, the region E, and randomly oriented area, for example, the region F. The horizontal oriented area is the water surface, while the randomly oriented area is disordered random grassland.

(f) The "Surface 1" can also be divided into uniformly and horizontally oriented scatter targets as water surface, for example, region G, and uniformly and vertically oriented targets as ordered grassland, for example, region H.

(g) There are not many surfaces classed as "Canopy 2" and "Surface 3".

Following the above orientation analysis, the terrain surfaces are further classified into subclasses and are identified by their types.

Appendix 5A: Matrix Transformations under Various Conventions

As shown in Figure 5.5, here we present the transformations between the scattering matrix, Mueller matrix, and coherency matrix, and their transformations under different conventions.

For the sake of simplicity, variables without specific subscript or superscript are by default defined under h/v and $\pm$, for example, $\mathbf{M} \equiv \mathbf{M}_{\pm}^{h/v}$, $\mathbf{T} \equiv \mathbf{T}_{\pm}^{h/v}$; meanwhile, variables defined under antenna coordinates (BSA) are denoted by subscript A, while those without subscript A are by default defined under wave coordinates (FSA).

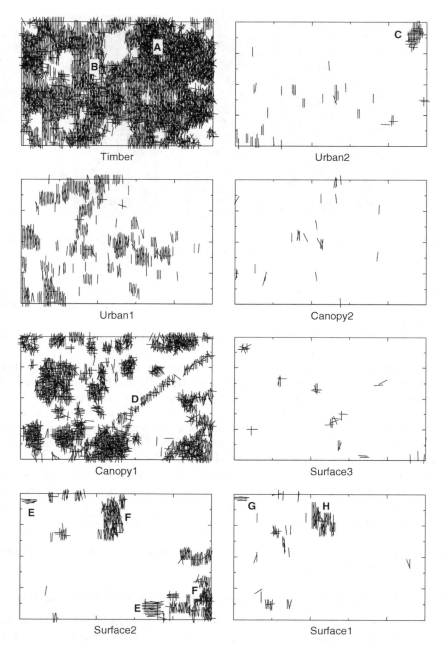

Figure 5.32 Orientation distributions of eight classes. © [2005] IEEE. Reprinted, with permission, from *IEEE Transactions on Geoscience and Remote Sensing*

5A.1 Transformation under Wave Coordinates (FSA)

The Mueller matrix $\mathbf{M}$ is expressed by the elements of the scattering matrix $\mathbf{S}$. The derivations of $\mathbf{M}_{\pm}^{h/v}$ are listed as follows:

$$M_{11} = \frac{1}{2}\left(\left\langle|S_{hh}|^2\right\rangle + \left\langle|S_{hv}|^2\right\rangle + \left\langle|S_{vh}|^2\right\rangle + \left\langle|S_{vv}|^2\right\rangle\right)$$

$$M_{12} = \frac{1}{2}\left(\left\langle|S_{hh}|^2\right\rangle - \left\langle|S_{hv}|^2\right\rangle + \left\langle|S_{vh}|^2\right\rangle - \left\langle|S_{vv}|^2\right\rangle\right)$$

$$M_{21} = \frac{1}{2}\left(\left\langle|S_{hh}|^2\right\rangle + \left\langle|S_{hv}|^2\right\rangle - \left\langle|S_{vh}|^2\right\rangle - \left\langle|S_{vv}|^2\right\rangle\right)$$

$$M_{22} = \frac{1}{2}\left(\left\langle|S_{hh}|^2\right\rangle - \left\langle|S_{hv}|^2\right\rangle - \left\langle|S_{vh}|^2\right\rangle + \left\langle|S_{vv}|^2\right\rangle\right)$$

$$M_{13} = \mathrm{Re}\left\langle S_{hh}S_{hv}^*\right\rangle + \mathrm{Re}\left\langle S_{vv}S_{vh}^*\right\rangle$$

$$M_{14} = \mathrm{Im}\left\langle S_{hh}S_{hv}^*\right\rangle - \mathrm{Im}\left\langle S_{vv}S_{vh}^*\right\rangle$$

$$M_{23} = \mathrm{Re}\left\langle S_{hh}S_{hv}^*\right\rangle - \mathrm{Re}\left\langle S_{vv}S_{vh}^*\right\rangle$$

$$M_{24} = \mathrm{Im}\left\langle S_{hh}S_{hv}^*\right\rangle + \mathrm{Im}\left\langle S_{vv}S_{vh}^*\right\rangle \qquad \text{(5A.1)}$$

$$M_{31} = \mathrm{Re}\left\langle S_{hh}S_{vh}^*\right\rangle + \mathrm{Re}\left\langle S_{vv}S_{hv}^*\right\rangle$$

$$M_{32} = \mathrm{Re}\left\langle S_{hh}S_{vh}^*\right\rangle - \mathrm{Re}\left\langle S_{vv}S_{hv}^*\right\rangle$$

$$M_{41} = -\mathrm{Im}\left\langle S_{hh}S_{vh}^*\right\rangle + \mathrm{Im}\left\langle S_{vv}S_{hv}^*\right\rangle$$

$$M_{42} = -\mathrm{Im}\left\langle S_{hh}S_{vh}^*\right\rangle - \mathrm{Im}\left\langle S_{vv}S_{hv}^*\right\rangle$$

$$M_{33} = \mathrm{Re}\left\langle S_{hh}S_{vv}^*\right\rangle + \mathrm{Re}\left\langle S_{hv}S_{vh}^*\right\rangle$$

$$M_{34} = \mathrm{Im}\left\langle S_{hh}S_{vv}^*\right\rangle - \mathrm{Im}\left\langle S_{hv}S_{vh}^*\right\rangle$$

$$M_{43} = -\mathrm{Im}\left\langle S_{hh}S_{vv}^*\right\rangle - \mathrm{Im}\left\langle S_{hv}S_{vh}^*\right\rangle$$

$$M_{44} = \mathrm{Re}\left\langle S_{hh}S_{vv}^*\right\rangle - \mathrm{Re}\left\langle S_{hv}S_{vh}^*\right\rangle$$

The conversions of the Mueller matrix under different definitions are:

$$\mathbf{M}_{vh}^{v/h} \leftrightarrow \mathbf{M}_{\pm}^{v/h}: \quad \mathbf{M}_{\pm}^{v/h} = \mathbf{U}_1^{-1}\cdot\mathbf{M}_{vh}^{v/h}\cdot\mathbf{U}_1, \quad \mathbf{U}_1 = \begin{bmatrix} 1 & 1 & 0 & 0 \\ 1 & -1 & 0 & 0 \\ 0 & 0 & 1 & 0 \\ 0 & 0 & 0 & 1 \end{bmatrix} \qquad \text{(5A.2)}$$

$$\mathbf{M}_{\pm}^{v/h} \leftrightarrow \mathbf{M}: \quad \mathbf{M} = \mathbf{U}_2^{-1}\cdot\mathbf{M}_{\pm}^{v/h}\cdot\mathbf{U}_2, \quad \mathbf{U}_2 = \begin{bmatrix} 1 & 0 & 0 & 0 \\ 0 & -1 & 0 & 0 \\ 0 & 0 & 1 & 0 \\ 0 & 0 & 0 & -1 \end{bmatrix} \qquad \text{(5A.3)}$$

The conversions between the Mueller matrix and the coherency matrix are:

$$
\begin{aligned}
\mathbf{M} \to \mathbf{T} : T_{11} &= (M_{11} + M_{22} + M_{33} + M_{44})/2 \\
T_{22} &= (M_{11} + M_{22} - M_{33} - M_{44})/2 \\
T_{33} &= (M_{11} - M_{22} + M_{33} - M_{44})/2 \\
T_{44} &= (M_{11} - M_{22} - M_{33} + M_{44})/2 \\
T_{12} &= (M_{12} + M_{21} - jM_{34} + jM_{43})/2 \\
T_{21} &= (M_{12} + M_{21} + jM_{34} - jM_{43})/2 \\
T_{34} &= (jM_{12} - jM_{21} + M_{34} + M_{43})/2 \\
T_{43} &= (-jM_{12} + jM_{21} + M_{34} + M_{43})/2 \\
T_{13} &= (M_{13} + M_{31} + jM_{24} - jM_{42})/2 \\
T_{31} &= (M_{13} + M_{31} - jM_{24} + jM_{42})/2 \\
T_{24} &= (-jM_{13} + jM_{31} + M_{24} + M_{42})/2 \\
T_{42} &= (jM_{13} - jM_{31} + M_{24} + M_{42})/2 \\
T_{14} &= (M_{14} - jM_{23} + jM_{32} + M_{41})/2 \\
T_{41} &= (M_{14} + jM_{23} - jM_{32} + M_{41})/2 \\
T_{23} &= (jM_{14} + M_{23} + M_{32} - jM_{41})/2 \\
T_{32} &= (-jM_{14} + M_{23} + M_{32} + jM_{41})/2
\end{aligned}
\tag{5A.4}
$$

$$
\begin{aligned}
\mathbf{T} \to \mathbf{M} : M_{11} &= \mathrm{Re}(T_{11} + T_{22} + T_{33} + T_{44})/2 \\
M_{22} &= \mathrm{Re}(T_{11} + T_{22} - T_{33} - T_{44})/2 \\
M_{33} &= \mathrm{Re}(T_{11} - T_{22} + T_{33} - T_{44})/2 \\
M_{44} &= \mathrm{Re}(T_{11} - T_{22} - T_{33} + T_{44})/2 \\
M_{12} &= \mathrm{Re}(T_{12} + T_{21} - jT_{34} + jT_{43})/2 \\
M_{21} &= \mathrm{Re}(T_{12} + T_{21} + jT_{34} - jT_{43})/2 \\
M_{34} &= \mathrm{Re}(jT_{12} - jT_{21} + T_{34} + T_{43})/2 \\
M_{43} &= \mathrm{Re}(-jT_{12} + jT_{21} + T_{34} + T_{43})/2 \\
M_{13} &= \mathrm{Re}(T_{13} + T_{31} + jT_{24} - jT_{42})/2 \\
M_{31} &= \mathrm{Re}(T_{13} + T_{31} - jT_{24} + jT_{42})/2 \\
M_{24} &= \mathrm{Re}(-jT_{13} + jT_{31} + T_{24} + T_{42})/2 \\
M_{42} &= \mathrm{Re}(jT_{13} - jT_{31} + T_{24} + T_{42})/2 \\
M_{14} &= \mathrm{Re}(T_{14} - jT_{23} + jT_{32} + T_{41})/2 \\
M_{41} &= \mathrm{Re}(T_{14} + jT_{23} - jT_{32} + T_{41})/2 \\
M_{23} &= \mathrm{Re}(jT_{14} + T_{23} + T_{32} - jT_{41})/2 \\
M_{32} &= \mathrm{Re}(-jT_{14} + T_{23} + T_{32} + jT_{41})/2
\end{aligned}
\tag{5A.5}
$$

The conversions between the coherency matrix and the covariance matrix are:

$$\mathbf{T} \leftrightarrow \mathbf{C} : \mathbf{C} = \frac{1}{2}\mathbf{U}_3^+ \cdot \mathbf{T} \cdot \mathbf{U}_3, \ \mathbf{U}_3 = \begin{bmatrix} 1 & 0 & 0 & 1 \\ 1 & 0 & 0 & -1 \\ 0 & 1 & 1 & 0 \\ 0 & j & -j & 0 \end{bmatrix} \quad (5A.6)$$

5A.2 Transformation under Antenna Coordinates (BSA)

From the scattering matrix, the Mueller matrix is calculated as follows:

$$\mathbf{S}_A \rightarrow \mathbf{M}_A : M_{A,11} = \frac{1}{2}\left(\left\langle\left|S_{hh}^A\right|^2\right\rangle + \left\langle\left|S_{hv}^A\right|^2\right\rangle + \left\langle\left|S_{vh}^A\right|^2\right\rangle + \left\langle\left|S_{vv}^A\right|^2\right\rangle\right)$$

$$M_{A,12} = \frac{1}{2}\left(\left\langle\left|S_{hh}^A\right|^2\right\rangle - \left\langle\left|S_{hv}^A\right|^2\right\rangle + \left\langle\left|S_{vh}^A\right|^2\right\rangle - \left\langle\left|S_{vv}^A\right|^2\right\rangle\right)$$

$$M_{A,21} = \frac{1}{2}\left(\left\langle\left|S_{hh}^A\right|^2\right\rangle + \left\langle\left|S_{hv}^A\right|^2\right\rangle - \left\langle\left|S_{vh}^A\right|^2\right\rangle - \left\langle\left|S_{vv}^A\right|^2\right\rangle\right)$$

$$M_{A,22} = \frac{1}{2}\left(\left\langle\left|S_{hh}^A\right|^2\right\rangle - \left\langle\left|S_{hv}^A\right|^2\right\rangle - \left\langle\left|S_{vh}^A\right|^2\right\rangle + \left\langle\left|S_{vv}^A\right|^2\right\rangle\right)$$

$$M_{A,13} = \text{Re}\left\langle S_{hh}^A S_{hv}^{A*}\right\rangle + \text{Re}\left\langle S_{vv}^A S_{vh}^{A*}\right\rangle$$

$$M_{A,14} = -\text{Im}\left\langle S_{hh}^A S_{hv}^{A*}\right\rangle + \text{Im}\left\langle S_{vv}^A S_{vh}^{A*}\right\rangle$$

$$M_{A,23} = \text{Re}\left\langle S_{hh}^A S_{hv}^{A*}\right\rangle - \text{Re}\left\langle S_{vv}^A S_{vh}^{A*}\right\rangle$$

$$M_{A,24} = -\text{Im}\left\langle S_{hh}^A S_{hv}^{A*}\right\rangle - \text{Im}\left\langle S_{vv}^A S_{vh}^{A*}\right\rangle \quad (5A.7)$$

$$M_{A,31} = \text{Re}\left\langle S_{hh}^A S_{vh}^{A*}\right\rangle + \text{Re}\left\langle S_{vv}^A S_{hv}^{A*}\right\rangle$$

$$M_{A,32} = \text{Re}\left\langle S_{hh}^A S_{vh}^{A*}\right\rangle - \text{Re}\left\langle S_{vv}^A S_{hv}^{A*}\right\rangle$$

$$M_{A,33} = \text{Re}\left\langle S_{hh}^A S_{vv}^{A*}\right\rangle + \text{Re}\left\langle S_{hv}^A S_{vh}^{A*}\right\rangle$$

$$M_{A,34} = -\text{Im}\left\langle S_{hh}^A S_{vv}^{A*}\right\rangle + \text{Im}\left\langle S_{hv}^A S_{vh}^{A*}\right\rangle$$

$$M_{A,41} = -\text{Im}\left\langle S_{hh}^A S_{vh}^{A*}\right\rangle + \text{Im}\left\langle S_{vv}^A S_{hv}^{A*}\right\rangle$$

$$M_{A,42} = -\text{Im}\left\langle S_{hh}^A S_{vh}^{A*}\right\rangle - \text{Im}\left\langle S_{vv}^A S_{hv}^{A*}\right\rangle$$

$$M_{A,43} = -\text{Im}\left\langle S_{hh}^A S_{vv}^{A*}\right\rangle - \text{Im}\left\langle S_{hv}^A S_{vh}^{A*}\right\rangle$$

$$M_{A,44} = -\text{Re}\left\langle S_{hh}^A S_{vv}^{A*}\right\rangle + \text{Re}\left\langle S_{hv}^A S_{vh}^{A*}\right\rangle$$

The conversions between the Mueller matrix and the coherency matrix are:

$$\mathbf{M}_A \rightarrow \mathbf{T}_A : T_{A,11} = (M_{A,11} + M_{A,22} + M_{A,33} - M_{A,44})/2$$

$$T_{A,22} = (M_{A,11} + M_{A,22} - M_{A,33} + M_{A,44})/2$$

$$T_{A,33} = (M_{A,11} - M_{A,22} + M_{A,33} + M_{A,44})/2$$

$$T_{A,44} = (M_{A,11} - M_{A,22} - M_{A,33} - M_{A,44})/2$$

$$T_{A,12} = (M_{A,12} + M_{A,21} + jM_{A,34} + jM_{A,43})/2$$

$$T_{A,21} = (M_{A,12} + M_{A,21} - jM_{A,34} - jM_{A,43})/2$$

$$T_{A,34} = (jM_{A,12} - jM_{A,21} - M_{A,34} + M_{A,43})/2$$

$$T_{A,43} = (-jM_{A,12} + jM_{A,21} - M_{A,34} + M_{A,43})/2 \qquad (5A.8)$$

$$T_{A,13} = (M_{A,13} + M_{A,31} - jM_{A,24} - jM_{A,42})/2$$

$$T_{A,31} = (M_{A,13} + M_{A,31} + jM_{A,24} + jM_{A,42})/2$$

$$T_{A,24} = (-jM_{A,13} + jM_{A,31} - M_{A,24} + M_{A,42})/2$$

$$T_{A,42} = (jM_{A,13} - jM_{A,31} - M_{A,24} + M_{A,42})/2$$

$$T_{A,14} = (-M_{A,14} - jM_{A,23} + jM_{A,32} + M_{A,41})/2$$

$$T_{A,41} = (-M_{A,14} + jM_{A,23} - jM_{A,32} + M_{A,41})/2$$

$$T_{A,23} = (-jM_{A,14} + M_{A,23} + M_{A,32} - jM_{A,41})/2$$

$$T_{A,32} = (jM_{A,14} + M_{A,23} + M_{A,32} + jM_{A,41})/2$$

$$\mathbf{T}_A \rightarrow \mathbf{M}_A : M_{A,11} = \mathrm{Re}(T_{A,11} + T_{A,22} + T_{A,33} + T_{A,44})/2$$

$$M_{A,22} = \mathrm{Re}(T_{A,11} + T_{A,22} - T_{A,33} - T_{A,44})/2$$

$$M_{A,33} = \mathrm{Re}(T_{A,11} - T_{A,22} + T_{A,33} - T_{A,44})/2$$

$$M_{A,44} = -\mathrm{Re}(T_{A,11} - T_{A,22} - T_{A,33} + T_{A,44})/2$$

$$M_{A,12} = \mathrm{Re}(T_{A,12} + T_{A,21} - jT_{A,34} + jT_{A,43})/2$$

$$M_{A,21} = \mathrm{Re}(T_{A,12} + T_{A,21} + jT_{A,34} - jT_{A,43})/2$$

$$M_{A,34} = -\mathrm{Re}(jT_{A,12} - jT_{A,21} + T_{A,34} + T_{A,43})/2$$

$$M_{A,43} = \mathrm{Re}(-jT_{A,12} + jT_{A,21} + T_{A,34} + T_{A,43})/2 \qquad (5A.9)$$

$$M_{A,13} = \mathrm{Re}(T_{A,13} + T_{A,31} + jT_{A,24} - jT_{A,42})/2$$

$$M_{A,31} = \mathrm{Re}(T_{A,13} + T_{A,31} - jT_{A,24} + jT_{A,42})/2$$

$$M_{A,24} = -\mathrm{Re}(-jT_{A,13} + jT_{A,31} + T_{A,24} + T_{A,42})/2$$

$$M_{A,42} = \mathrm{Re}(jT_{A,13} - jT_{A,31} + T_{A,24} + T_{A,42})/2$$

$$M_{A,14} = -\mathrm{Re}(T_{A,14} - jT_{A,23} + jT_{A,32} + T_{A,41})/2$$

$$M_{A,41} = \mathrm{Re}(T_{A,14} + jT_{A,23} - jT_{A,32} + T_{A,41})/2$$

$$M_{A,23} = \mathrm{Re}(jT_{A,14} + T_{A,23} + T_{A,32} - jT_{A,41})/2$$

$$M_{A,32} = \mathrm{Re}(-jT_{A,14} + T_{A,23} + T_{A,32} + jT_{A,41})/2$$

5A.3　Interconversion between Wave Coordinates (FSA) and Antenna Coordinates (BSA)

$$\mathbf{S} \leftrightarrow \mathbf{S}_A : \mathbf{S}_A = \begin{bmatrix} S_{hh}^A & S_{hv}^A \\ S_{vh}^A & S_{vv}^A \end{bmatrix} = \begin{bmatrix} 1 & 0 \\ 0 & -1 \end{bmatrix} \cdot \mathbf{S}^* = \begin{bmatrix} S_{hh}^* & S_{hv}^* \\ -S_{vh}^* & -S_{vv}^* \end{bmatrix} \tag{5A.10}$$

$$\mathbf{M} \leftrightarrow \mathbf{M}_A : \mathbf{M}_A = \mathbf{U}_4 \cdot \mathbf{M}, \quad \mathbf{U}_4 = \begin{bmatrix} 1 & 0 & 0 & 0 \\ 0 & 1 & 0 & 0 \\ 0 & 0 & -1 & 0 \\ 0 & 0 & 0 & 1 \end{bmatrix} \tag{5A.11}$$

References

Britton, M.C. (2000) Radio astronomical polarimetry and the Lorentz group. *Astrophysical Journal*, **532**, 1240–1244.

Cameron, W.L. and Leung, L.K. (1990) Feature motivated polarization scattering matrix decomposition. Radar Conference 1990, Record of the IEEE 1990 International Radar Conference.

Cameron, W.L., Youssef, N.N., and Leung, L.K. (1996) Simulated polarimetric signatures of primitive geometrical shapes. *IEEE Transactions on Geoscience and Remote Sensing*, **34**(3), 793–803.

Carrea, L. and Wanielik, G. (2001) Polarimetric SAR processing using the polar decomposition of the scattering matrix. IEEE Geoscience and Remote Sensing Symposium 2001.

Cloude, S.R. (1986) Group theory and polarization algebra. *Optik*, **75**(1), 26–36.

Cloude, S.R. (1992) Uniqueness of target decomposition theorems in radar polarimetry, in *Direct and Inverse Methods in Radar Polarimetry* (ed. W.M. Boerner), Kluwer Academic, Dordrecht, The Netherlands.

Cloude, S.R. (1995) An entropy based classification scheme for polarimetric SAR data. IEEE International Geoscience and Remote Sensing Symposium 1995.

Cloude, S.R. and Pottier, E. (1996) A review of target decomposition theorems in radar polarimetry. *IEEE Transactions on Geoscience and Remote Sensing*, **34**(2), 498–518.

Cloude, S.R. and Pottier, E. (1997) An entropy based classification scheme for land applications of polarimetric SAR. *IEEE Transactions on Geoscience and Remote Sensing*, **35**(1), 68–78.

Cloude, S.R., Papathanassiou, K.P., and Pottier, E. (2001) Radar polarimetry and polarimetric interferometry. *IEICE Transactions on Electronics*, **E84-C**(12), 1814–1822.

Davidovitz, M. and Boerner, W. (1986) Extension of Kennaugh's optimal polarization concept to the asymmetric scattering matrix case. *IEEE Transactions on Antennas and Propagation*, **34**(4), 569–574.

Ferro-Famil, L., Pottier, E., and Lee, J.S. (2000) Unsupervised classification of multifrequency and fully polarimetric SAR images based on the H/A/Alpha–Wishart classifier. *IEEE Transactions on Geoscience and Remote Sensing*, **39**(11), 2332–2342.

Freeman, A. and Durden, S.L. (1998) A three-component scattering model for polarimetric SAR data. *IEEE Transactions on Geoscience and Remote Sensing*, **36**, 963–973.

Fung, A.K. (1994) *Microwave Scattering and Emission Models and Their Applications*, Artech House, Norwood, MA.

Hellmann, M. (2001) *SAR Polarimetry Tutorial*, http://epsilon.nought.de/tutorials/polsmart/index.php (accessed 24 October 2012).

Huynen, J.R. (1970) *Phenomenological Theory of Radar Targets*, University of Technology, Delft, The Netherlands.

Jin, Y.Q. (1994) *Electromagnetic Scattering Modeling for Quantitative Remote Sensing*, World Scientific, Singapore.

Jin, Y.Q. and Cloude, S.R. (1994) Numerical eigenanalysis of the coherency matrix for a layer of random nonspherical scatterers. *IEEE Transactions on Geoscience and Remote Sensing*, **32**, 1179–1185.

Kang, X., Han, C.Z., and Xu, F. (2007) Clustering polarimetric SAR imagery under deorientation theory. Proceedings of the IEEE International Conference on Acoustics, Speech and Signal Processing, ICASSP2007, Hawaii.

Kennaugh, E.M. (1952) *Polarization Properties of Radar Reflections*, Antenna Laboratory, Ohio State University, Columbus, OH.

Kong, J.A. (2005) *Electromagnetic Wave Theory*, EMW Publishing, Cambridge, MA.

Kong, J.A., Yueh, S.H., Lim, H.H., Shin, R.T., and van Zyl, J.J. (1990) Classification of Earth terrain using polarimetric synthetic aperture radar images. *Progress in Electromagnetics Research*, **3**, 327–370.

Krogager E. (1990) New decomposition of the radar target scattering matrix. *Electronics Letters*, **26**(18), 1525–1527.

Lee, J.S., Grunes, M.R., Ainsworth, T.L. *et al.* (1999) Unsupervised classification using polarimetric decomposition and the complex Wishart classifier. *IEEE Transactions on Geoscience and Remote Sensing*, **37**, 2249–2258.

Lee, J.S., Grunes, M.R., Pottier, E., and Ferro-Famil, L. (2004) Unsupervised terrain classification preserving polarimetric scattering characteristics. *IEEE Transactions on Geoscience and Remote Sensing*, **42**, 722–731.

Lopez-Martinez, C., Ferro-Famil, L., and Pottier, E. (2005) *Polarimetry Tutorial*, http://earth.esa.int/printer_friendly/polsarpro/tutorial.html (accessed 24 October 2012).

Mischenko, M.I. (1992) Enhanced backscattering of polarized light from discrete random media: calculations in exactly the backscattering direction. *Journal of the Optical Society of America A*, **9**, 978–982.

Pottier, E. and Lee, J.S. (1999) Application of the H/A/alpha polarimetric decomposition theorem for unsupervised classification of fully polarimetric SAR data based on the Wishart distribution. Proceedings of Committee on Earth Observing Satellites SAR Workshop, Toulouse, France.

Tsang, L., Kong, J.A., and Shin, R. (1985) *Theory of Microwave Remote Sensing*, John Wiley & Sons, Inc., New York.

van Zyl, J.J. (1985) *On the Importance of Polarization in Radar Scattering Problems*, California Institute of Technology, Pasadena, CA.

van Zyl, J.J. (1989) Unsupervised classification of scattering behavior using radar polarimetry data. *IEEE Transactions on Geoscience and Remote Sensing*, **27**(1), 36–45.

Xu, F. and Jin, Y.Q. (2004) Polarimetric pulse echoes from a layer of non-spherical particles above rough surface. Proceedings of Progress in Electromagnetic Research Symposium (PIERS 2004), Nanjing, China.

Yang, J., Yamaguchi, Y., Yamada, H. *et al.* (2002) The characteristic polarization states and the equi-power curves. *IEEE Transactions on Geoscience and Remote Sensing*, **40**(2), 305–313.

Zebker, H.A., van Zyl, J.J., Durden, S.L., and Norikane, L. (1991) Calibrated imaging radar polarimetry: techniques, examples and applications. *IEEE Transactions on Geoscience and Remote Sensing*, **29**, 942–961.

6

Inversions from Polarimetric SAR Images

The fully polarimetric SAR image presents rich, qualitative, and quantitative information for the inversion of object characteristics. Two examples of inversions from the 2×2 dimensional complex scattering amplitude functions S_{pq} (p, $q = v$, h) and 4×4 dimensional real Mueller matrix M_{ij} ($i, j = 1, \ldots ,4$) are presented in this chapter.

From Equation 1.18, it can be seen that co-polarized (co-pol) or cross-polarized (cross-pol) backscattering signature is a function of the incidence wave with ellipticity angle χ and orientation angle ψ. When the terrain surface is flat, the polarimetric scattering signature has a maximum largely at the orientation angle $\psi = 0$. However, it has been shown that, as the surface is tilted, the orientation angle ψ at the maximum of co-pol or cross-pol signature can shift from $\psi = 0$. This shift can be applied to convert the surface slopes. Making use of the assumption of real co-pol and zero cross-pol scattering amplitude functions (Lee, Jansen, and Schuler, 1998), the ψ shift is expressed by the real scattering amplitude functions. Since both the range and azimuth angles are coupled, two- or multi-pass SAR image data are required for solving two unknowns of the surface slopes. This approach has been well demonstrated for inversion of digital elevation mapping (DEM) and terrain topography by using airborne SAR data (Schuler, Ainsworth, and Lee, 1998; Schuler *et al.*, 2000; Lee, Schuler, and Ainsworth, 2000).

However, the scattering signature is an ensemble average of echo power from random scatter media. Measurable Stokes parameters, such as the polarized scattering intensity, should be directly related to the ψ shift. In this chapter, as the first example, using the Mueller matrix solution, the ψ shift is newly derived as a function of three Stokes parameters, I_{vs}, I_{hs}, U_s, which are measurable by polarimetric SAR imagery. Using the Euler angle transformation between the principal and local coordinates, the orientation angle ψ is related with both the range and azimuth angles, β and γ, of the tilted surface pixel and radar viewing geometry. These results are consistent with Lee, Jansen, and Schuler (1998) and Schuler *et al.* (2000), but are more general.

It is proposed that the linear texture of tilted surface alignment is used to specify the azimuth angle γ. The adaptive threshold method and image morphological thinning algorithm

Polarimetric Scattering and SAR Information Retrieval, First Edition. Ya-Qiu Jin and Feng Xu.
© 2013 John Wiley & Sons Singapore Pte. Ltd. Published 2013 by John Wiley & Sons Singapore Pte. Ltd.

are applied to determine the azimuth angle γ from image linear textures. Thus, the range angle β is then solved, and both β and γ are utilized to obtain the azimuth slope and range slope. Then, the full multi-grid algorithm is employed to solve the Poisson equation of DEM and produce the terrain topography from a single-pass polarimetric SAR image (Jin and Luo, 2004).

The second example is the inversion of bridge height over water surface. Based on the mapping and projection algorithm (MPA) of Chapter 3, the polarimetric SAR image of a bridge over a water surface is simulated. It demonstrates that a bridge in a SAR image can be identified by three strips corresponding to single-, double-, and triple-order scattering, respectively. A set of parameters based on the deorientation theory of Chapter 5 is applied to locate the imaging positions of the single-, double-, and triple-order scattering of the bridge body. The thinning algorithm, clustering algorithm, and Hough transform are employed to detect the lines indicating different orders of scattering. Then, these lines are used to invert the bridge height. Fully polarimetric data of airborne Pi-SAR at X band are applied to the inversion of the height and width of the Naruto Bridge. Based on the same principles, space-borne ALOS PALSAR data of the Eastern Sea Bridge are also applied to the bridge height inversion. The results show the good feasibility of the bridge height inversion.

6.1 Inversion of Digital Elevation Mapping

6.1.1 The Shift of Orientation Angle

Consider a polarized electromagnetic wave incident upon the terrain surface at $(\pi - \theta_i, \varphi_i)$. The incident polarization is defined by the ellipticity angle χ and orientation angle ψ as described by Equation 1.5 and Figure 1.2b. The 2×2 complex scattering amplitude functions $\overline{\mathbf{S}}$ are obtained from polarimetric measurement as

$$\mathbf{E}_s(\theta_s, \varphi_s) = \frac{e^{ikr}}{4\pi r} \overline{\mathbf{S}}(\theta_s, \varphi_s; \pi - \theta_i, \varphi_i) \cdot \mathbf{E}_i(\pi - \theta_i, \varphi_i; \chi, \psi) \tag{6.1}$$

where the elements of the scattering amplitude matrix, S_{pq} ($p, q = v, h$), take account of volumetric and surface scattering from layered scatter media. In SAR imagery, $\overline{\mathbf{S}}$ in Equation 6.1 is measured and is used to construct the real Mueller matrix as follows:

$$\mathbf{I}_s(\theta_s, \varphi_s) = \overline{\mathbf{M}}(\theta_s, \varphi_s; \pi - \theta_i, \varphi_i) \cdot \mathbf{I}_i(\pi - \theta_i, \varphi_i; \chi, \psi) \tag{6.2}$$

where the normalized incident Stokes vector is written as

$$\begin{aligned} \mathbf{I}_i(\chi, \psi) &= [I_{vi}, I_{hi}, U_i, V_i]^{\mathrm{T}} \\ &= [0.5(1 - \cos 2\chi \cos 2\psi), \, 0.5(1 + \cos 2\chi \cos 2\psi), \, -\cos 2\chi \sin 2\psi, \, \sin 2\chi]^{\mathrm{T}} \end{aligned} \tag{6.3}$$

The elements of Equation 6.2, M_{ij} ($i, j = 1, \ldots, 4$), are expressed by the ensemble averages of the real or imaginary parts of $\langle S_{pq} S_{st}^* \rangle$ ($p, q, s, t = v, h$). It is noted that the co-pol terms $\langle |S_{pp}|^2 \rangle$ are always much larger than the cross-pol terms, such that $\langle S_{pq} S_{st}^* \rangle$ ($s \neq t$).

The co-pol backscattering coefficient $\sigma_c(\theta_i, \pi - \varphi_i; \pi - \theta_i, \varphi_i)$ can be obtained as

$$\sigma_c = 4\pi \cos \theta_i \, P_n \tag{6.4}$$

where

$$P_n = 0.5[I_{vs}(1 - \cos 2\chi \cos 2\psi) + I_{hs}(1 + \cos 2\chi \cos 2\psi) + U_s \cos 2\chi \sin 2\psi + V_s \sin 2\chi]$$
(6.5)

When the terrain surface is flat, co-pol backscattering σ_c versus the incidence polarization (χ, ψ) has a maximum at $\psi = 0$ (Lee, Jansen, and Schuler, 1998). However, it has been shown that, as the surface is tilted, the orientation angle ψ for the σ_c maximum is shifted from $\psi = 0$ (Lee, Jansen, and Schuler, 1998; Lee, Schuler, and Ainsworth, 2000; Schuler, Ainsworth, and Lee, 1998; Schuler *et al.*, 2000).

Let $\partial P_n / \partial \psi = 0$ at the maximum σ_c and $\chi = 0$ of the symmetric case. Then, from Equation 6.5, it yields

$$0 = -(I_{hs} - I_{vs})\sin 2\psi + U_s \cos 2\psi + 0.5(I_{hs} + I_{vs})' + 0.5U_s' \sin 2\psi + 0.5(I_{hs} - I_{vs})'\cos 2\psi$$
(6.6)

where the prime denotes $\partial / \partial \psi$. Equation 6.2 and comparison with the terms on the right-hand side of Equation 6.6 shows that

$$0.5(I_{hs} + I_{vs})' \sim (M_{13} + M_{23})\cos 2\psi \sim (\text{Re}\langle S_{vv}S_{vh}^*\rangle + \text{Re}\langle S_{hv}S_{hh}^*\rangle)\cos 2\psi \quad (6.7\text{a})$$

$$0.5U_s' \sim M_{33}\cos 2\psi \sim \text{Re}\langle S_{vv}S_{hh}^* + S_{vh}S_{hv}^*\rangle \cos 2\psi \quad (6.7\text{b})$$

$$(I_{hs} - I_{vs}) \sim 0.5(M_{11} + M_{22})\cos 2\psi \sim 0.5(\langle|S_{vv}|^2\rangle + \langle|S_{hh}|^2\rangle)\cos 2\psi \quad (6.7\text{c})$$

$$0.5(I_{hs} - I_{vs})' \sim 0.5(M_{13} - M_{23})\cos 2\psi \sim 0.5(\text{Re}\langle S_{vv}S_{vh}^*\rangle - \text{Re}\langle S_{hv}S_{hh}^*\rangle)\cos 2\psi \quad (6.8\text{a})$$

$$U_s \sim 0.5(M_{31} + M_{32}) \sim \text{Re}\langle S_{vv}S_{hv}^*\rangle + \text{Re}\langle S_{vh}S_{hh}^*\rangle \quad (6.8\text{b})$$

It can be seen that Equations 6.7a and 6.7b are much less than Equation 6.7c, and Equation 6.8a is much less than Equation 6.8b, so the last three terms on the right-hand side of Equation 6.6 are now neglected. Thus, it yields the ψ shift at the σ_c maximum expressed by the Stokes parameters as follows:

$$\tan 2\psi = \frac{U_s}{I_{hs} - I_{vs}}$$
(6.9)

It can be seen that the third Stokes parameter $U_s \neq 0$ causes the ψ shift. Note that, if both U_s and $I_{hs} - I_{vs}$ approach zero, for example, scattering from uniformly oriented scatterers or isotropic scatter media such as thick vegetation canopy, ψ cannot be well defined by Equation 6.9. Furthermore, from Equation 6.2, we obtain

$$U_s \sim 0.5(M_{31} + M_{32}) = \text{Re}(\langle S_{vv}S_{hv}^*\rangle + \langle S_{vh}S_{hh}^*\rangle) \quad (6.10\text{a})$$

$$I_{hs} - I_{vs} \sim 0.5(M_{22} - M_{11}) = 0.5(\langle|S_{hh}|^2\rangle - \langle|S_{vv}|^2\rangle) \quad (6.10\text{b})$$

It is interesting to see that if we let

$$\text{Re}(\langle S_{vv}S_{hv}^* \rangle + \langle S_{vh}S_{hh}^* \rangle) = 2S_{hv}S_0 \quad \text{and} \quad 0.5(\langle |S_{hh}|^2 \rangle - \langle |S_{vv}|^2 \rangle) \sim S_0(S_{hh} - S_{vv}),$$

where $S_0 \equiv S_{vv} = S_{hh}$ is assumed, Equation 6.9 can be reduced to the result of Schuler *et al.* (2000),

$$\tan 2\psi = \frac{2S_{vh}}{S_{hh} - S_{vv}} \tag{6.11}$$

Using the cross-pol backscattering coefficient σ_x, the same result can be obtained.

6.1.2 Range and Azimuth Angles from Euler Angle Transformation

The polarization vectors $\hat{h}_i, \hat{v}_i$ of the incident wave at $(\pi - \theta_i, \phi_i)$ in the principle coordinates $(\hat{x}, \hat{y}, \hat{z})$ are defined as described by Equation 1.2 as

$$\hat{h}_i = \frac{\hat{z} \times \hat{k}_i}{|\hat{z} \times \hat{k}_i|} = -\sin \phi_i \hat{x} + \cos \phi_i \hat{y} \quad \text{and} \quad \hat{v}_i = \hat{h}_i \times \hat{k}_i \tag{6.12}$$

where the incident wavevector

$$\hat{k}_i = \sin \theta_i \cos \phi_i \hat{x} + \sin \theta_i \sin \phi_i \hat{y} - \cos \theta_i \hat{z} \tag{6.13}$$

As the pixel surface has local slope as shown in Figure 6.1, the polarization vectors should be redefined following the local normal vector $\hat{z}_b$ as follows:

$$\hat{h}_b = \frac{\hat{z}_b \times \hat{k}_{ib}}{|\hat{z}_b \times \hat{k}_{ib}|} \quad \text{and} \quad \hat{v}_b = \hat{h}_b \times \hat{k}_{ib} \tag{6.14}$$

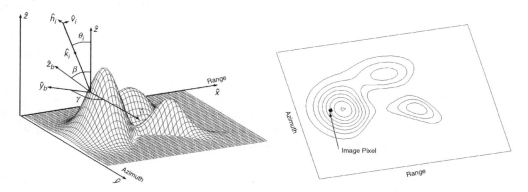

Figure 6.1 Incidence upon a pixel of a tilted surface and the contour to be inverted

By using the transformation of the Euler angles (β, γ) between the two coordinate systems $(\hat{x}, \hat{y}, \hat{z})$ and $(\hat{x}_b, \hat{y}_b, \hat{z}_b)$ (Jin, 1994), one has

$$\hat{x} = \cos\gamma\cos\beta\,\hat{x}_b + \sin\gamma\,\hat{y}_b - \cos\gamma\sin\beta\,\hat{z}_b \tag{6.15a}$$

$$\hat{y} = -\sin\gamma\cos\beta\,\hat{x}_b + \cos\gamma\,\hat{y}_b + \sin\gamma\sin\beta\,\hat{z}_b \tag{6.15b}$$

$$\hat{z} = \sin\beta\,\hat{x}_b + \cos\beta\,\hat{z}_b \tag{6.15c}$$

Substituting Equations 6.14 and (6.15a–6.15c) into Equations 6.12 and 6.13, it yields

$$\hat{h}_i = -\cos\beta\sin(\gamma+\phi_i)\hat{x}_b + \cos(\gamma+\phi_i)\hat{y}_b + \sin\beta\sin(\gamma+\phi)\hat{z}_b \tag{6.16}$$

$$\hat{k}_{ib} = (\cos\beta\sin\theta_i\cos(\gamma+\phi_i) - \sin\beta\cos\theta_i)\hat{x}_b + \sin\theta_i\sin(\gamma+\phi_i)\hat{y}_b$$
$$- [\sin\beta\sin\theta_i\cos(\gamma+\phi_i) + \cos\beta\cos\theta_i]\hat{z}_b \tag{6.17}$$

Substituting Equation 6.17 into Equation 6.14, the polarization vector $\hat{h}_b$ for local surface pixel is written as

$$\hat{h}_b = \frac{a\hat{x}_b + b\hat{y}_b}{\sqrt{a^2 + b^2}} \tag{6.18}$$

where

$$a = -\sin\theta_i\sin(\gamma+\phi_i) \quad\text{and}\quad b = \sin\theta_i\cos\beta\cos(\gamma+\phi_i) - \sin\beta\cos\theta_i \tag{6.19}$$

Thus, Equations 6.16 and 6.18 yield the orientation angle as

$$\cos\psi = \hat{h}_b \cdot \hat{h}_i = \frac{\cos\beta\sin\theta_i - \cos(\gamma+\phi_i)\sin\beta\cos\theta_i}{\sqrt{a^2 + b^2}} \tag{6.20a}$$

$$\tan\psi = \frac{\tan\beta\sin(\gamma+\phi_i)}{\sin\theta_i - \cos\theta_i\tan\beta\cos(\gamma+\phi_i)} \tag{6.20b}$$

Taking $\phi_i = 0$, that is, $\hat{x}$ is the range direction and $\bar{\hat{y}}$ is the azimuth direction, Equation 6.20b is reduced to the equation given by Schuler *et al.* (2000). Note that the azimuth angle in Lee, Jansen, and Schuler (1998) was defined by the angle between $\hat{\rho}$ and $\hat{y}$. Thus, the range and azimuth slopes of the pixel surface can be defined as in Lee, Jansen, and Schuler (1998) as

$$S_R = \tan\beta\cos\gamma$$
$$S_A = \tan\beta\sin\gamma \tag{6.21}$$

Since a single ψ shift cannot simultaneously determine two unknowns β and γ in Equation 6.20b, two- or multi-pass SAR image data are usually needed. This has been discussed in Lee, Jansen, and Schuler (1998).

6.1.3   The Azimuth Angle of Every Pixel in a SAR Image

If only the single-pass SAR image data are available, one of two unknown angles, β or γ, should be determined first. The azimuth alignment of tilted surface pixels can be visualized as a good indicator of the azimuth direction. We apply the adaptive threshold method and image morphological thinning algorithm (Castleman, 1996) to specify the azimuth angle in all SAR image pixels. The algorithm contains the following steps:

1. Carry out speckle filtering over the entire image.
2. Apply the adaptive threshold method (Chan *et al.*, 1998) to produce a binary image (see Figure 6.2a). The global threshold value is not adopted because of the heterogeneity of the image pixels.
3. Apply the image morphological processing for the binary image, remove those isolated pixels and fill small holes. Referring to regions of binary "1"s as foreground and binary "0"s as background, the edges from the foreground are extracted (see Figure 6.2b).
4. Each pixel on the edge is set as the center of a square 21×21 window (see Figure 6.2c), and a curve segment through the centered pixel is then obtained. Then, applying the polynomial algorithm for fitting a curve segment in the least-squares

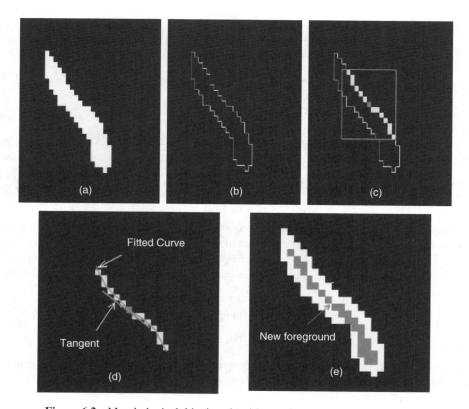

Figure 6.2   Morphological thinning algorithm to determine the azimuth angle

sense, the tangential slope of the centered pixel is obtained (see Figure 6.2d). It yields the azimuth angle of the centered pixel. Make a mark on that pixel so that it will not be calculated in the next turn.

5. Removing the edge in step 4 from the foreground, a new foreground is formed. Repeat step 4 until the azimuth angle of every pixel in the initial foreground is determined.

6. Make the complementary binary image, that is, the initial background now becomes the foreground (see the gray/red line in Figure 6.2e). Then, steps 4 and 5 are repeated for this complementary image until the azimuth angle of every pixel in the initial background is determined.

This approach provides supplementary information to first determine the angle γ over the whole image area if there is no other information available.

6.2 An Example of Algorithm Implementation

As an example for DEM inversion, the SAR image of total polarimetric power *hh*, *vv*, and *hv* data at L band on 4 October 1994 over Taiwan, China, measured by SIR-C of the Cal-Tech JPL is investigated, as shown in Figure 6.3. The center location is about (120.964°E, 23.487°N) under $\theta_i = 32.85°$, and the pixel size is 12.5 m × 12.5 m. Figure 6.3 is a 512 pixel × 512 pixel part of the original SAR image. The center location is (120.8626°E, 23.6659°N) situated in the Ali Mountain of Taiwan. Figure 6.4 is the image after processing via the morphological thinning algorithm in aforementioned steps 1–6 in the previous section.

To focus on the prominent topographic structure of DEM, some trivial undulations are omitted. Note that the azimuth angle γ is the angle between the slope direction $\hat{y}_b$ and the azimuth direction $\hat{y}$, as indicated in Figure 6.1.

As the azimuth angle γ of each pixel is obtained by the adaptive threshold method and thinning algorithm as described in the above steps, the β angle of each pixel can be

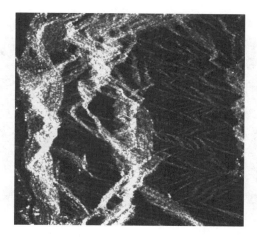

Figure 6.3 A SAR image

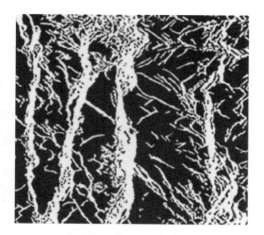

Figure 6.4   Morphological thinning processed image

determined by Equation 6.20b. Taking $\phi_i = 0$, this yields

$$\tan \beta = \frac{\tan \psi \sin \theta_i}{\sin \gamma + \tan \psi \cos \theta_i \cos \gamma} \tag{6.22}$$

where the orientation angle ψ is calculated using Equation 6.9, while the incident angle θ_i is determined by the SAR view geometry.

Substituting β and γ into Equation 6.21, the azimuth slope S_A and range slope S_R are obtained. Following the approach of Ghiglia and Romero (1994), Pritt and Shipman (1994), and Demmel (1997), we can utilize the two slopes to invert the terrain topography and DEM. The slopes S_A and S_R of all pixels in the SAR image are calculated as shown in Figure 6.5a and b.

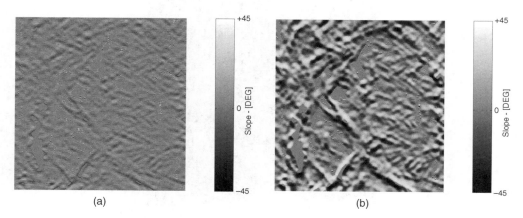

(a)   (b)

Figure 6.5   (a) Azimuth slope S_A. (b) Range slope S_R

The DEM can be generated by solving the Poisson equation for an $M \times N$ rectangular grid area (Schuler *et al.*, 2000). The approach is similar to the phase unwrapping of INSAR (Pritt, 1996; Takajo and Takahashi, 1998; Ghiglia and Romero, 1994). The Poisson equation can be written as

$$\nabla^2 \phi(x, y) = \rho(x, y) \tag{6.23}$$

where ∇^2 is the Laplace operator $\partial^2/\partial x^2 + \partial^2/\partial y^2$. The source function $\rho(x, y)$ consists of the surface curvature calculated by the slope $S(x, y)$. Taking the $\hat{x}$ direction as the range direction and $\hat{y}$ direction as the azimuth direction, the discrete values of the source function can be derived (the superscript A and R represent the azimuth and range directions, respectively) as follows:

$$\rho_{ij}^R = S_{ij}^R - S_{i-1,j}^R, \quad \rho_{ij}^A = S_{ij}^A - S_{i,j-1}^A \quad (0 \le i \le M, \, 0 \le j \le N) \tag{6.24}$$

Assuming that the errors in the input slopes are Gaussian random numbers, the integrated elevation values also have Gaussian errors. Therefore, χ^2 statistics is a good measure of matching the solution to the exact value (this is also the standard assumption of the phase unwrapping algorithm of INSAR); χ^2 is given by

$$\chi^2 = \sum_{i,j} (\phi_{ij} - \phi_{i-1,j} - S_{ij}^R)^2 + \sum_{i,j} (\phi_{ij} - \phi_{i,j-1} - S_{ij}^A)^2 \tag{6.25}$$

The differences between the partial derivatives of the solution ϕ_{ij} and the partial derivatives defined by Equation 6.24 should be minimized. Differentiating χ^2 with respect to ϕ_{ij} and setting the results equal to zero yields

$$(\phi_{i+1,j} - 2\phi_{ij} + \phi_{i-1,j}) + (\phi_{i,j+1} - 2\phi_{ij} + \phi_{i,j-1}) = \rho_{ij} \tag{6.26}$$

Equation 6.26 presents the Poisson equation to be solved.

The discrete source function ρ_{ij} is defined by

$$\rho_{ij} = (S_{i+1,j}^R - S_{ij}^R) + (S_{i,j+1}^A - S_{ij}^A) \tag{6.27}$$

In the INSAR approach (Lee, Jansen, and Schuler, 1998; Schuler, Ainsworth, and Lee, 1998), the alternating direction iteration algorithm (ADI algorithm) was applied to generate the DEM. In this chapter, the full multi-grid algorithm (FMG algorithm) (Pritt and Shipman, 1994; Demmel, 1997) is employed to solve the Poisson equation. The benefits of FMG are due to its rapid convergence, robustness, and low computation load. For an $M \times N$ rectangular grid, the FMG method has computations on the order of $(MN)\ln(MN)$. The detailed steps of FMG can be found in Pritt and Shipman (1994).

Figure 6.6 is the inverted DEM, Figure 6.7 is the contour map, and Figure 6.8 presents the DEM picture. It will not be a surprise that there is a similarity between the inverted topography (Figure 6.6) and the SAR image (Figure 6.3). Also, the inverted contour map has been validated in comparison to the map in the China Integrative Atlas (1990). But, because more accurate DEM as ground truth is not yet available, the inverted topography has only been roughly verified.

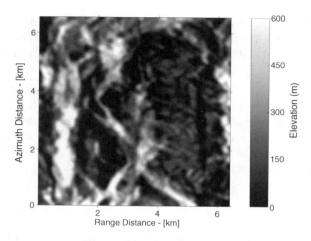

Figure 6.6 The DEM image

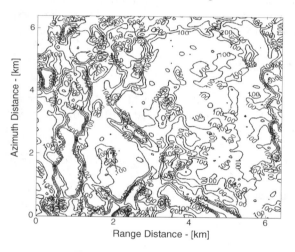

Figure 6.7 The contour map

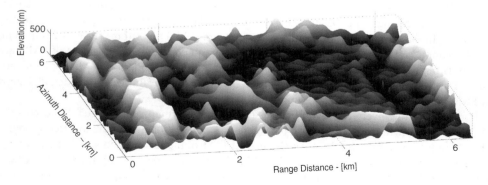

Figure 6.8 A picture of inverted DEM

6.3   Inversion of Bridge Height

Monitoring bridge security all day, all night, and in all weather (especially severe rain) is one of the extensive applications of synthetic aperture radar (SAR) technology. As the SAR resolution has much improved, for example, airborne SAR having been upgraded to the centimeter level, quantitative inversion of the target characteristics, for example, bridge height over water, has become feasible (Oliver and Quegan, 1998; Ouchi, 2004).

Airborne Pi-SAR presented fully polarimetric images with 5 m resolution. The Japanese ALOS PALSAR launched in January 2006 has resolution of about 5 m (JAXA, http://www .eorc.jaxa.jp/ALOS/about/palsar.htm) and the resolution of TerraSAR-X reaches 2 m (Roth, Eineder, and Schaettler, 2003). Airborne SAR has higher resolution, which is in the order of 10 cm to 1 m. High-resolution full polarimetric airborne Pi-SAR data (Kobayashi, Umehara, and Satake, 2000) and medium-resolution spaceborne ALOS-PALSAR data are used to develop the bridge height inversion algorithm in this chapter.

In this section, the scattering mechanism of a bridge object is first described by a set of geometric rays. Then an inversion method is presented. A simple bridge model is first used for SAR image simulation and analysis of different order scattering shown in the image. Then, a set of classification parameters from the deorientation theory of Chapter 5 are applied to indicate different order scattering appearing in the SAR image. Using the thinning and clustering algorithms, and Hough transform, these lines are clearly indicated for bridge height inversion. It is applied to height inversion of the Naruto Bridge from polarimetric Pi-SAR image (Wang, Xu, and Jin, 2009). Finally, the same principle is applied to height inversion of the Eastern Sea Bridge from a mono-polarized ALOS-PALSAR image.

6.3.1   Geometric Rays Projection for Analysis of Bridge Scattering

SAR is a side-looking radar, and target position in a SAR image is decided by the time delay of radar echoes, not simply by the pixel distance in the image, which is affected by geometric shadowing, foreshortening, layover, and so on. A steel bridge structure with a high reflectivity can produce strong scattering as bright signatures in the image. Geometric rays projection is used to describe the scattering mechanism of a bridge object in Figure 6.9 (Robalo and Lichtenegger, 2000), where the numbers 1, 2, 3 denote single-, double-, and triple-order

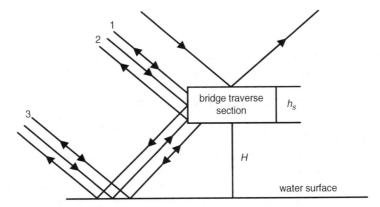

Figure 6.9   Single and multi-bounce scattering of a bridge; the number denotes scattering order

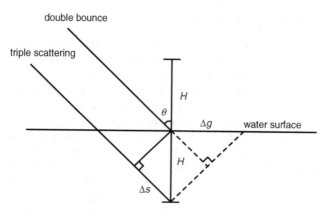

Figure 6.10   Distances of double and triple scattering, and the relation between the slant range and ground range

scattering, and H and h_s are the bridge height over the water surface and bridge body thickness, respectively. Single, double, and triple scattering of the bridge might produce three strips in the image due to different time delays of the echoes. In the case of the bridge azimuth direction not being parallel to the SAR flight, the strips due to double and triple scattering might become wider (Xu and Jin, 2006). Of course, more complicated bridge structures can produce much higher-order scattering and make the image more complicated.

To make the rays projection, Figure 6.10 presents a simple description of the travel distance of double- and triple-order scattering, which produces the relation of slant range and ground range in a SAR image. In Figure 6.10, θ is the incidence angle, and Δs and Δg are, respectively, the distance difference of double and triple scattering in the slant range and ground range. This yields the relation $\Delta s = \Delta g \sin \theta$. Thus, the bridge height over the water surface is inverted as

$$H = \Delta g \times \tan \theta \tag{6.28}$$

From Figure 6.9, it can be seen that the single scattering makes two strip lines due to the echoes from the bridge body with h_s, which is also estimated by Equation 6.22.

6.3.2   SAR Image Simulation of a Bridge Object

The MPA of Chapter 3 is employed to simulate a SAR image of a bridge object and validate the aforementioned analysis of different order scattering.

Figure 6.11 shows a picture of the artificial bridge, which is modeled as a lengthy cuboid with the side pillars over a rough water surface. Some geometric parameters of the bridge are taken as follows: bridge length 1 km, distance between the two pillars 800 m, bridge width 25 m, bridge height 45 m, bridge body thickness 10 m, pillar height 140 m, and angle between SAR flight and bridge azimuth 25°. Scatterings of the cuboid interfaces and water

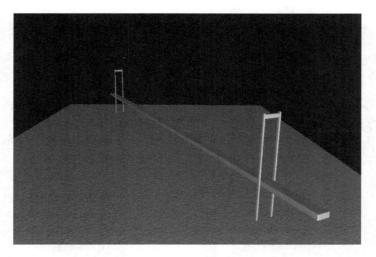

Figure 6.11   Bridge model for SAR image simulation

surface are calculated by the integral equation method (IEM) model (Fung, 1994). The scattering simulation of the bridge cuboid over the water surface follows the approach of scattering from a building on land (Xu and Jin, 2006). To enhance scattering from the bridge body, the roughness of the cuboid interfaces is especially increased.

The SAR image simulation for the bridge model of Figure 6.11 is shown in Figure 6.12. It can be seen, especially from the color image of Figure 6.12b, that the images from different order scattering are produced in different positions, which are predicted in the aforementioned analysis, that is, single-, double-, and triple-order scattering and bridge shadow are located in a sequence from near to far. Also, the pillars can be well recognized.

In Figure 6.12b, it can be seen that double scattering appears as two lines in gray/red color, which is caused by two converse paths between bridge and water surface. But, in a real SAR image, this clear separation will not be identified.

6.3.3   Inversion of Naruto Bridge Height using Classification Parameters

In Chapter 5, using deorientation theory, a set of parameters, u, v, w, and the entropy H have been studied and applied to surface classification (Xu and Jin, 2007). From Figure 5.21, it has been found that, as $v < -0.2$, it indicates dominant double scattering; as $-0.2 < v < 0.2$, it indicates higher-order or multiple scattering; and as $v > 0.2$, it is mainly caused by single scattering. The threshold 0.2 is empirically obtained from simulation of a canopy model. However, it can be identified that negative v indicates dominance of double scattering. Thus, the parameter v is employed to identify double scattering produced by the bridge.

Figure 6.13 is a photo of Naruto Bridge connecting Honshu Island and Shikoku Island in Japan. The bridge height is 45 m over the water surface, the length is 1629 m, the width is 25 m, with two pillars of height 144.3 m and separation 876 m.

Japanese airborne Pi-SAR with two fully polarimetric X band (9.55 GHz) antennas and a single fully polarimetric L band (1.27 GHz) antenna has operated under spatial resolution of

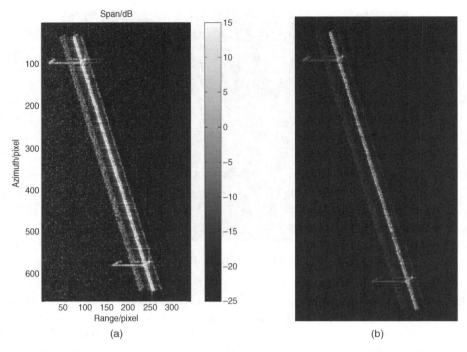

Figure 6.12   A SAR image simulation of the bridge: (a) total power image; (b) a pseudo-color SAR image composed by R = HH − VV, G = HV + VH, and B = HH + VV

1.5 m × 1.5 m at X band and 3 m × 3 m at L band. Figure 6.14a–c shows Pi-SAR images of HH, VV, and HV polarizations of Naruto Bridge on 7 November 2004. The azimuth direction of SAR flight is from the top to bottom and the range is from left to right. It can be seen that the bright dots of the bridge suspensions are well identified and might give the bridge width.

Figure 6.13   A photo of Naruto Bridge, which connects Honshu Island and Shikoku Island, Japan

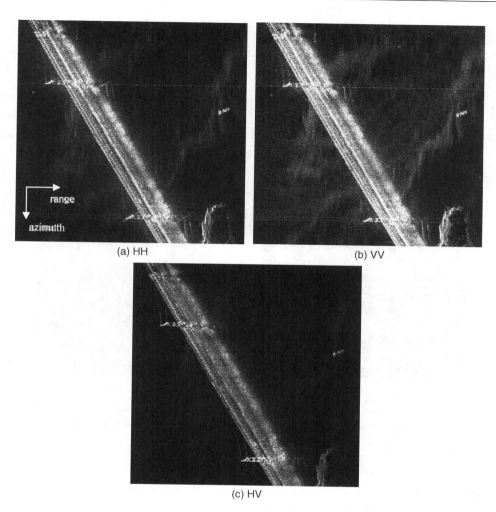

Figure 6.14 Pi-SAR images over Naruto Bridge, the center at (134°38′E, 34°14′N) under incidence angle at 41.8°

It is expected that co-pol echoes are generally higher than cross-pol ones, and VV is higher than HH. A ship object on the right middle and a land area on the right bottom can be identified. The brightness of the pillars image is especially strong, and the suspensions appear as bright dots, particularly in the HV image, which is caused by scattering of the joints structures. Some bright lines near the pillar positions of the co-pol image might be due to the side-lobe effect, while they are not identified in the HV image because of low intensity.

The parameters u, v, w of the Pi-SAR data are obtained as shown in Figure 6.15a–c. In Figure 6.15a for the parameter u, the triple-order scattering appears blue, but single- and double-order scattering are not very distinguishable. In Figure 6.15b for the parameter v, all of the single- and triple-order scattering (yellow and red colors) and the double-scattering indicated by negative v as blue color are well identified. Accordingly, the parameter v is

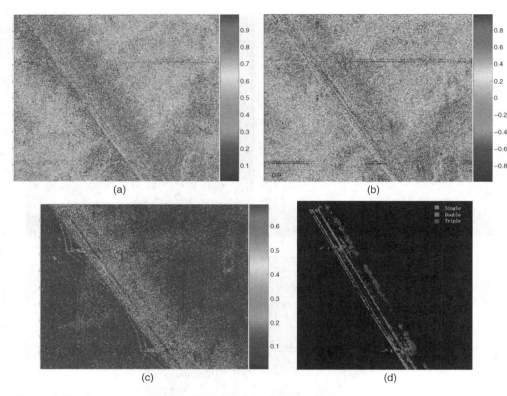

Figure 6.15 The parameters (a) u, (b) v, and (c) w of Pi-SAR data after deorientation. (d) Filtering and clustering of the parameter v image: green, red, and blue colors represents single, double, and triple scattering, respectively

employed to construct a result as shown in Figure 6.15d, and further invert the bridge height H. In Figure 6.15c for the parameter w, the slings can be seen clearly.

Figure 6.15b is further processed by filtering and clustering algorithms. The medium filtering is firstly utilized by moving a window with size of 3×3 pixels, and the clustering algorithm using a window of 5×5 pixels is then employed. The central pixel dots are confined to decide where the order scattering occurs most frequently. Finally, the images of single-, double-, and triple-order scatterings are clustered by green, red, and blue colors as shown in Figure 6.15d. Here, the ship and land area are simply erased.

Based on our analysis of SAR image simulation and deorientation analysis, there are three strip lines from near range to far range to indicate single-, double-, and triple-order scattering, respectively. The center lines of the strips are constructed from Figure 6.15d for inversion of the bridge height and width. Table 6.1 gives the inversion of H, h_s, d and comparison with the measurements. The inversion error of H is about 3.8 m, that is, about 8.4%, and the error of d is only 0.7 m, that is, about 2.8%. The errors are most likely due to (1) the coarse resolution of about 1.5 m (the resolution of airborne SAR can now reach the 10 cm level),

Table 6.1 Comparison of inversed height and ground truth data

	H (m)	h_s (m)	d (m)
Measurement	45	—	25
Estimation	48.8	10.1	25.7

and (2) the complicated structure of a real bridge making the scattering mechanism more complicated, not as well described as by Figure 6.9.

6.3.4 Inversion of the Height of Eastern Sea Bridge using ALOS SAR Data

Figure 6.16 is a photo of the Eastern Sea Bridge connecting the international deep water port to Shanghai. Its total length is about 32 km, and the width is about 31.5 m. The maximum height of the bridge is about 40 m above the water surface.

ALOS SAR, launched by the Japan Aerospace Exploration Agency (JAXA) in January 2006, carries L band PALSAR (phased array L band SAR) with three polarimetric modes. The resolution of single and dual polarization mode (HH + HV or VV + HV) is 7 m and 14 m, respectively, and the resolution of the fully polarimetric mode is 24 m. Since HH-pol PALSAR data are only available at this moment, these data are utilized to invert the height of the Eastern Sea Bridge following the aforementioned multi-order scattering analysis.

Figure 6.17 shows the ALOS-PALSAR HH image of Eastern Sea Bridge on 22 February 2007. The SAR flight was from the left to right with the center incidence angle 39.3°. Owing to limited resolution and mono-polarization, the bridge and suspensions are not identified as well as in Figure 6.14. The area marked by the white box is chosen for our inversion.

First, the thinning algorithm is applied to decrease noise and confine the lines (Xu and Jin, 2007). The intensity of a pixel is compared with its neighborhood to decide if this pixel is on the line or not.

The Hough transform is then employed to construct the strip lines of Figure 6.18. It can be seen that in Figure 6.18 there are three lines to indicate single-, double-, and triple-order

Figure 6.16 A photo of the Eastern Sea Bridge, Shanghai, China

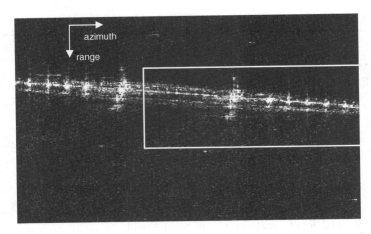

Figure 6.17   ALOS-PALSAR image of the Eastern Sea Bridge (the center is at 121°97′E and 30°73′N, under incidence angle of 39.3°)

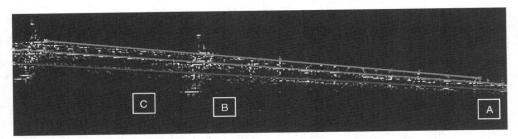

Figure 6.18   Lines detected from Figure 6.17 based on the thinning algorithm and Hough transform. The bridge heights at A, B, and C are inverted

scattering by green, red, and blue lines, respectively. Since the bridge height at position C between the two pillars is highest and gradually goes down to the sides, the confined lines are certainly not parallel.

Table 6.2 shows the height inversion of the Eastern Sea Bridge at the locations A, B, and C, and comparison with the measurement. The inversion error is on the order of a meter.

Table 6.2   Inverted bridge heights at A, B, and C

	A	B	C
Measurement (m)	15	35	40
Inversion (m)	17.8	38.6	44.6

References

Castleman, K.R. (1996) *Digital Image Processing*, Prentice-Hall, Englewood Cliffs, NJ.

Chan, F.H.Y., Lam, F.K., and Zhu, H. (1998) Adaptive thresholding by variational method. *IEEE Transactions on Geoscience and Remote Sensing*, **7** (3), 468–473.

China Map Publisher (1990) *China Integrative Atlas*, China Map Publisher, Beijing, p. 151.

Demmel, J. (1997) *Applied Numerical Linear Algebra*, Society for Industrial and Applied Mathematics, Philadelphia, PA.

Fung, A.K. (1994) *Microwave Scattering and Emission Models and Their Applications*, Artech House, Norwood, MA.

Ghiglia, D.C. and Romero, L.A. (1994) Robust two-dimensional weighted and unweighted phase unwrapping that uses fast transforms and iterative methods. *Journal of the Optical Society of America A*, **11** (1), 107–117.

Jin, Y.Q. (1994) *Electromagnetic Scattering Modelling for Quantitative Remote Sensing*, World Scientific, Singapore.

Jin, Y.Q. and Luo, L. (2004) Terrain topographic inversion using single-pass polarimetric SAR image data. *Science in China*, **47** (4), 490–500.

Kobayashi, T., Umehara, T., and Satake, M. (2000) Airborne dual-frequency polarimetric and interferometric SAR. *IEICE Transactions on Communications*, **E83-B** (9), 1945–1954.

Lee, J.S., Jansen, R.W., and Schuler, D.L. (1998) Polarimetric analysis and modeling of multi-frequency SAR signatures from Gulf Stream fronts. *IEEE Journal of Oceanic Engineering*, **23**, 322–333.

Lee, J.S., Schuler, D.L., and Ainsworth, T.L. (2000) Polarimetric SAR data compression for terrain azimuthal slope variation. *IEEE Transactions on Geoscience and Remote Sensing*, **38** (5), 2153–2163.

Oliver, C. and Quegan, S. (1998) *Understanding Synthetic Aperture Radar Images*, Artech House, London.

Ouchi, K. (2004) *Principles of Synthetic Aperture Radar for Remote Sensing*, Denki University Press, Tokyo (in Japanese).

Pritt, M.D. (1996) Phase unwrapping by means of multi-grid techniques for interferometric SAR. *IEEE Transactions on Geoscience and Remote Sensing*, **34**, 728–738.

Pritt, M.D. and Shipman, J.S. (1994) Least-squares two-dimensional phase unwrapping using FFTs. *IEEE Transactions on Geoscience and Remote Sensing*, **32** (3), 706–708.

Robalo, J. and Lichtenegger, J. (2000) ERS-SAR Images a Bridge, http://esapub.esrin.esa.it/eoq/eoq64/bridge.pdf.

Roth, A., Eineder, M., and Schaettler, B. (2003) SAR interferometry with TerraSAR-X, ESA Workshop Fringe.

Schuler, D.L., Ainsworth, T.L., and Lee, J.S. (1998) Topographic mapping using polarimetric SAR data. *International Journal of Remote Sensing*, **19** (1), 141–160.

Schuler, D.L., Lee, J.S., Ainsworth, T.L., and Grunes, M.R. (2000) Terrain topography measurements using multipass polarimetric synthetic aperture radar data. *Radio Science*, **35**, 813–832.

Takajo, H. and Takahashi, T. (1998) Least-squares phase estimation from phase difference. *Journal of the Optical Society of America A*, **5**, 416–425.

Wang, H., Xu, F., and Jin, Y.Q. (2009) Estimation of bridge height over water from polarimetric SAR image data using mapping and projection algorithm and de-orientation theory. *IEICE Transactions on Electronics*, **E92-B** (12), 3875–3882.

Xu, F. and Jin, Y.Q. (2005) Deorientation theory of polarimetric scattering targets and application to terrain surface classification. *IEEE Transactions on Geoscience and Remote Sensing*, **43** (10), 2351–2364.

Xu, F. and Jin, Y.Q. (2006) Imaging simulation of polarimetric SAR for a comprehensive terrain scene using the mapping and projection algorithm. *IEEE Transactions on Geoscience and Remote Sensing*, **44** (11), 3219–3234.

Xu, F. and Jin, Y.Q. (2007) Automatic reconstruction of building objects from multiaspect meter-resolution SAR images. *IEEE Transactions on Geoscience and Remote Sensing*, **45** (7), 2336–2353.

7

Automatic Reconstruction of Building Objects from Multi-Aspect SAR Images

The inversion and information retrieval studies in the previous chapters have been largely restricted to the scope of one single pixel or 1D pulse profile of radar return. The abundant spatial information conveyed by 2D SAR images in the form of mutual correlation among neighboring pixels is another important subject in remote sensing. In medium- to low-resolution SAR images, spatial information mostly reflects the rough geometric shape of the target being observed. As SAR technology, in particular the spaceborne SAR programs, has advanced toward the high-resolution era (meters to decimeters), the fine characteristics of the target become observable, and the rich information hidden among the tiny pixels is yet to be explored.

With the rapid advancement of synthetic aperture radar (SAR) technology and its operational services, such as ALOS, TerraSAR-X, RadarSat-2, Cosmo-Skymed, and so on, high-resolution SAR image data have become more accessible, in particular multi-aspect high-resolution images from left/right side-looking and descending/ascending orbits (Roth, 2003). This has greatly promoted new studies of surface monitoring and information retrieval for SAR applications (Roth, 2003), in particular in the areas of target extraction, parameter retrieval, and 3D object reconstruction.

A SAR image is produced through interactions of the radar wave with the objects and background scene. Owing to the complexity and heterogeneity of distributed objects and structures, most previous work on normal-resolution SAR imagery has been restricted to conventional image processing (Henderson, 1995), or interpretation of the double-scattering mechanism for built-up areas (Henderson, 1995; Lee *et al.*, 1999; Xu and Jin, 2005; Henderson and Xia, 1997; Xia and Henderson, 1997). A few studies have been devoted to the fundamental physics and modeling of the wave–building interaction process. Xia and Henderson (1997) studied the relationship between radar returns and the material and geometry of man-made objects. Franceschetti, Iodice, and Riccio (2002) proposed a scattering model of simple cuboid buildings based on geometrical and physical optics. They also derived a preliminary expression of the scattered wave as a function of building size and shape. Recently, Xu and Jin (2006)

Polarimetric Scattering and SAR Information Retrieval, First Edition. Ya-Qiu Jin and Feng Xu.
© 2013 John Wiley & Sons Singapore Pte. Ltd. Published 2013 by John Wiley & Sons Singapore Pte. Ltd.

proposed a computerized SAR image simulation method for a comprehensive terrain scene. It is based on the efficient mapping and projection algorithm (MPA) and physics-based models such as the integral equation method (IEM) for a rough surface and the vector radiative transfer (VRT) equation for the vegetation canopy. It successfully addressed, for the first time, the problem of SAR imaging simulation of both building and natural environments.

As an inverse problem, the reconstruction of 3D objects from SAR images has become a key issue of information retrieval for SAR imagery technology. The 3D reconstruction of a man-made object is usually based on interferometric techniques (Bolter and Leberl, 2000; Stilla, Soergel, and Thoennessen, 2001), fusion with other data resources, for example, optical and geographical information system (GIS) data (Tupin, 2004; Soergel *et al.*, 2005), or other modeling methods (Bolter, 2000; Quartulli and Datcu, 2004). Recently, some studies on the reconstruction of building objects from multi-incidence or multi-aspect high-resolution SAR images have been reported (Simonetto, Oriot, and Garello, 2005; Hamasaki *et al.*, 2005; Michaelsen, Soergel, and Thoennessen, 2005). For example, Simonetto, Oriot, and Garello (2005) use the profile line detection and height inversion for building reconstruction from spotlight SAR images with different incidences. Hamasaki *et al.* (2005) employ square-loop Pi-SAR data to detect man-made targets by evaluating the correlation of Pauli components between different aspects. Michaelsen, Soergel, and Thoennessen (2005) also proposed a reconstruction scheme for multi-aspect high-resolution SAR data.

However, fully exploiting high-resolution (meter or decimeter) polarimetric SAR data is still in the preliminary stage. How to carry out an automatic reconstruction of complicated or randomly distributed building objects from a SAR image at meter or decimeter resolution remains for further study.

It has been noted that scattering from man-made objects produces bright spots at decimeter resolution, but presents strip/block images at meter resolution. This difference is largely attributed to the difference in the respective imaging modes, that is, stripmap for meter resolution as opposed to the spotlight mode for decimeter resolution, which accumulates views in a much wider angle region. With the advancement of spaceborne meter-resolution SAR, it seems more important and practicable to develop a reconstruction approach at the level of meter resolution.

In this chapter, we present an automatic method for the detection and reconstruction of 3D objects from multi-aspect meter-resolution SAR images, that is, automatic multi-aspect reconstruction (AMAR). It consists of the following steps:

1. As *a priori* knowledge, the imaging features of the object are first generalized, such as geometrical profile and spatial distribution.
2. An edge-based approach is used to identify and extract the scattering image of the object from the SAR image, and retrieve the contained information.
3. Thereafter, a statistical description of the object image and its coherency to those of other aspects is given.
4. Finally, an automatic algorithm is designed to co-register object images in different aspects and further reconstruct the 3D geometry.

The first half of this chapter addresses the problem of how to extract strip-like building images from SAR imagery. It first employs the POL-CFAR technique to detect edges, which are then thinned by ridge filtering. After that, a parallel Hough transform is used to extract parallelogram-like façade images of building objects. The second half establishes a multi-aspect coherent model of building objects. It models the façade images using statistical geometry and reconstructs the building via maximum-likelihood estimation.

Throughout this chapter, both simulated SAR images and real Pi-SAR images are processed and used as examples to demonstrate the efficacy and necessity of each step. Reconstruction performance is evaluated by comparison to ground truth information. It is concluded that the reconstruction accuracy closely depends on the number of available aspects. In the end, a practicable scheme of 3D object reconstruction from satellite-borne SAR imagery is proposed.

7.1 Detection and Extraction of Object Image

7.1.1 Features of a Simple Building Object in a SAR Image with Meter Resolution

Following from the building scattering model introduced in Chapter 3, it is clear that scattering of a building object can be decomposed into different terms (Xu and Jin, 2006). As shown in Figure 3.12, the image of a simple building consists of direct scattering from the wall/ façade, roof, and edges, wall–ground double scattering and shadows projected, and so on. When the wall/façade directly faces the radar incidence, the wall–ground double bounce becomes dominant and thus is often used as a unique signature to identify built-up areas in low-resolution SAR images where each individual building may not be recognizable. However, in meter-resolution SAR, individual buildings become visible and their images often appear as bright strips/blocks that are produced by scattering from wall/façade, wall–ground bounce, and roof. The key to meter-resolution building reconstruction is to identify and extract these strips/ blocks. Note that this rule will not be valid for higher-resolution (centimeters) SAR images where buildings appear more like spots and finer structures are revealed. A different strategy should be used for submeter-resolution SAR images, which is beyond the scope of our book.

Our study focuses only on a simple building object (SBO), which is defined as follows: first, its geometry can be approximated as a cuboid; second, its spatial distribution is sparse so that no serious shadowing and superposition occurs between adjacent buildings.

The cuboid object is described by seven geometric parameters: 3D position, 3D size, and orientation. Besides, we also assume that the roof is flat and smooth so that its scattering is insignificant as compared to the edges and façades. This is generally true for cuboid SBOs because the edge and wall structures, for example, metal windows, often have more significant scattering than a flat roof. Thus, in high-resolution SAR images the façade images are dominant, whereas in aerial optical images the roofs dominate the appearance of buildings.

Figure 7.1a shows a model of the cuboid object, and Figure 7.1b shows its simulated SAR image using the MPA in Chapter 3 (Xu and Jin, 2006). Figure 7.1c and d give a photo of a real rectangular building and its image obtained by Pi-SAR observation, respectively. It can be seen that the radar echoes appear as strips or blocks, mostly coming from the façades, edges, or wall–ground.

On most occasions, scattering directly from façades and edges makes the most significant contribution to the object image, which yields the geometric profile information of the building. This is the basic rule upon which our approach is built. SBOs would have extremely dominant wall–ground double bounce only at the aspect angle exactly aligned with the radar incidence. In that extreme case, the spreading side-lobes of double bounce would mask the

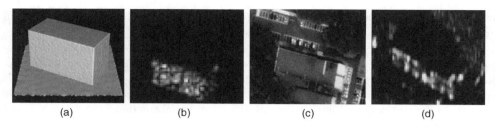

Figure 7.1 A cuboid object in meter-resolution SAR image (range directions are bottom-up): (a) the cuboid model; (b) corresponding simulated SAR image (span); (c) photo of a real building; (d) corresponding Pi-SAR image (span). © [2007] IEEE. Reprinted, with permission, from *IEEE Transactions on Geoscience and Remote Sensing*

remaining terms and thus result in the loss of useful geometric information. This issue of perpendicular aspect will be discussed later in this chapter.

7.1.2 CFAR Edge Detection and Ridge Filter Thinning

It can be seen from Figure 7.1 that the longer wall of the SBO, referred to as major wall henceforth, composes most of its image. Thus, our effort is mainly focused on the major wall image, with a possibility of taking advantage of the shorter wall as well. As a first step, edge detectors of constant false-alarm rate (CFAR) such as ratio gradient (Touzi, Lopes, and Bousquet, 1988) are used to delineate the profiles of wall images. Owing to the limited resources of high-resolution SAR data, it remains an open question whether a CFAR detector should be adapted particularly for multi-look HR-POLSAR imaging over urban areas. In our study, a standard detector, the POL-CFAR detector (Schou *et al.*, 2003), is employed.

The parameters of the POL-CFAR detector include the size of the two side windows, the gap between them, the number of directions, the false-alarm rate, and so on. The equivalent number of looks (ENL) is estimated by using the method of Foucher, Boucher, and Benie (2000). It is calculated based on statistics of an optimized window size for a homogeneous area. In this study, the parameters are taken as 7×3 window, one pixel gap, four directions, and false alarm equals 0.01. In our experience, the POL-CFAR detector derived from the complex Wishart distribution can fulfill the segmentation requirements of medium- and small-scale building images. Besides, the ridge filter applied after this step can accommodate segmentation error to some extent.

In order to improve the segmentation accuracy, the edge detected by the window operator needs to be further thinned. Using the edge intensity produced by POL-CFAR detection, an edge thinning method called ridge filtering is proposed. It is a simple morphological window operator. The name ridge filter is adopted because each edge pixel is eliminated or preserved, depending on whether it is located on the "ridge" of the edge intensity field.

Taking an eight-neighbor 3×3 window as an example, the center pixel is regarded as on the ridge if its intensity is higher than the pixels along two sides. Given a possible direction of the ridge, two neighboring pixels have to be located on the ridge too, and the other six should be along two sides. Define the filtering levels 2, 4, and 6 as the center pixel higher than at least two, four, and six pixels along two sides, respectively.

As depicted in Figure 7.2, assuming that the possible ridge is oriented as 4–8, the boundaries of the ridges in different levels are the dashed lines denoted as (2), (4), and (6). A higher level has a thinner ridge and yields finer filtering result. Considering the four different

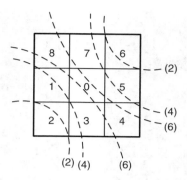

Figure 7.2   Ridge filtering. © [2007] IEEE. Reprinted, with permission, from *IEEE Transactions on Geoscience and Remote Sensing*

orientations, the center pixel is eliminated only if it is not on the ridge of any orientation. Denoting e_i as the edge intensity of the pixel i, the ridge filter can be expressed as

$$
\begin{aligned}
&\text{level 2:} \quad e_0 > \min_{i=1,2,3,4} \max(e_i, e_{i+4}) \\
&\text{level 4:} \quad e_0 > \min_{i=1,2,3,4} \max(e_i, e_{i'}, e_{i+4}, e_{i'+4}) \\
&\text{level 6:} \quad e_0 > \min_{i=1,2,3,4} \max(e_i, e_{i'}, e_{i''}, e_{i+4}, e_{i'+4}, e_{i''+4}) \\
&i' = [i+1]_4, \quad i'' = [i+2]_4
\end{aligned}
$$

(7.1)

In the following, the step of edge detection will be tested on both simulated and real SAR images. Figure 7.3a shows a scene for simulating the SAR image. Two cuboid buildings are

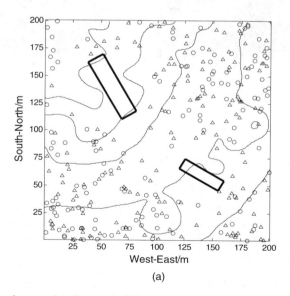

(a)

Figure 7.3   (a) A virtual scene, showing two buildings, evergreen (triangles) and deciduous (circles) trees, and contour lines. (b)–(e) The simulated SAR images (span) of different aspects (0°, 180°, 20°, and 200°, respectively). © [2007] IEEE. Reprinted, with permission, from *IEEE Transactions on Geoscience and Remote Sensing*

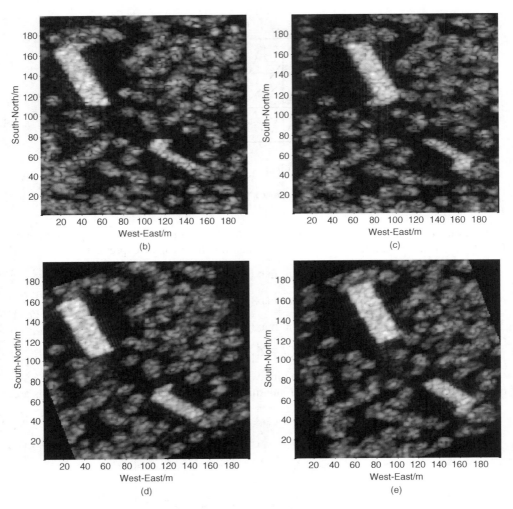

Figure 7.3 *(Continued)*

located with orientations 30° and 60° within forested ground, and there is about 2 m difference in ground elevation.

Four polarimetric SAR images of four aspects (0°, 180°, 20°, and 200°) are simulated using the MPA (Xu and Jin, 2006), as shown in Figure 7.3b–e. For example, the detected edge of Figure 7.3b is shown by Figure 7.4a, and the thinning results of the levels 2 and 4 ridge filtering are shown in Figure 7.4b and c, respectively. The level 2 is further de-spurred by using the spur filter (Pratt, 1991), as finally shown in Figure 7.4c.

Four-aspect Pi-SAR images acquired over Sendai, Japan, by a square-loop flying path (flight nos. 8708, 8709, 8710, 8711, X-band, pixel resolution 1.25 m) are taken as an example of a real SAR image study. The region of interest (ROI) is the Aobayama campus of Tohoku University. Figure 7.5a–e give the optical picture and four-aspect Pi-SAR images, respectively. The results of the edge detection and thinning are shown in Figure 7.6a–c.

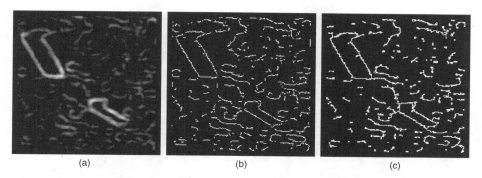

Figure 7.4 Edge detection of simulated SAR image: (a) edge detected from Figure 7.3b by POL-CFAR; (b) the level 4 ridge; (c) the level 2 ridge de-spurred. © [2007] IEEE. Reprinted, with permission, from *IEEE Transactions on Geoscience and Remote Sensing*

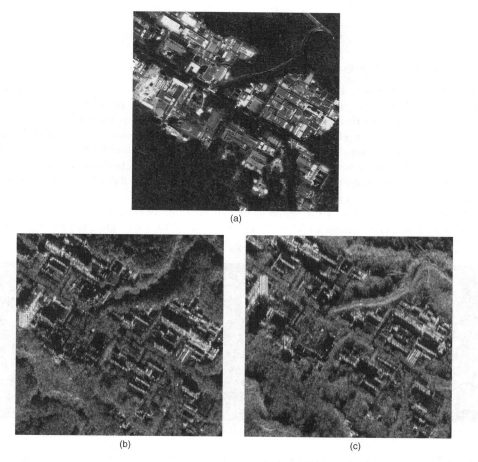

Figure 7.5 SAR image over the Aobayama campus of Tohoku University, Japan: (a) optical picture of ROI (resolution 0.6 m); (b)–(e) are four-aspect Pi-SAR images of ROI (90°, 270°, 0°, 180°, respectively), with R = HH, G = HV, and B = VV. © [2007] IEEE. Reprinted, with permission, from *IEEE Transactions on Geoscience and Remote Sensing*

(d)   (e)

Figure 7.5   (*Continued*)

7.1.3   Building Image Extraction via Hough Transform

The most distinct feature of an SBO in a SAR image is parallel lines. The Hough transform is employed to detect straight segments from the thinned edges, and parallelogram outlines of the façade images can then be extracted. The Hough transform is not sensitive to short segments and has high complexity of $o(N^2M)$, where $N \times N$ is the size of the spatial domain and M is the size of one dimension, for example, offsets, of the transform domain, respectively. For these two reasons, it is carried out in tiling manner in this study, that is, the original picture is partitioned into blocks, each of which is detected independently via Hough transform.

To preserve continuity of the segment detection, some overlapped areas between two neighboring blocks are retained. As illustrated in Figure 7.7a, the original edge image is first

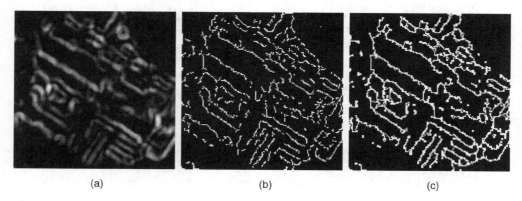

(a)   (b)   (c)

Figure 7.6   Edge detection of Pi-SAR images: (a) edge detection of part of Figure 7.5b; (b) level 4 ridge; (c) level 2 ridge de-spurred. © [2007] IEEE. Reprinted, with permission, from *IEEE Transactions on Geoscience and Remote Sensing*

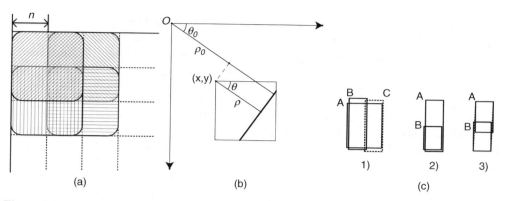

Figure 7.7 Hough transform for parallel line segment pairs: (a) block partitioning of edge image for Hough transform; (b) block local coordinates versus principle coordinates; (c) post-processing. © [2007] IEEE. Reprinted, with permission, from *IEEE Transactions on Geoscience and Remote Sensing*

equally divided into small blocks of $n \times n$ size, and then all four small neighboring blocks form the sub-images, where the Hough transform is performed. Hence, those segments contained within one small block can be fully detected. To make sure that most façade images can be contained within one or two blocks, the size of the small block is selected according to the length of objective building images. The number of sub-images is $(N/n - 1)^2$ and the size of the transform domain reduces proportionally to $M \cdot 2n/N$. Hence, the complexity of the Hough transform after tiling is reduced to

$$\frac{(N-n)^2}{n^2} \times 4n^2 \times \frac{2n}{N} M = \frac{8n}{N}(N-n)^2 M = o(NM) \tag{7.2}$$

as well as the storage requirement. The resource saving effect is significant when the size N of the image is large.

The edge of a strip-like building image is a pair of parallel line segments. Each point (ρ, θ) in the Hough transform domain corresponds to a unique straight line in the spatial domain. Accordingly, a pair of parallel lines can be described by two points (ρ_1, θ) and (ρ_2, θ). Further defining $\rho = \min(\rho_1, \rho_2)$, $\Delta = |\rho_1 - \rho_2|$, the traditional 2D transform domain can be extended to 3D, that is, (ρ, θ, Δ), where each point corresponds to a unique parallel line pair.

The Hough transform $h(\rho, \theta)$ is defined as the accumulation of the points along the corresponding line. Likewise, an extended Hough transform for parallel line pairs can be defined in the domain (ρ, θ, Δ) as the mean of point accumulations of the two corresponding parallel lines, that is

$$b(\rho, \theta, \Delta) = \sqrt{h(\rho, \theta) \cdot h(\rho + \Delta, \theta)} \tag{7.3}$$

The detection steps of parallel line segments are: (i) find bright spots in the transform domain with a minimum distance between every pair of them so as to avoid redetection; (ii) search for the segments consisting of successive points along the corresponding parallel lines in the spatial domain, which is longer than a minimum length, and the distance between two successive points is shorter than a maximum gap; and (iii) only the pairs of points lying on two lines and directly facing are taken into account in segment searching.

Here two different levels (4 and 2) of thinned edges are used for the Hough transform and segment searching, respectively, to prevent either low efficiency of bright spot finding or the loss of line segments in searching.

As depicted in Figure 7.7b, there is a transform from local coordinates to principal coordinates after line segments are detected in each block. Besides, overlapping or broken parallelograms detected from neighboring blocks should be unified in post-processing steps. For two overlapped and parallel parallelograms, there might be three different situations as represented in Figure 7.7c.

1. Wide "A" contains narrow "B", and both of them have comparable lengths. Then, "A" should be removed, because the building image is supposed to be a homogeneous area without edges in the middle. In this occasion, there always exists another parallelogram "C" equal to "A minus B". However, the image of a large building might reveal some textures, which is detected as edges. For this reason, "A" should be preserved if the median levels of scattering power in the areas "B" and "C" do not satisfy the CFAR criterion for edge detection, which indicates the possibility of "A" as a wrongly divided large building image.
2. A long "A" can contain a short "B", and both of them have comparable widths. Thus, "B" should be removed, because "B" occurs in the boundary between two neighboring blocks.
3. "A" and "B" are partly overlapped, and have comparable lengths. This also may result from tiling. "A" and "B" should be combined into a longer parallelogram, which takes both far-side end points of "A" and "B" as its new ends.

Note that only information of the long lateral of the parallelogram is considered in the previous steps. The position of short laterals should be determined using the originally CFAR-detected edge intensity. First, the long laterals are elongated to one end and the short lateral is moved to a new position, correspondingly. Meanwhile, the edge intensities along the adjusted short laterals are summed up and recorded. The elongating stops when both end points of two long laterals are not edge pixels. The final position of the short lateral is determined where the sum of the edge intensities reaches its maximum. The other end is adjusted in the same way. Note that post-processing, for example, combining, should be executed once again after the short lateral has been determined.

Another issue is the edge offset caused by the intrinsic inter-pixel correlation of SAR images, that is, the side-lobe effect of the focusing algorithm adopted in SAR raw data processing. This introduces error into image extraction and object reconstruction. Suppose that two homogeneous areas divided by a real edge are a strong and a weak scattering area, respectively. The side-lobe effect causes strong scattering spreading across the real edge and yields an offset from strong side to weak side. This offset is determined by the side-lobe dampening rate of the point target response. It can be estimated as the position where the dampened side-lobe level can be detected by the edge detector from the main lobe. As an example, for ideal sinc point target response, given the resolution 1.5 m and pixel spacing 1.25 m of Pi-SAR (Toshihiko *et al.*, 2002), the estimated edge offset is about 1.25 pixels.

As shown in Figure 7.8, the boundary of a building image (ABCD) might be detected as a round-angled parallelogram ($A_1B_1C_1D_1$), where S_1 and S_2 denote the edge offsets in normal directions of its long lateral and short lateral, respectively. Here, the building image is

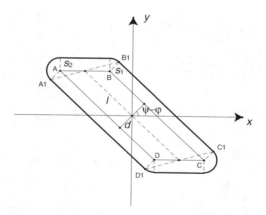

Figure 7.8 Detected boundary of building image taking into account the edge offset. © [2007] IEEE. Reprinted, with permission, from *IEEE Transactions on Geoscience and Remote Sensing*

assumed stronger than the surrounding area. Let us describe the building image (parallelogram ABCD) by five parameters, namely, the length of long lateral ℓ, the distance between them d, their normal direction toward the sensor ψ, and the center position x, y; and indicate φ as the range direction toward the sensor ($\varphi = 0$ is assumed in Figure 7.8 for simplicity, that is, $+x$ is the range direction). Given the edge offsets, all these parameters of ABCD can be easily obtained from the detected parallelogram $A_1B_1C_1D_1$. Notice that it would produce a severe error in the case of large $\psi - \varphi_i$ without taking into account the edge offset.

Figure 7.9a–c shows the detection result in each step of the simulated SAR image in Figure 7.4. Figure 7.10a–c shows the detection from Pi-SAR image given in Figure 7.5. A few building images are not detected and some false images are wrongly produced. In fact, there is a compromise between over-detection and incomplete detection. We prefer to preserve more detected building images, so as to control the non-detected rate, while the false images are expected to be eliminated through the subsequent step of auto-selection of effective images for reconstruction. However, there always remain some undetectable images,

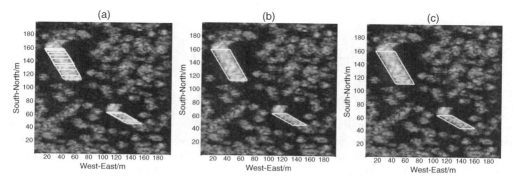

Figure 7.9 Extraction of strip-like building images from the simulated SAR image of Figure 7.3b: (a) first result via Hough transform; (b) after post-processing; (c) after end adjusting. © [2007] IEEE. Reprinted, with permission, from *IEEE Transactions on Geoscience and Remote Sensing*

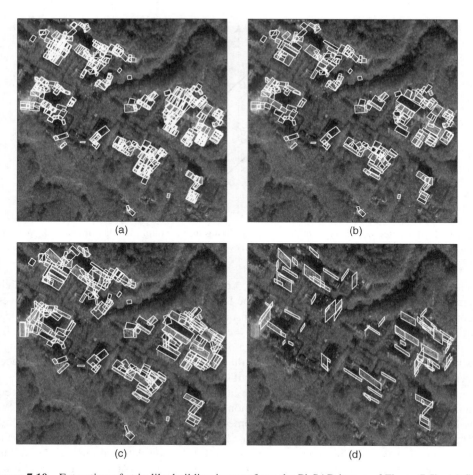

(a)   (b)

(c)   (d)

Figure 7.10 Extraction of strip-like building images from the Pi-SAR image of Figure 7.5b: (a) first result via Hough transform; (b) after post-processing; (c) after end adjusting; (d) final building images after offset removal and classification. © [2007] IEEE. Reprinted, with permission, from *IEEE Transactions on Geoscience and Remote Sensing*

attributable to shadowing of the tree canopy, overlapping of nearby buildings, or with too complicated wall structures.

7.1.4   Classification of Building Images

The detected parallelogram of a homogeneous scattering area could be direct scattering from the façade, wall–ground double scattering, a combination of direct and double scatterings, the projected shadow of the building, or even strong scattering of strip-like objects (e.g., metal fence or metal awning, which is not considered in this study).

Shadowing is identified via the criterion that the scattering power of that area is much lower than that of the near vicinity. Specifically, first set up two parallel equal-length strip windows on its two sides and then calculate the median scattering powers of the three regions. If the middle one is weaker than the two sides, it is classified as shadow instead of building image. Figure 7.10d shows the building images after offset removal and classification.

The wall–ground double scattering can be differentiated from direct scattering based on polarimetric information. In this study, the deorientation parameter v (Henderson and Xia, 1997) is used to identify double scattering. The whole area is first proportionally divided into the direct scattering and double scattering sub-areas (Xu and Jin, 2006). If the median value of v is less than the threshold value -0.7 in the double scattering sub-area and larger than 0 in the direct scattering sub-area, the whole area is identified as a combination of two different scatterings. Otherwise, only if the median value of v in the whole area is less than -0.7 is the area classified as double scattering; otherwise, the area is seen as direct scattering.

It should be noted that polarimetric calibration of Pi-SAR data is necessary before it can be used for classification. Owing to the difficulty of installing a reference target for different view angles (Toshihiko *et al.*, 2002), the acquired multi-aspect Pi-SAR data need polarimetric calibration using known terrain objects. The main purpose is to compensate the imbalance between horizontally polarized and vertically polarized channels for building image classification. A calibration procedure for multi-aspect Pi-SAR is proposed in Section 7.5 (Xu and Jin, 2008). However, this channel imbalance does not affect the POL-CFAR edge detection.

7.2 Building Reconstruction from a Multi-Aspect Image

7.2.1 Probabilistic Description of Building Image

Uncertainty in object image detection is inherited from the complexity of distributed objects, the randomness of terrain surfaces, and the SAR image speckle. We propose a parameterized probabilistic model to describe the detected multi-aspect object images, which is critically important for automatic reconstruction. For convenience, the detected building images are parameterized in rotated coordinates (ρ, τ) where the normal direction becomes $+\rho$. As depicted in Figure 7.11a, ρ_1, ρ_2 denotes the positions of the two long laterals, respectively, while τ_1, τ_2 are for the short laterals. Then, $\ell = \tau_2 - \tau_1$ and $d = \rho_2 - \rho_1$. The center (ρ, τ) corresponds to (x, y) in the original coordinate system.

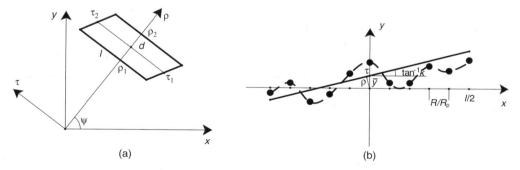

(a) (b)

Figure 7.11 The detected edge pixels around the theoretical boundary of an object: (a) parameterization of building image; (b) the detected edge. © [2007] IEEE. Reprinted, with permission, from *IEEE Transactions on Geoscience and Remote Sensing*

Generally, the edge pixels detected by CFAR are randomly situated around the real (or, as we say, theoretical) boundary of the object. In the simple case shown in Figure 7.11b, the theoretical boundary lies on the x axis and the detected edge points (after thinning) are randomly distributed around the x axis. Through the method presented in the previous section, the detected straight boundary (the tilted line) will deviate from the real one. It is reasonable to presume that the deviation distances of the edge points follow a normal distribution $N(0, \sigma^2)$.

Owing to the inter-pixel correlation of SAR images, in order to be considered as independent, two edge points must be further apart from each other than the intrinsic resolution of the sensor. Thus, the original edge can be equivalently treated as a set of independent edge points with intervals equal to R/R_p pixels, where R and R_p denote the intrinsic resolution and pixel spacing, respectively. Given the length of edge ℓ, the number of equivalently independent edge points is $N_\ell = R_p \hat{\ell}/R$. Note that symbols with a hat (e.g., $\hat{\ell}, \hat{\rho}, \hat{d}$) are used to represent the true value (deterministic), while symbols without a hat represent the measured value (probabilistic).

The line detection approach presented in the previous section is considered as an equivalent linear regression, that is, line fitting to random sample points. According to linear regression using independent normal samples (Hannan and Deistler, 1987), the slope of the fitted line follows the normal distribution:

$$k \sim N(0, \sigma^2/N_k), \quad N_k = 2(R/R_p)^2 \sum_{n=-N_\ell/2}^{N_\ell/2} n^2 = \frac{R^2 N_\ell(N_\ell + 1)(N_\ell + 2)}{6R_p^2} \tag{7.4}$$

where the factor "2" denotes both edges of two long laterals. Obviously, we have $N_k \gg 1$ and $k \ll 1$. Thus, the orientation angle, $\psi - \pi/2 = \tan^{-1}k$, follows the same distribution as k, which is the probabilistic characteristic of the first parameter of a building image.

According to linear regression (Xu and Jin, 2007), the mean position of the edge points $\bar{y}$ follows $\bar{y} \sim N(0, \sigma^2/N_\ell)$ and k is independent of ψ. Under the condition $k \ll 1$, the detected line position $\rho_1 = \bar{y} \cos \psi = \bar{y}\sqrt{1 + k^2} \approx \bar{y}$ follows the same distribution as $\bar{y}$, where the small variations of τ and ℓ caused by the variance of ψ are ignored. Likewise, ρ_2 follows the same distribution and is independent of ρ_1. Therefore, we have the third parameter $d = \rho_2 - \rho_1$ obeying the distribution $d \sim N(\hat{d}, 2\sigma^2/N_\ell)$, the fourth parameter $\rho = (\rho_2 + \rho_1)/2$ with the distribution $\rho \sim N(\hat{\rho}, \sigma^2/(2N_\ell))$, and the correlation coefficient between them $r_{d,\rho} = 0$, that is, ρ independent of d.

Likewise, the positions of the two short laterals τ_1, τ_2 also follow a normal distribution with variance σ^2/N_d, where N_d denotes the number of independent edge points around the short lateral. Note that the short laterals are adjusted to the most likely position (with most edge points), which means that the edge points lying on the short laterals are non-deterministic. The best situation is $N_d = R_p\hat{d}/R$. The worst case is when no edge point can be found lying on the short laterals. Under this circumstance, the factor to determine the position of the short laterals is the end points of the two long laterals, that is, $N_d = 2$. Taking the probabilistic average of these two cases as an approximation yields $N_d = 4R_p\hat{d}/(R_p\hat{d} + 2R)$. Finally, we obtain the distribution of the last two parameters, $\ell \sim N(\hat{\ell}, 2\sigma^2/N_d)$ and $\tau \sim N(\hat{\tau}, \sigma^2/(2N_d))$. They are mutually independent as well.

Since all estimations in detection are regarded as unbiased, the aforementioned case can be easily extended to the general case via substituting the means of their distributions by their theoretical values. Further transforming the center coordinates (ρ, τ) back to (x, y), that is, $[x, y]^{\mathrm{T}} = \mathbf{R}(-\hat{\psi}) \cdot [\rho, \tau]^{T}$, where $\mathbf{R}$ is the rotation matrix, yields that x, y follow a bivariate joint normal distribution.

To sum up, the parameters of a detected building image in any aspect (x, y, d, ℓ, ψ) are distributed around their theoretical values $(\hat{x}, \hat{y}, \hat{d}, \hat{\ell}, \hat{\psi})$ as

$$
\begin{aligned}
&\psi \sim N\left(\hat{\psi}, \frac{\sigma^2}{N_k}\right), \quad N_k = \frac{R^2 N_\ell (N_\ell + 1)(N_\ell + 2)}{6 R_p^2} \\[2mm]
&d \sim N\left(\hat{d}, \frac{2\sigma^2}{N_\ell}\right), \quad N_\ell = R_p \hat{\ell}/R \\[2mm]
&\ell \sim N\left(\hat{\ell}, \frac{2\sigma^2}{N_d}\right), \quad N_d = \frac{4 R_p \hat{d}}{R_p \hat{d} + 2R} \\[2mm]
&x \sim N\left(\hat{x}, \frac{R_{11}^2 \sigma^2}{2 N_\ell} + \frac{R_{12}^2 \sigma^2}{2 N_d}\right), \quad R_{11} = \cos\hat{\psi}, \quad R_{12} = -\sin\hat{\psi} \\[2mm]
&y \sim N\left(\hat{y}, \frac{R_{21}^2 \sigma^2}{2 N_\ell} + \frac{R_{22}^2 \sigma^2}{2 N_d}\right), \quad R_{21} = \sin\hat{\psi}, \quad R_{22} = \cos\hat{\psi} \\[2mm]
&r = \frac{N_d R_{11} R_{21} + N_\ell R_{12} R_{22}}{\sqrt{(N_d R_{11}^2 + N_\ell R_{12}^2)(N_d R_{21}^2 + N_\ell R_{22}^2)}}
\end{aligned}
\tag{7.5}
$$

where r is the correlation coefficient between x and y.

All parameters have normal distributions and their variances are determined by the variance of the edge point deviation σ^2, which can be estimated via calculating statistics over the detected results. After counting the deviation of the edge points in the vicinity of each lateral of all detected building images and making its histogram, the variance σ^2 can be determined through a minimum square error (MSE) fitting of the histogram using the normal probability density function (PDF). The histogram and fitted PDF of the four-aspect Pi-SAR images are plotted in Figure 7.12. They match very well and, in addition, prove the normal

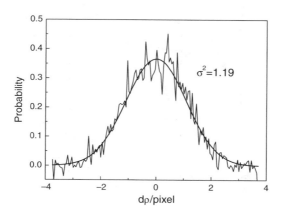

Figure 7.12   Estimation of σ^2 via histogram fitting on Pi-SAR images. © [2007] IEEE. Reprinted, with permission, from *IEEE Transactions on Geoscience and Remote Sensing*

distribution presumption of the edge point deviation. It should be pointed out that individual histograms of images of different aspects also seem identical and suggest the aspect independence of this distribution.

One more issue has to be addressed here. Owing to the sheltering and overlapping of adjacent terrain objects or the reflected texture of large building walls, building images might not be accurately detected, which leads to uncertainty in the distribution centers, that is, the "theoretical" parameters have randomness themselves. The influences resulting from this uncertainty might be much larger than the randomness. However, it is impossible to quantify the corresponding probabilistic property due to the unknown situation in the imaged multi-building area. It subsequently becomes one of the major sources of reconstruction errors.

7.2.2 Multi-Aspect Coherence of Building Images

The imaging geometry of a wall/façade is described in Figure 7.13. In the principal coordinates (x, y, z), xOy is the reference plane of zero elevation and also defined as the ground-range imaging plane for all aspects. Obviously, SAR images of all aspects can be fitted to the xOy plane via translation and rotation.

Define the wall parameters as $\mathbf{w} = [\psi, \ell, h, x, y, e]$, where ψ denotes the outward normal direction, e is the elevation of the center, h is the height, ℓ is the length, and (x, y) are the coordinates of the center projected in the xOy plane.

Given the incident angle (θ_i, φ_i) of the ith aspect, the gray and deep gray areas on the ground-range plane in Figure 7.13 correspond to the wall direct scattering and wall–ground double scattering images, respectively. The scattering images are projected on the imaging plane from the wall toward the sensor. The direct scattering part is on the far side. The long laterals of the images are parallel with the wall, while the short laterals are parallel with the

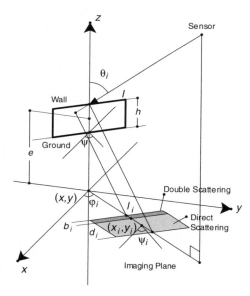

Figure 7.13 Geometry of a wall/façade imaging in multi-aspect SAR observation. © [2007] IEEE. Reprinted, with permission, from *IEEE Transactions on Geoscience and Remote Sensing*

range direction, which makes them parallelograms. Note that the direction of the short laterals with respect to the wall itself is irrelevant.

To avoid confusion with the geometric parameters of the wall, the parameters of the building image in the ith aspect are redefined with the subscript i, namely, length ℓ_i, width of direct scattering part d_i, width of double scattering part b_i, center coordinates x_i, y_i, and normal direction toward the sensor ψ_i. Since b_i can be determined by d_i and ψ_i according to the imaging geometry (i.e., $b_i = d_i \tan^2 \theta_i \sin^2(\varphi_i - \psi)$), the remaining five independent parameters $\boldsymbol{\omega}_i = [\psi_i, \ell_i, d_i, x_i, y_i]$ are able to completely represent the building image of the ith aspect. As for those detected double scattering building images, d_i can be obtained via b_i.

In the light of Equation 7.5, the observed building image $\boldsymbol{\omega}_i = [\psi_i, \ell_i, d_i, x_i, y_i]$ is a set of normal random variables, whose expectations are their theoretical values expressed as $\hat{\boldsymbol{\omega}}_i = [\hat{\psi}_i, \hat{\ell}_i, \hat{d}_i, \hat{x}_i, \hat{y}_i]$. Obviously, the theoretical parameter $\hat{\boldsymbol{\omega}}_i$ has an inherent relationship with the wall $\mathbf{w}$. Following the imaging geometry, the relationship between $\hat{\boldsymbol{\omega}}_i$ and $\mathbf{w}$ is written as

$$
\begin{aligned}
\hat{\psi}_i &= \psi, \quad \hat{\ell}_i = \ell \\
\hat{d}_i &= h \cos(\varphi_i - \psi)/\tan \theta_i = h\lambda_i \\
\hat{b}_i &= \hat{d}_i \tan^2 \theta_i \sin^2(\varphi_i - \psi) \\
\hat{x}_i &= x + e \cos \varphi_i/\tan \theta_i = x + e\mu_i \\
\hat{y}_i &= y + e \sin \varphi_i/\tan \theta_i = y + e\nu_i
\end{aligned}
\tag{7.6}
$$

As mentioned above, a simple building object (SBO) has seven degrees of freedom, six of which can be determined by one side major wall, and the last one is the building width a, that is, the distance between the two major walls. Defining $\mathbf{w}^*$ as the opposite wall of $\mathbf{w}$, its normal direction is $\psi^* = \psi + \pi$, its center is located at a distance a away from $\mathbf{w}$ along the direction of ψ^*, and the other parameters are the same as those for $\mathbf{w}$. Consequently, an SBO can be represented by a set of parameters $\mathbf{c} = [\psi, \ell, h, a, x, y, e]$, where $[x, y]_{\mathbf{c}} = 0.5([x, y]_{\mathbf{w}} + [x, y]_{\mathbf{w}^*})$ is defined as the center of the whole building.

A wall might be "visible" or "not visible" depending on the wall orientation, the aspect, and so on. Hence, a set of relevant concepts are defined as follows:

(a) "visible aspects" of wall $\mathbf{w}$, $i \in V$, which satisfy $|\varphi_i - \psi| < \pi/2$ and belong to the effective aspects;
(b) "non-visible aspects" of wall $\mathbf{w}$, that is, visible aspects of the opposite wall $\mathbf{w}^*$, $j \in V^*$, which satisfy $|\varphi_j - \psi^*| < \pi/2$ and belong to the effective aspects;
(c) "effective aspects" of wall $\mathbf{w}$, $k \in V \cup V^*$, where the images of either $\mathbf{w}$ or $\mathbf{w}^*$ are properly detected; and
(d) "ineffective aspects" of wall $\mathbf{w}$, $\overline{V \cup V^*}$, which do not belong to the effective aspects.

Some conclusions are obvious. For examples, two opposite walls cannot be visible from the same aspect, that is, $V \cap V^* = \emptyset$. In the effective aspects, the images of $\mathbf{w}$ or $\mathbf{w}^*$ are correctly detected and can be used for reconstruction. A special case of the ineffective aspect is $\varphi_i = \psi$ or $\varphi_i = \psi^*$, when the wall–ground scattering is extremely strong and its theoretical width tends to $b_i = 0$, which means that it reflects no height information of the building.

Examples of ineffective aspects include situations where complete wall images cannot be detected due to shadowing and interference.

As an estimation problem, the wall $\mathbf{w}$ and building $\mathbf{c}$ can be estimated from their multi-aspect images. Given the observed images $\boldsymbol{\omega}_i$, $i \in V$, the maximum-likelihood estimation of $\mathbf{w}$ is written as

$$\mathbf{w}_{ML} = \arg \max_{\mathbf{w}} \prod_{i \in V} p_i(\boldsymbol{\omega}_i | \mathbf{w}) \tag{7.7}$$

where the conditional probability $p_i(\boldsymbol{\omega}_i|\mathbf{w})$ is the PDF of the observed image $\boldsymbol{\omega}_i$ under the premise of a known $\mathbf{w}$. Likewise, the building $\mathbf{c}$ can be estimated from the observed two side walls $\boldsymbol{\omega}_i$, $i \in V$ and $\boldsymbol{\omega}_j^*$, $j \in V^*$, that is

$$\mathbf{c}_{ML} = \arg \max_{\mathbf{c}} \prod_{i \in V} p_i(\boldsymbol{\omega}_i | \mathbf{c}) \prod_{j \in V^*} p_j(\boldsymbol{\omega}_j^* | \mathbf{c}) \tag{7.8}$$

Note that the conditional probabilities of different aspects are considered independent. Following from the previous subsection, it can be seen that the factor influencing $p_i(\boldsymbol{\omega}_i|\mathbf{w})$ or $p_i(\boldsymbol{\omega}_i|\mathbf{c})$ is the random deviation of the edge pixels. Such randomness should be independent for every pair of edge pixels with large enough separation, because the inter-pixel relevance is attributed to the focusing effect in raw data compression. Therefore, it is justified to assume the edge pixel deviation of different observations to be independent.

Following Equation 7.8, the conditional probability $p_i(\boldsymbol{\omega}_i|\mathbf{w})$ can be expressed as

$$p_i(\boldsymbol{\omega}_i|\mathbf{w}) = p_N(\psi_i; \hat{\psi}_i, \sigma_\psi^2) \cdot p_N(\ell_i; \hat{\ell}_i, \sigma_{\ell_i}^2) \cdot p_N(d_i; \hat{d}_i, \sigma_d^2) \cdot p_{N^2}(x_i, y_i; \hat{x}_i, \sigma_{x_i}^2, \hat{y}_i, \sigma_{y_i}^2, r_i) \tag{7.9}$$

where $p_N(w; \hat{w}, \sigma_w^2)$ stands for the PDF of the normal distribution $N(\hat{w}, \sigma_w^2)$ and $p_{N^2}(\cdot)$ for the bivariate joint normal distribution with correlation coefficient r_i. All the expectations are given in Equation 7.6, and the variances and correlation coefficients are given in Equation 7.8. Note that only the variances σ_ψ^2, σ_d^2 are aspect-independent.

It can be found from Equation 7.9 that ψ affects not only the probability of ψ_i, but also d_i by determining its expectation $\hat{d}_i$ in Equation 7.6. However, the contribution of d_i to the estimation of ψ_{ML} can be neglected, because the variance of ψ_i is negligible to d_i. Therefore, ψ_{ML} is estimated from ψ_i first.

Moreover, note that the variance and correlation coefficient in Equation 7.8 are determined from $\hat{\ell}_i, \hat{d}_i, \hat{\psi}_i$. During the estimation process, the estimated $\ell_{ML}, h_{ML}, \psi_{ML}$ are used, instead of $\hat{\ell}_i, \hat{d}_i, \hat{\psi}_i$, to calculate the variance and correlation coefficient. Thus, ψ, ℓ, h can be estimated, sequentially and independently. By maximizing the logarithm of Equation 7.7, we obtain

$$\psi_{ML} = \sum_{i \in V} \psi_i / I = E(\psi_i)$$

$$h_{ML} = \sum_{i \in V} \lambda_i d_i \bigg/ \sum_{i \in V} \lambda_i^2 = E(\lambda_i d_i) / E(\lambda_i^2) \tag{7.10}$$

$$\ell_{ML} = \sum_{i \in V} \sigma_{\ell_i}^{-2} \ell_i \bigg/ \sum_{i \in V} \sigma_{\ell_i}^{-2} = E(\sigma_{\ell_i}^{-2} \ell_i) / E(\sigma_{\ell_i}^{-2})$$

where I denotes the number of aspects in V, and $E(w_i)$ indicates the average of w_i over all aspects in V. The sequence for calculating Equation 7.10 is first to estimate ψ_{ML}, then use ψ_{ML} to calculate λ_i, and further estimate h_{ML}. After that, $\hat{d}_i$ can be calculated using h_{ML}, as well as $\sigma_{\ell_i}^2$, and finally ℓ_{ML} can be estimated.

However, the other three parameters x, y, e can only be jointly estimated from x_i, y_i, that is

$$[x, y, e]_{ML} = \arg\min_{x,y,e} \sum_{i \in V} \frac{1}{2(1 - r_i^2)}$$
$$\left[\frac{(x_i - x - e\mu_i)^2}{\sigma_{x_i}^2} - 2r_i \frac{(x_i - x - e\mu_i)(y_i - y - ev_i)}{\sigma_{x_i}\sigma_{y_i}} + \frac{(y_i - y - ev_i)^2}{\sigma_{y_i}^2} \right] \quad (7.11)$$

This yields

$$[x_{ML}, y_{ML}, e_{ML}]^{\mathrm{T}} = \mathbf{A}_{3\times3}^{-1} \cdot \mathbf{B}_{3\times1}$$

$$A_{11} = E(q_i \sigma_{x_i}^{-2}), \quad A_{22} = E(q_i \sigma_{y_i}^{-2})$$

$$A_{12} = -E(q_i r_i \sigma_{x_i}^{-1} \sigma_{y_i}^{-1}), \quad A_{13} = E(q_i \mu_i \sigma_{x_i}^{-2} - q_i r_i v_i \sigma_{x_i}^{-1} \sigma_{y_i}^{-1})$$

$$A_{23} = E(q_i v_i \sigma_{y_i}^{-2} - q_i r_i \mu_i \sigma_{x_i}^{-1} \sigma_{y_i}^{-1})$$

$$A_{33} = E(q_i \mu_i^2 \sigma_{x_i}^{-2} + q_i v_i^2 \sigma_{y_i}^{-2} - 2q_i r_i \mu_i v_i \sigma_{x_i}^{-1} \sigma_{y_i}^{-1}) \quad (7.12)$$

$$B_1 = E(q_i \sigma_{x_i}^{-2} x_i - q_i r_i \sigma_{x_i}^{-1} \sigma_{y_i}^{-1} y_i), \quad B_2 = E(q_i \sigma_{y_i}^{-2} y_i - q_i r_i \sigma_{x_i}^{-1} \sigma_{y_i}^{-1} x_i)$$

$$B_3 = E(q_i \mu_i \sigma_{x_i}^{-2} x_i + q_i v_i \sigma_{y_i}^{-2} y_i - q_i r_i \mu_i \sigma_{x_i}^{-1} \sigma_{y_i}^{-1} y_i - q_i r_i v_i \sigma_{x_i}^{-1} \sigma_{y_i}^{-1} x_i)$$

$$q_i = \frac{1}{(1 - r_i^2)}, \quad i \in V$$

where $\mathbf{A} = \mathbf{A}^{\mathrm{T}}$ is symmetric. Note that, if the aspect number in V is less than 2, that is, $I < 2$, it yields $|\mathbf{A}| = 0$ and $\mathbf{A}^{-1}$ is not defined, which means that it is unsolvable (three unknowns) if we have only one visible aspect (two observed samples). In this case, we have first to give up e and estimate x, y by eliminating the third dimension of $\mathbf{A}, \mathbf{B}$. Then e is obtained via other means, for example, using the estimated elevation of neighboring buildings, and at last x, y are compensated.

In the case of building estimation, ℓ, h, ψ are estimated similarly, but extending the observed samples to all visible aspects $i \in V, j \in V^*$, that is

$$\psi_{ML} = E(\psi_k'), \quad \psi_k' = \psi_i \text{ or } \psi_j - \pi, \quad i \in V, j \in V^*, k \in V \cup V^*$$

$$h_{ML} = E(\lambda_k d_k) / E(\lambda_k^2) \quad (7.13)$$

$$\ell_{ML} = E(\sigma_{\ell_k}^{-2} \ell_k) / E(\sigma_{\ell_k}^{-2})$$

and x, y, e, a jointly estimated as

$$
[x, y, e, a]_{ML} = \arg\min_{x,y,e,a} \sum_{k \in V \cup V^*} \frac{1}{2(1 - r_k^2)} \left[\frac{(x_k - x - a\delta_k - e\mu_k)^2}{\sigma_{x_k}^2} \right.
$$

$$
\left. -2r_k \frac{(x_k - x - a\delta_k - e\mu_k)(y_k - y - a\Delta_k - ev_k)}{\sigma_{x_k}\sigma_{y_k}} + \frac{(y_k - y - a\Delta_k - ev_k)^2}{\sigma_{y_k}^2} \right]
$$

$$(7.14)$$

where $2\delta_i = \cos\hat{\psi}$, $2\Delta_i = \sin\hat{\psi}$, $i \in V$, $2\delta_j = -\cos\hat{\psi}$, $2\Delta_j = -\sin\hat{\psi}$, $j \in V^*$. This yields

$$
[x_{ML}, y_{ML}, e_{ML}, a_{ML}]^{\mathrm{T}} = \overline{A}_{4\times4}^{-1} \cdot \mathbf{B}_{4\times1}
$$

$$
A_{14} = E(q_k\delta_k\sigma_{x_k}^{-2} - q_k r_k \Delta_k \sigma_{x_k}^{-1}\sigma_{y_k}^{-1})
$$

$$
A_{24} = E(q_k\Delta_k\sigma_{y_k}^{-2} - q_k r_k \delta_k \sigma_{x_k}^{-1}\sigma_{y_k}^{-1})
$$

$$
A_{34} = E(q_k\delta_k\mu_k\sigma_{x_k}^{-2} + q_k\Delta_k v_k\sigma_{y_k}^{-2} - q_k r_k \Delta_k \mu_k \sigma_{x_k}^{-1}\sigma_{y_k}^{-1} - q_k r_k \delta_k v_k \sigma_{x_k}^{-1}\sigma_{y_k}^{-1}) \qquad (7.15)
$$

$$
A_{44} = E(q_k\delta_k^2\sigma_{x_k}^{-2} + q_k\Delta_k^2\sigma_{y_k}^{-2} - 2q_k r_k \delta_k \Delta_k \sigma_{x_k}^{-1}\sigma_{y_k}^{-1})
$$

$$
B_4 = E(q_k\delta_k x_k\sigma_{x_k}^{-2} + q_k\Delta_k y_k\sigma_{y_k}^{-2} - q_k r_k \Delta_k x_k \sigma_{x_k}^{-1}\sigma_{y_k}^{-1} - q_k r_k \delta_k y_k \sigma_{x_k}^{-1}\sigma_{y_k}^{-1})
$$

$$
k \in V \cup V^*
$$

where the first three dimensions of $\mathbf{A}, \mathbf{B}$ are the same as given in Equation 7.12 but with all the subscripts i substituted by k. Note that δ_k, Δ_k are calculated by using ψ_{ML}.

Note in Equation 7.15 that, if the aspect number in either V or V^* is 0, it yields $|\mathbf{A}| = 0$ and $\mathbf{A}^{-1}$ is not defined, which means that a cannot be obtained from only information of one side wall. By eliminating the fourth dimension of $\mathbf{A}, \mathbf{B}$, it reduces to Equation 7.12 for wall estimation. In this case, one possible solution is to invert a from the detected shadow behind the façade image. In the case of no overlapping with adjacent objects, the width of the shadow d_i' behind the façade image preserves the relationships with d_i and a as

$$
d_i' = a + d_i \tan^2\theta_i \qquad (7.16)
$$

Thus, the building can still be reconstructed if only images of one side façade and the shadow behind are detected.

After obtaining the estimated values, the corresponding maximum-likelihood probabilities can be calculated as

$$
p_{ML} = \prod_{i \in V} p_i(\omega_i | \mathbf{w}_{ML}) \quad \text{or} \quad p_{ML} = \prod_{i \in V} p_i(\omega_i | \mathbf{c}_{ML}) \prod_{j \in V^*} p_j(\omega_j^* | \mathbf{c}_{ML}) \qquad (7.17)
$$

Thus far, given a group of building images detected from multi-aspect SAR images, the most likely wall or building can be reconstructed, and the corresponding maximum-likelihood probability can be further used as an assessment of the coherence among this group of multi-aspect images. A large maximum-likelihood probability indicates a strong coherence among the group of multi-aspect building images, and vice versa.

7.2.3 Multi-Aspect Co-Registration

In the previous sections, SAR images of all aspects were assumed to be co-registered to the global imaging plane. However, multi-aspect co-registration, as a critical pre-processing step, is necessary when dealing with real SAR data.

Given the specification of a ROI, for example, the longitudes and latitudes, the corresponding area in a SAR image of each aspect can be coarsely delimited according to its orbit parameters. It can be regarded as a coarse co-registration step. Manual intervention is necessary if the orbit parameters are not accurate enough.

The delimited sub-images from all aspects have to be transformed from slant to ground range. Thereafter, building images are detected through the chain of processing given in the previous section. Eventually, a set of parameterized building images is obtained from the SAR image of each aspect.

After that, only the parameters of the building images need to be co-registered to the global coordinates, rather than the original SAR images. This is regarded as a fine co-registration step. In this study, only linear co-registrations are considered. Thus, it includes a rotation of ψ_i and x, y and a translation of x, y. The former can be easily done by first rotating x, y and then compensating ψ_i according to the azimuth angle of each aspect.

As for the translation of x, y, a simple and effective method should be manually choosing ground control points. In general, a featured terrain object or building is first selected as the reference of zero elevation, and then the locations of different aspects of it are pinpointed. However, the manual method becomes inconvenient when the aspect number increases.

An automatic co-registration scheme for multi-aspect SAR images is proposed here. As depicted in Figure 7.14, all aspects are sorted as 1 to N. The first round of co-registration is along,

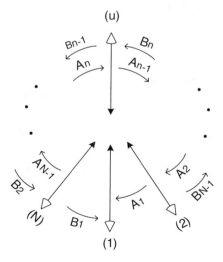

Figure 7.14 Multi-aspect co-registration. © [2007] IEEE. Reprinted, with permission, from *IEEE Transactions on Geoscience and Remote Sensing*

say, the counter-clockwise direction. At each step, the next aspect is co-registered according to the former aspect by using a one-to-one co-registration method for two neighboring aspects. For instance, aspect 2 is co-registered to aspect 1 at step A_1, aspect 3 to aspect 2 at step A_2, and so forth, till the last aspect N. The second round is along the opposite direction. The aspect N is co-registered to 1 at step B_1, the aspect $N-1$ to N at step B_2, and so on, till the aspect 2. Repeat several rounds until the adjustments through the whole round are small enough.

For the co-registration of two neighboring aspects, a new method is provided based on the automatic matching of multi-aspect building images. From each aspect, a set of parameterized building images, that is, $\omega_{i,m}$, $m \in U_i$, $i \in V$, where U_i is the set of building images of the ith aspect, is obtained. Given two aspects, namely, i, j, with building images $\omega_{i,m}$, $m \in U_i$, $\omega_{j,n}$, $n \in U_j$, each building image $\omega_{i,m}$ has its counterpart $\omega_{j,n}$ for the same building wall. The distance between the two building images of an image pair is restricted in order to prevent exhaustive searching.

Since there is a position bias $\Delta x, \Delta y$ between two aspects, the multi-aspect coherence of the parameters x, y is lost. But the other three parameters, ℓ, d, ψ, can still be used in maximum-likelihood estimation of the building wall. Hence, for each image pair $\omega_{i,m}$ and $\omega_{j,n}$, a maximum-likelihood probability, $p_{ML1}(\omega_{i,m}, \omega_{j,n} | \ell_{ML}, d_{ML}, \psi_{ML})$, is calculated, and is used as a measure of the probability of the occurrence of the event that $\omega_{i,m}$ and $\omega_{j,n}$ are actually images of the same building, that is

$$p_{ML1}(\omega_{i,m}, \omega_{j,n} | \ell_{ML}, d_{ML}, \psi_{ML}) = p_t(\langle m, n \rangle) \tag{7.18}$$

where $\langle m, n \rangle$ stands for the event that two images paired correctly. Such a probability can be calculated for all image pairs.

Now, consider a wall $\mathbf{w}$ and its two images $\omega_{i,m}, \omega_{j,n}$. Given the position bias $\Delta x, \Delta y$, the parameters $x_{j,n}, y_{j,n}$ of the second image $\omega_{j,n}$ can be calibrated, that is, $\tilde{x}_{j,n}, \tilde{y}_{j,n}$ and $\tilde{\omega}_{j,n}$. Then the coherence of x, y is restored. In the same way, a maximum-likelihood probability can be estimated from each image pair after calibration, that is, $p_{ML2}(\omega_{i,m}, \tilde{\omega}_{j,n} | x_{ML}, y_{ML}, e_{ML})$, which is considered as the probability of occurrence of the event $\langle m, n \rangle$ under the condition of bias $\Delta x, \Delta y$, that is

$$p_{ML2}(\omega_{i,m}, \tilde{\omega}_{j,n} | x_{ML}, y_{ML}, e_{ML}) = p(\langle m, n \rangle | \Delta x, \Delta y) \tag{7.19}$$

Thus, for all events $\langle m, n \rangle \in \langle U_i, U_j \rangle$, we have a measure of their probability of occurring, that is, $p_t(\langle m, n \rangle)$, and the conditional probability given a known bias $\Delta x, \Delta y$, that is, $p(\langle m, n \rangle | \Delta x, \Delta y)$. Using $p_t(\langle m, n \rangle)$ as the weights, a likelihood function of $\Delta x, \Delta y$ can be constructed, that is

$$L(\Delta x, \Delta y) = \sum_{\langle m, n \rangle} p_t(\langle m, n \rangle) \cdot \ln p(\langle m, n \rangle | \Delta x, \Delta y) \tag{7.20}$$

Subsequently, the maximum-likelihood estimation of $\Delta x, \Delta y$ can be obtained

$$[\Delta x, \Delta y]_{ML} = \arg \max_{\Delta x, \Delta y} L(\Delta x, \Delta y) \tag{7.21}$$

$$\Delta x_{ML} = \sum_{m,n} p_t(m, n)(x_{j,n} - x_{i,m}) \bigg/ \sum_{m,n} p_t(m, n)$$

$$\Delta y_{ML} = \sum_{m,n} p_t(m, n)(y_{j,n} - y_{i,m}) \bigg/ \sum_{m,n} p_t(m, n) \tag{7.22}$$

The estimated bias is used to compensate the jth aspect, and this procedure can be iterated till the estimated bias tends to zero. This means that the two neighboring aspects are co-registered.

Nevertheless, there is a disadvantage of this one-to-one co-registration method. Since this method is based on the information of common visible walls, a number of walls must be visible in both the two neighboring aspects, and the average elevation of the visible buildings from all aspects should be similar. If these two conditions are not satisfied for some aspects, co-registration has to be performed manually.

7.3 Automatic Multi-Aspect Reconstruction (AMAR)

Coherence among multi-aspect building images is the basis of automatic multi-aspect reconstruction. Given a group of building images (at most one for each aspect), an SBO can be estimated from Equation 7.8, and its maximum-likelihood probability is the measure of the probability that such an SBO truly exists. On the other hand, it can also be regarded as a measure of the probability that such a group of building images actually belong to the same building.

An important issue not addressed yet is the reliability of reconstruction based on different numbers of effective aspects. Following the previous discussion, different levels of reconstruction are summarized here.

1. If V, V^* both include at least one aspect, a building can be reconstructed from Equation 7.15.
2. If V includes at least two aspects and V^* is empty, only one side wall can be reconstructed from Equation 7.12; if a shadow is detected behind the façade images, a can be further inverted from Equation 7.16 and then the building is finally reconstructed.
3. If V includes only one aspect and V^* is empty, one side façade can be reconstructed with its elevation e obtained from adjacent buildings, and the building can be further reconstructed if its shadow is detected.

Obviously, the reliabilities of these three levels of reconstruction, sequentially, descend as $1 > 2 > 3$. As for the same level of reconstruction, more effective aspects used in estimation means more observed samples, which lead to more accurate estimation. Therefore, a priority criterion is defined as follows: for different levels, $1 > 2 > 3$; for the same level, the one with more effective aspects has higher priority; for the same number of effective aspects, the one with higher maximum-likelihood probability has higher priority. In other words, more reliable reconstruction has higher priority.

However, because this priority criterion cannot be used to judge whether a group of building images is wrongly matched or not, a prerequisite condition of the criterion is that the group of building images used for reconstruction must truly belong to the same building. As a simple case, suppose that the building A is rightly reconstructed from its images of three aspects, and another building B is wrongly reconstructed from the same three images plus one more image that does not belong to A. This yields that B has a higher priority than A according to the criterion.

In order to test whether a group of building images is matched correctly or not, a quasi-CFAR method is proposed here. Following Equations 7.9 and 7.17, if $\boldsymbol{\omega}_k$ is represented using five independent scalars, that is, $w_{k,q}$, $q = 1, \ldots, 5$, and, moreover, the variance of each scalar $w_{k,q}$ is approximately regarded as aspect-independent, that is, $\sigma_{q,k}^2 \equiv \sigma_q^2$, then the maximum-likelihood estimations should be equal to the average of all aspects, that is, $w_{q,ML} = \overline{w}_q$. Consequently, the maximum-likelihood probability can be written as

$$-\ln p_{ML} = -\sum_q \ln p_{ML}^{(q)}$$

$$-\ln p_{ML}^{(q)} + C_q = \sigma_q^{-2} \sum_{k \in V \cup V^*} \left(w_{k,q} - \overline{w}_q \right)^2$$

(7.23)

where C_q is a constant determined by σ_q^2. According to the property of normal samples, we have

$$-\ln p_{ML}^{(q)} + C_q \sim \chi^2(K - 1)$$

(7.24)

where K is the number of effective aspects in $V \cup V^*$. Further, we have

$$-\ln p_{ML} + \sum_q C_q \sim \chi^2(5K - 5)$$

(7.25)

where 5 is the number of independent random variables.

Considering the null hypothesis "H_0: building images are correctly matched", it is examined by

$$P\left\{ -\ln p_{ML} + \sum_q C_q > \chi_{1-P_f}^2(5K - 5) \right\} = P_f$$

(7.26)

where P_f is the false-alarm rate. If the condition within $\{\cdots\}$ is satisfied, H_0 is rejected, that is, wrongly matched, and vice versa. However, since this method is derived under a certain approximation and, to some extent, is also affected by the accuracy of σ_q^2 of Equation 7.5, it is proposed as a quasi-CFAR method. Therefore, the false-alarm rate P_f should be determined according to experience in practical operation.

Finally, the automatic multi-aspect reconstruction algorithm shown in Figure 7.15 is designed based on the priority criterion of reconstruction reliability. Briefly speaking, first take a building image as a reference, and then one possible matched building image is selected from each aspect (or none is selected when it is considered as an ineffective aspect) to compose a candidate group of building images. A set of all potential candidate groups is established and the corresponding buildings are reconstructed. When those buildings reconstructed from the groups are judged as wrongly matched by Equation 7.26, they are eliminated. Finally, only buildings with higher priorities are selected as the final results. In the case that buildings conflict with each other, for example, the same building images are used for reconstruction or overlapped in the 3D space.

The notation of "condition for potential same-side/opposite-side building images" in Figure 7.15 describes some presumed restrictions on the difference between the reference building image and the selected possible matched building image. For example, those building images located too far away from each other cannot belong to the same building. This

```
initialization;
for (next aspect)
    for (next building image, referred as BI, of this aspect)
        // take this BI as a reference, search for all candidate BIs
        for (other aspects)
            for (next BI of this aspect)
                if (condition for potential same-side BI is satisfied)
                    select this BI as a candidate same-side BI;
                elseif (condition for potential opposite-side BI is satisfied)
                    select this BI as a candidate opposite-side BI;
                end
            end
            add an empty BI (stand for ineffective aspect) as a candidate BI;
        end
        // reconstruction
        while (all combinations are exhausted)
            form a new combination of candidate BIs from all aspects as a group;
            perform reconstruction using all effective BIs of this group;
            if (this group are matched correctly)
                save the reconstructed building;
                set up connections between the building and its BIs;
            end
        end
    end
end
// select buildings with high priorities
sort all reconstructed buildings as priority-descending;
check the parameters of buildings, if reasonable, label as valid, or as invalid;
for (next valid building)
    if (it is overlapped with a higher-priority valid building)
        label as invalid; continue;
    end
    establish the list of all BIs connected with this building;
    for (the next BI on the list)
        label all buildings connected with this BI as invalid (except this one);
    end
end
// buildings labeled as valid are the final reconstruction results
```

Figure 7.15 Automatic reconstruction algorithm for multi-aspect images. © [2007] IEEE. Reprinted, with permission, from *IEEE Transactions on Geoscience and Remote Sensing*

restriction is necessary to prevent an exhaustive search and to keep the algorithm at a high level of efficiency.

7.4 Results and Discussion

7.4.1 Analysis of Reconstruction Results

The two building objects are reconstructed from the four-aspect simulated SAR images. Table 7.1 gives a comparison of the reconstruction with the true value. The inversion accuracy is exceptional. Figure 7.16 shows the 3D reconstructed objects on the true digital elevation mapping (DEM). Note that two co-registered aspects of SAR images are simultaneously displayed on the colorful ground plane in red and green colors, respectively. It seems that the inverted elevations also match well with the true DEM.

In a statistical sense, the precision of reconstruction can be improved as the aspect number increases. However, for reconstruction from a real SAR image, complicated buildings and the background scene may cause many uncertainties and reduce the feasibility of detection. A possible solution is to increase the observation aspects or strengthen the robustness of object image detection.

Reconstruction is carried out on Pi-SAR images. Figure 7.17a gives the 3D view of reconstructed buildings and its comparison with a cartoon of this area (Figure 7.17b) given on the

Table 7.1 Reconstruction from simulated SAR images versus true values[a]

	Orientation	Length	Width	Height	Center position		Elevation
Object A	**60°**	**40**	**10**	**10**	**140**	**60**	**1.94**
	59.55°	41.96	12.45	10.00	138.84	61.54	3.16
Object B	**30°**	**60**	**15**	**15**	**60**	**140**	—
	30.16°	58.69	16.86	13.87	58.83	140.25	—

[a]True values are in **bold**; unit, meter.

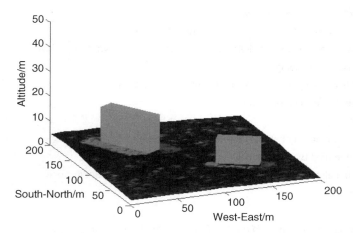

Figure 7.16 Reconstruction from simulated SAR images. © [2007] IEEE. Reprinted, with permission, from *IEEE Transactions on Geoscience and Remote Sensing*

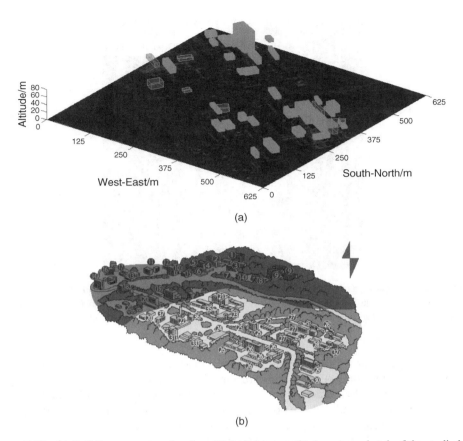

(a)

(b)

Figure 7.17   (a) Building reconstruction from Pi-SAR images. (b) A cartoon sketch of the studied site (the shadowed area is outside of the SAR images). © [2007] IEEE. Reprinted, with permission, from *IEEE Transactions on Geoscience and Remote Sensing*

website of Tohoku University (http://www.tohoku.ac.jp), where the shadowed area is outside of the SAR images. Note that the cartoon is not perfectly accurate. About 80% reconstructed SBOs have real corresponding buildings, and those transparent SBOs are false results with no corresponding building in reality. The false-reconstruction rate is defined as the proportion of false results. It is about 20%. Moreover, a "reconstructable building" is defined as a truly existing building (observed in the satellite photo), which has building images (not covered by surroundings) in SAR images of at least two aspects. Then, the reconstruction rate can be defined as the ratio of correctly reconstructed buildings to reconstructable buildings. It is about 72%.

Owing to the difficulty for authors to collect ground truth data, a high-resolution satellite optical picture (0.6 m QuickBird data) is used as a reference. The geometric parameters of each building manually measured from the picture are taken as ground truth data to evaluate the accuracy of reconstruction. The satellite picture is co-registered to SAR images and shown in Figure 7.18, where the reconstructed SBOs and corresponding real buildings are

Figure 7.18   Comparison of reconstructed buildings with real buildings on optical photo (dashed lines are the roofs of real buildings). © [2007] IEEE. Reprinted, with permission, from *IEEE Transactions on Geoscience and Remote Sensing*

labeled. Histograms of the inversion errors of SBO parameters (ψ, x, y, l, a, h) are given in Figure 7.19, with the corresponding standard variances and average relative errors.

The error of inverted elevation is not given due to the lack of true DEM data. In fact, the elevation is jointly inverted with the position and width in Equation 7.12. The accuracy of the position and width can somewhat reflect the error level of the elevation.

There is a trade-off between the false-reconstruction rate and the reconstruction rate. If we increase the false-alarm rate of edge detection, relax the requirements of building image detection, and/or increase the false-alarm rate of judging a correctly matched group, this will raise the reconstruction rate but also boost the false-reconstruction rate. Efficiency will be worsened if the false-reconstruction rate goes too high. On the contrary, the false-reconstruction rate can be reduced and the accuracy of reconstructed buildings can be improved, but the reconstruction rate will also decline.

A critical factor confining the reconstruction precision should be the number of effective aspects, hereafter referred as the effective aspect number (EAN). The relationship between the average error level and EAN is plotted in Figure 7.20, where EAN = 1 means that only one side wall is reconstructed and the width is inverted with the aid of the detected shadow. Basically, the precision of reconstruction has an improving tendency as EAN increases. This implies that the reconstruction result will become better if more SAR images of different aspects are available.

The average EAN is 2.59 in the reconstruction on four-aspect Pi-SAR dataset. For an arbitrarily oriented building, there inevitably are, in general, two to four ineffective aspects that have intersection angles near 0° or 90° with wall orientation. Suppose that 10 to 12 aspects are used in reconstruction. Then, it is anticipated that the EANs of most buildings will be at least 4, and the accuracy would be upgraded (errors < 10 m). Nevertheless, the relationship

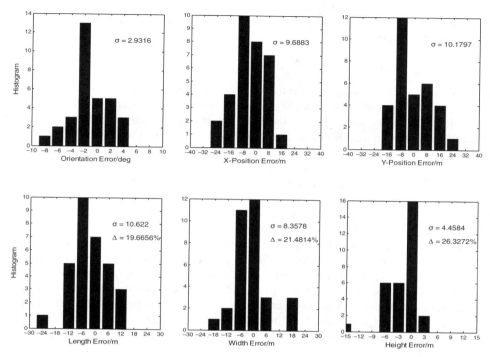

Figure 7.19 Error statistics of reconstruction on Pi-SAR images (σ = standard variance, Δ = average relative error). © [2007] IEEE. Reprinted, with permission, from *IEEE Transactions on Geoscience and Remote Sensing*

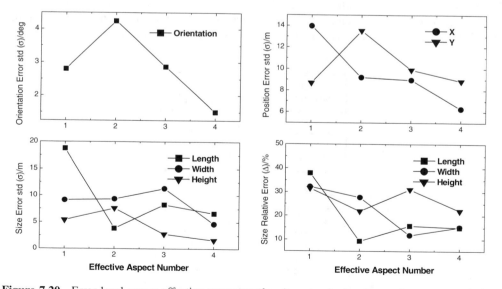

Figure 7.20 Error level versus effective aspect number (σ = standard variance, Δ = average relative error). © [2007] IEEE. Reprinted, with permission, from *IEEE Transactions on Geoscience and Remote Sensing*

of EAN versus total number of aspects needs to be further studied when image data of more aspects become available.

7.4.2 Discussion

The main error in reconstruction is caused by the boundary deviation of detected building images, which originates from complicated scattering and interactions of spatially distributed objects and backgrounds. In a probabilistic description of detected building images, the presented boundary in a SAR image is regarded as the same as in reality. However, the real detected building image might be biased or even partly lost due to obstacles and overlapping.

In addition, it is noted that large-scale buildings might not be well reconstructed. The reason is partly attributed to the premise of the Wishart distribution of the POL-CFAR detector. Since the images of large-scale buildings might reveal more detailed information about the textural features and heterogeneity, this deteriorates the performance of the edge detector. Based on a certain specific speckle model for high-resolution images, a new edge detector can be developed to improve the results. Another more feasible way is to employ a multi-scale analysis, through which building images of different scales can present Gaussian properties in their own scale levels. Of course, the expense of multi-scale analysis might be a loss in precision.

Another issue to be addressed is the exploitation of multi-aspect polarimetric scattering information. In a heterogeneous urban area, the terrain objects appear distinctively in different aspects, which gives rise to a very low coherence between their multi-aspect scatterings. Hence, polarimetric scattering information may not be a good option for the fusion of multi-aspect SAR images over an urban area.

Figure 7.21 summarizes the whole reconstruction process. After the ROI is coarsely chosen in each aspect image, edge detection and object image extraction are carried out subsequently. Then the object images are parameterized and finely co-registered. As long as multi-aspect object images are automatically matched, 3D objects are reconstructed at the same time.

The merits of this approach are process automation with few manual interventions, the full utilization of all aspect information, and the high efficiency for computer processing. Making

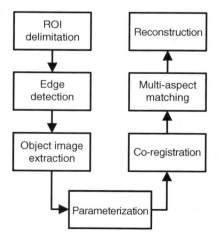

Figure 7.21 Flowchart of building reconstruction based on multi-aspect SAR dataset. © [2007] IEEE. Reprinted, with permission, from *IEEE Transactions on Geoscience and Remote Sensing*

a reconstruction on a four-aspect Pi-SAR dataset (500×500) takes less than 10 min CPU time (at CPU frequency of 3 GHz). The complexity of this approach is about $o(KN^2)$, where K is the aspect number and N is the size of the SAR images, that is, $K = 4$, $N = 500$ in this case.

It is tractable to extend this approach to reconstruction of other kinds of objects. For other types of primary objects, given the *a priori* knowledge, new image detection methods have to be developed. It is possible to treat more complicated buildings as combinations of different primary objects, which can be reconstructed separately and then put together.

Based on multi-aspect meter-resolution airborne or spaceborne SAR datasets, simple building targets can be automatically and efficiently reconstructed. This is applicable to built-up areas where buildings have nearly rectangular shapes and are not too dense. The reconstruction accuracy depends on the number of observation aspects.

Considering a spaceborne SAR with the functions of left/right side-looking (e.g., Radarsat2 SAR) and forward/backward oblique side-looking (e.g., PRISM in ALOS), six different aspect settings are available for ascending orbit. There is a 20° angle between ascending and descending flights for sun-synchronous orbit (Yuan, 2003). Therefore, it can observe from 12 different aspects. Supposing the sensor uses different aspects in each visit and the repeat period is 15 days, a set of SAR images acquired from 12 aspects can be obtained over a three-month observation period. Then, application of SBO reconstruction with acceptable precision can be realized.

7.5 Calibration and Validation of Multi-Aspect SAR Data

To retrieve information accurately from SAR observations, polarimetric calibration of SAR data is a prerequisite. It mainly includes three parts: calibration of cross-talk effects, correction for channel imbalance, and radiometric calibration. Usually, trihedral corner reflectors are installed in the imaging area for channel imbalance and radiometric calibration, whereas the cross-talk effect is corrected from the SAR data itself (van Zyl, 1990).

Dealing with a set of four-aspect Pi-SAR data (Xu and Jin, 2008) for target reconstruction, the data are acquired from different viewing angles. It is usually difficult to install calibration targets, especially in urban areas, for calibration of multi-aspect flights. A simple method is first to choose a natural object, which is believed to have identical polarimetric scattering for all viewing angles (aspects). Then, the channel imbalance can be estimated for all aspects for the phase difference and amplitude ratio of co-horizontally (*hh*) and co-vertically (*vv*) polarized echoes. Calibration of the cross-talk effect is performed in the conventional way. The calibration fidelity is demonstrated by comparing the polarimetric signatures and the deorientation parameters of the SAR images before and after calibration. These calibrated multi-aspect SAR data have been implemented for target reconstruction (Xu and Jin, 2007).

7.5.1 *Method*

A set of four-aspect Pi-SAR images in a square-loop flying path at X band with 1.25 m spatial resolution over Sendai City, Japan (the flights are indicated by the numbers 8708, 8709, 8710, and 8711) was studied. The airborne Pi-SAR was developed by the Communication Research Laboratory (CRL) and Japan Aerospace Exploration Agency (JAXA) (Toshihiko *et al.*, 2002) for polarimetric and interferometric observations at X and L bands.

Usually, some manual reference targets are installed in the imaging area for data calibration. However, it has been reported that installing those reference targets for different viewing angles is actually not practicable or is not convenient (Toshihiko *et al.*, 2002), especially in crowded urban areas. Furthermore, the phase information errors might be significant because antennas with different positions are used for different polarizations and their roll angles are variable (Toshihiko *et al.*, 2002). In fact, such abnormal behaviors of the original Pi-SAR data did catch the authors' attention in their reconstruction study. Those polluted data might become meaningless to fulfill reconstruction. Therefore, calibration of multi-aspect data is the first issue to be resolved.

The measured symmetric scattering matrix is simply described as

$$\overline{\mathbf{Z}} = Ae^{j\phi}\overline{\mathbf{R}}_x \cdot \overline{\mathbf{R}}_c \cdot \overline{\mathbf{S}} \cdot \overline{\mathbf{R}}_c^{\mathrm{T}} \cdot \overline{\mathbf{R}}_x^{\mathrm{T}} \tag{7.27}$$

where **S** is the desired scattering matrix, which is also symmetric due to the reciprocity. The superscript T means "transpose". The coefficient A and the phase ϕ are the overall absolute amplitude and phase factor, respectively. $\mathbf{R}_x$ and $\mathbf{R}_c$ denote the cross-talk effect and the channel imbalance, respectively, which can be further expressed as

$$\overline{\mathbf{R}}_x = \begin{bmatrix} 1 & \delta_2/f \\ \delta_1 & 1 \end{bmatrix}, \quad \overline{\mathbf{R}}_c = \begin{bmatrix} 1 & 0 \\ 0 & f \end{bmatrix} \tag{7.28}$$

where δ_1, δ_2 are the cross-talk coefficients, and $\bar{f}$ is the channel imbalance factor. As indicated by van Zyl (1990), the cross-talk effect can be corrected independently without external calibration devices. An iterative method was applied to the calibration of $\mathbf{R}_x$. Since this study focuses on the polarimetric information, absolute radiometric calibration is not the major concern here. Thereafter, only the channel imbalance is left for calibration, that is

$$\overline{\mathbf{W}} = \overline{\mathbf{R}}_c \cdot \overline{\mathbf{S}} \cdot \overline{\mathbf{R}}_c^{\mathrm{T}} = \begin{bmatrix} S_{hh} & fS_{hv} \\ fS_{hv} & f^2 S_{vv} \end{bmatrix} \tag{7.29}$$

where **W** is the scattering matrix after calibration of the cross-talk effect, and the absolute amplitude and phase are ignored.

It is assumed that one aspect has been calibrated with the assistance of external calibration devices. A simple natural target (e.g., flat bare ground) is chosen as the reference target, which has identical polarimetric scattering for all aspects. The centers of the distribution of the phase difference and amplitude ratio of hh and vv polarizations are estimated. Then, by comparison with the calibrated aspect, the channel imbalance factors of other aspects can be retrieved.

The distribution of the phase difference ϕ and amplitude ratio r from uniformly distributed scatterers can be determined by the correlation coefficient ρ_c and the intensity ratio τ as (Joughin, Winebrenner, and Percival, 1994; Lee *et al.*, 1994)

$$p_\phi(\phi) = \frac{(1 - |\rho_c|^2)[(1 - \beta^2)^{1/2} + \beta(\pi - \cos^{-1}\beta)]}{2\pi(1 - \beta^2)^{3/2}}, \quad \beta = |\rho_c|\cos(\phi - \theta) \tag{7.30}$$

$$p_r(r) = \frac{2\tau(1 - |\rho_c|^2)(\tau + r^2)r}{[(\tau + r^2)^2 - 4\tau|\rho_c|^2 r^2]^{3/2}} \tag{7.31}$$

where

$$\rho_c = \frac{E(S_{hh}S_{vv}^*)}{\sqrt{E(S_{hh}S_{hh}^*)E(S_{vv}S_{vv}^*)}} = |\rho_c|e^{j\theta} \tag{7.32}$$

$$\tau = \frac{E(S_{hh}S_{hh}^*)}{E(S_{vv}S_{vv}^*)} \tag{7.33}$$

Note that single-look data are assumed. Here, co-polarization is chosen because cross-polarization is more likely to suffer from system noise. Following Equation 7.29, these distributions have the same forms for uncalibrated data $\mathbf{W}$, but ρ_c and τ are replaced by another ρ_c' and τ' as follows:

$$\rho_c' = \frac{E(W_{hh}W_{vv}^*)}{\sqrt{E(W_{hh}W_{hh}^*)E(W_{vv}W_{vv}^*)}} = |\rho_c|e^{j\theta}e^{-2j\varphi} \tag{7.34}$$

$$\tau' = \frac{E(W_{hh}W_{hh}^*)}{E(W_{vv}W_{vv}^*)} = \tau|f|^{-4} \tag{7.35}$$

where φ denotes the phase of f. Therefore, ρ_c' and τ' can be calculated directly from Equations 7.34 and 7.35 via averaging the system over the area. However, the directly calculated parameters are not accurate enough, partly due to the influence of strong scatterers or bright spots, which are very likely to appear in homogeneous areas. Here, these parameters are further optimized through curve fitting of the histograms of phase difference and amplitude ratio using the probability distribution functions (PDFs) in Equations 7.30 and 7.31, whereas the directly calculated values are taken as initial values for optimization of PDF fitting. The fitting step is realized via conventional optimization, for example, Nelder–Mead downhill simplex method, where the input parameters of the PDF are adjusted step by step with a goal function indicating the disparity between the PDF and the histogram, for example, the squared sum of errors.

Under the assumption that the reference target has identical scattering for all aspects, the estimated ρ_c' and τ' of all aspects are expressed as

$$\begin{aligned}
\rho_{c,0}' &= |\rho_c|e^{j\theta}, \quad \rho_{c,i}' = |\rho_c|e^{j\theta}e^{-2j\varphi_i} \\
\tau_0' &= \tau, \quad \tau_i' = \tau|f_i|^{-4}, \quad i \in U
\end{aligned} \tag{7.36}$$

where the subscript 0 denotes the calibrated aspect, $i \in U$ denotes other aspects. Hence, the channel imbalance factors are obtained as

$$f_i^2 = \sqrt{\tau_0'/\tau_i'} \cdot \rho_{c,0}'/\rho_{c,i}', \quad i \in U \tag{7.37}$$

which are used to compensate the channel imbalances of all aspects.

To preserve constant polarimetric scattering in all aspects, a reference target with an upright symmetric axis of rotation is suitable for calibration, because it can be seen as identical from any aspect. Another condition is that the incident angles should not differ too much. Normally, a flat ground area located in the image center is chosen as the reference target, which should also be large enough to assure sufficient samples for the estimation.

However, some side-lobes of strong scatterers in the vicinity might be involved and contaminate the reference target, especially in a crowded urban area. Besides, some moving targets might be displaced and overlap with the reference target. It is necessary to avoid these cases. Moreover, elevation of the reference target is a factor to be taken into account, as well as its range distances, to ensure that its incidence angles in multi-aspect do not vary too much.

7.5.2 Experiments

Following the aforementioned criteria, a sports field located in the center of the Pi-SAR image is chosen as the reference, as shown in Figure 7.22. This sports field is regarded as a flat surface with moderate roughness to keep almost the same isotropic scattering for all aspects, and the incidence angles of all aspects are similar, that is, $48.9°$, $47.5°$, $48.6°$, $48.0°$. From Figure 7.22, no severe vicinity contamination is observed, but some spatial inhomogeneity due to possible strong scatterers does appear. In this case, about 2500 pixels within the selected area were taken as samples for the estimation.

The first aspect, No. 8078, is seen as a calibrated one via validation of the polarimetric signature predicted by the integral equation method (IEM) model (Fung, 1994). The input parameters of the IEM model (Gaussian) are assumed as: roughness $k\sigma = 1$, correlation length $kL = 4$, and dielectric constant $\varepsilon_r = 5 + 0.5i$. The co-polarized backscattering

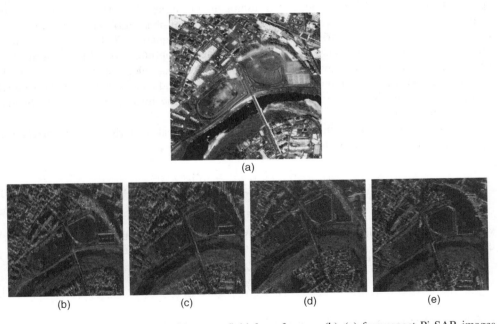

Figure 7.22 (a) Scenario photo with sports field for reference; (b)–(e) four-aspect Pi-SAR images (radar look directions are bottom–top, top–bottom, left–right, and right–left, respectively). © [2007] IEEE. Reprinted, with permission, from *IEEE Transactions on Geoscience and Remote Sensing*

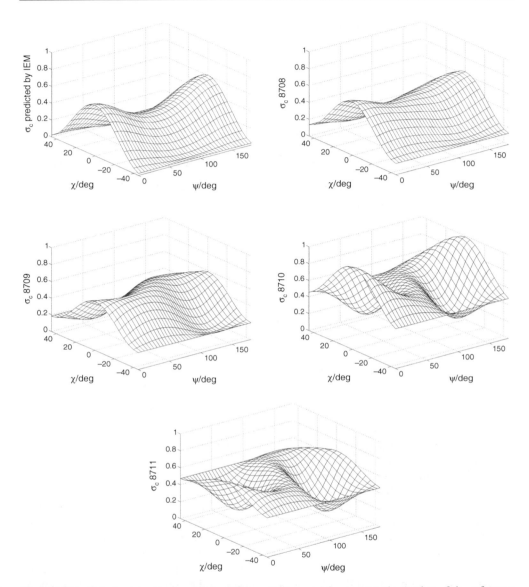

Figure 7.23 Polarimetric signature of the IEM prediction; and four-aspect observation of the reference target. © [2007] IEEE. Reprinted, with permission, from *IEEE Transactions on Geoscience and Remote Sensing*

signature σ_c predicted by the Mueller matrix solution of the IEM and those averaged over the sports field region in four aspects are plotted in Figure 7.23. Note that cross-talk effects have been corrected. It can be seen that the other three aspects have great channel imbalances.

The phase difference and amplitude ratio of S_{hh}/S_{vv} was counted over all the sports field area and the normalized histograms were employed to run a PDF curve fitting process. The fitting-optimized PDFs match with the histograms very well (Figure 7.24). During this fitting

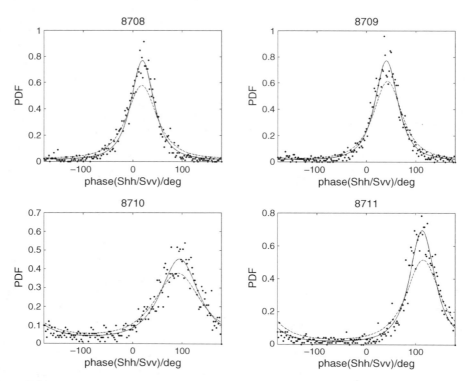

Figure 7.24 The fitted probability distribution functions of phase differences (dashed line – initial PDF). © [2007] IEEE. Reprinted, with permission, from *IEEE Transactions on Geoscience and Remote Sensing*

process, a severe misalignment of distribution centers of phase differences among the four-aspect data is observed.

The channel imbalance factors are calculated from Equation 7.37 and then used for calibration. The polarimetric signatures of the reference target after calibration are shown in Figure 7.25. It seems that No. 8710 still has some discrepancy compared with the other aspects. In fact, the statistics and PDF of No. 8710 reveal a wider spread of phase difference (Figure 7.24), which indicates a large variety of double/multiple scatterings observed in this aspect. Two possible reasons for the abnormal polarimetric behavior of No. 8710 are inhomogeneous features of rough ground surface and scattering contribution from grassland. Nevertheless, a wider spread distribution will not affect the correct shifting of distribution center, which guarantees the accuracy of calibration here.

Using the channel imbalance factors obtained in the aforementioned steps, all SAR images were calibrated. Calibration error is a critical issue. However, we are unable to evaluate the accuracy of this method quantitatively due to the fact that no artificial calibrator was installed during the collection of the data. To confirm the calibration reliability, we choose an indirect way instead, that is, comparing the polarimetric retrievals against ground truth information. Another region of the same image set, the Aobayama campus of Tohoku University, Sendai, Japan, was also studied. Figures 7.26a and b show the optical photo and one aspect of the Pi-SAR image, respectively.

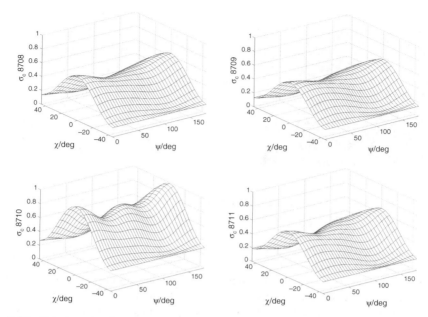

Figure 7.25 Polarimetric signatures of the reference target of four-aspect data after calibration.
© [2007] IEEE. Reprinted, with permission, from *IEEE Transactions on Geoscience and Remote Sensing*

Figure 7.26 The photo and SAR image of Aobayama campus of Tohoku University, Japan: (a) optical photo (resolution 0.6 m); (b) Pi-SAR image (No. 8708), with R = HH, G = HV, and B = VV. © [2007] IEEE. Reprinted, with permission, from *IEEE Transactions on Geoscience and Remote Sensing*

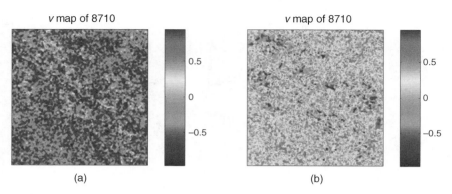

Figure 7.27 The parameter v of left–right aspect (No. 8710 image): (a) before and (b) after calibration.
© [2007] IEEE. Reprinted, with permission, from *IEEE Transactions on Geoscience and Remote Sensing*

The classification parameters of the deorientation approach (Xu and Jin, 2005), v and ψ, are adopted for comparison of before and after calibration. The regions, $v > 0.2$, $v < -0.2$, or other cases, classify scattering mechanisms as single, double or multiple scattering, respectively. The deorientation angle ψ indicates the orientation of scattering targets.

The area for this study is a place full of rectangular buildings and surrounding forest. Since these buildings do not directly face the radar, the wall–ground double scattering is not significant as usual. Figures 7.27a and b show maps of the classification parameter v (median filtered by a 7×7 boxcar) of No. 8710 data before and after calibration, respectively. Apparently, owing to the channel imbalances, most of these areas are misclassified as dominated by double scattering before calibration.

The deorientation angle ψ provides the orientation information of the symmetry axis of the scattering target in the coordinates of the polarization basis (Xu and Jin, 2005). For example, the normal direction of the building wall surface is regarded as its symmetry axis. From the geometry of the vertical wall surfaces, the angle η, defined as the deviation of the normal direction of the wall surface from vertical polarization, can be expressed by the wall orientation Ω, that is, the angle between the wall normal direction and the range direction, and the local incidence angle θ_i as

$$\eta = \tan^{-1}(\tan \Omega / \cos \theta_i), \quad \eta \in [0°, 45°] \tag{7.38}$$

Since most targets have two mutually perpendicular axes of symmetry, there is a 90° ambiguity in the definition of the deorientation angle ψ (Xu and Jin, 2005). If the deorientation angle ψ is equivalently limited within $[0°, 45°]$ as

$$\psi_m = 45 - \big|[\psi]_{90} - 45\big| \tag{7.39}$$

where $[y]_x$ denotes the reminder of y divided by x, then ψ_m should theoretically equal η. Both of them mean the angle between the symmetric axis and the nearest polarization basis.

As shown in Figure 7.28a, the parallelograms are the detected wall images from No. 8710 data (Xu and Jin, 2007). Orientation of each wall can be obtained from the detected parallelogram. Hence, the deviation angle η of each wall can be calculated from Equation 7.38. At the same time, the deorientation parameter ψ_m is calculated via averaging the Mueller matrix over the region of each parallelogram. Figure 7.28b shows the relationship between

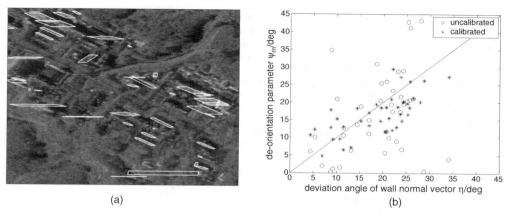

(a)   (b)

Figure 7.28 Validation via deorientation angle: (a) wall images detected from left–right aspect (No. 8710 image); (b) deorientation parameter of wall image versus wall orientation. © [2007] IEEE. Reprinted, with permission, from *IEEE Transactions on Geoscience and Remote Sensing*

ψ_m and η. Clearly, they have a much closer relationship after calibration, which demonstrates the reliability of the calibration approach.

Calibration of multi-aspect SAR images is a key issue for data exploration, such as surface classification and 3D object reconstruction. A simple method of polarimetric calibration for multi-aspect SAR data is presented. First, a natural object is chosen as a reference target, which retains identical scattering for all aspects. Then, the channel imbalance factors are estimated from the statistical distribution of the phase difference and amplitude ratio of hh and vv polarizations of the reference target. With respect to the calibrated aspect, other aspects are then calibrated. The experiments on four-aspect Pi-SAR data show the fidelity of this calibration approach.

However, this approach requires that at least one aspect is calibrated with the assistance of external devices or by other means. Another major limitation of this method is the selection of a natural reference target. It is restricted by many conditions, such as the need to be aspect-invariant, to have a large number of pixels, and to be free of strong scatterer/moving target contamination.

Our calibration has been applied to target reconstruction from multi-aspect SAR data (Xu and Jin, 2008). A potential extended application of this approach is calibration of descending-orbit data using calibrated ascending-orbit data, or vice versa.

References

Bolter, R. (2000) Reconstruction of man-made objects from high resolution SAR images. Proceedings of the 2000 IEEE Aerospace Conference, Vol. 3, pp. 287–292.

Bolter, R. and Leberl, F. (2000) Detection and reconstruction of human scale features from high resolution interferometric SAR data. Proceedings of the 15th International Conference on Pattern Recognition, 2000, Vol. 4, pp. 291–294.

Foucher, S., Boucher, J.M., and Benie, G.B. (2000) Maximum likelihood estimation of the number of looks in SAR images. Proceedings of the 13th International Conference on Microwaves, Radar and Wireless Communications, 2000, MIKON-2000, Vol. 2, pp. 657–660.

Franceschetti, G., Iodice, A., and Riccio, D. (2002) A canonical problem in electromagnetic backscattering from buildings. *IEEE Transactions on Geoscience and Remote Sensing*, **40**(8), 1787–1801.

Fung, A.K. (1994) *Microwave Scattering and Emission Models and Their Applications*, Artech House, Norwood, MA.

Hamasaki, T., Sato, M., Ferro-Famil, L., and Pottier, E. (2005) Polarimetric SAR stereo using Pi-SAR square loop path. Proceedings Geoscience and Remote Sensing Symposium, 2005, IGARSS'05, Vol. 1, p. 4.

Hannan, E.J. and Deistler, M. (1987) *The Statistical Theory of Linear Systems*, John Wiley & Sons, Inc., New York.

Henderson, F.M. (1995) An analysis of settlement characterization in central Europe using SIR-B radar imagery. *Remote Sensing of Environment*, **54**(1), 61–70.

Henderson, F.M. and Xia, Z.G. (1997) SAR applications in human settlement detection, population estimation and urban land use pattern analysis: a status report. *IEEE Transactions on Geoscience and Remote Sensing*, **35**(1), 79–85.

Joughin, I.R., Winebrenner, D.P., and Percival, D.B. (1994) Probability density functions for multilook polarimetric signatures. *IEEE Transactions on Geoscience and Remote Sensing*, **32**(3), 562–574.

Lee, J.S., Hoppel, K.W., Mango, S.A., and Miller, A.R. (1994) Intensity and phase statistics of multilook polarimetric and interferometric SAR imagery. *IEEE Transactions on Geoscience and Remote Sensing*, **32**(5), 1017–1028.

Lee, J.S., Grunes, M.R., Ainsworth, T.L. *et al.* (1999) Unsupervised classification using polarimetric decomposition and the complex Wishart classifier. *IEEE Transactions on Geoscience and Remote Sensing*, **37**(5), 2249–2258.

Michaelsen, E., Soergel, U., and Thoennessen, U. (2005) Potential of building extraction from multi-aspect high resolution amplitude SAR data, in Proceedings of the ISPRS Workshop CMRT 2005: Object Extraction for 3D City Models, Road Databases and Traffic Monitoring – Concepts, Algorithms and Evaluation, 29–30 August 2005, Vienna, Austria (eds U. Stilla, F. Rottensteiner, and S. Hinz), Vol. XXXVI, Part 3/W24.

Pratt, W.K. (1991) *Digital Image Processing*, John Wiley & Sons, Inc., New York.

Quartulli, M. and Datcu, M. (2004) Stochastic geometrical modeling for built-up area understanding from a single SAR intensity image with meter resolution. *IEEE Transactions on Geoscience and Remote Sensing*, **42**(9), 1996–2003.

Roth, A. (2003) TerraSAR-X: a new perspective for scientific use of high resolution spaceborne SAR data. Proceedings of the 2nd GRSS/ISPRS Joint Workshop on Remote Sensing and Data Fusion over Urban Areas, URBAN 2003, pp. 22–23.

Schou, J., Skriver, H., Nielsen, A.A., and Conradsen, K. (2003) CFAR edge detector for polarimetric SAR images. *IEEE Transactions on Geoscience and Remote Sensing*, **41**(1), 20–32.

Simonetto, E., Oriot, H., and Garello, R. (2005) Rectangular building extraction from stereoscopic airborne radar images. *IEEE Transactions on Geoscience and Remote Sensing*, **43**(10), 2386–2395.

Soergel, U., Michaelsen, E., Thoennessen, U., and Stilla, U. (2005) Potential of high-resolution SAR images for urban analysis. Proceedings of the ISPRS Joint Conference (URS 2005), Tempe, AZ.

Stilla, U., Soergel, U., and Thoennessen, U. (2001) Potential and limits of InSAR data for the reconstruction of buildings. Remote Sensing and Data Fusion over Urban Areas, IEEE/ISPRS Joint Workshop, 2001, pp. 64–68.

Toshihiko, U., Seiho, U., Tatsuharu, K. *et al.* (2002) Development of airborne high-resolution multi-parameter imaging radar SAR, Pi-SAR. *Journal of the Communications Research Laboratory*, **49**(2), 109–126.

Touzi, R., Lopes, A., and Bousquet, P. (1988) A statistical and geometrical edge detector for SAR images. *IEEE Transactions on Geoscience and Remote Sensing*, **26**(6), 764–773.

Tupin, F. (2004) Merging of SAR and optical features for 3D reconstruction in a radargrammetric framework. Proceedings of the IEEE International Geoscience and Remote Sensing Symposium, 2004, IGARSS'04, Vol. 1, p. 92.

van Zyl, J. (1990) Calibration of polarimetric radar images using only image parameters and trihedral corner reflector responses. *IEEE Transactions on Geoscience and Remote Sensing*, **28**(3), 337–348.

Xia, Z.G. and Henderson, F.M. (1997) Understanding the relationships between radar response patterns and the bio- and geophysical parameters of urban areas. *IEEE Transactions on Geoscience and Remote Sensing*, **35**(1), 93–101.

Xu, F. and Jin, Y.Q. (2005) Deorientation theory of polarimetric scattering targets and application to terrain surface classification. *IEEE Transactions on Geoscience and Remote Sensing*, **43**(10), 2351–2364.

Xu, F. and Jin, Y.Q. (2006) Imaging simulation of polarimetric synthetic aperture radar for comprehensive terrain scene using the mapping and projection algorithm. *IEEE Transactions on Geoscience and Remote Sensing*, **44**(11), 3219–3234.

Xu, F. and Jin, Y.Q. (2007) Automatic reconstruction of building objects from multi-aspect metric-resolution SAR images. *IEEE Transactions on Geoscience and Remote Sensing*, **45**(7), 2336–2353.

Xu, F. and Jin, Y.Q. (2008) Calibration and validation of a set of multiaspect airborne polarimetric Pi-SAR data. *International Journal of Remote Sensing*, **29**(23), 6801–6810.

Yuan, X.K. (2003) *Introduction to the Spaceborne Synthetic Aperture Radar*, Defense Industry Press, Beijing, chap. 2, pp. 48–56.

8

Faraday Rotation on Polarimetric SAR Image at UHF/VHF Bands

The moisture profiles in the subcanopy and subsurface (0–5 m) are key parameters for studies of global change, land hydrological processes, the global carbon balance, and so on. To obtain these profiles is the main purpose of microwave observation of subcanopy and subsurface (MOSS) (Dobson *et al.*, 1992; Rignot *et al.*, 1995; Moghaddam, Saatchi, and Cuenca, 2000; Moghaddam *et al.*, 2003; Freeman, Durden, and Zimmermann, 1992; Pierce and Moghaddam, 2005). In the general case, SAR technology employs the L, C, X or Ku bands, which can monitor the non-dense vegetation canopy (with $\sim$4 kg m^{-2} of biomass) and penetrate through the land media to a depth of order about 10 cm or so, even at lowest L band. It needs UHF/VHF bands (435 MHz, 135 MHz) to penetrate through the dense subcanopy ($\sim$20 kg m^{-2} and more) with little scattering and reach the subsurface.

It is known that the Earth's ionosphere extends from approximately 50 km above the Earth to several Earth radii (the mean Earth radius is about 6371 km) with the maximum in ionization density at about 300 km. As the polarized electromagnetic wave propagates through the Earth's ionosphere, the polarization vectors are rotated through the anisotropic ionosphere and the action of the geomagnetic field. This rotation is called the Faraday rotation (FR) effect. The FR depends on the wavelength, electron density, geomagnetic field, the angle between the direction of wave propagation and the geomagnetic field, and the wave incidence angle (Le Vine and Abraham, 2002). The FR may decrease the difference between co-polarized (*vv* and *hh*) backscattering, enhance cross-polarized echoes, and mix different polarized terms. Since the FR is proportional to the square power of the wavelength, it has an especially serious impact on SAR observations operating at a frequency lower than L band. The FR angle at P band can reach dozens of degrees. Even for mono-polarized measurement with a known FR degree, the true scattering vector cannot be inverted from the observed SAR data (Wright *et al.*, 2003).

Also, in the deep-space remote sensing program of the Mars Advanced Radar for Subsurface and Ionospheric Sounding (MARSIS) on the Mars Express (MEX) spacecraft, numerous measurements (at 1–5 MHz) of radar echoes from the Martian surface and subsurface were

Polarimetric Scattering and SAR Information Retrieval, First Edition. Ya-Qiu Jin and Feng Xu.
© 2013 John Wiley & Sons Singapore Pte. Ltd. Published 2013 by John Wiley & Sons Singapore Pte. Ltd.

taken. All of these measurements were distorted by the Mars ionospheric FR. Some techniques were developed to compensate for the ionospheric distortions (Mouginota *et al.*, 2008).

In a study of the FR issues, Freeman (2004) derived the polarimetric 2×2 (dimensional) complex scattering matrix without FR from SAR data with FR. But he pointed out that there existed a $\pm\pi/2$ ambiguity error to fully determine the scattering matrix without FR. This ambiguity error remains to be resolved.

In this chapter, an example shows how the P band data with FR is distorted and why the data cannot be directly applied to terrain surface classification. Then, the 4×4 Mueller matrix without FR is inverted from one with FR, and it shows that the remaining $\pm\pi/2$ ambiguity error does not affect the classification parameters u, v, H, α, A, which have been discussed in Chapter 5. This Mueller matrix without FR can be directly applied to surface classification, regardless of the $\pm\pi/2$ ambiguity error. Based on an intuitive assumption of gradual change of FR degree with geographical position, a method – an alternative approach to Freeman (2004) – to eliminate the $\pm\pi/2$ ambiguity error is designed. Our example shows the polarimetric SAR without FR at low frequency, such as P band, can be fully inverted (Qi and Jin, 2007).

8.1 Faraday Rotation Effect on Terrain Surface Classification

8.1.1 Faraday Rotation in Plasma Media

The electromagnetic wavevector $\hat{k}$ is always perpendicular to the electric displacement vector $\mathbf{D}$ and magnetic flux density $\mathbf{B}$. In anisotropic media the constitutive relation $\mathbf{D} = \bar{\varepsilon} \cdot \mathbf{E}$ might make the electric field vector $\mathbf{E}$ not simply on the plane of DB. Thus, using $\mathbf{D}$, instead of $\mathbf{E}$, to define the polarization of a plane electromagnetic wave is more convenient. This is called the kDB system (Kong, 1985).

Assume that the Earth's magnetic field $\mathbf{B}$ is in the $\hat{z}$ direction. The constitutive relation in the ionosphere can usually be written as

$$\bar{\varepsilon} = \begin{bmatrix} \varepsilon & i\varepsilon_g & 0 \\ -i\varepsilon_g & \varepsilon & 0 \\ 0 & 0 & \varepsilon_z \end{bmatrix} \tag{8.1}$$

where

$$\varepsilon = \varepsilon_0 \left[1 - \frac{\omega_p^2}{(\omega^2 - \omega_B^2)} \right], \quad \varepsilon_g = \varepsilon_0 \left[\frac{-\omega_p^2 \omega_B}{\omega(\omega^2 - \omega_B^2)} \right], \quad \varepsilon_z = \varepsilon_0 (1 - \omega_p^2/\omega^2)$$

$\omega_B = qB/m_e$ is the cyclotron frequency, $\omega_p^2 = Nq^2/(m_e\varepsilon_0)$ is the plasma frequency, and q, m_e, N are electron charge, mass, and number per unit volume, respectively. The electron density N varies from approximately $10^7 \, \text{m}^{-3}$ to $10^{12} \, \text{m}^{-3}$.

In the kDB system, let $\bar{\kappa} = \bar{\varepsilon}{-1}$, and write the constitutive relation as $\mathbf{E} = \bar{\kappa} \cdot \mathbf{D}$ with

$$\bar{\kappa} = \bar{\varepsilon}{-1} = \begin{bmatrix} \kappa & i\kappa_g & 0 \\ -i\kappa_g & \kappa & 0 \\ 0 & 0 & \kappa_z \end{bmatrix} \tag{8.2}$$

where

$$\kappa = \frac{\varepsilon}{\varepsilon^2 - \varepsilon_g^2}, \quad \kappa_g = \frac{-\varepsilon_g}{\varepsilon^2 - \varepsilon_g^2}, \quad \kappa_z = 1/\varepsilon_z$$

Writing the Maxwell equations in the kDB system, it yields the following equation (Kong, 1985):

$$\begin{bmatrix} u^2 - \kappa/\mu & i\kappa_g\cos\theta/\mu \\ -i\kappa_g\cos\theta/\mu & u^2 - (\kappa\cos^2\theta + \kappa_z\sin^2\theta)/\mu \end{bmatrix} \begin{bmatrix} D_1 \\ D_2 \end{bmatrix} = 0 \tag{8.3}$$

where the phase velocity $u = \omega/k$, and the third $D_3 = 0$ in the kDB system. To satisfy Equation 8.3, it requires that there are two orthogonal characteristic waves propagating in the θ direction through the ionosphere plasma, and the phase velocity must satisfy

$$u^2 = \frac{1}{2\mu}\left[\kappa(1 + \cos^2\theta) + \kappa_z\sin^2\theta \pm \sqrt{(\kappa - \kappa_z)^2\sin^4\theta + 4\kappa_g^2\cos^2\theta}\right] \tag{8.4}$$

It gives two phase velocities of the characteristic waves, and the ratio of two elliptically polarized characteristic waves is

$$\frac{D_1}{D_2} = \frac{2i\kappa_g\cos\theta}{(\kappa - \kappa_z)\sin^2\theta \pm \sqrt{(\kappa - \kappa_z)^2\sin^4\theta + 4\kappa_g^2\cos^2\theta}} \tag{8.5}$$

Using the typical values of the Earth's ionosphere, this gives $(\kappa - \kappa_z)\sin^2\theta \ll 2\kappa_g\cos\theta$. Thus, it yields $D_1/D_2 = \pm i$. It means that two characteristic waves through the ionosphere are circularly polarized. Neglecting collisions, the Appleton–Hartree formula of the refractive index can be written as

$$n_\pm^2 = \frac{c^2}{u^2} = 1 - \frac{f_{pn}^2}{1 - \left[\frac{f_{Bn}^2\sin^2\theta}{2(1 - f_{pn}^2)}\right] \pm \sqrt{f_{Bn}^2\cos^2\theta + \frac{f_{Bn}^4\sin^4\theta}{4(1 - f_{pn}^2)^2}}} \tag{8.6}$$

where $f_{Bn} = \omega_B/\omega = f_B/f$, $f_{pn} = \omega_p/\omega = f_p/f$. Using typical parameters of the Earth's ionosphere, it gives $f_{Bn}^2\cos^2\theta \gg f_{Bn}^4\sin^4\theta/[4(1 - f_{pn}^2)^2]$, and Equation 8.6 is simplified to

$$n_\pm^2 \approx 1 - f_{pn}^2\left(1 \mp \frac{1}{2}f_{Bn}\cos\theta\right) \equiv (1 - f_{pn}^2) \pm \Delta \tag{8.7}$$

where $\Delta \equiv \frac{1}{2}f_{pn}^2 f_{Bn}\cos\theta$. It means that two characteristic waves travel through the ionosphere with different refractive indices, and their difference is Δ.

It can be proved that any polarized electromagnetic wave can be expressed in terms of two orthogonal linearly polarized waves, and any linearly polarized wave can be expressed in terms of two orthogonal circularly polarized characteristic waves (Kong, 1985). Thus, as the electromagnetic waves propagate though the ionosphere by dr, the phase of the synthetic wave combination of these two characteristic waves is rotated by $k_0\Delta \cdot dr$, that is, the polarization base turns by $k_0 \Delta \cdot dr$. This is called Faraday rotation (FR). Thus, as a polarized electromagnetic wave passes through the ionosphere, the polarization vectors are rotated because of the FR effect (Le Vine and Abraham, 2002):

$$\Omega = 2 \int k_0 \Delta \, dr = \left(\frac{\pi}{cf^2} \right) \int f_p^2(s) f_B(s) \cos \Theta_B(s) \, ds \tag{8.8}$$

where f is the electromagnetic wave frequency, f_B is the electronic gyro-frequency, f_p is the plasma frequency, and Θ_B is the angle between the directions of electromagnetic wave propagation and the geomagnetic field. Note that as $\Theta_B = 90°$, it gives FR $= 0$. So, Equation 8.8 is not restricted by $\Theta_B \neq 90°$.

Assuming that the propagation direction is not changed on passing through the homogeneous ionosphere, $ds = \sec \theta_i \, dz$ (where $\hat{z}$ is the normal to the surface), and that the geomagnetic field remains constant at the value at 400 km altitude (Le Vine and Abraham, 2000, 2002, see also http://omniweb.gsfc.nasa.gov/vitmo/igrf_vitmo.html), the FR angle of Equation 8.8 is written and calculated as

$$\Omega_F \approx -2620 \rho_e B(400) \lambda^2 \cos \Theta_B \sec \theta_i \quad \text{[radians]} \tag{8.9}$$

where ρ_e is the total electron content per unit area (10^{16} electron/m^3), and $B(400)$ is the intensity of the geomagnetic field at 400 km altitude. The intensity of the geomagnetic field is generally reported in nanoteslas (nT). It ranges from about 25 000 to 65 000 nT. It is greatest near the poles and weakest near the equator.

8.1.2 Mueller Matrix with Faraday Rotation

As a SAR system experiences two Faraday rotations when completing a round trip, the scattering matrix with FR (indicated by superscript F) is written in terms of the scattering matrix without FR as follows (Freeman and Saatchi, 2004):

$$\begin{bmatrix} S_{hh}^F & S_{hv}^F \\ S_{vh}^F & S_{vv}^F \end{bmatrix} = \begin{bmatrix} \cos \Omega & \sin \Omega \\ \sin \Omega & \cos \Omega \end{bmatrix} \cdot \begin{bmatrix} S_{hh} & S_{hv} \\ S_{vh} & S_{vv} \end{bmatrix} \cdot \begin{bmatrix} \cos \Omega & \sin \Omega \\ \sin \Omega & \cos \Omega \end{bmatrix} \tag{8.10a}$$

that is

$$\begin{aligned}
S_{hh}^F &= S_{hh} \cos^2 \Omega - S_{vv} \sin^2 \Omega \\
S_{hv}^F &= S_{hv} + (S_{hh} + S_{vv}) \sin \Omega \cos \Omega \\
S_{vh}^F &= S_{hv} - (S_{hh} + S_{vv}) \sin \Omega \cos \Omega \\
S_{vv}^F &= -S_{hh} \sin^2 \Omega + S_{vv} \cos^2 \Omega
\end{aligned} \tag{8.10b}$$

where backscattering reciprocity $S_{hv} = S_{vh}$ is adopted as usual.

Based on the assumption of reflection symmetry ($\langle S_{hh} S_{hv}^* \rangle = \langle S_{hv} S_{vv}^* \rangle = 0$), Freeman (2004) derived a relationship between the Mueller matrices with FR and without FR.

However, as the scatterers are not uniformly oriented in the azimuthal direction, the assumption of so-called reflection symmetry is not true. Actually, keeping these two items, we derived the following components of the Mueller matrix with FR:

$$\langle S_{hh}S_{hh}^*\rangle^F = \langle S_{hh}S_{hh}^*\rangle\cos^4\Omega + \langle S_{vv}S_{vv}^*\rangle\sin^4\Omega - 2\mathrm{Re}\langle S_{hh}S_{vv}^*\rangle\sin^2\Omega\cos^2\Omega \qquad (8.11\mathrm{a})$$

$$\langle S_{vv}S_{vv}^*\rangle^F = \langle S_{vv}S_{vv}^*\rangle\cos^4\Omega + \langle S_{hh}S_{hh}^*\rangle\sin^4\Omega - 2\mathrm{Re}\langle S_{hh}S_{vv}^*\rangle\sin^2\Omega\cos^2\Omega \qquad (8.11\mathrm{b})$$

$$\begin{aligned}\langle S_{hv}S_{hv}^*\rangle^F &= \langle S_{hv}S_{hv}^*\rangle + \big(\mathrm{Re}\langle S_{hh}S_{hv}^*\rangle + \mathrm{Re}\langle S_{hv}S_{vv}^*\rangle\big)\sin 2\Omega \\ &\quad + S_{hh}S_{vv}^*\big(2\mathrm{Re}\langle S_{hh}S_{vv}^*\rangle + \langle S_{hh}S_{hh}^*\rangle + \langle S_{vv}S_{vv}^*\rangle\big)\sin^2\Omega\cos^2\Omega\end{aligned} \qquad (8.11\mathrm{c})$$

$$\begin{aligned}\langle S_{vh}S_{vh}^*\rangle^F &= \langle S_{hv}S_{hv}^*\rangle - \big(\mathrm{Re}\langle S_{hh}S_{hv}^*\rangle + \mathrm{Re}\langle S_{hv}S_{vv}^*\rangle\big)\sin 2\Omega \\ &\quad + \big(2\mathrm{Re}\langle S_{hh}S_{vv}^*\rangle + \langle S_{hh}S_{hh}^*\rangle + \langle S_{vv}S_{vv}^*\rangle\big)\sin^2\Omega\cos^2\Omega\end{aligned} \qquad (8.11\mathrm{d})$$

$$\begin{aligned}\mathrm{Re}\langle S_{hh}S_{hv}^*\rangle^F &= \mathrm{Re}\langle S_{hh}S_{hv}^*\rangle\cos^2\Omega + \langle S_{hh}S_{hh}^*\rangle\sin\Omega\cos^3\Omega \\ &\quad + \frac{1}{4}\mathrm{Re}\langle S_{hh}S_{vv}^*\rangle\sin 4\Omega - \langle S_{vv}S_{vv}^*\rangle\sin^3\Omega\cos\Omega - \mathrm{Re}\langle S_{hv}S_{vv}^*\rangle\sin^2\Omega\end{aligned} \qquad (8.11\mathrm{e})$$

$$\mathrm{Im}\langle S_{hh}S_{hv}^*\rangle^F = \mathrm{Im}\langle S_{hh}S_{hv}^*\rangle\cos^2\Omega + \frac{1}{2}\mathrm{Im}\langle S_{hh}S_{vv}^*\rangle\sin 2\Omega + \mathrm{Im}\langle S_{hv}S_{vv}^*\rangle\sin^2\Omega \qquad (8.11\mathrm{f})$$

$$\begin{aligned}\mathrm{Re}\langle S_{hh}S_{vv}^*\rangle^F &= -\langle S_{hh}S_{hh}^*\rangle\sin^2\Omega\cos^2\Omega + \mathrm{Re}\langle S_{hh}S_{vv}^*\rangle\cos^4\Omega \\ &\quad + \mathrm{Re}\langle S_{hh}S_{vv}^*\rangle\sin^4\Omega - \langle S_{vv}S_{vv}^*\rangle\sin^2\Omega\cos^2\Omega\end{aligned} \qquad (8.11\mathrm{g})$$

$$\mathrm{Im}\langle S_{hh}S_{vv}^*\rangle^F = \mathrm{Im}\langle S_{hh}S_{vv}^*\rangle\cos 2\Omega \qquad (8.11\mathrm{h})$$

$$\begin{aligned}\mathrm{Re}\langle S_{hv}S_{vv}^*\rangle^F &= \mathrm{Re}\langle S_{hv}S_{vv}^*\rangle\cos^2\Omega - \mathrm{Re}\langle S_{hh}S_{hv}^*\rangle\sin^2\Omega - \langle S_{hh}S_{hh}^*\rangle\sin^3\Omega\cos\Omega \\ &\quad + \frac{1}{4}\mathrm{Re}\langle S_{hh}S_{vv}^*\rangle\sin 4\Omega + \langle S_{vv}S_{vv}^*\rangle\sin\Omega\cos^3\Omega\end{aligned} \qquad (8.11\mathrm{i})$$

$$\mathrm{Im}\langle S_{hv}S_{vv}^*\rangle^F = \mathrm{Im}\langle S_{hv}S_{vv}^*\rangle\cos^2\Omega + \mathrm{Im}\langle S_{hh}S_{hv}^*\rangle\sin^2\Omega + \mathrm{Im}\langle S_{hh}S_{vv}^*\rangle\sin\Omega\cos\Omega \qquad (8.11\mathrm{j})$$

$$\begin{aligned}\mathrm{Re}\langle S_{hh}S_{vh}^*\rangle^F &= \mathrm{Re}\langle S_{hh}S_{hv}^*\rangle\cos^2\Omega - \langle S_{hh}S_{hh}^*\rangle\sin\Omega\cos^3\Omega - \frac{1}{4}\mathrm{Re}\langle S_{hh}S_{vv}^*\rangle\sin 4\Omega \\ &\quad + \langle S_{vv}S_{vv}^*\rangle\sin^3\Omega\cos\Omega - \mathrm{Re}\langle S_{hv}S_{vv}^*\rangle\sin^2\Omega\end{aligned} \qquad (8.11\mathrm{k})$$

$$\mathrm{Im}\langle S_{hh}S_{vh}^*\rangle^F = \mathrm{Im}\langle S_{hh}S_{hv}^*\rangle\cos^2\Omega - \frac{1}{2}\mathrm{Im}\langle S_{hh}S_{vv}^*\rangle\sin 2\Omega + \mathrm{Im}\langle S_{hv}S_{vv}^*\rangle\sin^2\Omega \qquad (8.11\mathrm{l})$$

$$\begin{aligned}\mathrm{Re}\langle S_{vh}S_{vv}^*\rangle^F &= \mathrm{Re}\langle S_{hv}S_{vv}^*\rangle\cos^2\Omega - \mathrm{Re}\langle S_{hh}S_{hv}^*\rangle\sin^2\Omega + \langle S_{hh}S_{hh}^*\rangle\sin^3\Omega\cos\Omega \\ &\quad - \frac{1}{4}\mathrm{Re}\langle S_{hh}S_{vv}^*\rangle\sin 4\Omega - \langle S_{vv}S_{vv}^*\rangle\sin\Omega\cos^3\Omega\end{aligned} \qquad (8.11\mathrm{m})$$

$$\mathrm{Im}\langle S_{vh}S_{vv}^*\rangle^F = \mathrm{Im}\langle S_{hv}S_{vv}^*\rangle\cos^2\Omega + \mathrm{Im}\langle S_{hh}S_{hv}^*\rangle\sin^2\Omega - \mathrm{Im}\langle S_{hh}S_{vv}^*\rangle\sin\Omega\cos\Omega \qquad (8.11\mathrm{n})$$

$$\mathrm{Re}\langle S_{hv}S_{vh}^*\rangle^F = \langle S_{hv}S_{hv}^*\rangle - \big(\langle S_{hh}S_{hh}^*\rangle + \langle S_{vv}S_{vv}^*\rangle + 2\mathrm{Re}\langle S_{hh}S_{vv}^*\rangle\big)\sin^2\Omega\cos^2\Omega \qquad (8.11\mathrm{o})$$

Figure 8.1 A photo of the area of Tabernash, Colorado. © [2007] IEEE. Reprinted, with permission, from *IEEE Transactions on Geoscience and Remote Sensing*

$$\mathrm{Im}\langle S_{hv}S_{vh}^*\rangle^F = \left(\mathrm{Im}\langle S_{hh}S_{hv}^*\rangle - \mathrm{Im}\langle S_{hv}S_{vv}^*\rangle\right)\sin 2\Omega \qquad (8.11p)$$

where $\langle\cdots\rangle$ denotes the ensemble average. It can be seen that each term of the Mueller matrix with FR is composed of different polarized elements of the Mueller matrix without FR; and that the FR periodicity is 180°.

Figure 8.1 gives a photo of Tabernash, Colorado (center location 39°98′N, 105°50′W). Figure 8.2 is the JPL airborne AIRSAR image at P band in that location (21 February 2002). It is an airborne SAR image without any FR.

Now we suppose that we move this SAR antenna to the satellite and impose the FR effect on the polarimetric AIRSAR data following Equations 8.11a–p. In this example, the satellite is supposed be at 675 km altitude with looking to the east, total electron content follows the 2001 list of the International Reference Ionosphere (IRI) (see http://modelweb.gsfc.nasa.gov/models/iri.html) and the geomagnetic field is based on the International Geomagnetic

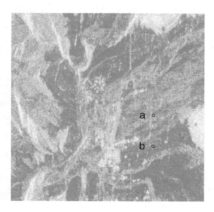

Figure 8.2 An AIRSAR image at P band (from Google Earth). © [2007] IEEE. Reprinted, with permission, from *IEEE Transactions on Geoscience and Remote Sensing*

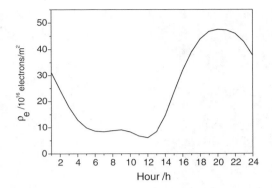

Figure 8.3 Temporal change of total electron content (ρ_e) per unit area. © [2007] IEEE. Reprinted, with permission, from *IEEE Transactions on Geoscience and Remote Sensing*

Reference Field (IGRF) (see http://modelweb.gsfc.nasa.gov/models/igrf.html). Figure 8.3 shows the temporal change of total electron content. Geomagnetic induction intensity $B(400)$ at this location (latitude and longitude) is taken as 44 413.1 nT with magnetic declination degree 10.2° and magnetic inclination degree 66.9°.

From Equation 8.8 it can be found that Θ_B is calculated as

$$\cos \Theta_B = \cos \theta_i \sin \Theta + \sin \theta_i \cos \Theta \sin \Phi \tag{8.12}$$

where Θ is the magnetic inclination angle, and Φ is the magnetic declination angle. Following ρ_e of Figure 8.3, that is, $(6\text{–}47.5) \times 10^{16}$ electron/m^2, the FR degree is 16.4–129.6° ($\lambda = 0.857$ m). All the terms of Equation 8.11aa–p, $\langle S_{pq} S_{rt}^* \rangle^F$, are calculated with FR = 30°. Figure 8.2 now with FR becomes Figure 8.4.

An example of an application that is strongly affected by FR is land cover classification. As a conventional approach, the parameters α, A, and the entropy H have been applied to surface classification, which can be read in Chapter 5. The deorientation concept was introduced in

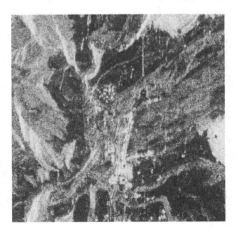

Figure 8.4 The AIRSAR image of Figure 8.2 is distorted by an FR angle of 30°. © [2007] IEEE. Reprinted, with permission, from *IEEE Transactions on Geoscience and Remote Sensing*

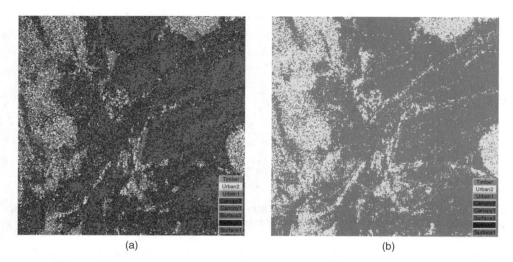

(a)   (b)

Figure 8.5 Terrain surface classification from (a) original (Figure 8.2) and (b) FR-distorted (Figure 8.4) SAR images. © [2007] IEEE. Reprinted, with permission, from *IEEE Transactions on Geoscience and Remote Sensing*

Chapter 5 to reduce the influence of randomly fluctuating orientation and shows the prominence of targets' generic characteristics for terrain surface classification. Making the target orientation into such a fixed state with minimization of cross-polarization, a set of parameters u, v, ψ is defined from deorientation of the principal scattering vector of random scatter targets. All these parameters as well as the entropy H are applied to terrain surface classification.

Using these parameters u, v, H for terrain surface classification, Figure 8.5a is obtained from Figure 8.2 without FR. As a comparison, Figure 8.5b is a classification from Figure 8.4 with FR. These two classifications are totally different. For instance, the flat soil land with weak scattering in Figure 8.5a is misclassified as urban area with strong scattering in Figure 8.5b.

Figure 8.6 shows change of the classification parameters u, v, H with FR degree at the point b of Figure 8.2. It can be seen that the set of parameters ($u < 0.3$, $v > 0.2$, $H < 0.5$)

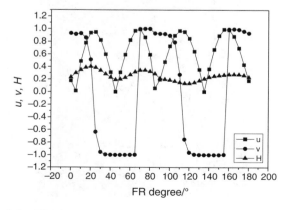

Figure 8.6 Change of classification parameters (u, v, H) with the FR degree. © [2007] IEEE. Reprinted, with permission, from *IEEE Transactions on Geoscience and Remote Sensing*

without FR is for the class of Surface 3 (flat soil land), but FR 30° makes the parameters $(u > 0.7, v < -0.2, H < 0.5)$ and leads to misclassification as Urban 1. The parameters u, v are very sensitive to FR degree. The consequence is that the P-band polarimetric data with FR cannot be directly applied to terrain surface classification.

8.2 Recovering the Mueller Matrix with Ambiguity Error $\pm\pi/2$

From Equation 8.10b, Freeman (2004) obtained

$$\Omega = \frac{1}{2}\tan^{-1}\left[\frac{S_{hv}^F - S_{vh}^F}{S_{hh}^F + S_{vv}^F}\right] \tag{8.13}$$

The scattering matrix without FR, $[\tilde{S}_{pq}]$ $(p, q = h, v)$, is obtained from $[S_{pq}^F]$ with FR as follows:

$$\begin{bmatrix} \tilde{S}_{hh} & \tilde{S}_{hv} \\ \tilde{S}_{vh} & \tilde{S}_{vv} \end{bmatrix} = \begin{bmatrix} \cos\Omega & -\sin\Omega \\ \sin\Omega & \cos\Omega \end{bmatrix} \begin{bmatrix} S_{hh}^F & S_{hv}^F \\ S_{vh}^F & S_{vv}^F \end{bmatrix} \begin{bmatrix} \cos\Omega & -\sin\Omega \\ \sin\Omega & \cos\Omega \end{bmatrix} \tag{8.14}$$

It yields

$$\tilde{S}_{hh} = S_{hh}^F\cos^2\Omega - S_{vv}^F\sin^2\Omega + \frac{1}{2}\left(S_{hv}^F - S_{vh}^F\right)\sin 2\Omega \tag{8.15a}$$

$$\tilde{S}_{hv} = S_{hv}^F\cos^2\Omega + S_{vh}^F\sin^2\Omega - \frac{1}{2}\left(S_{hh}^F + S_{vv}^F\right)\sin 2\Omega \tag{8.15b}$$

$$\tilde{S}_{vh} = S_{vh}^F\cos^2\Omega + S_{hv}^F\sin^2\Omega + \frac{1}{2}\left(S_{hh}^F + S_{vv}^F\right)\sin 2\Omega \tag{8.15c}$$

$$\tilde{S}_{vv} = S_{vv}^F\cos^2\Omega - S_{hh}^F\sin^2\Omega + \frac{1}{2}\left(S_{hv}^F - S_{vh}^F\right)\sin 2\Omega \tag{8.15d}$$

According to Equation 8.10a, an alternative method to recover the cross-polarized items is given as:

$$S_{hv} = S_{vh} = \frac{S_{hv}^F + S_{vh}^F}{2} \tag{8.16}$$

If only measured Mueller matrix data are available, the FR degree can be equally well derived from Equation 8.11a–p as

$$\Omega = \frac{1}{2}\tan^{-1}\left[\frac{\text{Im}\langle S_{hh}S_{hv}^*\rangle^F - \text{Im}\langle S_{hh}S_{vh}^*\rangle^F}{\text{Im}\langle S_{hh}S_{vv}^*\rangle^F}\right] \tag{8.17}$$

Two other techniques, based on combinations of the Mueller matrix terms, were also presented in equations (12)–(15) of Freeman (2004).

The polarimetric Mueller matrix without FR can be inverted as

$$
\begin{aligned}
\langle S_{hh}S_{hh}^*\rangle = {}& \langle S_{hh}S_{hh}^*\rangle^F \cos^4\Omega + \langle S_{vv}S_{vv}^*\rangle^F \sin^4\Omega - 2\mathrm{Re}\langle S_{hh}S_{vv}^*\rangle^F \sin^2\Omega\cos^2\Omega \\
&+ 2\left(\mathrm{Re}\langle S_{hh}S_{hv}^*\rangle^F - \mathrm{Re}\langle S_{hh}S_{vh}^*\rangle^F\right)\sin\Omega\cos^3\Omega \\
&- 2\left(\mathrm{Re}\langle S_{hv}S_{vv}^*\rangle^F - \mathrm{Re}\langle S_{vh}S_{vv}^*\rangle^F\right)\sin^3\Omega\cos\Omega \\
&+ \frac{1}{4}\left(\langle S_{hv}S_{hv}^*\rangle^F - 2\mathrm{Re}\langle S_{hv}S_{vh}^*\rangle^F + \langle S_{vh}S_{vh}^*\rangle^F\right)\sin^2 2\Omega
\end{aligned}
$$
(8.18a)

$$
\begin{aligned}
\langle S_{vv}S_{vv}^*\rangle = {}& \langle S_{vv}S_{vv}^*\rangle^F \cos^4\Omega + \langle S_{hh}S_{hh}^*\rangle^F \sin^4\Omega - 2\mathrm{Re}\langle S_{hh}S_{vv}^*\rangle^F \sin^2\Omega\cos^2\Omega \\
&+ 2\left(\mathrm{Re}\langle S_{hv}S_{vv}^*\rangle^F - \mathrm{Re}\langle S_{vh}S_{vv}^*\rangle^F\right)\sin\Omega\cos^3\Omega \\
&- 2\left(\mathrm{Re}\langle S_{hh}S_{hv}^*\rangle^F - \mathrm{Re}\langle S_{hh}S_{vh}^*\rangle^F\right)\sin^3\Omega\cos\Omega \\
&+ \frac{1}{4}\left(\langle S_{hv}S_{hv}^*\rangle^F - 2\mathrm{Re}\langle S_{hv}S_{vh}^*\rangle^F + \langle S_{vh}S_{vh}^*\rangle^F\right)\sin^2 2\Omega
\end{aligned}
$$
(8.18b)

$$
\langle S_{hv}S_{hv}^*\rangle = \langle S_{hv}S_{vh}^*\rangle = \langle S_{vh}S_{vh}^*\rangle = \frac{1}{4}\left(\langle S_{hv}S_{hv}^*\rangle^F + 2\mathrm{Re}\langle S_{hv}S_{vh}^*\rangle^F + \langle S_{vh}S_{vh}^*\rangle^F\right)
$$
(8.18c)

$$
\begin{aligned}
\mathrm{Re}\langle S_{hh}S_{vv}^*\rangle = {}& \mathrm{Re}\langle S_{hh}S_{vv}^*\rangle^F\left(\sin^4\Omega + \cos^4\Omega\right) - \frac{1}{4}\left(\langle S_{hh}S_{hh}^*\rangle^F + \langle S_{vv}S_{vv}^*\rangle^F\right)\sin^2 2\Omega \\
&+ \frac{1}{4}\left(\mathrm{Re}\langle S_{hh}S_{hv}^*\rangle^F + \mathrm{Re}\langle S_{hv}S_{vv}^*\rangle^F - \mathrm{Re}\langle S_{hh}S_{vh}^*\rangle^F - \mathrm{Re}\langle S_{vh}S_{vv}^*\rangle^F\right)\sin 4\Omega \\
&+ \frac{1}{4}\left(\langle S_{hv}S_{hv}^*\rangle^F - 2\mathrm{Re}\langle S_{hv}S_{vh}^*\rangle^F + \langle S_{vh}S_{vh}^*\rangle^F\right)\sin^2 2\Omega
\end{aligned}
$$
(8.18d)

$$
\begin{aligned}
\mathrm{Im}\langle S_{hh}S_{vv}^*\rangle = {}& \mathrm{Im}\langle S_{hh}S_{vv}^*\rangle^F \cos 2\Omega \\
&+ \frac{1}{2}\left(\mathrm{Im}\langle S_{hh}S_{hv}^*\rangle^F + \mathrm{Im}\langle S_{hv}S_{vv}^*\rangle^F - \mathrm{Im}\langle S_{hh}S_{vh}^*\rangle^F - \mathrm{Im}\langle S_{vh}S_{vv}^*\rangle^F\right)\sin 2\Omega
\end{aligned}
$$
(8.18e)

$$
\begin{aligned}
\mathrm{Re}\langle S_{hh}S_{hv}^*\rangle = {}& \mathrm{Re}\langle S_{hh}S_{vh}^*\rangle = \frac{1}{2}\left(\mathrm{Re}\langle S_{hh}S_{hv}^*\rangle^F + \mathrm{Re}\langle S_{hh}S_{vh}^*\rangle^F\right)\cos^2\Omega \\
&- \frac{1}{2}\left(\mathrm{Re}\langle S_{hv}S_{vv}^*\rangle^F + \mathrm{Re}\langle S_{vh}S_{vv}^*\rangle^F\right)\sin^2\Omega + \frac{1}{4}\left(\langle S_{hv}S_{hv}^*\rangle^F - \langle S_{vh}S_{vh}^*\rangle^F\right)\sin 2\Omega
\end{aligned}
$$
(8.18f)

$$
\begin{aligned}
\mathrm{Im}\langle S_{hh}S_{hv}^*\rangle = {}& \mathrm{Im}\langle S_{hh}S_{vh}^*\rangle = \frac{1}{2}\left(\mathrm{Im}\langle S_{hh}S_{hv}^*\rangle^F + \mathrm{Im}\langle S_{hh}S_{vh}^*\rangle^F\right)\cos^2\Omega \\
&+ \frac{1}{2}\left(\mathrm{Im}\langle S_{hv}S_{vv}^*\rangle^F + \mathrm{Im}\langle S_{vh}S_{vv}^*\rangle^F\right)\sin^2\Omega + \frac{1}{2}\mathrm{Im}\langle S_{hv}S_{vh}^*\rangle^F \sin 2\Omega
\end{aligned}
$$
(8.18g)

$$\mathrm{Re}\langle S_{hv}S_{vv}^*\rangle = \mathrm{Re}\langle S_{vh}S_{vv}^*\rangle = \frac{1}{2}\left(\mathrm{Re}\langle S_{hv}S_{vv}^*\rangle^F + \mathrm{Re}\langle S_{vh}S_{vv}^*\rangle^F\right)\cos^2\Omega$$

$$-\frac{1}{2}\left(\mathrm{Re}\langle S_{hh}S_{hv}^*\rangle^F + \mathrm{Re}\langle S_{hh}S_{vh}^*\rangle^F\right)\sin^2\Omega + \frac{1}{4}\left(\langle S_{hv}S_{hv}^*\rangle^F - \langle S_{vh}S_{vh}^*\rangle^F\right)\sin 2\Omega$$

$$(8.18h)$$

$$\mathrm{Im}\langle S_{hv}S_{vv}^*\rangle = \mathrm{Im}\langle S_{vh}S_{vv}^*\rangle = \frac{1}{2}\left(\mathrm{Im}\langle S_{hv}S_{vv}^*\rangle^F + \mathrm{Im}\langle S_{vh}S_{vv}^*\rangle^F\right)\cos^2\Omega$$

$$(8.18i)$$

$$+\frac{1}{2}\left(\mathrm{Im}\langle S_{hh}S_{hv}^*\rangle^F + \mathrm{Im}\langle S_{hh}S_{vh}^*\rangle^F\right)\sin^2\Omega - \frac{1}{2}\mathrm{Im}\langle S_{hv}S_{vh}^*\rangle^F\sin 2\Omega$$

As shown in Equation 8.6, Freeman (2004) pointed out that, because the values of $\tan^{-1}$ lie between $\pm\pi/2$, the FRs Ω estimated by Equation 8.6 should be between $\pm\pi/4$. This means that the Ω values can only be estimated modulo $\pi/2$, which is not sufficient. It leaves a $\pm\pi/2$ ambiguity error in Ω because the periodicity of the FR effect is $180°$.

The $\pm\pi/2$ ambiguity error introduces an undetermined inversion as (assuming that the true FR degree is $\Omega + n\pi/2$):

$$\tilde{S}_{vv} = S_{vv} \quad \text{(when } n \text{ is even) or}$$

$$= -S_{hh} \quad \text{(when } n \text{ is odd)}$$

$$\tilde{S}_{hh} = S_{hh} \quad \text{(when } n \text{ is even) or} \qquad (8.19)$$

$$= -S_{vv} \quad \text{(when } n \text{ is odd)}$$

$$\tilde{S}_{hv} = S_{hv}$$

Similarly, there exists a $\pm\pi/2$ ambiguity error of Ω in Equation 8.10a. Thus, the coherency matrix $\overline{\mathbf{T}}^{\pm}$ indicated by the undetermined sign $\pm$ (corresponding to even or odd n in Equation 8.19) is written as

$$\overline{\mathbf{T}}^{\pm} = \begin{bmatrix} T_{11} & \pm T_{12} & \pm T_{13} \\ \pm T_{21} & T_{22} & T_{23} \\ \pm T_{31} & T_{32} & T_{33} \end{bmatrix} \qquad (8.20)$$

where T_{ij} denotes the element of the coherency matrix without FR and ambiguity. Correspondingly, the eigenvectors of $\overline{\mathbf{T}}^{\pm}$ are written as

$$\mathbf{k}_{Pi}^{\pm} = \begin{bmatrix} \pm k_{Pi,1} \\ k_{Pi,2} \\ k_{Pi,3} \end{bmatrix} \qquad (8.21)$$

It is easy to find that the eigenvalues λ_i of $\overline{\mathbf{T}}^{\pm}$ ($i = 1, 2, 3$) are the same as those of $\overline{\mathbf{T}}$. Thus, the entropy H and anisotropy A of Equations 5.44 and 5.45 derived from the eigenvalues λ_i are not affected by the $\pm\pi/2$ ambiguity error.

Taking the principal eigenvector $\mathbf{k}_{P1}^{\pm}$, the classification parameter α (Pottier and Lee, 1999) is calculated as

$$\alpha^{\pm} = \arctan\left(\frac{\sqrt{|k_{P1,2}|^2 + |k_{P1,3}|^2}}{|\pm k_{P1,1}|}\right) = \alpha \tag{8.22}$$

Thus, the set of classification parameters (α, H, A) is also not affected by the $\pm\pi/2$ ambiguity error.

Performing a deorientation of the eigenvector $\mathbf{k}_{P1}^{\pm}$, one obtains

$$\mathbf{k}_{P1}^{\pm\prime} = \begin{bmatrix} \pm k_{P1,1} \\ k_{P1,2}\cos 2\varphi_m + k_{P1,3}\sin 2\varphi_m \\ -k_{P1,2}\sin 2\varphi_m + k_{P1,3}\cos 2\varphi_m \end{bmatrix} \tag{8.23}$$

where φ_m is the deorientation angle of the polarization base to minimize the cross-polarized scattering (Xu and Jin, 2005) in Chapter 5. We obtain each element of the scattering matrix from Equation 8.23:

$$S_{hh}^{\pm} = \frac{\sqrt{2}}{2}(\pm k_{P1,1} + k_{P1,2}\cos 2\varphi_m + k_{P1,3}\sin 2\varphi_m) = S_{hh} \text{ or } = -S_{vv}$$
$$\tag{8.24}$$
$$S_{vv}^{\pm} = \frac{\sqrt{2}}{2}(\pm k_{P1,1} - k_{P1,2}\cos 2\varphi_m - k_{P1,3}\sin 2\varphi_m) = S_{vv} \text{ or } = -S_{hh}$$

It yields

$$a^{\pm} = \tan^{-1}\left(\frac{|S_{vv}^{\pm}|}{|S_{hh}^{\pm}|}\right) = a \quad \text{or} \quad = \frac{\pi}{2} - a \tag{8.25}$$

$$b^{\pm} = \frac{1}{2}\arg\left(\frac{S_{vv}^{\pm}}{S_{hh}^{\pm}}\right) = \pm b \tag{8.26}$$

The classification parameters, u, v are found as

$$u^{\pm} = \sin c \cos 2a^{\pm} = \pm u \tag{8.27}$$

$$v^{\pm} = \sin c \sin 2a^{\pm} \cos 2b = v \tag{8.28}$$

$$\psi^{\pm} = \begin{cases} \varphi_m, & u^{\pm} > 0 \\ \varphi_m + \dfrac{\pi}{2}, & u^{\pm} < 0 \end{cases} = \begin{cases} \varphi_m, & \varphi_m + \dfrac{\pi}{2}u > 0 \\ \varphi_m + \dfrac{\pi}{2}, & \varphi_m u < 0 \end{cases} \tag{8.29}$$

In deorientation classification, the parameter set uses $(|u|, v, H)$. Thus, the $\pm\pi/2$ ambiguity error and the undetermined sign $\pm$ of $\overline{\mathbf{T}}^{\pm}$ do not affect the classification by the parameter set

(u, v, H) or (α, H, A). But the orientation angle $\psi^{\pm}$ of Equation 5.63 is affected by the $\pm\pi/2$ ambiguity error.

8.3   Method to Eliminate the $\pm\pi/2$ Ambiguity Error

Freeman (2004) proposed four approaches to eliminate the $\pm\pi/2$ ambiguity error:

1. Deploy a calibration target (e.g., a transponder) within the scene with polarization signature

$$\begin{pmatrix} 1 & 0 \\ 0 & 0 \end{pmatrix} \quad \text{or} \quad \begin{pmatrix} 0 & 0 \\ 0 & 1 \end{pmatrix}$$

2. Examine the polarization signature of a slightly rough surface, for which the vv backscattering is typically greater than or equal to the hh. This might be easily implemented for a land surface, simply by initiating or concluding each data-take over the ocean and using that as a benchmark. On the SRTM (Shuttle Radar Topography Mission), all data-takes started and finished over the ocean to provide a calibration reference for surface topography.
3. Constrain Ω itself by observing when the ionosphere is relatively quiet, for example, at night during solar minimum, and Ω may be less than $45°$ at both L and P bands.
4. Correlate the dataset with an earlier one, obtained by the same radar under identical viewing conditions, but without FR rotation.

From the calculation in Equation 8.8, Figure 8.7a shows a global distribution of the FR degree at 6 a.m. on 1 June 1989 based on total electron content and geomagnetic field at that time. Generally, it can be seen that, due to the orientation of the geomagnetic field, the FR degree is negative in the northern hemisphere, and is positive in the southern hemisphere. It also can be seen that both the geomagnetic field and total electron content gradually change with geographical location, which produces a correspondingly smooth change of the FR degree.

The $\pm\pi/2$ ambiguity comes from the calculation of $\Omega' = 0.5\tan^{-1}[\tan 2\Omega]$. We recalculate $0.5\tan^{-1}[\tan 2\Omega]$ for each pixel of Figure 8.7a, and obtain a global distribution of FR degree with $\pm\pi/2$ ambiguity error as shown in Figure 8.7b. It can be seen that Figure 8.7b differs totally from Figure 8.7a. The solid and dashed lines in Figure 8.8 are the true FR and the FR with $\pm\pi/2$ ambiguity error along the longitude $150°E$ of Figure 8.7a and b, respectively.

The problem of $\pm\pi/2$ ambiguity is similar to the phase wrapping in interferometric SAR measurement. As an alternative approach to correct the $\pm\pi/2$ ambiguity error, a one-dimensional phase unwrapping method (integral of phase difference) can be employed to recover the true FR degree.

To recover the true FR degree, a benchmark is defined as the zero FR as shown by the white line in Figure 8.7a, where the angle between the directions of wave propagation and the geomagnetic field reaches $90°$ by Equation 8.17 and zero FR by Equation 8.2. Then, a method similar to phase unwrapping is employed to recover the true FR degree on each pixel starting from the points with zero FR.

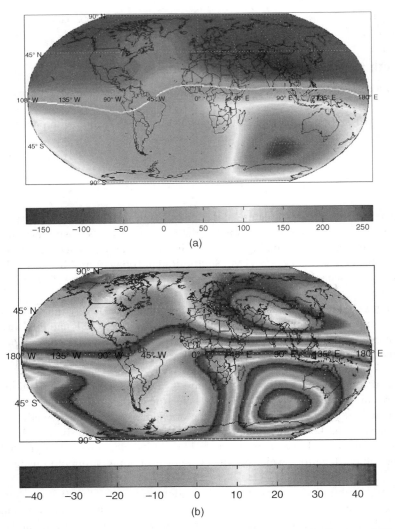

Figure 8.7 (a) Global distribution of FR degree at P band. (b) Global distribution of FR degree at P band with $\pm \pi/2$ ambiguity error. © [2007] IEEE. Reprinted, with permission, from *IEEE Transactions on Geoscience and Remote Sensing*

Similar to the unambiguous phase difference condition (UPDC) (Fornaro *et al.*, 2003), it is assumed that the difference of the FR degree from its adjacent pixels should be $|\Delta\Omega| < \pi/4$. The true FR degree along the longitude (column) is recovered by the following process:

$$\Omega_n = \sum_{i=s}^{n} (\Delta\Omega_i + k_i \pi/2) + \Omega_s \tag{8.30}$$

$$k_i = \begin{cases} 0 & (|\Delta\Omega| \leq \pi/4) \\ -1 & (\Delta\Omega > \pi/4) \\ 1 & (\Delta\Omega < -\pi/4) \end{cases} \tag{8.31}$$

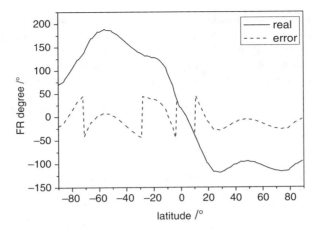

Figure 8.8 The FR degree along the longitudes 150°E. © [2007] IEEE. Reprinted, with permission, from *IEEE Transactions on Geoscience and Remote Sensing*

where Ω_s is the FR degree on the benchmark, $\Delta\Omega_i$ is the difference of FR degree from adjacent pixels (when $n > s$, $\Delta\Omega_i = \Omega_i - \Omega_{i-1}$; when $n < s$, $\Delta\Omega_i = \Omega_i - \Omega_{i+1}$; when $n = s$, $\Delta\Omega_i = 0$).

Since each column contains a benchmark, the true FR degree of each pixel can be recovered sequentially. Recovered FR degrees from the dashed FR in Figure 8.8 are exactly returned to the solid line; and the global distribution of FR degree without $\pm\pi/2$ ambiguity error (identical with Figure 8.7a) can be recovered from Figure 8.7b.

Similarly, some phase unwrapping methods (such as the integral track or least square methods) can be employed to eliminate noise or other disturbances of SAR measurements, which may break FR values.

This method involving finding a benchmark area close to the equator might become restrictive because polarimetric SAR data rates tend to be much higher than that of conventional single-polarization SAR, and operation in full polarimetric SAR mode is often restricted to a few minutes per orbit. In practice, this would limit the full correction capability to tropical areas only.

However, if there is no zero-FR available, any single point in the image can be chosen as the benchmark. The FR degree of other pixels can be inverted by the two-dimensional phase unwrapping method. These results might not be true, but are only different from the true values by integer multiples of $\pi/2$ in the same way as the benchmark. These inverted polarimetric data might be correct, or might all keep the same $\pm\pi/2$ ambiguity error. Further, based on the criterion $\left\langle |S_{vv}|^2 \right\rangle > \left\langle |S_{hh}|^2 \right\rangle$ in some known areas such as the ocean, the $\pm\pi/2$ ambiguity in those points can be resolved, as suggested by Freeman (2004). Those points can also be chosen as the benchmark for the phase unwrapping:

1. Polarimetric SAR measurement with FR at low frequency such as P band cannot be directly applied to surface classification.
2. The scattering and Mueller matrices can be derived with the $\pm\pi/2$ ambiguity error from those with FR, and applicable to surface classifications by using the parameters u, v, H, α, A.

3. Based on gradual change of FR degree and zero FR as a starting point, similar to the phase unwrapping method, then the $\pm\pi/2$ ambiguity error can be eliminated. Thus, the true FR degree and polarimetric data without any FR effect can be recovered.

4. The two-dimensional phase unwrapping method with benchmark at some well-known areas can be employed to resolve the $\pm\pi/2$ ambiguity error when no zero FR is contained in SAR image data.

5. All these discussions and this approach can be applied to even lower VHF bands.

References

Dobson, M.C. *et al.* (1992) Dependence of radar backscatter on conifer forest biomass. *IEEE Transactions on Geoscience and Remote Sensing*, **30**, 412–415.

Fornaro, G. *et al.* (2003) Phase difference based multiple acquisition phase unwrapping. Proceedings of IGARSS'03, Toulouse, France, Vol. **2**, pp. 948–950.

Freeman, A. (2004) Calibration of linearly polarized polarimetric SAR data subject to Faraday rotation. *IEEE Transactions on Geoscience and Remote Sensing*, **42**(8), 1617–1624.

Freeman, A. and Saatchi, S.S. (2004) On the detection of Faraday rotation in linearly polarized L-band SAR backscatter signatures. *IEEE Transactions on Geoscience and Remote Sensing*, **40**(8), 1607–1616.

Freeman, A., Durden, S.L., and Zimmermann, R. (1992) Mapping subtropical vegetation using multifrequency, multipolarization SAR data. Proceedings of International Symposium of Geoscience and Remote Sensing (IGARSS'92), Houston, TX, pp. 1986–1689.

Kong, J.A. (1985) *Electromagnetic Wave Theory*, Wiley-Interscience, New York.

Le Vine, D.M. and Abraham, S. (2000) Faraday rotation and passive microwave remote sensing of soil moisture from space, in *Microwave Radiometry and Remote Sensing of the Earth's Surface and Atmosphere* (eds P. Pampaloni and S. Paloscia), VSP, Zeist, The Netherlands, pp. 89–96.

Le Vine, D.M. and Abraham, S. (2002) The effect of the ionosphere on remote sensing of sea surface salinity from space: absorption and emission at L band. *IEEE Transactions on Geoscience and Remote Sensing*, **40**(4), 771–782.

Moghaddam, M., Saatchi, S., and Cuenca, R.H. (2000) Estimating subcanopy soil moisture with radar. *Journal of Geophysical Research*, **105**(D11), 14899–14911.

Moghaddam, M., Rodriguez, E., Rahmat-Samii, Y. *et al.* (2003) Microwave Observation of Subcanopy and Subsurface (MOSS): a low-frequency radar for global deep soil moisture measurements. Proceedings of International Symposium of Geoscience and Remote Sensing (IGARSS'03), Vol. **1**, pp. 500–502.

Mouginota, J. *et al.* (2008) Correction of the ionospheric distortion on the MARSIS surface sounding echoes. *Planetary and Space Science*, **56**, 917–926.

Pierce, L. and Moghaddam, M. (2005) A tower-based prototype UHF/VHF radar for subcanopy and subsurface soil moisture. Proceedings of International Symposium of Geoscience and Remote Sensing (IGARSS'05), Seoul, Korea.

Pottier, E. and Lee, J.S. (1999) Application of the H/A/alpha polarimetric decomposition theorem for unsupervised classification of fully polarimetric SAR data based on the Wishart distribution. Proceedings of Committee on Earth Observing Satellites SAR Workshop, Toulouse, France, 26–29 October.

Qi, R. and Jin, Y.Q. (2007) Analysis of the effects of Faraday rotation on spaceborne polarimetric SAR observation at P band. *IEEE Transactions on Geoscience and Remote Sensing*, **45**(5), 1115–1122.

Rignot, E., Zimmermann, R., van Zyl, J., and Oren, R. (1995) Spaceborne applications of a P-band imaging radar for mapping of forest biomass. *IEEE Transactions on Geoscience and Remote Sensing*, **33**(5), 1162–1169.

Wright, P.A. *et al.* (2003) Faraday rotation effects on L-band spaceborne SAR data. *IEEE Transactions on Geoscience and Remote Sensing*, **41**(12), 383–390.

Xu, F. and Jin, Y.Q. (2005) Deorientation theory of polarimetric scattering targets and application to terrain surface classification. *IEEE Transactions on Geoscience and Remote Sensing*, **43**(10), 2351–2364.

9

Change Detection from Multi-Temporal SAR Images

Multi-temporal observations of spaceborne remote sensing imagery provide a fast and practical technical means for surveying and assessing changes of terrain surfaces over the course of years. Some direct applications to detect and classify such surface changes can be found in Rignot and van Zyl (1993), Singh (1989), Bruzzone and Prieto (2000), Merril and Liu (1998), and Kasetkasem and Varshney (2002), among others. The SAR imagery technology, such as SIR-C SAR, ERS-1,2, and RADARSAT, during the past 20 years has provided high-resolution and all-weather image data. Such data are very helpful to retrieve accurate information on terrain surface changes.

It would be laborious to make an intuitive assessment for a huge amount of multi-temporal image data over a vast area. Such assessment based on qualitative gray-level analysis is not accurate and might lose some important information. How to detect, collect, and automatically analyze change information is one of the most important issues in remote sensing applications. There are two usual methods: the supervised and unsupervised approaches. The former needs a mass of multi-temporal ground truth data for training samples in a classification procedure. The latter performs change detection by making a direct comparison of the multi-temporal remote sensing images without relying on any additional information. Our discussion in this chapter adopts the unsupervised approach.

Based on Bayes theorem and Markov models, Bruzzone and Prieto (2000) proposed an automatic algorithm known as expectation maximum (EM) and Markov random field (MRF) analysis. Two classifications are made: changed and unchanged regimes. Their approach, however, has not been validated by using real SAR observations. Moreover, so-called changes might have different meanings. For example, backscattering is reduced when an area of roughly distributed buildings becomes flat grassland, or vice versa; backscattering is increased when a flat surface turns into a group of buildings or many trees. A plus or minus decibels change in backscattering might contain different content.

In this chapter, the two-thresholds EM–MRF algorithm (2EM-MRF) is developed to detect the change direction of enhanced, reduced, or unchanged regimes of backscattering from the SAR difference image (Jin, Chen, and Luo, 2004; Jin and Wang, 2009). Two

Polarimetric Scattering and SAR Information Retrieval, First Edition. Ya-Qiu Jin and Feng Xu.
© 2013 John Wiley & Sons Singapore Pte. Ltd. Published 2013 by John Wiley & Sons Singapore Pte. Ltd.

examples are presented. One is for change in an urban area. Two ERS-2 SAR images of Shanghai City in 1996 and 2002 are analyzed. The results illustrate well the direction of change in the Shanghai area during those years. Another example is change detection following damage resulting from the earthquake in Wenchuan, China, on 12 May 2008. Two images of ALOS PALSAR before and after the earthquake are applied for the detection and classification of the terrain surface changes caused by the earthquake.

9.1 The 2EM-MRF Algorithm

9.1.1 The EM Algorithm

Consider two co-registered images X_1 and X_2 with size $I \times J$ at two different times (t_1, t_2). Their difference image is $X = (X_2 - X_1)$; it could also take the form $X = X_2/X_1$. Let X be a random variable to represent the value of $I \times J$ pixels in $X_D = \{X(i,j); 1 \leq i \leq I, 1 \leq j \leq J\}$.

Change detection and analysis for a difference image is usually to distinguish two opposite classes: the unchanged class ω_n and the changed class ω_c. The probability density function p (X) of the pixel value in the difference image X_D was modeled as a mixture distribution of two density components associated with ω_n and ω_c, that is,

$$p(X) = p(X \mid \omega_n)P(\omega_n) + p(X \mid \omega_c)P(\omega_c) \tag{9.1}$$

where $P(\omega_n), P(\omega_c)$ of classes ω_n and ω_c are the *a priori* probabilities. The conditional probability density function is $p(X \mid \omega_k)$ $(k = n, c)$. The unsupervised estimates of $p(X|\omega_n)$, $p(X \mid \omega_c)$, $P(\omega_n)$ and $P(\omega_c)$ can be performed by the expectation maximum algorithm (EM algorithm).

The EM algorithm is a general approach to maximum-likelihood estimation (ML estimation) for incomplete data problems (Fukunaga, 1990; Dempster, Laird, and Rubin, 1977; Moon, 1996; Render and Walker, 1984; Shahshahani and Landgrebe, 1994). It consists of an expectation step and a maximization step. The expectation step is computed with respect to the unknown underlying variable using the estimates of the parameters conditioned by observations. The maximization step provides new estimates of the parameters. Assume that both $p(X \mid \omega_n)$ and $p(X \mid \omega_c)$ are modeled by Gaussian distributions. The density function associated with the class ω_n is described by the mean μ_n and the variance σ_n^2, and the density function associated with the class ω_c is described by μ_c and σ_c^2. It can be proved that the equations for estimating the statistical parameters and the *a priori* probability for the class ω_n are as follows (Bruzzone and Prieto, 2000; Render and Walker, 1984):

$$P^{t+1}(\omega_n) = \frac{\displaystyle\sum_{X(i,j)\in X_D} \frac{P^t(\omega_n)p^t(X(i,j)|\omega_n)}{p^t(X(i,j))}}{IJ} \tag{9.2}$$

$$\mu_n^{t+1} = \frac{\displaystyle\sum_{X(i,j)\in X_D} \frac{P^t(\omega_n)p^t(X(i,j)|\omega_n)}{p^t(X(i,j))}X(i,j)}{\displaystyle\sum_{X(i,j)\in X_D} \frac{P^t(\omega_n)p^t(X(i,j)|\omega_n)}{p^t(X(i,j))}} \tag{9.3}$$

$$(\sigma_n^2)^{t+1} = \frac{\displaystyle\sum_{X(i,j)\in X_D} \frac{P^t(\omega_n)p^t(X(i,j)|\omega_n)}{p^t(X(i,j))}[X(i,j)-\mu_n^t]^2}{\displaystyle\sum_{X(i,j)\in X_D} \frac{P^t(\omega_n)p^t(X(i,j)|\omega_n)}{p^t(X(i,j))}} \tag{9.4}$$

where the superscripts t and $t+1$ denote the current and next iterations.

Analogously, these equations can also be used to estimate the *a priori* probability, the mean, and the variation of the conditional probability density function for the class ω_c.

The estimates are computed starting from the initial values by iterating the above equations until convergence. It can be proved that the estimated parameters provide an increase in the log-likelihood function $L(\theta) = \ln p(X_D | \theta)$ after each iteration, where $\theta = \{P(\omega_n), P(\omega_c), \mu_n, \mu_c, \sigma_n^2, \sigma_c^2\}$. At convergence, a local maximum of the log-likelihood function is reached (Dempster, Laird, and Rubin, 1977; Moon, 1996).

The initial values of the estimated parameters can be obtained by the analysis of the histogram of the difference image. A pixel subset S_n likely belonging to ω_n and a pixel subset S_c likely belonging to ω_c can be obtained by first selecting two-threshold T_n and T_c on the histogram. These T_n and T_c are expressed as $T_n = M_D(1-\gamma)$ and $T_c = M_D(1+\gamma)$, where $M_D = [\max\{X_D\} + \min\{X_D\}]/2$ as the middle value of $h(X)$, and $\gamma \in (0,1)$ is a parameter that defines the range around M_D in which the pixels cannot be easily identified. The two subsets $S_n = \{X(i,j) | X(i,j) < T_n\}$ and $S_c = \{X(i,j) | X(i,j) > T_c\}$ are used to compute the initial estimates associated with the two classes ω_n and ω_c, respectively.

Under the assumption of inter-pixel independence and according to the Bayes rule for minimum error, each pixel $X(i,j)$ in the difference image X_D should be assigned to the class ω_k that maximizes the posterior conditional probability, that is,

$$\omega_k = \arg\max_{\omega_i \in \{\omega_n, \omega_c\}} \{P(\omega_i | X(i,j))\} = \arg\max_{\omega_i \in \{\omega_n, \omega_c\}} \{P(\omega_i)p(X(i,j) | \omega_i)\} \tag{9.5}$$

This is equivalent to solving the ML boundary T_o for the two classes ω_n and ω_c on the difference image. The optional threshold T_o requires

$$\frac{P(\omega_c)}{P(\omega_n)} = \frac{p(X | \omega_n)}{p(X | \omega_c)} \tag{9.6}$$

Under the assumption of a Gaussian distribution, Equation 9.6 yields

$$(\sigma_n^2 - \sigma_c^2)T_o^2 + 2(\mu_n\sigma_c^2 - \mu_c\sigma_n^2)T_o + \mu_c^2\sigma_n^2 - \mu_n^2\sigma_c^2 - 2\sigma_n^2\sigma_c^2 \ln\left[\frac{\sigma_c P(\omega_n)}{\sigma_n P(\omega_c)}\right] = 0 \tag{9.7}$$

to obtain the optimal T_o.

9.1.2 Two-Thresholds EM Algorithm

The pixel values of the SAR image represent the echo power (backscattering σ^0) of the radar observation. The difference image (i.e., $X = \sigma_D^0 = \sigma_2^0 - \sigma_1^0$) can show where the backscattering is enhanced ($\sigma_D^0 > 0$), unchanged ($\sigma_D^0 \approx 0$), or reduced ($\sigma_D^0 < 0$). This classification is applicable to information retrieval of the change in terrain surfaces. The pixels of the difference image are classified into three classes: σ_D^0 for enhanced ω_{c1}, σ_D^0 for unchanged ω_n, and σ_D^0 for reduced ω_{c2}. That is, there are three subsets of pixels with $\sigma_D^0 > 0$, $\sigma_D^0 \approx 0$, and $\sigma_D^0 < 0$.

Thus, the probability density function $p(X)$ of X_D is modeled as a mixture density distribution consisting of three components:

$$p(X) = p(X \mid \omega_{c1})P(\omega_{c1}) + p(X \mid \omega_n)P(\omega_n) + p(X \mid \omega_{c2})P(\omega_{c2}) \tag{9.8}$$

First, T_{o1} is obtained by application of the EM algorithm to the enhanced class ω_{c1} and non-enhanced class ω_{n1} (ω_n and ω_{c2}). Then, T_{o2} is obtained by applying the EM to the reduced class ω_{c2} and non-reduced class ω_{n2} (ω_n and ω_{c1}), where $\omega_{n1} = \omega_n \cup \omega_{c2}$ and $\omega_{n2} = \omega_n \cup \omega_{c1}$. Finally, the two results are superimposed to get the final classification.

In Figure 9.1a, the subset S_{n1} likely belonging to class ω_{n1} and the subset S_{c1} likely belonging to class ω_{c1} are obtained by applying two thresholds T_{n1} and T_{c1} in the enhanced regime

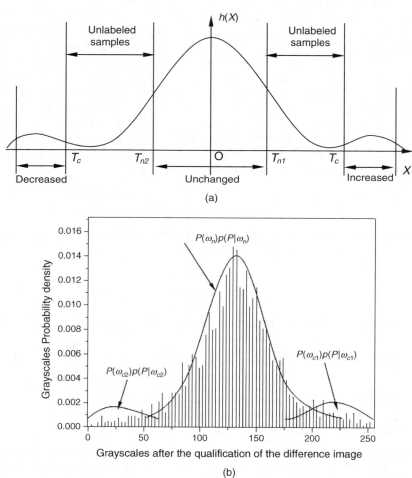

Figure 9.1 (a) The histogram of the difference image and thresholds. (b) The normalized histogram for three classes from real difference image data

$(X \geq 0)$ of the histogram $h(X)$. These T_{n1} and T_{c1} are expressed as $T_{n1} = M_{D1}(1 - \gamma)$ and $T_{c1} = M_{D1}(1 + \gamma)$, where M_{D1} is the mean value of the enhanced regime in $h(X)$, that is, $M_{D1} = [\max\{X_D\}/2]$. Analogously, the subset S_{n2} likely belonging to the class ω_{n2} and the subset S_{c2} likely belonging to the class ω_{c2} are obtained in the reduced regime ($X < 0$) of $h(X)$, where the mean value of the reduced regime in $h(X)$ is $M_{D2} = [\min\{X_D\}/2]$. In the present case, $\gamma = 0.5$ is chosen as a middle value of $(0,1)$.

Figure 9.1b shows the grayscale histogram of the difference image (shown in Figure 9.5) from the ERS-2 SAR image data around the Yan Zhong greenbelt of Shanghai downtown center. Three classes of probability distributions are outlined in this figure. The histogram is normalized in accordance with the probability density distributions.

The initial statistical parameters of the subsets $S_{n1} = \{X(i,j) \mid 0 < X(i,j) < T_{n1}\}$ and $S_{c1} = \{X(i,j) \mid X(i,j) > T_{c1}\}$ are first derived, and then the parameters of $S_{n2} = \{X(i,j) \mid T_{n2} < X(i,j) < 0\}$ and $S_{c2} = \{X(i,j) \mid X(i,j) < T_{c2}\}$ are derived. The initial statistical parameters are related to the classes ω_{n1}, ω_{c1} and the classes ω_{n2}, ω_{c2}, respectively.

Then, Equations 9.2–9.4 are sequentially used to perform the iterations on the four classes, that is, the *a priori* probability and other statistical parameters $[P(\omega_{n1}), \mu_{n1}, \sigma_{n1}^2]$, $[P(\omega_{c1}), \mu_{c1}, \mu_{c1}]$, $[P(\omega_{n2}), \mu_{n2}, \sigma_{n2}^2]$, and $[P(\omega_{c2}), \mu_{c2}, \sigma_{c2}^2]$. The thresholds T_{o1} and T_{o2} are solved one by one from Equation 9.7.

9.1.3 Spatial–Textual Classification based on MRF

As Bruzzone and Prieto (2000) discussed for the MRF approach, the pixels of the image are spatially correlated, that is, a pixel belonging to the class ω_k is likely to be surrounded by pixels belonging to the same class. To take account of spatial texture may yield more reliable and accurate change detection.

Let the set $C = \{C_\ell, 1 \leq l \leq L\}$ represent the possible sets of the labels in the difference image, where $C_\ell = \{C_\ell(i,j), 1 \leq i \leq I, 1 \leq j \leq J\}$, $C_\ell(i,j) \in \{\omega_{c1}, \omega_{c2}, \omega_n\}, L = 3^{IJ}, I, J$, are the height and width of the difference image, respectively. According to Bayes rule for minimum error, the classification result should maximize the posterior conditional probability,

$$C_k = \arg \max_{C_\ell \in C} \{P(C_\ell \mid X_D)\} = \arg \max_{C_\ell \in C} \{P(C_\ell)p(X_D \mid C_\ell)\} \qquad (9.9)$$

where X_D is the difference image, $P(C_\ell)$ is the prior model for the class labels, and $p(X_D \mid C_\ell)$ is the joint density function of the pixel values in the difference image given the set of labels C_ℓ. The solution of Equation 9.9 requires the MRF approach to estimate both $P(C_\ell)$ and $p(X_D \mid C_\ell)$.

In order to formulate the problem by using the MRF, it is necessary to introduce the concept of a spatial neighborhood system defining N. Considering a pixel (i,j), its neighborhood can be expressed as

$$N(i,j) = \{(i,j) + (\nu, \varsigma), (\nu, \varsigma) \in N\} \qquad (9.10)$$

In our case, $N = \{(\pm 1, 0), (0, \pm 1), (1, \pm 1), (-1, \pm 1)\}$, that is, the neighborhood of one pixel is composed of the eight pixels surrounding it.

Let $C_\ell(i,j)$ represent the class label of (i,j), $C_{\ell N}$ represent the class labels of the neighborhood pixels of (i,j), and $C_{\ell|(i,j)}$ represent the class labels of the pixels in the difference image except (i,j). If C_ℓ is taken as an MRF, C_ℓ has the following two properties:

$$P(C_\ell) > 0, \quad \forall C_\ell(i,j) \in \{\omega_{c1}, \omega_{c2}, \omega_n\} \tag{9.11}$$

$$P(C_\ell(i,j) \mid C_{\ell|(i,j)}) = P(C_\ell(i,j) \mid C_{\ell N}) \tag{9.12}$$

The second-order neighborhood system and cliques are shown in Figure 9.2.

Cliques refer to the connected subgraph of neighboring pixels, including the center pixel in the neighborhood system. In our work, only single-pixel and two-pixel cliques are considered. As shown in Figure 9.2b and c, the single-pixel clique has one form, and the two-pixel clique has four forms. The mathematical expressions are, respectively,

$$L_1 = (i,j), \quad L_2 = \{\{(i,j), (g,h)\} \mid (g,h) \in N_i(i,j)\} \tag{9.13}$$

According to the Hammersley–Clifford theorem, MRF is equivalent to the Gibbs distribution. Therefore, the conditional probability of the class label of (i,j) given the pixel labels elsewhere is expressed as (Bruzzone and Prieto, 2000)

$$\begin{aligned}
P(C_\ell(i,j) \mid &\{C_\ell(g,h), (g,h) \neq (i,j)\}) \\
&= P(C_\ell(i,j) \mid \{C_\ell(g,h), (g,h) \in N(i,j)\}) \\
&= \frac{1}{Z}\exp[-U(C_\ell(i,j) \mid \{C_\ell(g,h), (g,h) \in N(i,j)\})]
\end{aligned} \tag{9.14}$$

where $U(\cdot)$ is the Gibbs energy function, and Z is the normalizing factor,

$$Z = \sum_{(g,h)\in\{(i,j)\cup N(i,j)\}} \exp[-U(C_\ell(i,j) \mid \{C_\ell(g,h)\})] \tag{9.15}$$

and

$$\begin{aligned}
U(C_\ell \mid X_D) &= \sum_{1\leq i\leq I}\sum_{1\leq j\leq J} U(C_\ell(i,j) \mid C_\ell, X_D) \\
&= \sum_{1\leq i\leq I}\sum_{1\leq j\leq J} U(C_\ell(i,j) \mid C_{\ell N}, X_D)
\end{aligned} \tag{9.16}$$

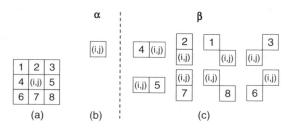

Figure 9.2 The second-order neighborhood system and cliques: (a) second-order neighborhood system; (b) single-pixel clique L_1 and its weight α; (c) two-pixel clique L_2 and its weight β

In the Gaussian case

$$
U(C_\ell(i,j) \mid C_{\ell N}, X_D) = \frac{1}{2} \ln \left| 2\pi\sigma^2_{C_\ell(i,j)} \right| + \frac{(X(i,j) - \mu_{C_\ell(i,j)})^2}{2\sigma^2_{C_\ell(i,j)}}
$$
$$
+ v_1(C_\ell(i,j)) + \sum_{(g,h)\in N(i,j)} v_2(C_\ell(i,j), C_\ell(g,h))
$$
(9.17)

Here

$$
v_1(C_\ell(i,j)) = \alpha, \quad (i,j) \in L_1
$$
(9.18)

where α is the weight value of a single-pixel clique, and $\alpha = 0$ is assumed. Finally

$$
v_2(C_\ell(i,j), C_\ell(g,h)) = \begin{cases} -\beta, & C_\ell(i,j) = C_\ell(g,h), \{(i,j),(g,h)\} \in L_2 \\ \beta, & C_\ell(i,j) \neq C_\ell(g,h), \{(i,j),(g,h)\} \in L_2 \end{cases}
$$
(9.19)

where β is the weight value of a two-pixel clique, $0.7 < \beta < 1.5$, and in our calculation $\beta = 1.0$.

According to Equation 9.9, the generation of the final change detection map involves the labeling of all the pixels in the difference image so that the posterior probability is maximized. In terms of Markov models, this is equivalent to the minimization of the following energy function:

$$
U(C_\ell \mid X_D) = \sum_{1 \leq i \leq I} \sum_{1 \leq j \leq J} [U_I(X(i,j) \mid C_\ell(i,j)) + U_C(C_\ell(i,j) \mid \{C_\ell(g,h), (g,h) \in N(i,j)\})]
$$
(9.20)

Here $U_C(\cdot)$ describes the inter-pixel class dependence, which is given by the last two terms in Equation 9.17; and $U_I(\cdot)$ represents the energy function under the assumption of conditional independence, as defined by the first two terms in Equation 9.17:

$$
U_I(X(i,j) \mid C_\ell(i,j)) = \frac{1}{2} \ln \left| 2\pi\sigma^2_{C_\ell(i,j)} \right| + \frac{1}{2}(X(i,j) - \mu_{C_\ell(i,j)})^2 [\sigma^2_{C_\ell(i,j)}]^{-1}
$$
(9.21)

where $\sigma^2_{C_\ell(i,j)} \in \{\sigma^2_{c1}, \sigma^2_{c2}, \sigma^2_n\}$ and $\mu_{C_\ell(i,j)} \in \{\mu_{c1}, \mu_{c2}, \mu_n\}$ have been obtained by the EM algorithm.

Generally, the solution of Equation 9.21 requires an iterative algorithm, for example, the simulated annealing algorithm. We employ the iterated conditional modes algorithm (ICM algorithm) (Besag, 1974, 1986), which has been proved to be computationally economical and converges rapidly to a local minimum of the energy function. The mathematical expression of the ICM algorithm can be written as

$$
C_\ell = ICM(\alpha, \beta, \mu_{C_\ell(i,j)}, \sigma^2_{C_\ell(i,j)}, X_D, C_{\ell 0})
$$
(9.22)

where $C_{\ell 0}$ is the initial value of C_ℓ. The value C_ℓ is obtained by the following steps:

(a) obtain $C_{\ell 0}$ by the EM algorithm, and initialize C_ℓ with $C_{\ell 0}$;
(b) calculate the new $C_\ell(i,j)$ that minimizes the energy function in Equation 9.20; and
(c) repeat step (b) until convergence is reached.

The convergence condition of the ICM algorithm is either that the number of iterative cycles exceeds 30 (actual convergence is reached much more quickly) or that fewer than $\sqrt{IJ}/20$ pixels change. The details of the ICM algorithm can be found in Besag (1986).

9.2 The 2EM-MRF for Change Detection in an Urban Area

Figures 9.3 and 9.4 present two images from data produced by ERS-2 SAR on 4 June 1996 and 9 April 2002, respectively, over Shanghai, China ($\sim$31.3°N, 121.4°E). The image resolution is 12.5 m $\times$ 12.5 m. The images have been co-registered, and the noise is simply reduced by applying mean filtering (3 $\times$ 3 window size). The size of the images given in Figures 9.3 and 9.4 is 580 $\times$ 980 pixels. Figure 9.5 shows the difference image. The white pixels of Figure 9.5 indicate enhanced scattering and the black ones reduced backscattering.

Figure 9.3 ERS-2 SAR image on 4 June 1996, Shanghai

Figure 9.4 ERS-2 SAR image on 9 April 2002, Shanghai

Figure 9.5   The difference image between Figures 9.3 and 9.4

The two-thresholds EM algorithm is applied to the difference image to obtain iteratively the *a priori* probability and other statistical parameters. The iterative process of the first class is shown in Figure 9.6a–c, and the iteration of the second class is very similar. The two thresholds T_{o1} and T_{o2} for surface change detection are then computed. The present example

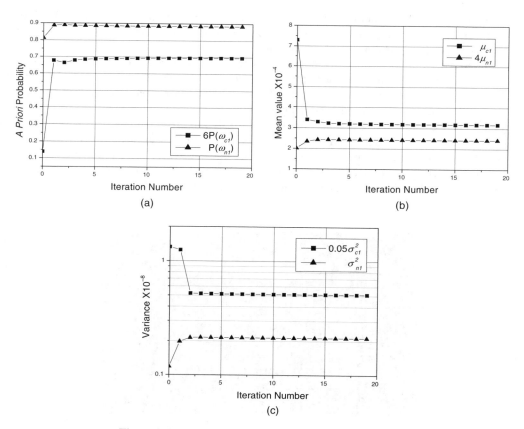

Figure 9.6   The parameters after each iteration ($\gamma = 0.5$)

Figure 9.7   Change detection using two-thresholds EM

yields $T_{o1} = 0.946 \times 10^{-4}$ and $T_{o2} = -1.013 \times 10^{-4}$ (the maximum $\sigma_D^0|_{max} = 5.202 \times 10^{-4}$ and the minimum $\sigma_D^0|_{min} = -3.381 \times 10^{-4}$).

Figure 9.7 shows the automatic detection of change direction by utilizing the EM algorithm. The white pixels indicate the σ^0 enhanced places, the black pixels show the σ^0 reduced places, and the gray area is σ^0 unchanged.

Figure 9.8 shows the spatially contextual change based on the MRF-ICM algorithm. Compared with Figure 9.7, it can be seen that the pixels of the same class are more merged and that isolated pixels become much fewer. In Figure 9.8, the black areas marked by odd numbers (1, 3, 5) are the σ^0 reduced areas, and the white areas marked by even numbers (2, 4) are the σ^0 enhanced areas. For example, area 1 is the Lu Jia Zui greenbelt, area 3 is the Pu Dong Park, and area 5 is the Yan Zhong greenbelt, where the flat green surface causes less scattering. But area 2 is the commercial center at the Lu Jia Zui, where there are many different buildings causing more electromagnetic wave double reflection and enhanced scattering. Area 4 is the location of the new buildings of the International Exhibition Center of

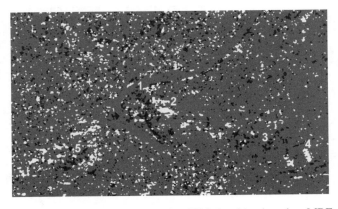

Figure 9.8   Change detection using ICM algorithm based on MRF

Figure 9.9   Change detection superimposed on the Shanghai tourist map

Shanghai. The areas 1–5 were newly rebuilt after 1996 and are shown well in the change detection data of 2002 minus 1996.

It is interesting to overlap Figure 9.8 on the tourist map of Shanghai and to identify detailed changes along the city streets during these years as indicated on Figure 9.9.

9.3   Change Detection after the 2008 Wenchuan Earthquake

On 12 May 2008, a major earthquake, measuring 8.0 on the Richter scale, jolted the Wenchuan area of southwestern China's Sichuan Province. The earthquake destroyed thousands of buildings in the area, and thousands of people died. This earthquake was the worst natural disaster to hit China over at least 30 years. Bad weather, for example, rain and cloud, following the occurrence of a natural disaster, causes serious difficulties for large-scale rescue and aerial photography. All-weather and all-day/all-night satellite-borne microwave remote sensing plays a key role to monitor such disasters.

To evaluate the damages and terrain surface changes caused by the earthquake, multi-temporal ALOS PALSAR image data before and after the earthquake were applied to the detection and classification of the terrain surface changes. The 2EM-MRF is developed for automatic detection of surface changes with scattering enhanced, reduced, and not changed (Jin and Wang, 2009). Using the tools of Google Earth for surface mapping, the surface change situation after the earthquake overlapping the DEM topography can be shown in multi-azimuth views as an animated cartoon. The detection and classification can also be compared with the optical photographs.

Figure 9.10a and b are the ALOS PALSAR images (L band, HH polarization) on 17 February and 19 May 2008, respectively, before and after the earthquake in Beichuan county, Wenchuan area. The spatial resolution is 4.7 m × 4.5 m.

Figure 9.11 is the difference image between Figure 9.10a and b. From Figure 9.11, it seems very difficult to evaluate the terrain surface changes only using the human eye,

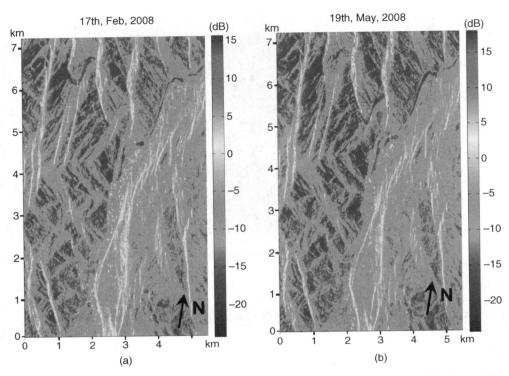

Figure 9.10 ALOS PALSAR images over Beichuan county, Wenchuan area: (a) on 17 February 2008, before the earthquake; (b) on 19 May 2008, after the earthquake. © [2009] IEEE. Reprinted, with permission, from *IEEE Geoscience and Remote Sensing Letters*

especially to classify accurately the change classes on a large scale. Figure 9.12 shows the grayscale histogram of a small part chosen from the difference image in Figure 9.11. The three classes of probability distributions are outlined in this figure. The histogram is normalized in accordance with the probability density distributions.

The EM algorithm is applied to the difference image to obtain iteratively the *a priori* probability and other statistical parameters. The iterative process is shown in Figure 9.13a–d. Solving Equation 9.7, the thresholds $T_{o1} = 3.1643$ dB and $T_{o2} = -2.8381$ dB are obtained.

Figure 9.14 is the change detection using the 2EM algorithm. It shows three classes of change in the difference image: the red color indicates the area with scattering enhanced, blue color indicates the area with scattering reduced, and green color denotes no change.

The MRF algorithm is employed with a spatial neighborhood of 5×5 pixels to take account of spatial texture. The generation of the final change detection map involves the labeling of all the pixels in the difference image so that the posterior probability of Equation 9.8 is maximized.

Following Bruzzone and Prieto (2000), in terms of Markov models, the MRF algorithm is equivalent to the minimization of the Gibbs energy function. It requires an iterative algorithm, for example, the simulated annealing algorithm or the ICM, which are seen as "hard decisions", that is, either one or the other. The 2EM algorithm is given as a good initial value for iteration. Alternatively, to implement the renewing of category labels and the parameter estimation, the

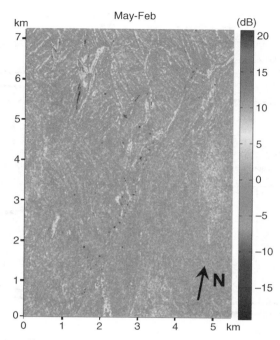

Figure 9.11   Difference image of ALOS PALSAR before and after the earthquake. © [2009] IEEE. Reprinted, with permission, from *IEEE Geoscience and Remote Sensing Letters*

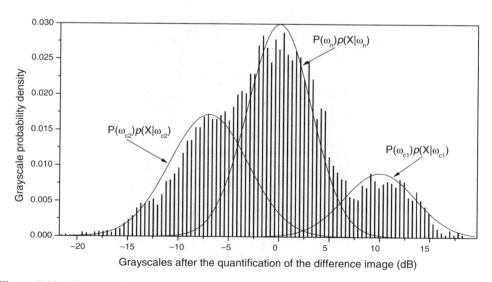

Figure 9.12   The normalized histogram for three classes from real difference image data. © [2009] IEEE. Reprinted, with permission, from *IEEE Geoscience and Remote Sensing Letters*

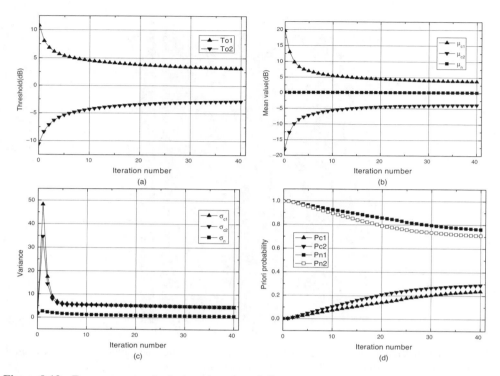

Figure 9.13   Four parameters in the iterations ($\gamma = 0.5$). © [2009] IEEE. Reprinted, with permission, from *IEEE Geoscience and Remote Sensing Letters*

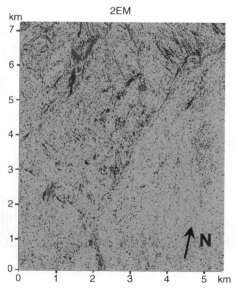

Figure 9.14   Detection and classification of terrain surface changes using the 2EM algorithm. © [2009] IEEE. Reprinted, with permission, from *IEEE Geoscience and Remote Sensing Letters*

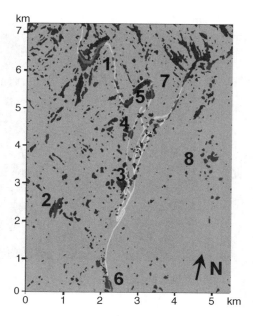

Figure 9.15 Automatic detection and classification of terrain surface changes using 2EM-MRF. © [2009] IEEE. Reprinted, with permission, from *IEEE Geoscience and Remote Sensing Letters*

pixels are given a membership of a certain category. Following the steps of Bruzzone and Prieto (2000) and Zhang, Modestino, and Langan (1994), the MRF algorithm is carried out iteratively.

Figure 9.15 presents the final result of 2EM-MRF for detection and classification of the terrain surface changes from the difference image in Figure 9.14. The numbers 1–8 indicate some areas with typical changes. It can be seen that, for example, in area 1 the river was largely blocked by a landslide; in area 2, there were landslides causing large-scale blockages; area 3 is Beichuan town, where the terrain surface was significantly undulated or roughened due to landslides and building collapse; in areas 4 and 5 the river was blocked due to land-slides along river courses; in area 6 the highway was significantly blocked by a landslide; in area 7 there was a collection of landslides almost toward one azimuth direction, and the flat area with reduced scattering along the river might be due to seasonally higher water level in May than in February; finally, in area 8 collapse of mountain blocks caused undulation of the terrain surface to enhance stronger scattering.

Figure 9.16 gives the topography and contour lines of the same area as Figure 9.15 from the Google website. It is useful to locate where and which kind of changes are happening. It might be seen that the landslides, for example, in areas 1 and 7, are correlated with the direction and magnitude of slopes. A quantitative analysis to assess the landslides and surface damage can be further developed.

Figure 9.17 is a part of the photograph distributed to the public from the website of the Ministry of State Resources of China. The numbers in the figure indicate the same areas as the numbers in Figure 9.15.

For example, Figure 9.18 specifically presents comparisons of the detections of ALOS PALSAR and optical photographs in regions 3 and 4, where the left figures are from

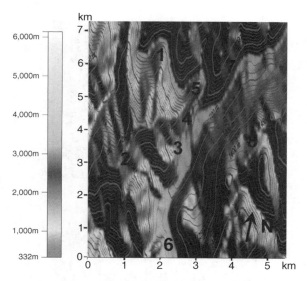

Figure 9.16 The topography of the same area as Figure 9.15. © [2009] IEEE. Reprinted, with permission, from *IEEE Geoscience and Remote Sensing Letters*

Figure 9.15, and the right figures are from Figure 9.17. It can be clearly understood in both figures that A1 and A2 show rough surfaces, and A2 shows the river blocked; B1, B2, and B3 show flat surfaces; and B2 shows the road blocked. It can be seen that optical shadowing actually does not confuse these classifications.

Using the tools of Google Earth mapping with our 2EM-MRF results, the terrain surface change situation classified by three types overlapping the DEM topography can be shown in multi-azimuth views as an animated cartoon. Figure 9.19a–d shows four such views. Since

Figure 9.17 A photograph of the area shown in Figures 9.15 and 9.16 after the earthquake. © [2009] IEEE. Reprinted, with permission, from *IEEE Geoscience and Remote Sensing Letters*

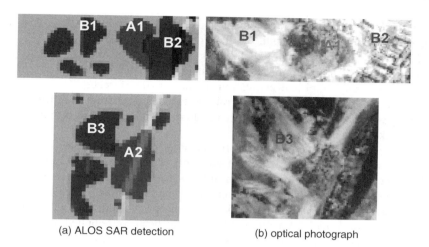

(a) ALOS SAR detection   (b) optical photograph

Figure 9.18 Comparison of (a) detections of ALOS PALSAR and (b) optical photographs in the regions 3 and 4 of Figures 9.15 and 9.17. © [2009] IEEE. Reprinted, with permission, from *IEEE Geoscience and Remote Sensing Letters*

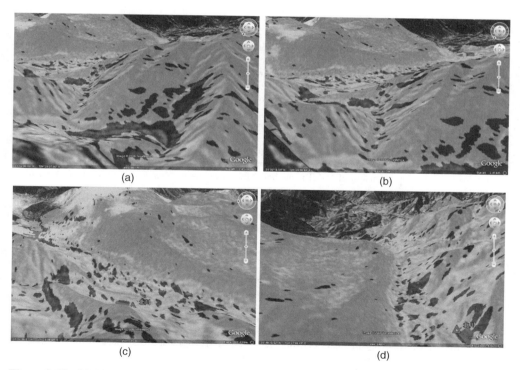

Figure 9.19 Multi-azimuth views in an animated cartoon showing terrain surface changes on Google Earth mapping

all the processes of 2EM-MRF are automatic and are carried out in real time, this should be helpful, especially for commanding the rescue works in the disaster area.

The 2EM-MRF method is applied to automatic detection and classification of the terrain changes from multi-temporal SAR images. This method can be carried out in real time without any empirical estimation or manual interference. The two-thresholds EM and MRF algorithm classified the terrain surface changes from the difference image into three types: enhanced scattering, reduced scattering, and no change. Spatially correlated textures are taken into account.

It can be seen that information retrieval from a single amplitude image (scattered power) with one frequency, mono-polarization, at one time is very limited. In fully polarimetric technology, the physical status of the terrain surfaces, such as vegetation canopy, road, building, river, and so on, can be well classified and would be of great help to change detection. It is feasible to show the surface slope variation and anisotropy using the four Stokes parameters of the polarimetric measurement, and INSAR has been well applied to the retrieval of surface digital elevation. All of this progress shows superiority over a single-amplitude image analysis and manual estimation by the human eye. It is also helpful to include the constant false-alarm rate (Schou *et al.*, 2003) for detection of the object, for example, edge and block, from a SAR image.

The 2EM-MRF can be further extended to utilization of the principal characteristic parameters of the difference image, such as VV + HH, VV – HH, entropy, and angles α, β, γ, Ψ, as described in Chapter 5, for change detection and classification.

References

Besag, J. (1974) Spatial interaction and the statistical analysis of lattice systems. *Journal of the Royal Statistical Society, Series B*, **36**(2), 192–326.

Besag, J. (1986) On the statistical analysis of dirty pictures. *Journal of the Royal Statistical Society, Series B*, **48**, 259–302.

Bruzzone, L. and Prieto, D.F. (2000) Automatic analysis of the difference image for unsupervised change detection. *IEEE Transactions on Geoscience and Remote Sensing*, **38**(3), 1171–1182.

Dempster, A.P., Laird, N.M., and Rubin, D.B. (1977) Maximum likelihood from incomplete data via the EM algorithm. *Journal of the Royal Statistical Society*, **39**(1), 1–38.

Fukunaga, K. (1990) *Introduction to Statistical Pattern Recognition*, 2nd edn. Academic Press, London.

Jin, Y.Q. and Wang, D. (2009) Automatic detection of terrain surface changes after Wenchuan earthquake, May 2008 from ALOS SAR images using 2EM-MRF method. *IEEE Geoscience and Remote Sensing Letters*, **6**(2), 344–348.

Jin, Y.Q., Chen, F., and Luo, L. (2004) Automatic detection of change direction of multi-temporal ERS-2 SAR images using two-threshold EM and MRF algorithms. *Imaging Science Journal*, **52**, 234–241.

Kasetkasem, T. and Varshney, P.K. (2002) An image change detection algorithm based on Markov random field models. *IEEE Transactions on Geoscience and Remote Sensing*, **40**(8), 1815–1823.

Merril, K.R. and Liu, J. (1998) A comparison of four algorithms for change detection in an urban environment. *Remote Sensing of Environment*, **63**, 95–100.

Moon, T.K. (1996) The expectation-maximization algorithm. *Signal Processing Magazine*, **13**(6), 47–60.

Render, A.P. and Walker, H.F. (1984) Mixture densities, maximum likelihood and the EM algorithm. *SIAM Review*, **26**(2), 195–239.

Rignot, E. and van Zyl, J. (1993) Change detection techniques for ERS-1 SAR data. *IEEE Transactions on Geoscience and Remote Sensing*, **31**(4), 896–906.

Schou, J., Skriver, H., Nielsen, A.A., and Conradsen, K. (2003) CFAR edge detector for polarimetric SAR images. *IEEE Transactions on Geoscience and Remote Sensing*, **41**(1), 20–32.

Shahshahani, B.M. and Landgrebe, D.A. (1994) The effect of unlabeled samples in reducing the small size problem and mitigating the Hughes phenomenon. *IEEE Transactions on Geoscience and Remote Sensing*, **32**, 1087–1095.

Singh, A. (1989) Digital change detection techniques using remotely-sensed data. *International Journal of Remote Sensing*, **10**(6), 989–1003.

Zhang, J., James, W.M., and David, A.L. (1994) Maximum-likelihood parameter estimation for unsupervised stochastic model based on image segmentation. *IEEE Transactions on Image Processing*, **3**(4), 404–420.

10

Temporal Mueller Matrix for Polarimetric Scattering

With the advance of radar polarimetry and imagery of SAR technology (Zebker and van Zyl, 1991; Boerner *et al.*, 1997), the study of completely polarimetric scattering of the terrain surface has become of great importance. The Mueller matrix solution of vector radiative transfer has been studied for polarimetric scattering from a layer of random non-spherical particles (Jin, 1994). However, as the incidence is a plane pulse, it leads to the problem of pulse echoes for polarimetric scattering from random scattering media and the temporal Mueller matrix solution.

There are a few works on the pulse echo from random media. For example, using the two-frequency mutual coherence function, Hong and Ishimaru (1977), Liu and Yeh (1978), and Ito (1980) studied the scalar pulse wave propagation in random media. Oguchi and Ito (1990) obtained numerical solutions of the two-frequency vector radiative transfer equation in the case where a circularly polarized plane pulse was normally incident upon a half-space of spherical raindrops. Whitman *et al.* (1996) investigated pulse propagation in a strongly forward scattering random medium. All the aforementioned studies are limited to discussion of homogeneous random media. Using the ray tracing method and range calculation, Sun and Ranson (2000) presented a model of the forest canopy with vertical structure and calculated scattering for simulation of lidar returns. However, the completely polarimetric scattering echo from inhomogeneous random media remains to be further studied.

This chapter presents a theoretical model of an inhomogeneously layered random medium and derives the temporal Mueller matrix solution of vector radiative transfer for polarimetric scattering of random non-spherical scatterers. Co-pol and cross-pol bistatic and backscattering are numerically calculated. Variations of the shape and intensity of the polarimetric echoes well depict the inhomogeneous fraction profile, and the functional dependence upon layer thickness, incident angle, and other parameters are discussed (Jin, Chen, and Chang, 2004; Chang and Jin, 2002).

Polarimetric Scattering and SAR Information Retrieval, First Edition. Ya-Qiu Jin and Feng Xu.
© 2013 John Wiley & Sons Singapore Pte. Ltd. Published 2013 by John Wiley & Sons Singapore Pte. Ltd.

10.1   Radiative Transfer in Inhomogeneous Random Scattering Media

The N-layer model is shown in Figure 10.1. The fractional volume of non-spherical scatterers in the nth layer is assumed to be f_{sn} ($n = 1, 2, \ldots, N$). The layer $(N+1)$ is assumed be the air region above the underlying surface, with dielectric permittivity ε_b. Note that the dashed lines in the model of Figure 10.1 indicate no solid interfaces. Let a polarized plane pulse be incident at an angle of $(\pi - \theta_0, \varphi_0)$. The time-dependent vector radiative transfer (VRT) equations for the up-going wave $\mathbf{I}_n(\theta, \varphi, z, t)$ and the down-going wave $\mathbf{I}_n(\pi - \theta, \varphi, z, t)$ in layer n are written as ($0 \leq \theta < \pi/2$)

$$\cos\theta \frac{\partial}{\partial z}\mathbf{I}_n(\theta, \varphi, z, t) = -\overline{\boldsymbol{\kappa}}_{en}(\theta, \varphi) \cdot \mathbf{I}_n(\theta, \varphi, z, t) + \mathbf{S}_n(\theta, \varphi, z, t) - \frac{1}{c_0}\frac{\partial}{\partial t}\mathbf{I}_n(\theta, \varphi, z, t) \quad (10.1\text{a})$$

$$\begin{aligned}
-\cos\theta &\frac{\partial}{\partial z}\mathbf{I}_n(\pi - \theta, \varphi, z, t) \\
&= -\overline{\boldsymbol{\kappa}}_{en}(\pi - \theta, \varphi) \cdot \mathbf{I}_n(\pi - \theta, \varphi, z, t) + \mathbf{W}_n(\theta, \varphi, z, t) - \frac{1}{c_0}\frac{\partial}{\partial t}\mathbf{I}_n(\pi - \theta, \varphi, z, t)
\end{aligned} \quad (10.1\text{b})$$

Here the subscript n denotes the nth layer, $z \in [d_{n-1}, d_n]$, $\overline{\boldsymbol{\kappa}}_{en}$ is the extinction matrix in the nth layer, c_0 is the light velocity, the multiple scattering function is

$$\begin{aligned}
\mathbf{S}_n(\theta, \varphi, z, t) = \int_0^{\pi/2} d\theta' \int_0^{2\pi} d\varphi' \sin\theta' &[\overline{\mathbf{P}}_n(\theta, \varphi; \theta', \varphi') \cdot \mathbf{I}_n(\theta', \varphi', z, t) \\
&+ \overline{\mathbf{P}}_n(\theta, \varphi; \pi - \theta', \varphi') \cdot \mathbf{I}_n(\pi - \theta', \varphi', z, t)]
\end{aligned} \quad (10.2)$$

the function $\mathbf{W}_n(\theta, \varphi, z, t)$ is the same as $\mathbf{S}_n(\theta, \varphi, z, t)$ with all θ replaced by $\pi - \theta$, $\overline{\mathbf{P}}_n$ is the phase matrix, and (θ, φ) and (θ', φ') are the incident and scattered direction angles in the nth layer, respectively.

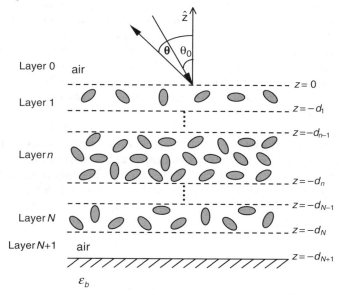

Figure 10.1   Model of inhomogeneous layered random media

The boundary conditions are written as

$$\mathbf{I}_1(\pi - \theta, \varphi, z = 0, t) = \mathbf{I}_0(t)\delta(\cos \theta - \cos \theta_0)\delta(\varphi - \varphi_0) \tag{10.3a}$$

$$\mathbf{I}_{n+1}(\pi - \theta, \varphi, z = -d_n, t) = \mathbf{I}_n(\pi - \theta, \varphi, z = -d_n, t) \quad (n = 1, \ldots, N-1) \tag{10.3b}$$

$$\mathbf{I}_N(\theta, \varphi, z = -d_N, t) = \overline{\mathbf{R}}(\theta) \cdot \mathbf{I}_N\left(\pi - \theta, \varphi, z = -d_N, t - \frac{2\Delta d_{N+1}\sec \theta}{c_0}\right) \tag{10.3c}$$

$$\mathbf{I}_n(\theta, \varphi, z = -d_n, t) = \mathbf{I}_{n+1}(\theta, \varphi, z = -d_n, t) \quad (n = N-1, \ldots, 1) \tag{10.3d}$$

where $\mathbf{I}_0(t)$ is the incident Stokes vector, $\overline{\mathbf{R}}(\theta)$ is the reflectivity matrix of the underlying surface at $z = -d_{N+1}$, and $\Delta d_{N+1} (= d_{N+1} - d_N)$ is the thickness of the layer $(N+1)$.

Introducing the exponential Fourier series representations for the time dependence of Equations 10.1a–10.3d

$$\mathbf{I}(\theta, \varphi, z, t) = \frac{1}{2\pi}\int_{-\infty}^{\infty} \mathbf{I}(\theta, \varphi, z, \omega)e^{j\omega t}d\omega \tag{10.4}$$

it yields the VRT equations (Equations 10.1a and 10.1b) as

$$\cos \theta \frac{\partial}{\partial z}\mathbf{I}_n(\theta, \varphi, z, \omega) = -\left[\overline{\boldsymbol{\kappa}}_{en}(\theta, \varphi) + \frac{j\omega}{c_0}\overline{\mathbf{U}}\right] \cdot \mathbf{I}_n(\theta, \varphi, z, \omega) + \mathbf{S}_n(\theta, \varphi, z, \omega) \tag{10.5a}$$

$$-\cos \theta \frac{\partial}{\partial z}\mathbf{I}_n(\pi - \theta, \varphi, z, \omega) = -\left[\overline{\boldsymbol{\kappa}}_{en}(\pi - \theta, \varphi) + \frac{j\omega}{c_0}\overline{\mathbf{U}}\right] \cdot \mathbf{I}_n(\pi - \theta, \varphi, z, \omega) + \mathbf{W}_n(\theta, \varphi, z, \omega) \tag{10.5b}$$

where $\overline{\mathbf{U}}$ is the 4×4 (dimensional) unit matrix, and Equation 10.2 becomes

$$\mathbf{S}_n(\theta, \varphi, z, \omega) = \int_0^{\pi/2} d\theta' \int_0^{2\pi} d\varphi' \sin \theta' \left[\overline{\mathbf{P}}_n(\theta, \varphi; \theta', \varphi') \cdot \mathbf{I}_n(\theta', \varphi', z, \omega)\right.$$
$$\left. + \overline{\mathbf{P}}_n(\theta, \varphi; \pi - \theta', \varphi') \cdot \mathbf{I}_n(\pi - \theta', \varphi', z, \omega)\right] \tag{10.6}$$

and $\mathbf{W}_n(\theta, \varphi, z, \omega)$ is obtained in the same way. It can be seen that the term of $j\omega\mathbf{U}/c$ takes into account the time delay of wave propagation through the random media in the Fourier domain.

The boundary conditions of Equations 10.3a–10.3d are rewritten as

$$\mathbf{I}_1(\pi - \theta, \varphi, z = 0, \omega) = \mathbf{I}_0(\omega)\delta(\cos \theta - \cos \theta_0)\delta(\varphi - \varphi_0) \tag{10.7a}$$

$$\mathbf{I}_{n+1}(\pi - \theta, \varphi, z = -d_n, \omega) = \mathbf{I}_n(\pi - \theta, \varphi, z = -d_n, \omega) \quad (n = 1, \ldots, N-1) \tag{10.7b}$$

$$\mathbf{I}_N(\theta, \varphi, z = -d_N, \omega) = \overline{\mathbf{R}}(\theta) \cdot \mathbf{I}_N(\pi - \theta, \varphi, z = -d_N, \omega) \cdot \exp\left(-j\omega\frac{2\Delta d_{N+1}\sec \theta}{c_0}\right) \tag{10.7c}$$

$$\mathbf{I}_n(\theta, \varphi, z = -d_n, \omega) = \mathbf{I}_{n+1}(\theta, \varphi, z = -d_n, \omega) \quad (n = N-1, \ldots, 1) \tag{10.7d}$$

Given that the eigenvalues of $\overline{\boldsymbol{\kappa}}_{en}(\pi - \theta, \varphi)$ are $\xi_{ni}(\pi - \theta, \varphi)$, $i = 1, 2, 3, 4$, and that the corresponding eigenmatrix is $\overline{\mathbf{E}}(\pi - \theta, \varphi)$ (see Equation 2.39 of Chapter 2; Tsang, Kong, and

Ding, 2000, pp. 37–40; Ito, 1980), it is easy to understand that the eigenvalues $\zeta_{ni}(\pi - \theta, \varphi)$ of $[\overline{\mathbf{K}}_{en}(\pi - \theta, \varphi) + j\omega\overline{\mathbf{U}}/c_0]$ are equal to $(\xi_{ni}(\pi - \theta, \varphi) + j\omega/c_0)$ and the eigenmatrix is also $\overline{\mathbf{E}}(\pi - \theta, \varphi)$.

The formal general solution of Equation 10.5b can be written as (Tsang, Kong, and Ding et al., 2000, p. 124)

$$\mathbf{I}_n(\pi - \theta, \varphi, z, \omega) = \overline{\mathbf{E}}(\pi - \theta, \varphi) \cdot \overline{\mathbf{D}}[\zeta_n(\pi - \theta, \varphi)z \sec \theta] \cdot \mathbf{C}_n + \sec \theta \, \overline{\mathbf{E}}(\pi - \theta, \varphi)$$

$$\cdot \int_z^{-d_{n-1}} \overline{\mathbf{D}}[\zeta_n(\pi - \theta, \varphi)(z - z') \sec \theta] \cdot \overline{\mathbf{E}}^{-1}(\pi - \theta, \varphi) \cdot \mathbf{W}_n(\theta, \varphi, z', \omega) \, dz'$$

$$(10.8)$$

where $\overline{\mathbf{D}}$ is a 4×4 diagonal matrix with the iith element $\exp[\zeta_{ni}(\pi - \theta)z \sec \theta]$, and $\mathbf{C}_n$ are the unknown column vectors to be determined by the boundary conditions in Equations 10.7a and 10.7b.

Let the eigenvalues of the matrix $[\overline{\mathbf{K}}_{en}(\theta, \varphi) + j\omega\overline{\mathbf{U}}/c_0]$ be $\zeta_{ni}(\theta, \varphi) = \xi_{ni}(\theta, \varphi) + j\omega/c_0$, $i = 1, 2, 3, 4$ (where $\xi_{ni}(\theta, \varphi)$ are the eigenvalues of the matrix $\overline{\mathbf{K}}_{en}(\theta, \varphi)$) and the eigenmatrix $\overline{\mathbf{E}}(\theta, \varphi)$, respectively. Then the formal general solution of Equations 10.5a is obtained as

$$\mathbf{I}_n(\theta, \varphi, z, \omega) = \overline{\mathbf{E}}(\theta, \varphi) \cdot \overline{\mathbf{D}}[-\zeta_n(\theta, \varphi)z \sec \theta] \cdot \mathbf{C}_n$$

$$+ \sec \theta \, \overline{\mathbf{E}}(\theta, \varphi) \cdot \int_{-d_n}^z \overline{\mathbf{D}}[-\zeta_n(\theta, \varphi)(z - z') \sec \theta] \cdot \overline{\mathbf{E}}^{-1}(\theta, \varphi) \cdot \mathbf{S}_n(\theta, \varphi, z', \omega) \, dz'$$

$$(10.9)$$

where $\mathbf{C}_n$ are the unknown column vectors to be determined by the boundary conditions in Equations 10.7c and 10.7d.

As $\mathbf{C}_1$ is first obtained by the boundary condition in Equation 10.7a, others $\mathbf{C}_2, \ldots, \mathbf{C}_N$ are then determined in turn by the boundary conditions in Equation 10.7b. Next, $\mathbf{C}'_N$ is first determined by Equation 10.7c after $\mathbf{C}_N$ is obtained, and the others $\mathbf{C}'_{N-1}, \ldots, \mathbf{C}'_1$ are then in turn determined by Equation 10.7d.

As $\mathbf{C}_n$ and $\mathbf{C}'_n$, $n = 1, \ldots, N$, are now determined, the formal solutions of $\mathbf{I}_n(\theta, \varphi, z, \omega)$ and $\mathbf{I}_n(\pi - \theta, \varphi, z, \omega)$ of Equations 10.8 and 10.9 are obtained.

Taking the integral-independent terms of the formal solutions of Equations 10.8 and 10.9 as the zeroth-order solutions $\mathbf{I}_n^{(0)}(\theta, \varphi, z, \omega)$ and $\mathbf{I}_n^{(0)}(\pi - \theta, \varphi, z, \omega)$, and substituting them into the multiple scattering terms $\mathbf{W}_1(\theta, \varphi, z, \omega), \ldots, \mathbf{W}_N(\theta, \varphi, z, \omega)$ and $\mathbf{S}_1(\theta, \varphi, z, \omega), \ldots, \mathbf{S}_N$ $(\theta, \varphi, z, \omega)$ of $\mathbf{I}_1(\theta, \varphi, z, \omega)$ (where $z \in [-d_1, 0]$), the first-order solution $\mathbf{I}_1^{(1)}(\theta, \varphi, z, \omega)$ at $z = 0$ can be obtained as

$$\mathbf{I}_1^{(1)}(\theta, \varphi, z = 0, \omega) = \sum_{n=1}^N \left\{ \sec \theta \, \overline{\mathbf{E}}(\theta, \varphi) \cdot \overline{\mathbf{D}}\left[-\sum_{m=1}^n \xi_m(\theta, \varphi)\Delta d_m \sec \theta \right] \right.$$

$$\cdot \int_{-d_n}^{-d_{n-1}} dz' \, D[\xi_n(\theta, \varphi)(z' + d_n) \sec \theta] \, \overline{\mathbf{E}}^{-1}(\theta, \varphi)$$

$$\cdot \overline{\mathbf{P}}_n(\theta, \varphi; \pi - \theta_0, \varphi_0) \cdot \overline{\mathbf{E}}(\pi - \theta_0, \varphi_0)$$

$$\left. \cdot \overline{\mathbf{D}}\left[-\sum_{m=1}^n \xi_m(\pi - \theta_0, \varphi_0)\Delta d_m \sec \theta_0 + \xi_n(\pi - \theta_0, \varphi_0)(z' + d_n) \sec \theta_0 \right] \right.$$

$$\cdot \overline{\mathbf{E}}^{-1}(\pi - \theta_0, \varphi_0) \cdot \mathbf{I}_0(\omega) \exp\left[j\omega \frac{z'(\sec\theta + \sec\theta_0)}{c_0}\right]\Bigg\}$$

$$+ \sum_{n=1}^{N}\Bigg\{\sec\theta\, \overline{\mathbf{E}}(\theta, \varphi) \cdot \overline{\mathbf{D}}\left[-\sum_{m=1}^{n}\xi_m(\theta, \varphi)\Delta d_m \sec\theta\right]$$

$$\cdot \int_{-d_n}^{-d_{n-1}} dz'\, \overline{\mathbf{D}}[\xi_n(\theta, \varphi)(z' + d_n)\sec\theta] \cdot \overline{\mathbf{E}}^{-1}(\theta, \varphi) \cdot \overline{\mathbf{P}}_n(\theta, \varphi; \theta_0, \varphi_0) \cdot \overline{\mathbf{E}}(\theta_0, \varphi_0)$$

$$\cdot \overline{\mathbf{D}}\left[-\sum_{m=n}^{N}\xi_m(\theta_0, \varphi_0)\Delta d_m \sec\theta_0 - \xi_n(\theta_0, \varphi_0)(z' + d_{n-1})\sec\theta_0\right]$$

$$\cdot \overline{\mathbf{E}}^{-1}(\theta_0, \varphi_0) \cdot \overline{\mathbf{R}}(\theta_0) \cdot \overline{\mathbf{E}}(\pi - \theta_0, \varphi_0) \cdot \overline{\mathbf{D}}\left[-\sum_{m=1}^{N}\xi_m(\pi - \theta_0, \varphi_0)\Delta d_m \sec\theta_0\right]$$

$$\cdot \overline{\mathbf{E}}^{-1}(\pi - \theta_0, \varphi_0) \cdot \mathbf{I}_0(\omega)\exp\left[-j\omega \frac{2d_{N+1}\sec\theta_0 + z'(\sec\theta_0 - \sec\theta)}{c_0}\right]\Bigg\}$$

$$+ \sum_{n=1}^{N}\Bigg\{\sec\theta\, \overline{\mathbf{E}}(\theta, \varphi) \cdot \overline{\mathbf{D}}\left[-\sum_{m=1}^{N}\xi_m(\theta, \varphi)\Delta d_m \sec\theta\right]$$

$$\cdot \overline{\mathbf{E}}^{-1}(\theta, \varphi) \cdot \overline{\mathbf{R}}(\theta) \cdot \overline{\mathbf{E}}(\pi - \theta, \varphi)$$

$$\cdot \overline{\mathbf{D}}\left[-\sum_{m=n}^{N}\xi_m(\pi - \theta, \varphi)\Delta d_m \sec\theta\right]\int_{-d_n}^{-d_{n-1}} dz'\, \overline{\mathbf{D}}\left[-\xi_n(\pi - \theta, \varphi)(z' + d_{n-1})\sec\theta\right]$$

$$\cdot \overline{\mathbf{E}}^{-1}(\pi - \theta, \varphi) \cdot \overline{\mathbf{P}}_n(\pi - \theta, \varphi; \pi - \theta_0, \varphi_0) \cdot \overline{\mathbf{E}}(\pi - \theta_0, \varphi_0)$$

$$\cdot \overline{\mathbf{D}}\left[-\sum_{m=1}^{n}\xi_m(\pi - \theta_0, \varphi_0)\Delta d_m \sec\theta_0 + \xi_n(\pi - \theta_0, \varphi_0)(z' + d_n)\sec\theta_0\right]$$

$$\cdot \overline{\mathbf{E}}^{-1}(\pi - \theta_0, \varphi_0) \cdot \mathbf{I}_0(\omega)\exp\left[-j\omega \frac{2d_{N+1}\sec\theta + z'(\sec\theta - \sec\theta_0)}{c_0}\right]\Bigg\}$$

$$+ \sum_{n=1}^{N}\Bigg\{\sec\theta\, \overline{\mathbf{E}}(\theta, \varphi) \cdot \overline{\mathbf{D}}\left[-\sum_{m=1}^{N}\xi_m(\theta, \varphi)\Delta d_m \sec\theta\right]$$

$$\cdot \overline{\mathbf{E}}^{-1}(\theta, \varphi) \cdot \overline{\mathbf{R}}(\theta) \cdot \overline{\mathbf{E}}(\pi - \theta, \varphi)$$

$$\cdot \overline{\mathbf{D}}\left[-\sum_{m=n}^{N}\xi_m(\pi - \theta, \varphi)\Delta d_m \sec\theta\right]\int_{-d_n}^{-d_{n-1}} dz'\, \overline{\mathbf{D}}[-\xi_n(\pi - \theta, \varphi)(z' + d_{n-1})\sec\theta]$$

$$\cdot \overline{\mathbf{E}}^{-1}(\pi - \theta, \varphi) \cdot \overline{\mathbf{P}}_n(\pi - \theta, \varphi; \theta_0, \varphi_0) \cdot \overline{\mathbf{E}}(\theta_0, \varphi_0)$$

$$\cdot \overline{\mathbf{D}}\left[-\sum_{m=n}^{N}\xi_m(\theta_0, \varphi_0)\Delta d_m \sec\theta_0 - \xi_n(\theta_0, \varphi_0)(z' + d_{n-1})\sec\theta_0\right]$$

$$\cdot \overline{\mathbf{E}}^{-1}(\theta_0, \varphi_0) \cdot \overline{\mathbf{R}}(\theta_0) \cdot \overline{\mathbf{E}}(\pi - \theta_0, \varphi_0) \cdot \overline{\mathbf{D}}\left[-\sum_{m=1}^{N}\xi_m(\pi - \theta_0, \varphi_0)\Delta d_m \sec\theta_0\right]$$

$$\cdot \overline{\mathbf{E}}^{-1}(\pi - \theta_0, \varphi_0) \cdot \mathbf{I}_0(\omega)\exp\left[-j\omega \frac{(z' + 2d_{N+1})(\sec\theta + \sec\theta_0)}{c_0}\right]\Bigg\}\Bigg\}$$

$$(10.10)$$

Taking the inverse Fourier transformation of $\mathbf{I}_1^{(1)}(\theta, \varphi, z = 0, \omega)$ in Equation 10.10, the first-order scattering intensity observed in layer 0 can be obtained as follows:

$$\mathbf{I}_1^{(1)}(\theta, \varphi, z = 0, t) = \sum_{n=1}^{N} \mathbf{I}_n^{PS} + \sum_{n=1}^{N} \mathbf{I}_n^{BP} + \sum_{n=1}^{N} \mathbf{I}_n^{PB} + \sum_{n=1}^{N} \mathbf{I}_n^{BPB} \qquad (10.11)$$

The four terms on the right-hand side of Equation 10.11 represent, respectively, the four scattering processes illustrated in Figure 10.2 (which also appeared in Equation 10.10 in the ω domain): $\mathbf{I}_n^{PS}$ describes the particle scattering (PS) in the nth layer; $\mathbf{I}_n^{BP}$ indicates a single boundary reflection and particle scattering (BP); $\mathbf{I}_n^{PB}$ takes account of the particle scattering and boundary reflection (PB); and $\mathbf{I}_n^{BPB}$ includes the boundary reflection, particle scattering, and boundary reflection again (BPB). These terms are explicitly formulated as follows:

$$\mathbf{I}_n^{PS}(\theta, \varphi, z = 0, t) = \sec \theta\, \overline{\mathbf{E}}(\theta, \varphi) \cdot \overline{\mathbf{D}}\left[-\sum_{m=1}^{n} \xi_m(\theta, \varphi)\Delta d_m \sec \theta\right]$$

$$\times \int_{-d_n}^{-d_{n-1}} dz'\, \overline{\mathbf{D}}\left[\xi_n(\theta, \varphi)(z' + d_n)\sec \theta\right]$$

$$\times \overline{\mathbf{E}}^{-1}(\theta, \varphi) \cdot \overline{\mathbf{P}}_n(\theta, \varphi; \pi - \theta_0, \varphi_0) \cdot \overline{\mathbf{E}}(\pi - \theta_0, \varphi_0)$$

$$\cdot \overline{\mathbf{D}}\left[-\sum_{m=1}^{n} \xi_m(\pi - \theta_0, \varphi_0)\Delta d_m \sec \theta_0 + \xi_n(\pi - \theta_0, \varphi_0)(z' + d_n)\sec \theta_0\right]$$

$$\cdot \overline{\mathbf{E}}^{-1}(\pi - \theta_0, \varphi_0) \cdot \mathbf{I}_0[t + z'(\sec \theta + \sec \theta_0)/c_0] \qquad (10.12a)$$

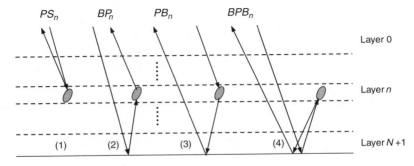

Figure 10.2 Four scattering processes in the respective layers

$$
\mathbf{I}_n^{BP}(\theta,\varphi,z=0,t) = \sec\theta\,\overline{\mathbf{E}}(\theta,\varphi)\cdot\overline{\mathbf{D}}\left[-\sum_{m=1}^{n}\xi_m(\theta,\varphi)\Delta d_m\sec\theta\right]\int_{-d_n}^{-d_{n-1}}dz'\,\overline{\mathbf{D}}\left[\xi_n(\theta,\varphi)(z'+d_n)\sec\theta\right]
$$
$$
\cdot\,\overline{\mathbf{E}}^{-1}(\theta,\varphi)\cdot\overline{\mathbf{P}}_n(\theta,\varphi;\theta_0,\varphi_0)\cdot\overline{\mathbf{E}}(\theta_0,\varphi_0)
$$
$$
\cdot\,\overline{\mathbf{D}}\left[-\sum_{m=n}^{N}\xi_m(\theta_0,\varphi_0)\Delta d_m\sec\theta_0-\xi_n(\theta_0,\varphi_0)(z'+d_{n-1})\sec\theta_0\right]
$$
$$
\cdot\,\overline{\mathbf{E}}^{-1}(\theta_0,\varphi_0)\cdot\overline{\mathbf{R}}(\theta_0)\cdot\overline{\mathbf{E}}(\pi-\theta,0\varphi_0)\cdot\overline{\mathbf{D}}\left[-\sum_{m=1}^{N}\xi_m(\pi-\theta_0,\varphi_0)\Delta d_m\sec\theta_0\right]
$$
$$
\cdot\,\overline{\mathbf{E}}^{-1}(\pi-\theta_0,\varphi_0)\cdot\mathbf{I}_0\{t-[2d_{N+1}\sec\theta_0+z'(\sec\theta_0-\sec\theta)]/c_0\}
$$

$$(10.12\mathrm{b})$$

$$
\mathbf{I}_n^{PB}(\theta,\varphi,z=0,t) = \sec\theta\,\overline{\mathbf{E}}(\theta,\varphi)\cdot\overline{\mathbf{D}}\left[-\sum_{m=1}^{N}\xi_m(\theta,\varphi)\Delta d_m\sec\theta\right]\cdot\overline{\mathbf{E}}^{-1}(\theta,\varphi)\cdot\overline{\mathbf{R}}(\theta)\cdot\overline{\mathbf{E}}(\pi-\theta,\varphi)
$$
$$
\cdot\,\overline{\mathbf{D}}\left[-\sum_{m=n}^{N}\xi_m(\pi-\theta,\varphi)\Delta d_m\sec\theta\right]\int_{-d_n}^{-d_{n-1}}dz'\,\overline{\mathbf{D}}[-\xi_n(\pi-\theta,\varphi)(z'+d_{n-1})\sec\theta]
$$
$$
\cdot\,\overline{\mathbf{E}}^{-1}(\pi-\theta,\varphi)\cdot\overline{\mathbf{P}}_n(\pi-\theta,\varphi;\pi-\theta_0,\varphi_0)\cdot\overline{\mathbf{E}}(\pi-\theta_0,\varphi_0)
$$
$$
\cdot\,\overline{\mathbf{D}}\left[-\sum_{m=1}^{n}\xi_m(\pi-\theta_0,\varphi_0)\Delta d_m\sec\theta_0+\xi_n(\pi-\theta_0,\varphi_0)(z'+d_n)\sec\theta_0\right]
$$
$$
\cdot\,\overline{\mathbf{E}}^{-1}(\pi-\theta_0,\varphi_0)\cdot\mathbf{I}_0\{t-[2d_{N+1}\sec\theta+z'(\sec\theta-\sec\theta_0)]/c_0\}
$$

$$(10.12\mathrm{c})$$

$$
\mathbf{I}_n^{BPB}(\theta,\varphi,z=0,t) = \sec\theta\,\overline{\mathbf{E}}(\theta,\varphi)\cdot\overline{\mathbf{D}}\left[-\sum_{m=1}^{N}\xi_m(\theta,\varphi)\Delta d_m\sec\theta\right]\cdot\overline{\mathbf{E}}^{-1}(\theta,\varphi)\cdot\overline{\mathbf{R}}(\theta)\cdot\overline{\mathbf{E}}(\pi-\theta,\varphi)
$$
$$
\cdot\,\overline{\mathbf{D}}\left[-\sum_{m=n}^{N}\xi_m(\pi-\theta,\varphi)\Delta d_m\sec\theta\right]\int_{-d_n}^{-d_{n-1}}dz'\,\overline{\mathbf{D}}[-\xi_n(\pi-\theta,\varphi)(z'+d_{n-1})\sec\theta]
$$
$$
\cdot\,\overline{\mathbf{E}}^{-1}(\pi-\theta,\varphi)\cdot\overline{\mathbf{P}}_n(\pi-\theta,\varphi;\theta_0,\varphi_0)\cdot\overline{\mathbf{E}}(\theta_0,\varphi_0)
$$
$$
\cdot\,\overline{\mathbf{D}}\left[-\sum_{m=n}^{N}\xi_m(\theta_0,\varphi_0)\Delta d_m\sec\theta_0-\xi_n(\theta_0,\varphi_0)(z'+d_{n-1})\sec\theta_0\right]
$$
$$
\cdot\,\overline{\mathbf{E}}^{-1}(\theta_0,\varphi_0)\cdot\overline{\mathbf{P}}(\theta_0)\cdot\overline{\mathbf{E}}(\pi-\theta_0,\varphi_0)\cdot\overline{\mathbf{D}}\left[-\sum_{m=1}^{N}\xi_m(\pi-\theta_0,\varphi_0)\Delta d_m\sec\theta_0\right]
$$
$$
\cdot\,\overline{\mathbf{E}}^{-1}(\pi-\theta_0,\varphi_0)\cdot\mathbf{I}_0[t-(z'+2d_{N+1})(\sec\theta+\sec\theta_0)/c_0]
$$

$$(10.12\mathrm{d})$$

where $\Delta d_m = d_m - d_{m-1}$ denotes the thickness of the mth layer, $m = 1, 2, \ldots, N$, and $d_0 = 0$.

10.2 Time-Dependent Mueller Matrix for Inhomogeneous Random Media

Let the incident plane pulse be a Gaussian with polarization (χ,ψ) as follows:

$$
\mathbf{I}_0(t) = I_0(t)\mathbf{I}_{i0}(\chi,\psi) \tag{10.13}
$$

where

$$I_0(t) = \frac{1}{\sqrt{2\pi}\sigma} \exp\left(-\frac{t^2}{2\sigma^2}\right) \tag{10.14}$$

with $\sigma = T_w/\sqrt{2\ln 2}$, and $2T_w$ is the half-power pulse width. In Equation 10.13, $\mathbf{I}_{i0}(\chi, \psi)$ is the normalized incident Stokes vector given by

$$\mathbf{I}_{i0} = [(1 - \cos 2\chi\cos 2\psi)/2, \quad (1 + \cos 2\chi \cos 2\psi)/2, \quad -\cos 2\chi \sin 2\psi, \quad \sin 2\chi]^{\mathrm{T}} \tag{10.15}$$

where $\chi \in [-45°, 45°]$ and $\psi \in [0°, 180°]$ are the elliptic and orientation angles of the incident polarization, respectively. When $\chi = 0°$ and $\psi = 0°$ or $180°$, this denotes a horizontally polarized wave; when $\chi = 0°$ and $\psi = 90°$, it is vertically polarized; and when $\chi = \pm 45°$, it is circularly polarized. All other cases are elliptic polarization.

Substituting Equations (10.13)–(10.15) into Equation 10.11 and using the integrand (Abramowitz and Stegun, 1972, p. 303)

$$\int dx \exp(-ax^2 - 2bx - c) = \frac{1}{2}\sqrt{\frac{\pi}{a}}\exp\left(\frac{b^2 - ac}{a}\right)\mathrm{erf}\left(\sqrt{a}x + \frac{b}{\sqrt{a}}\right) + \mathrm{const.} \tag{10.16}$$

the first Mueller matrix solution is obtained as

$$\mathbf{I}_s(\theta, \varphi, z = 0, t) = \overline{\mathbf{M}}(\theta, \varphi; \pi - \theta_0, \varphi_0; t) \cdot \mathbf{I}_{i0}(\chi, \psi) \tag{10.17}$$

where the time-dependent $\overline{\mathbf{M}}(\theta, \varphi; \pi - \theta_0, \varphi_0; t)$ includes the contributions from all N layers of scatterers and takes into account the four scattering processes. Its ℓjth elements $(\ell, j = 1, \ldots, 4)$ are derived as

$$M_{\ell j}(\theta, \varphi; \pi - \theta_0, \varphi_0; t) = \sum_{n=1}^{N} PS_n + \sum_{n=1}^{N} BP_n + \sum_{n=1}^{N} PB_n + \sum_{n=1}^{N} BPB_n \tag{10.18}$$

where the functions PS_n, BP_n, PB_n, and BPB_n are contributed by the scattering processes described in Figure 10.2:

$$PS_n = \sec\theta \sum_{k,i} E_{\ell k}(\theta, \varphi) \cdot \left\{ \overline{\mathbf{D}}\left[-\sum_{m=1}^{n} \xi_m(\theta, \varphi)\Delta d_m \sec\theta + \xi_n(\theta, \varphi)d_n \sec\theta \right] \right.$$

$$\cdot \overline{\mathbf{E}}^{-1}(\theta, \varphi) \cdot \overline{\mathbf{P}}_n(\theta, \varphi; \pi - \theta_0, \varphi_0) \cdot \overline{\mathbf{E}}(\pi - \theta_0, \varphi_0)$$

$$\cdot \overline{\mathbf{D}}\left[-\sum_{m=1}^{n} \xi_m(\pi - \theta_0, \varphi_0)\Delta d_m \sec\theta_0 + \xi_n(\pi - \theta_0, \varphi_0)d_n \sec\theta_0 \right]_{ki}$$

$$\times \frac{1}{2R_1}\left\{ \exp\left[\frac{2\sigma^2 B_{n1}^2}{R_1^2} - C_1\right] \times \left[\mathrm{erf}\left(\frac{\sqrt{2}\sigma B_{n1}}{R_1} - \frac{R_1 d_{n-1}}{\sqrt{2}\sigma}\right) - \mathrm{erf}\left(\frac{\sqrt{2}\sigma B_{n1}}{R_1} - \frac{R_1 d_n}{\sqrt{2}\sigma}\right)\right] \right\}_{ki}$$

$$\cdot \left\{\overline{\mathbf{E}}^{-1}(\pi - \theta_0, \varphi_0)\right\}_{ij}$$

$$\tag{10.19a}$$

$$BP_n = \sec\theta \sum_{k,i} E_{\ell k}(\theta,\varphi) \cdot \left\{ \overline{\mathbf{D}}\left[-\sum_{m=1}^{n} \xi_m(\theta,\varphi)\Delta d_m \sec\theta + \xi_n(\theta,\varphi)d_n \sec\theta \right] \right.$$

$$\cdot \overline{\mathbf{E}}^{-1}(\theta,\varphi) \cdot \overline{\mathbf{P}}_n(\theta,\varphi;\theta_0,\varphi_0) \cdot \overline{\mathbf{E}}(\theta_0,\varphi_0)$$

$$\cdot \overline{\mathbf{D}}\left[-\sum_{m=n}^{N} \xi_m(\theta_0,\varphi_0)\Delta d_m \sec\theta_0 - \xi_n(\theta_0,\varphi_0)d_{n-1} \sec\theta_0 \right]\bigg\}_{ki}$$

$$\times \frac{1}{2|R_2|} \left\{ \exp\left[\frac{2\sigma^2 B_{n2}^2}{R_2^2} - C_2 \right] \times \left[\text{erf}\left(\frac{\sqrt{2}\sigma B_{n2}}{|R_2|} - \frac{|R_2|d_{n-1}}{\sqrt{2}\sigma} \right) - \text{erf}\left(\frac{\sqrt{2}\sigma B_{n2}}{|R_2|} - \frac{|R_2|d_n}{\sqrt{2}\sigma} \right) \right] \right\}_{ki}$$

$$\cdot \left\{ \overline{\mathbf{E}}^{-1}(\theta_0,\varphi_0) \cdot \overline{\mathbf{R}}(\theta_0) \cdot \overline{\mathbf{E}}(\pi-\theta_0,\varphi_0) \right.$$

$$\cdot \overline{\mathbf{D}}\left[-\sum_{m=1}^{N} \xi_m(\pi-\theta_0,\varphi_0)\Delta d_m \sec\theta_0 \right] \cdot \overline{\mathbf{E}}^{-1}(\pi-\theta_0,\varphi_0) \bigg\}_{ij} \tag{10.19b}$$

$$PB_n = \sec\theta \sum_{k,i} \left\{ \overline{\mathbf{E}}(\theta,\varphi) \cdot \overline{\mathbf{D}}\left[-\sum_{m=1}^{N} \xi_m(\theta,\varphi)\Delta d_m \sec\theta \right] \cdot \overline{\mathbf{E}}^{-1}(\theta,\varphi) \cdot \overline{\mathbf{R}}(\theta) \cdot \overline{\mathbf{E}}(\pi-\theta,\varphi) \right\}_{\ell k}$$

$$\cdot \left\{ \overline{\mathbf{D}}\left[-\sum_{m=n}^{N} \xi_m(\pi-\theta,\varphi)\Delta d_m \sec\theta - \xi_n(\pi-\theta,\varphi)d_{n-1} \sec\theta \right] \right.$$

$$\cdot \overline{\mathbf{E}}^{-1}(\pi-\theta,\varphi) \cdot \overline{\mathbf{P}}_n(\pi-\theta,\varphi;\pi-\theta_0,\varphi_0) \cdot \overline{\mathbf{E}}(\pi-\theta_0,\varphi_0)$$

$$\cdot \overline{\mathbf{D}}\left[-\sum_{m=1}^{n} \xi_m(\pi-\theta_0,\varphi_0)\Delta d_m \sec\theta_0 + \xi_n(\pi-\theta_0,\varphi_0)d_n \sec\theta_0 \right]\bigg\}_{ki}$$

$$\times \frac{1}{2|R_2|} \left\{ \exp\left[\frac{2\sigma^2 B_{n3}^2}{R_2^2} - C_3 \right] \times \left[\text{erf}\left(\frac{\sqrt{2}\sigma B_{n3}}{|R_2|} - \frac{|R_2|d_{n-1}}{\sqrt{2}\sigma} \right) - \text{erf}\left(\frac{\sqrt{2}\sigma B_{n3}}{|R_2|} - \frac{|R_2|d_n}{\sqrt{2}\sigma} \right) \right] \right\}_{ki}$$

$$\cdot \left\{ \overline{\mathbf{E}}^{-1}(\pi-\theta_0,\varphi_0) \right\}_{ij} \tag{10.19c}$$

$$BPB_n = \sec\theta \sum_{k,i} \left\{ \overline{\mathbf{E}}(\theta,\varphi) \cdot \overline{\mathbf{D}}\left[-\sum_{m=1}^{N} \xi_m(\theta,\varphi)\Delta d_m \sec\theta \right] \cdot \overline{\mathbf{E}}^{-1}(\theta,\varphi) \cdot \overline{\mathbf{R}}(\theta) \cdot \overline{\mathbf{E}}(\pi-\theta,\varphi) \right\}_{\ell k}$$

$$\cdot \left\{ \overline{\mathbf{D}}\left[-\sum_{m=n}^{N} \xi_m(\pi-\theta,\varphi)\Delta d_m \sec\theta - \xi_n(\pi-\theta,\varphi)d_{n-1} \sec\theta \right] \right.$$

$$\cdot \overline{\mathbf{E}}^{-1}(\pi-\theta,\varphi) \cdot \overline{\mathbf{P}}_n(\pi-\theta,\varphi;\theta_0,\varphi_0) \cdot \overline{\mathbf{E}}(\theta_0,\varphi_0)$$

$$\cdot \overline{\mathbf{D}}\left[-\sum_{m=n}^{N} \xi_m(\theta_0,\varphi_0)\Delta d_m \sec\theta_0 - \xi_n(\theta_0,\varphi_0)d_{n-1} \sec\theta_0 \right]\bigg\}_{ki}$$

$$\times \frac{1}{2R_1} \left\{ \exp\left[\frac{2\sigma^2 B_{n4}^2}{R_1^2} - C_4\right] \times \left[\mathrm{erf}\left(\frac{\sqrt{2}\sigma B_{n4}}{R_1} - \frac{R_1 d_{n-1}}{\sqrt{2}\sigma}\right) - \mathrm{erf}\left(\frac{\sqrt{2}\sigma B_{n4}}{R_1} - \frac{R_1 d_n}{\sqrt{2}\sigma}\right)\right]\right\}_{ki}$$

$$\cdot \left\{ \overline{\mathbf{E}}^{-1}(\theta_0, \varphi_0) \cdot \overline{\mathbf{R}}(\theta_0)\overline{\mathbf{E}}(\pi - \theta_0, \varphi_0)\right.$$

$$\left.\cdot \overline{\mathbf{D}}\left[-\sum_{m=1}^{N} \xi_m(\pi - \theta_0, \varphi_0)\Delta d_m \sec\theta_0\right]\overline{\mathbf{E}}^{-1}(\pi - \theta_0, \varphi_0)\right\}_{ij}$$

$$(10.19\mathrm{d})$$

where i, j, ℓ, k take values 1,2,3,4. In these equations, $\mathrm{erf}(\cdot)$ is the error function. The eigen-matrix $\overline{\mathbf{E}}$, its inverse $\overline{\mathbf{E}}^{-1}$, and the diagonal matrix $\overline{\mathbf{D}}$ are related to the forward scattering amplitude functions of the scatterers. The phase matrices $\overline{\mathbf{P}}_n$ are constructed from functions of $N_{0n}\langle f_{pq}f_{st}^*\rangle$, where $p, q, s, t = v, h$, and N_{0n} is the number of particles per unit volume in the nth layer. The functions f_{pq} and f_{st}^* indicate the scattering amplitude functions of the particles and its conjugation, and the angular brackets $\langle\cdot\rangle$ denote the averaging of Euler angles orientation. The explicit expressions of $\overline{\mathbf{E}}$, $\overline{\mathbf{D}}$, and $\overline{\mathbf{P}}$ of small particles can be found in Tsang, Kong, and Ding (2000, pp. 16, 39, 40). Finally, the functions in Equations 10.19a–10.19d, are denoted as follows:

$$R_1 = (\sec\theta + \sec\theta_0)/c_0 \tag{10.20a}$$

$$R_2 = (\sec\theta - \sec\theta_0)/c_0 \tag{10.20b}$$

$$B_{n1} = [R_1 t/\sigma^2 - \xi_{nk}(\theta, \varphi)\sec\theta - \xi_{ni}(\pi - \theta_0, \varphi_0)\sec\theta_0]/2 \tag{10.21a}$$

$$B_{n2} = [R_2(t - 2d_{N+1}\sec\theta_0/c_0)/\sigma^2 - \xi_{nk}(\theta, \varphi)\sec\theta + \xi_{ni}(\theta_0, \varphi_0)\sec\theta_0]/2 \tag{10.21b}$$

$$B_{n3} = -[R_2(t - 2d_{N+1}\sec\theta/c_0)/\sigma^2 - \xi_{nk}(\pi - \theta, \varphi)\sec\theta + \xi_{ni}(\pi - \theta_0, \varphi_0)\sec\theta_0]/2 \tag{10.21c}$$

$$B_{n4} = -[R_1(t - 2d_{N+1}R_1)/\sigma^2 - \xi_{nk}(\pi - \theta, \varphi)\sec\theta - \xi_{ni}(\theta_0, \varphi_0)\sec\theta_0]/2 \tag{10.21d}$$

$$C_1 = t^2/(2\sigma^2) \tag{10.21e}$$

$$C_2 = (t - 2d_{N+1}\sec\theta_0/c_0)^2/(2\sigma^2) \tag{10.21f}$$

$$C_3 = (t - 2d_{N+1}\sec\theta/c_0)^2/(2\sigma^2) \tag{10.21g}$$

$$C_4 = (t - 2d_{N+1}R_1)^2/(2\sigma^2) \tag{10.21h}$$

As the scattering angle $\theta \to \theta_0$ (e.g., in the specular or backward direction), it yields $R_2 \to 0$, and the related terms of Equations 10.19b and 10.19c have to be rewritten as (see Appendix 10.A)

$$\lim_{|R_2|\to 0} \frac{1}{2|R_2|} \exp\left[\frac{2\sigma^2 B_{n2}^2}{R_2^2} - C_2\right] \times \left[\mathrm{erf}\left(\frac{\sqrt{2}\sigma B_{n2}}{|R_2|} - \frac{|R_2|d_{n-1}}{\sqrt{2}\sigma}\right) - \mathrm{erf}\left(\frac{\sqrt{2}\sigma B_{n2}}{|R_2|} - \frac{|R_2|d_n}{\sqrt{2}\sigma}\right)\right]$$

$$= \frac{1}{\sqrt{2\pi}\sigma} \exp\left[-\frac{(t - 2d_{N+1}\sec\theta_0/c_0)^2}{2\sigma^2}\right] \times \{\exp\{-d_{n-1}[\xi_{nk}(\theta,\varphi)\sec\theta - \xi_{ni}(\theta_0,\varphi_0)\sec\theta_0]\}$$

$$-\frac{\exp\{-d_n[\xi_{nk}(\theta,\varphi)\sec\theta - \xi_{ni}(\theta_0,\varphi_0)\sec\theta_0]\}}{\xi_{nk}(\theta,\varphi)\sec\theta - \xi_{ni}(\theta_0,\varphi_0)\sec\theta_0}\}$$

$$(10.22a)$$

$$\lim_{|R_2|\to 0} \frac{1}{2|R_2|} \exp\left[\frac{2\sigma^2 B_{n3}^2}{R_2^2} - C_3\right] \times \left[\mathrm{erf}\left(\frac{\sqrt{2}\sigma B_{n3}}{|R_2|} - \frac{|R_2|d_{n-1}}{\sqrt{2}\sigma}\right) - \mathrm{erf}\left(\frac{\sqrt{2}\sigma B_{n3}}{|R_2|} - \frac{|R_2|d_n}{\sqrt{2}\sigma}\right)\right]$$

$$= \frac{1}{\sqrt{2\pi}\sigma} \exp\left[-\frac{(t - 2d_{N+1}\sec\theta/c_0)^2}{2\sigma^2}\right]$$

$$\times \{\exp\{-d_{n-1}[\xi_{ni}(\pi - \theta_0,\varphi_0)\sec\theta_0 - \xi_{nk}(\pi - \theta,\varphi)\sec\theta]\}$$

$$-\frac{\exp\{-d_n[\xi_{ni}(\pi - \theta_0,\varphi_0)\sec\theta_0 - \xi_{nk}(\pi - \theta,\varphi)\sec\theta]\}}{\xi_{ni}(\pi - \theta_0,\varphi_0)\sec\theta_0 - \xi_{nk}(\pi - \theta,\varphi)\sec\theta}\}$$

$$(10.22b)$$

As $t \to 0$, $c_0 \to \infty$ (i.e., a plane pulse becomes a continuous plane wave), and $N = 1$, $d_{N+1} = d_N$ (i.e., inhomogeneous random media degenerate to the homogeneous case), the temporal Mueller matrix of Equation 10.18 can return to the conventional Mueller matrix in Tsang, Kong, and Ding (2000, p. 126) (see Appendix 10.A).

10.3 Polarimetric Bistatic and Backscattering Pulse Responses

Using the temporal Mueller matrix solution, polarimetric bistatic and backscattering from multi-layered random media of non-spherical scatterers under a Gaussian plane pulse incidence can be numerically simulated. The averaging result of many realizations in real field measurements statistically presents the scattered Stokes parameters. In our simulations of the Mueller solution, only scattering from the terrain canopy is taken into account – the propagation and fading effect through the turbid atmosphere are not yet discussed.

The polarized bistatic scattering coefficient is defined as

$$\gamma_{pq}(\theta_s,\varphi_s;\pi - \theta_0,\varphi_0) = 4\pi\frac{\cos\theta_s\, I_{sp}(\theta_s,\varphi_s)}{\cos\theta_0\, I_{i0q}}, \quad p,q = v,h \tag{10.23}$$

where I_{i0q} is the q-polarized incident intensity, and I_{sp} is the p-polarized scattered intensity. The polarized backscattering coefficient is defined as

$$\sigma_{pq}(\theta_0,\pi + \varphi_0;\pi - \theta_0,\varphi_0) = \cos\theta_0\,\gamma_{pq}(\theta_0,\pi + \varphi_0;\pi - \theta_0,\varphi_0)$$

$$= 4\pi\cos\theta_0\frac{I_{sp}(\theta_0,\pi + \varphi_0)}{I_{i0q}} \tag{10.24}$$

In the following calculations, the pulse echoes from four-layer media (i.e., $N+1=4$, and layer 4 is an air region) of random disk-like particles are numerically simulated. The carrier frequency is 5.3 GHz, and the Gaussian pulse has a half-power width 2.0 ns. For convenient programming, all particles in the media are simply assumed to be identical in shape and size (radius 1.0 cm, and thickness 0.025 cm) with the dielectric constant $(14.1+j4.5)\varepsilon_0$ and uniform spatial orientation with Euler angles $\beta \in (0°, 180°)$, $\gamma \in (0°, 360°)$. In fact, the scattering particles in the layered media can have different shapes and sizes, and can be spatially oriented with a non-uniform orientation distribution or around a particular direction. In such cases, it is believed that the polarimetric scattering from the media under a pulse incidence can be of great help in identifying the particular locations of those scattering particles.

The dielectric constant of the underlying medium ε_b is assumed as $(6+i0.5)\varepsilon_0$. The incident pulse is horizontally polarized ($\chi = 0°$, $\psi = 0°$) and $\theta_0 = 0°$, $\varphi_0 = 0°$. The particles are sparsely distributed and their sizes are small in the low-frequency limit. The generalized Rayleigh–Gans approximation is applied to derive the scattering amplitude functions of random small particles (Jin, 1994).

In the following figures, the solid curves denote the σ_c of inhomogeneous three-layer media and the dashed curves are the σ_c from an overall homogeneous layer of random disks (i.e., $N=1$). The dotted lines indicate the fractional volume profiles of random disks. To show the shape of the pulse echoes and compare with the fraction profile in the same figure, the time delay $t(s)$ of the echoes is transformed by the height as

$$\text{height (m)} = d_4 \text{ (m)} - t \text{ (s)} \times c_0 \text{ (m s}^{-1})/2 \tag{10.25}$$

The fractional volume of an overall homogeneous layer is averaged as

$$f_s = \sum_{n=1}^{N}(d_n - d_{n-1})f_{sn}/d_N \tag{10.26}$$

Figure 10.3a–d simulate temporal σ_c versus height with comparison of four different layering profiles of f_{s1}, f_{s2}, f_{s3}: (a) 0.0015, 0.0010, 0.0005; (b) 0.0010, 0.0015, 0.0005; (c) 0.0005, 0.0015, 0.0010; and (d) 0.0005, 0.0010, 0.0015, respectively. The layer depths are $d_1 = 0.5$ m, $d_2 = 1.0$ m, $d_3 = 1.5$ m, and $d_4 = 2.5$ m, respectively, that is, total layer thickness is 2.5 m.

It can be seen that the patterns of pulse echoes with height depict well the vertical fraction profiles of random disks. The separate peak comes from the reflection of the underlying surface. The return from the underlying surface can be strong due to the sparse scatterers and the thin layer depth. If the fractional volumes become larger, the peak due to the bottom reflection would be shadowed. The results for the cross-polarized case are similar, but the values are far smaller than those of the co-polarized case. Pulse distortion is due to multipath interaction described by PS_n, BP_n, PB_n, and BPB_n in Equations 10.18 and (10.19a), (10.19b),(10.19c),(10.19d), which certainly depend upon the parameters of the media, not only the fraction profile, but also the shape, size, and orientation of the scattering particles, dielectric properties, layer thickness, underlying surface, and so on.

Figure 10.4a–d show σ_c versus height for total layer thickness 5.0 m, where there are four different fraction profiles (corresponding to Figure 10.3a–d and $d_1 = 1.0$ m,

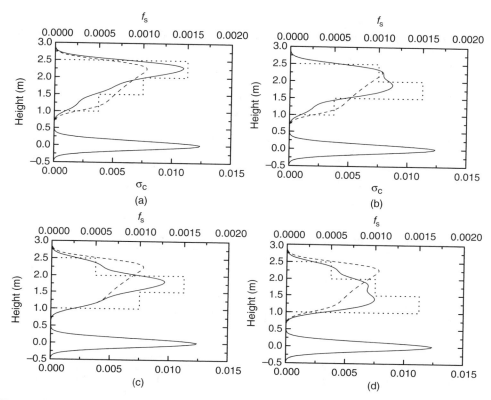

Figure 10.3 Plots of σ_c versus height with indication of layering fraction profile of random disks ($N = 3$, $d_1 = 0.5$ m, $d_2 = 1.0$ m, $d_3 = 1.5$ m, $d_4 = 2.5$ m, that is, total thickness = 2.5 m)

$d_2 = 2.0$ m, $d_3 = 3.0$ m, and $d_4 = 5.0$ m, respectively. Compared with Figure 10.3, the pulse echoes shift to the top layer due to the presence of more scatterers in the top crown, and the greater number of scatterers can darken the bottom returns.

Figure 10.5 presents the σ_c for a total thickness 10 m where $d_1 = 2.0$ m, $d_2 = 4.0$ m, $d_3 = 6.0$ m, and $d_4 = 10.0$ m, respectively. The fractional volumes of Figure 10.5a–d also correspond to Figure 10.3a–d, respectively. It can be seen that the pulse echoes are mainly contributed by the scatterers in the top crown and the bottom return is significantly shadowed.

Figure 10.6 shows the σ_c for different incident angles $\theta_0 = 0°$, $40°$, $60°$, respectively. All the parameters are the same as in Figure 10.4d (total layer thickness = 5.0 m). It can be seen that, as θ_0 increases, the layered media become thicker for wave penetration and the bottom return becomes weaker. The time delay difference Δt between the bottom returns for different incident angles θ_{10}, θ_{20} can be derived as a function of the total layer thickness d:

$$\Delta t = t_2 - t_1 = \frac{2d(\sec \theta_{20} - \sec \theta_{10})}{c_0} \tag{10.27}$$

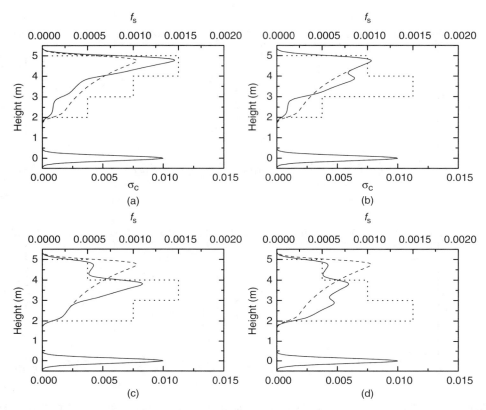

Figure 10.4 Plots of σ_c versus height with indication of layering fraction profile of random disks ($N = 3$, $d_1 = 1.0$ m, $d_2 = 2.0$ m, $d_3 = 3.0$ m, $d_4 = 5.0$ m, that is, total thickness = 5.0 m). Reprinted from *Journal of Quantitative Spectroscopy and Radiative Transfer*, with permission from Elsevier Science

In Figure 10.6, $t_1 = 43.6$ ns for $\theta_{10} = 40°$ and $t_2 = 66.6$ ns for $\theta_{20} = 60°$. It is interesting to see that Equation 10.27 can be inverted to give $d = 4.964$ m to 5 m.

Figure 10.7 presents the pulse echoes of bistatic scattering at $\theta_s = 0°$, $40°$, $60°$ with incident $\theta_0 = 0°$. All the parameters are the same as in Figure 10.4d. As θ_s becomes larger, the pulse shape is broadened since the media seem thicker for wave penetration. It is noted that the bistatic scattering shows two peak tails due to the interactions of random disks and underlying surface. The first tail is due to single reflection from the surface to particles and scattering. The second tail results from scattering of the particles to the surface and reflection. In backscattering, these two scattering processes take the same path and have the same time delay. So only one peak tail appears as $\theta_s = \theta_0$ ($= 0°$) in Figure 10.7.

Figure 10.8a and b show the co-pol and cross-pol backscattering coefficients σ_c, σ_x versus (χ, t), respectively, with the orientation angle of the incident polarization $\psi = 0°$. All the parameters are the same as in Figure 10.5d. It demonstrates temporal variations of the σ_c and σ_x for different incidence polarization (e.g., horizontal $\chi = 0°$, circular $\chi = \pm 45°$, and elliptic $\chi =$ others).

Figure 10.9a–c show numerical results of co-pol backscattering coefficients, σ_{hh} and σ_{vv}, and cross-pol backscattering, σ_{hv}, from four layers of random disk-like particles

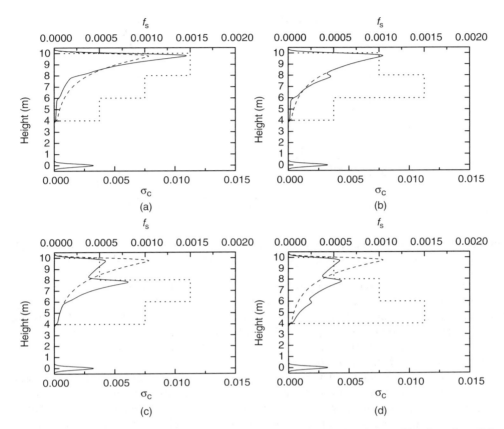

Figure 10.5 Plots of σ_c versus height with indication of layering fraction profile of random disks ($N = 3$, $d_1 = 2.0$ m, $d_2 = 4.0$ m, $d_3 = 6.0$ m, $d_4 = 10.0$ m, that is, total thickness $= 10.0$ m). Reprinted from *Journal of Quantitative Spectroscopy and Radiative Transfer*, with permission from Elsevier Science

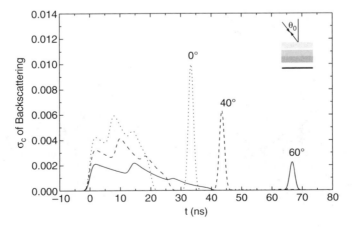

Figure 10.6 Temporal variation of backscattering σ_c for different incidence θ_0 $(0°, 40°, 60°)$. © [2004] IEEE. Reprinted, with permission, from *IEEE Transactions on Geoscience and Remote Sensing*

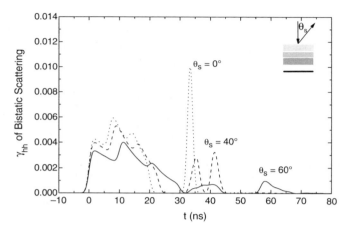

Figure 10.7   Temporal variation of bistatic scattering ($\theta_s = 0°, 40°, 60°$) coefficients γ_{hh} at $\theta_0 = 0°$. © [2004] IEEE. Reprinted, with permission, from *IEEE Transactions on Geoscience and Remote Sensing*

(i.e., $N + 1 = 4$, and layer 4 is an air region). Taking $m = 15$ for the height interval, the effective heights for three different fraction profiles are calculated. The total canopy thickness is 5 m, with three different fraction profiles of f_{s1}, f_{s2}, f_{s3}: (1) 0.004, 0.002, 0.001; (2) 0.002, 0.004, 0.001; and (3) 0.001, 0.002, 0.004, respectively. The layer thicknesses are $\Delta d_1 = 1.0$ m, $\Delta d_{21} = 1.0$ m, $\Delta d_3 = 1.0$ m, and $\Delta d_4 = 2.0$ m, respectively, that is, total layer thickness is $d_{N+1} = 5$ m.

It can be seen that temporal variation of the echo intensity depicts well the different inhomogeneous fraction profiles. Meanwhile, we can see that the effective heights are related to the respective fraction profiles and are all lower than the real heights (5 m). Different canopies can have different effective heights because an electromagnetic wave propagating through random media of sparsely distributed scatterers will have different scattering, absorption, and extinction coefficients. So the depth of penetration and the temporal pattern of the echoes are both changed.

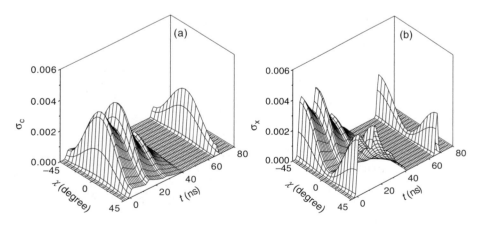

Figure 10.8   Plots of (a) σ_c and (b) σ_x versus (χ, t) at an orientation angle of $\psi = 0°$. © [2004] IEEE. Reprinted, with permission, from *IEEE Transactions on Geoscience and Remote Sensing*

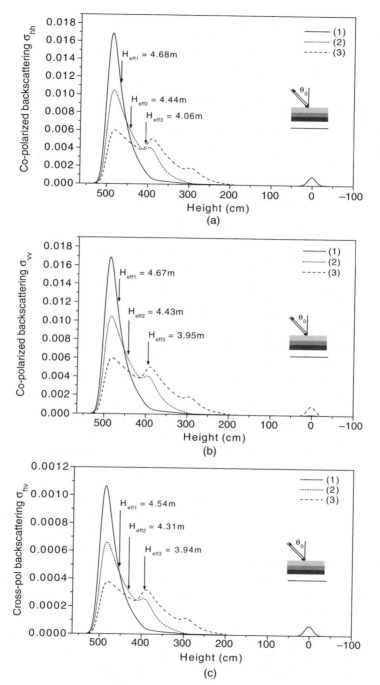

Figure 10.9 The effective heights of the vegetation canopy with different fraction profiles: inversion of the effective height from (a) σ_{hh}, (b) σ_{vv}, and (c) σ_{hv}. © [2004] IEEE. Reprinted, with permission, from *IEEE Transactions on Geoscience and Remote Sensing*

Co-pol and cross-pol backscattering present very similar effective heights from Figure 10.9a–c. Generally, linear co-pol σ_{hh} or σ_{vv} is always much larger than cross-pol σ_{hv}. Which one, σ_{hh} or σ_{vv}, is more significant depends on the shape of the scatterers, for example, whether they are disks or needles.

In our model of Figure 10.1, the underlying surface is simply smooth ground, and there is no direct ground–surface scattering in the VRT solutions, Equations 10.11 and 10.12a (in the next section, the model of a rough surface will be treated). However, even though a rough surface might, to a greater or lesser degree, diffuse or distort the final scattering echo pattern, the contribution from direct ground scattering can be identified and would not cause an essential difference from the above approach to invert the effective canopy height.

10.4 Pulse Echoes from Lunar Regolith Layer

The lunar regolith layer, the uppermost layer of the Moon's surface, preserves the geological history of the Moon, and knowledge of its structure, composition, and distribution might provide important information concerning lunar geology and resources for future lunar exploration. The regolith layer thickness varies from several meters to tens of meters. Low-frequency microwaves can penetrate through the regolith to considerable depth.

In the 1980s, co-pol and depol radar echoes from the lunar nearside were observed using the 430 MHz radar at Arecibo Observatory, Puerto Rico (Thompson, 1987). Some radar and optical data have been utilized for mapping lunar regolith layer thickness for the lunar nearside (Shkuratov and Bondarenko, 2001; Bondarenko and Shkuratov, 1998). Dual-polarized ground-based radars at 2.38 GHz and with 125 m resolution have also been utilized in the search for ice deposits in those shadowed areas of the lunar poles (Stacyn, Campbell, and Ford, 1997). During the Clementine-1 mission a bistatic radar experiment was also carried out for observation of the lunar poles at 2.273 GHz (Nozette *et al.*, 1996). All of these investigations demonstrated great utility of lunar surface probing by using radar technology.

In order to explore the potential utilities of lower-frequency (L band) polarimetric pulse radar sensing for exploration of the Moon, a theoretical model of a stratified lunar regolith media with scattering inhomogeneity and rough interfaces under polarimetric radar pulse incidence (Jin, Chen, and Chang, 2004) is developed in our study. Temporal echoes might display the regolith structure profile (Jin, 2005; Jin, Chen, and Chang, 2004, Jin and Chen, 2003). The time-domain Mueller matrix solution is applied to numerical simulation of polarimetric radar pulse echoes from stratified random scattering media. In this section, the Mueller matrix solution is derived and found to contain seven scattering mechanisms of the stratified media: surface scattering from the rough top and bottom interfaces, respectively, volumetric scattering from random stone scatterers, and their multiple interactions.

Model parameters are set according to the study of the lunar regolith structure (Fa and Jin, 2006; Shkuratov and Bondarenko, 2001; Bondarenko and Shkuratov, 1998). The temporal characteristics and structure of the polarimetric echo profile are analyzed. It displays a prominent peak due to the top boundary scattering and a complex tail due to the bottom interface and stone volumetric scattering. The contributions of different scattering mechanisms, that is, surface, volumetric scattering and interactions, are analyzed, and their dependence on model parameters such as the layer thickness and the content of $FeO + TiO_2$ are also discussed. Simulation of the pulse echoes displays an image of subsurface structures. It reveals information about the depth and other properties of the lunar regolith layer.

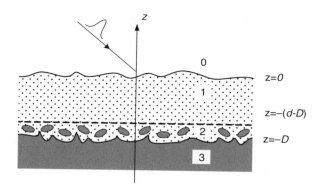

Figure 10.10 A model of the lunar regolith layer. Reprinted from *Radio Science*, from AGU

This approach has also been applied to monitoring and early warning of debris flows and landslides based on high-frequency radar echoes from multi-layered and very wet scattering media (Jin and Xu, 2011).

As shown in Figure 10.10, the lunar regolith layer is a lossy uniform medium (1) with thickness D, into which a layer of stone/rock scatterers (random oblate spheroids) with thickness d is embedded (2), which both overlay the lunar rocky media (3). The interfaces of the regolith layer (at $z = 0$ and $z = -D$) are randomly rough and tilted surfaces.

10.4.1 Mueller Matrix Solution for Seven Scattering Mechanisms

As a polarized pulse $\mathbf{I}_0(t)$ is incident upon and through the top interface at $z = 0$, the up-going scattered Stokes vector in medium (1) is written as

$$\mathbf{I}(t, \Theta) = \overline{\mathbf{R}}_{01}(\Theta; \Omega_i) \cdot \mathbf{I}_0(t) + \int d\Theta' \cdot \overline{\mathbf{T}}_{10}(\Theta; \Theta') \cdot \exp(-\tau_1/\mu') \cdot \int d\Omega'' \, \overline{\mathbf{M}}(t - t_1', \Theta'; \Omega'')$$
$$\cdot \exp(-\tau_1/\mu'') \cdot \overline{\mathbf{T}}_{01}(\Omega''; \Omega_i) \cdot \mathbf{I}_0(t - t_1' - t_1'')$$

$$(10.28)$$

Here, the first term on the right-hand side is the scattering from the top rough surface, where $\mathbf{R}_{01}$ is the surface scattering matrix (Fung, 1994). The second term is the Mueller matrix to describe the scattering and transmission of the pulse in the media 1 and 2 (including two rough interfaces at the top and bottom). In the second term, $\overline{\mathbf{T}}_{01}$, $\overline{\mathbf{T}}_{10}$ are the transmission matrices. The subscripts 01 or 10 indicate "from layer 0 to layer 1", or "from layer 1 to layer 0", respectively. The optical depth is denoted by $\tau_1 = \kappa_1(D - d)$, κ_1 is the absorption coefficient of the layer 1, $\mu = \cos\theta$, $\mu' = \cos\theta'$, and $\mu'' = \cos\theta''$ are defined, and $t_1 = (D - d)/(\mu c_0)$, $t_1' = (D - d)/(\mu' c_0)$, and $t_1'' = (D - d)/(\mu'' c_0)$ are the time delays in different propagation paths, respectively. The following notations are used for the up-going wave $\Theta = (\theta, \phi)$ and down-going wave $\Omega = (\pi - \theta, \phi)$, respectively.

Referring to Equations 64, 65, 66, 67, 68, 69 of Chapter 2, the 4×4 dimensional Mueller matrix $\overline{\mathbf{M}}$ contains the surface scattering from the rough surface at $z = -D$, volumetric scattering from stone scatterers (random oblate spheroids in this model), and interactions of the surface–volumetric scatterers. It is derived according to Xu and Jin (2006) as:

$$\overline{\mathbf{M}}(t,\Theta;\Omega_i) \cdot \mathbf{I}_0(t,\Omega_i) = \exp(-\tau_2/\mu - \tau_2/\mu_i) \cdot \overline{\mathbf{R}}_{23}(\Theta;\Omega_i) \cdot \mathbf{I}_0(t - t_2 - t_{2i}, \Omega_i)$$

$$+ \mu^{-1} \int_{-d}^{0} dz' \exp(\kappa_2 z'/\mu + \kappa_2 z'/\mu_i) \cdot \overline{\mathbf{P}}(\Theta;\Omega_i) \cdot \mathbf{I}_0[t + z'/(\mu c_0) + z'/(\mu_i c_0)]$$

$$+ \mu^{-1} \int_{-d}^{0} dz' \exp(\kappa_2 z'/\mu) \cdot \int d\Theta' \,\overline{\mathbf{P}}(\Theta;\Theta') \cdot \exp(-\tau_2/\mu' - \kappa_2 z'/\mu') \cdot \overline{\mathbf{R}}_{23}(\Theta';\Omega_i)$$

$$\cdot \exp(-\tau_2/\mu_i) \cdot \mathbf{I}_0[t + z'/(\mu c_0) - (d + z')/(\mu' c_0) - t_{2i}, \Omega_i]$$

$$+ \mu^{-1} \cdot \exp(-\tau_2/\mu) \cdot \int d\Omega' \,\overline{\mathbf{R}}_{23}(\Theta;\Omega') \cdot \int_{-d}^{0} dz' \exp(-\tau_2/\mu' - \kappa_2 z'/\mu') \cdot \overline{\mathbf{P}}(\Omega';\Omega_i)$$

$$\cdot \exp(\kappa_2 z'/\mu_i) \cdot \mathbf{I}_0[t - d/(\mu c_0) - d/(\mu' c_0) - z'/(\mu' c_0) + z'/(\mu_i c_0), \Omega_i]$$

$$+ \mu^{-1} \exp(-\tau_2/\mu) \cdot \int d\Omega'' \,\overline{\mathbf{R}}_{23}(\Theta;\Omega'') \cdot \int_{-d}^{0} dz' \exp(-\tau_2/\mu'' - \kappa_2 z'/\mu'')$$

$$\cdot \int d\Theta' \,\overline{\mathbf{P}}(\Omega'';\Theta') \cdot \exp(-\tau_2/\mu' - \kappa_2 z'/\mu') \cdot \overline{\mathbf{R}}_{23}(\Theta';\Omega_i) \cdot \exp(-\tau_2/\mu_i)$$

$$\cdot \mathbf{I}_0[t - d/(\mu c_0) - d/(\mu'' c_0) - z'/(\mu'' c_0) - d/(\mu' c_0) - z'/(\mu' c_0) - d/(\mu_i c_0), \Omega_i]$$

$$(10.29)$$

where $\overline{\mathbf{R}}_{23}$ is the scattering matrix of the bottom rough surface at $z = -D$, $\overline{\mathbf{P}}$ is the phase matrix of the stone scatterers, and $\tau_2 = \kappa_2 d$, $t_{2i} = d/(\mu_i c_0)$. Substituting the first four terms of Equation 10.29 into Equation 10.28, it is found that

$$\mathbf{I}(t,\Theta) = \overline{\mathbf{R}}_{01}(\Theta;\Omega_i) \cdot \mathbf{I}_0(t)$$

$$+ \int d\Theta' \,\overline{\mathbf{T}}_{10}(\Theta;\Theta') \cdot \int d\Omega''' \,\overline{\mathbf{R}}_{23}(\Theta';\Omega''') \cdot \overline{\mathbf{T}}_{01}(\Omega''';\Omega_i)$$

$$\cdot \exp(-\tau_1/\mu' - \tau_1/\mu''' - \tau_2/\mu' - \tau_2/\mu''') \cdot \mathbf{I}_0(t - t_1' - t_1''' - t_2' - t_2''')$$

$$+ \int d\Theta' \,\overline{\mathbf{T}}_{10}(\Theta;\Theta') \cdot 1/\mu' \cdot \int d\Omega''' \,\overline{\mathbf{P}}(\Theta';\Omega''') \cdot \overline{\mathbf{T}}_{01}(\Omega''';\Omega_i)$$

$$\cdot \int_{-d}^{0} dz' \exp(-\tau_1/\mu' + \kappa_2 z'/\mu' + \kappa_2 z'/\mu''' - \tau_1/\mu''')$$

$$\cdot \mathbf{I}_0\left[t + z'/(\mu' c_0) + z'/(\mu''' c_0) - t_1' - t_1'''\right]$$

$$+ \int d\Theta' \,\overline{\mathbf{T}}_{10}(\Theta;\Theta') \cdot 1/\mu' \cdot \int d\Theta'' \,\overline{\mathbf{P}}(\Theta';\Theta'') \cdot \int d\Omega''' \,\overline{\mathbf{R}}_{23}(\Theta'';\Omega''') \cdot \overline{\mathbf{T}}_{01}(\Omega''';\Omega_i)$$

$$\cdot \int_{-d}^{0} dz' \exp(-\tau_1/\mu' - \tau_1/\mu''' - \tau_2/\mu'' - \tau_2/\mu''' + \kappa_2 z'/\mu' - \kappa_2 z'/\mu'')$$

$$\cdot \mathbf{I}_0\left[t + z'/(\mu' c_0) - d/(\mu'' c_0) - z'/(\mu'' c_0) - d/(\mu''' c_0) - t_1' - t_1'''\right]$$

$$+ \int d\Theta' \,\overline{\mathbf{T}}_{10}(\Theta;\Theta') \cdot 1/\mu' \cdot \int d\Omega'' \,\overline{\mathbf{R}}_{23}(\Theta';\Omega'') \cdot \int d\Omega''' \,\overline{\mathbf{P}}(\Omega'';\Omega''') \cdot \overline{\mathbf{T}}_{01}(\Omega''';\Omega_i)$$

$$\cdot \int_{-d}^{0} dz' \exp(-\tau_1/\mu' - \tau_1/\mu''' - \tau_2/\mu' - \tau_2/\mu'' - \kappa_2 z'/\mu'' + \kappa_2 z'/\mu''')$$

$$\cdot \mathbf{I}_0\left[t - d/(\mu' c_0) - d/(\mu'' c_0) - z'/(\mu'' c_0) + z'/(\mu''' c_0) - t_1' - t_1'''\right] \qquad (10.30)$$

It can be seen that Equation 10.30 contains three runs of integration (totally fivefold integrals) resulting in lengthy and complicated computation. Those angular integrations in Equation 10.30 are generated due to diffused scattering and transmission through the randomly rough interfaces.

In fact, scattering and transmission can be decomposed into coherent and incoherent parts (Nozette *et al.*, 1996), that is, $\overline{\mathbf{R}}_{01}$, $\overline{\mathbf{T}}_{01}$, $\overline{\mathbf{T}}_{10}$, $\overline{\mathbf{R}}_{23}$ can be written as

$$
\begin{cases}
\overline{\mathbf{R}}(\Theta;\Omega_i) = \overline{\mathbf{R}}^n(\Theta;\Omega_i) + \overline{\mathbf{R}}^c(\Theta;\Omega_i) \cdot \delta(\Omega - \Omega_i) \\
\overline{\mathbf{T}}(\Theta;\Theta_i) = \overline{\mathbf{T}}^n(\Theta;\Theta_i) + \overline{\mathbf{T}}^c(\Theta;\Theta_i) \cdot \delta(\Theta - \Theta_i^+)
\end{cases}
\tag{10.31}
$$

where $\delta(\Omega - \Omega_i)$, $\delta(\Theta - \Theta_i^+)$ denote coherent scattering or transmission in specular or diffraction directions, Θ_i^+ indicates the diffraction angle of incident Θ_i, and Θ^- indicates the diffraction angle of Θ; the superscripts c, n denote coherent and incoherent components, respectively.

It can be seen that coherent scattering is dominant for large-scale surface perturbations (Fung, 1994). Because of the large-scale smoothly perturbed lunar surface and small permittivity ($\varepsilon \sim 3$) of the lunar surface, incoherent scattering of rough surfaces and stone scatterers, compared with coherent scattering, is negligible at low frequencies. It is assumed that, as Equation 10.31 is substituted into Equation 10.30, only those incoherent scattering, transmission or volumetric scattering in the specific direction described by Figure 10.2 are taken into account. Thus, no integral of Θ' or Θ'' is involved, and Equation 10.30 can be explicitly rewritten as

$$
\begin{aligned}
\mathbf{I}(t,0^+,\Theta) = {} & \overline{\mathbf{R}}_{01}^c(\Theta;\Omega_i) \cdot \mathbf{I}_0(t) \\
& + \overline{\mathbf{T}}_{10}^c(\Theta;\Theta^-) \cdot \overline{\mathbf{R}}_{23}^c(\Theta^-;\Omega^-) \cdot \overline{\mathbf{T}}_{01}^{-n}(\Omega^-;\Omega_i) \\
& \cdot \exp(-2\tau_1/\mu^- - 2\tau_2/\mu^-) \cdot \mathbf{I}_0(t - 2t_1^- - 2t_2^-) \\
& + \overline{\mathbf{T}}_{10}^c(\Theta;\Theta^-) \cdot \overline{\mathbf{R}}_{23}^n(\Theta^-;\Omega_i^+) \cdot \overline{\mathbf{T}}_{01}^c(\Omega_i^+;\Omega_i) \\
& \cdot \exp(-\tau_1/\mu^- - \tau_1/\mu_i^+ - \tau_2/\mu^- - \tau_2/\mu_i^+) \cdot \mathbf{I}_0(t - t_1^- - \tau_1/\mu_i^+ - t_2^- - \tau_2/\mu_i^+) \\
& + \overline{\mathbf{T}}_{10}^n(\Theta;\Theta_i^+) \cdot \overline{\mathbf{R}}_{23}^c(\Theta_i^+;\Omega_i^+) \cdot \overline{\mathbf{T}}_{01}^c(\Omega_i^+;\Omega_i) \\
& \cdot \exp(-2\tau_1/\mu_i^+ - 2\tau_2/\mu_i^+) \cdot \mathbf{I}_0(t - 2t_{1i}^+ - 2t_{2i}^+) \\
& + \overline{\mathbf{T}}_{10}^c(\Theta;\Theta^-) \cdot \overline{\mathbf{P}}(\Theta^-;\Omega_i) \cdot \overline{\mathbf{T}}_{01}^c(\Omega_i^+;\Omega_i) \\
& \cdot 1/\mu^- \int_{-d}^{0} dz' \exp(-\tau_1/\mu^- - \tau_1/\mu_i^+ + \kappa_2 z'/\mu^- + \kappa_2 z'/\mu_i^+) \\
& \cdot \mathbf{I}_0\left[t + z'/(\mu^- c_0) + z'/(\mu^+ c_0) - t_1^- - t_{1i}^+\right] \\
& + \overline{\mathbf{T}}_{10}^c(\Theta;\Theta^-) \cdot \overline{\mathbf{R}}_{23}^c(\Theta_i^+;\Omega_i^+) \cdot \overline{\mathbf{T}}_{01}^c(\Omega_i^+;\Omega_i) \\
& \cdot 1/\mu^- \int_{-d}^{0} dz' \exp(-\tau_1/\mu^- - \tau_1/\mu_i^+ - \tau_2/\mu_i^+ + \kappa_2 z'/\mu^- - \tau_2/\mu_i^+ - \kappa_2 z'/\mu_i^+) \\
& \cdot \mathbf{I}_0\left[t + z'/(\mu^- c_0) - d/(\mu_i^+ c_0) - z'/(\mu_i^+ c_0) - d/(\mu_i^+ c_0) - t_1^- - t_{1i}^+\right] \\
& + \overline{\mathbf{T}}_{10}^c(\Theta;\Theta^-) \cdot \overline{\mathbf{R}}_{23}^c(\Theta^-;\Omega^-) \cdot \overline{\mathbf{P}}(\Omega^-;\Omega_i^+) \cdot \overline{\mathbf{T}}_{01}^c(\Omega_i^+;\Omega_i) \\
& \cdot 1/\mu^- \int_{-d}^{0} dz' \exp(-\tau_1/\mu^- - \tau_1/\mu_i^+ - \tau_2/\mu^- - \tau_2/\mu^- - \kappa_2 z'/\mu^- + \kappa_2 z'/\mu_i^+) \\
& \cdot \mathbf{I}_0\left[t - d/(\mu^- c_0) - d/(\mu^- c_0) - z'/(\mu^- c_0) + z'/(\mu_i^+ c_0) - t_1^- - t_{1i}^+\right]
\end{aligned}
\tag{10.32}
$$

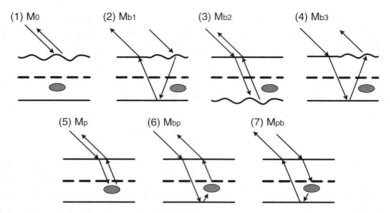

Figure 10.11 Scattering mechanism of the seven terms. Reprinted from *Radio Science*, from AGU

The seven terms on the right-hand side of Equation 10.32 are described in Figure 10.11, with the assumption of a less dense distribution of small oblate spheroidal scatterers, where the wavy and straight lines indicate the occurrence of incoherent and coherent scattering, respectively. The first term M_0 is due to diffuse scattering from the top rough surface, the second, third, and fourth terms M_{b1}, M_{b2}, M_{b3} result from scattering from the top and bottom interfaces, respectively, the fifth term M_p is volumetric scattering of rock scatterers, and the sixth and seventh terms, M_{bp}, M_{pb}, are the interactions of scatterers and interfaces.

We redefine

$$\sin\theta^- = \sqrt{\varepsilon_0/\varepsilon_1}\sin\theta, \quad \sin\theta_i^+ = \sqrt{\varepsilon_0/\varepsilon_1}\sin\theta_i$$

$$\mu^- = 1/\cos\theta^-, \quad \mu_i^+ = 1/\cos\theta_i^+$$

$$\tau_1 = \kappa_1(D-d), \quad \tau_2 = \kappa_2 d$$

$$t_1^- = (D-d)/(\mu^- c_0), \quad t_2^- = d/(\mu^- c_0)$$

$$t_{1i}^+ = (D-d)/(\mu_i^+ c_0), \quad t_{2i}^+ = d/(\mu_i^+ c_0)$$

$$\overline{\mathbf{T}}_{nm}^c(\Theta;\Theta^-) = \exp[-\sigma_{nm}^2(\kappa_n'/\mu - \kappa_m'/\mu^-)^2]\cdot\overline{\mathbf{T}}_{nm}^0(\Theta;\Theta^-)$$

$$\overline{\mathbf{R}}_{nm}^c(\Theta;\Omega) = \exp[-2\sigma_{nm}^2(\kappa_n'/\mu)^2]\cdot\overline{\mathbf{R}}_{nm}^0(\Theta;\Omega) \tag{10.33}$$

$$\kappa_1 = 2\kappa'', \quad \kappa_2 = \kappa_s + \kappa_{as} + (1-f_s)\kappa$$

Here ε_0, ε_1, ε_3 are, respectively, the dielectric permittivities of layers 0, 1, 3; ε_2 is the dielectric permittivity of rock scatterers; the background permittivity of layer 2 is the same as that of layer 1; $\overline{\mathbf{R}}^0$ and $\overline{\mathbf{T}}^0$ are the Fresnel reflection and transmission coefficients of the flat surface; σ_{nm} is the roughness variance of the interface between layers n and m; κ_n' and κ_n'' are the real and imaginary parts of the wavenumber in layer n (n, $m = 1, 2, 3$), respectively; the scattering and absorption coefficients of the scatterers (small oblate spheroids with assumed known solution) in layer 2 are κ_s and κ_{as}, respectively; the phase matrix is $\overline{\mathbf{P}}$; the scattering matrices are denoted as $\overline{\mathbf{R}}^n$ and $\overline{\mathbf{R}}^c$; and the transmission matrices are $\overline{\mathbf{T}}^n$ and $\overline{\mathbf{T}}^c$, respectively. Their analytic formulations are derived in Fung (1994).

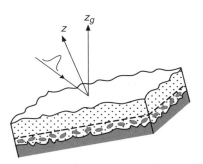

Figure 10.12   Tilted surface of the regolith layer. Reprinted from *Radio Science*, from AGU

It is noted that, for non-spherical scatterers embedded in layer 2, the extinction coefficient of layer 2, $\overline{\kappa}_2$, is a matrix form, that is, the extinction $\overline{\kappa}_e$ as defined in Equation 2.36. In that case, the exponential terms involving $\overline{\kappa}_e$ should first be expanded into the product of its eigenvector matrix and the exponential of its eigenvalues, as discussed in Chapter 2.

It is noted that in the backscattering direction, $\theta^- = \theta_i^+$, and the integration of the last two terms in Equation 10.32 is neglected because we have assumed the medium to be less dense. Also, the time delays of the second, third, fourth, sixth, and seventh terms are assumed to be equal.

As discussed in Chapter 1, Equations 1.115–1.119, to take account of large-scale perturbations of the tilted surface, the local coordinates $(\hat{x}, \hat{y}, \hat{z})$ of the lunar surface are described by the Euler angles (α, β, γ) in the principal coordinates $(\hat{x}_g, \hat{y}_g, \hat{z}_g)$, as shown in Figure 10.12. In the local coordinates, the time-domain Mueller matrix is expressed by Equation 10.32 with transformation of incident and scattering angles. In addition, the polarization bases (i.e., polarization base vectors $\hat{v}$ and $\hat{h}$) of the Mueller matrix in local coordinates are rotated in order to obtain the Mueller matrix, expressed in the principal coordinates, as follows:

$$\overline{M}(\Theta_s; \Omega_i) = \overline{U}(\Delta_s) \cdot \overline{M}^0(\Theta_s^0; \Omega_i^0) \cdot \overline{U}^{-1}(\Delta_i) \tag{10.34}$$

Here the superscript 0 denotes the Mueller matrix, incident and scattered angles are in local coordinates, Δ_i, Δ_s are rotation angles of incident and scattered bases, and $\overline{U}$ is the rotation matrix of the polarization bases of the Stokes vector:

$$\overline{U}(\Delta) = \begin{bmatrix} \cos^2 \Delta & \sin^2 \Delta & \sin 2\Delta & 0 \\ \sin^2 \Delta & \cos^2 \Delta & -\sin 2\Delta & 0 \\ -\frac{1}{2}\sin 2\Delta & \frac{1}{2}\sin 2\Delta & \cos 2\Delta & 0 \\ 0 & 0 & 0 & 1 \end{bmatrix} \tag{10.35}$$

10.4.2   Numerical Simulation of Pulse Echoes

Low-frequency 1.4 GHz (L band) sensing is assumed in this section for predicted full penetration through the regolith layer (Thompson, 1987). The incident Gaussian plane wave pulse is given as

$$I_0(t) = \exp[-b^2(t - T_0)^2] \tag{10.36}$$

where $b = \sqrt{\ln 2}/T_w$, $2T_w$ is the half-power width (12 ns), T_0 is the time of the pulse echo peak (18 ns), and $B_w = 2\ln 2/T_w = 231$ MHz.

The platform scenario of the pulse radar is described in Figure 10.13. The following points should be noted:

(a) To penetrate the dissipated regolith layer with thickness of 10 m or so, the carrier frequency should be lower, for example, L band, at least (Jin, 2005).
(b) To obtain good range resolution, the pulse width needs to be 10 ns, at least. In addition, a large bandwidth of the receiver and a high sampling rate are preferred.
(c) Dispersion effects are not taken into account in this section, which requires that the carrier frequency should not be too low, and the pulse width should not be too narrow.
(d) To see a uniform area illuminated by the incident wave, high spatial resolution is required. This leads to a demanding narrow beam and low flight altitude.
(e) From simulation, the dynamic range of the receiver is generally larger than 80 dB for probing a deep regolith layer of approximate depth of 20 m.

The parameters are designed, based on the Moon's topography (Fa and Jin, 2006), as follows: regolith layer thickness $D = 500$ cm; layer thickness of stone scatterers $d = 150$ cm; oblate spheroid semi-axes 2.5 cm and 1 cm; fractional volume 0.02; and spatial orientation over the ranges $\beta \in (0°, 60°)$, $\gamma \in (0°, 360°)$. Furthermore, the dielectric permittivity of the regolith layer is assumed to be $3 + i0.01$ (corresponding to FeO + TiO$_2$ content of 6%), and both the dielectric permittivity of the stone scatterers and underlying rock media are chosen as $8 + i0.5$. The rough interfaces are Gaussian, and the roughness variance and correlation length are 4 cm and 20 cm for the top interface, and 2 cm and 40 cm for the bottom interface, respectively. The incident angle in all examples is $\theta_i = 30°$. In the examples of Figures 10.14–10.16, the interfaces are assumed not to be tilted.

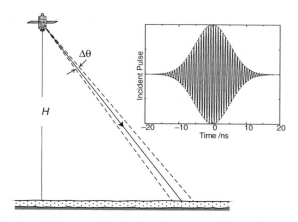

Figure 10.13   Geometry of the pulse incidence. Reprinted from *Radio Science*, from AGU

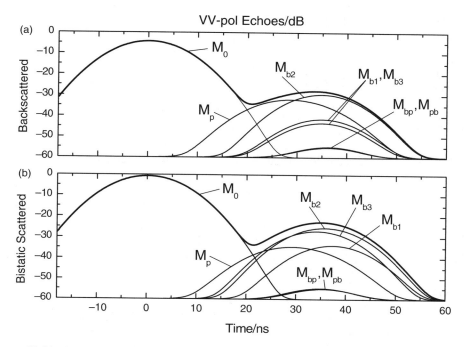

Figure 10.14 VV-pol scattering of the seven terms: (a) backscattering with incident angle 30°; (b) bistatic scattering with incident angle 20°, scattering direction 50°. Reprinted from *Radio Science*, from AGU

Figure 10.14a and b present co-pol (*vv*) backscattering and bistatic scattering contributed by the seven terms of Equation 10.32, respectively. The temporal behavior of the *hh*-pol case is also similar to the *vv* case. It can be seen that the top surface scattering M_0 is most dominant. In backscattering, the term M_{b2}, which is due to coherent transmission through the top interface, seems to dominate the other three terms coming from the bottom interface. The scattering from the stone scatterers M_{bp} and M_{pb} are the same because the time delay and attenuation of these two terms are the same. However, in bistatic scattering, these terms should be separated because the incident and scattered directions are different. Thus, M_{bp} and M_{pb} are being separated.

It is expected that, if there exist non-uniform and random stratified structures in the regolith layer (1), random fluctuations and multiple reflections can happen during the propagation time, for example, 0–25 ns. Also, if the structure of the layered media is anisotropic, different polarized echoes will display more significantly different responses and might reveal more information about the media.

Figure 10.15 shows pulse echoes for different $FeO + TiO_2$ contents, regolith layer thickness, and stone scattering layer thickness. It can be seen that high $FeO + TiO_2$ content increases the imaginary part of the dielectric permittivity and causes more attenuation; and the scattering from the underlying rock media is then shadowed. Increasing the regolith layer thickness increases the time delay of the pulse echoes from the bottom interface at $z = -D$ and results in more attenuation. Also, an increase of the layer thickness of rock scatterers makes the tails of the pulse echoes more significant.

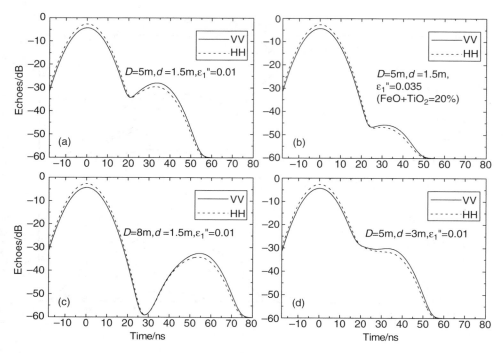

Figure 10.15 Pulse echoes from different lunar regolith layers. Reprinted from *Radio Science*, from AGU

The influence of the peaks due to the lunar parameters such as the $FeO + TiO_2$ content, layer thickness, fraction of rock scatterers, stone scattering layer thickness, and so on are shown in Figure 10.16. It can be seen that higher $FeO + TiO_2$ content or a thicker regolith layer can obscure scattering from the underlying stone scatterers and the bottom interface; and then scattering is mainly contributed by the top interface. A thicker scattering layer or denser scatterers enhances volumetric scattering. Roughness of the interfaces also increases scattering.

10.4.3 Pulse Echo Images of Lunar Layered Media

The regolith is the result of continual impacts of large and small meteoroids during the long history of the Moon's geology, and its average thickness is about 4–5 m for the maria and about 10–15 m for the highlands (Heiken, Vaniman, and French, 1991). In general, the regolith layer thickness correlates well with the lunar surface age: the greater the age, the thicker the regolith deposit. We propose to construct the lunar regolith layer thickness proportional to the altitude in the lunar digital elevation map (DEM) as a tentatively chosen distribution (Fa and Jin, 2006).

Simulation of the pulse echoes can show the stratified structures of the lunar regolith media. In radar probing with high resolution, the echoes from aligned pixels can compose an image, as illustrated in Figure 10.17. As an example, the parameters of the assumed DEM, regolith layer thickness, and $FeO + TiO_2$ content (Fa and Jin, 2006; Shkuratov and

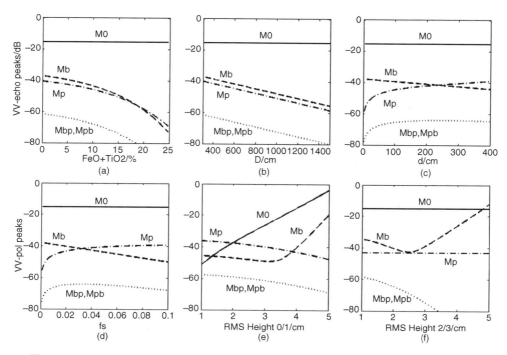

Figure 10.16 Echo peaks due to different parameters. Reprinted from *Radio Science*, from AGU

Bondarenko, 2001; Bondarenko and Shkuratov, 1998) at the lunar location along the latitude S57.5° from the longitude W22.5° to E2.5° (at Clavius) are chosen, as shown in Figure 10.18, and surface roughness is assumed to be of the order of 10% random fluctuation.

To take account of the tilted surface perturbation, the normal vector of any local area (i.e., local coordinate $\hat{z}$) can be obtained as

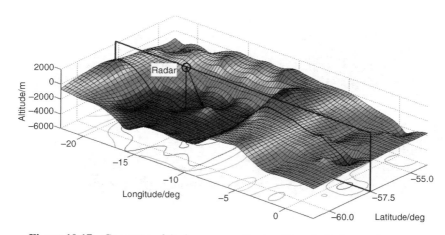

Figure 10.17 Geometry of the image area. Reprinted from *Radio Science*, from AGU

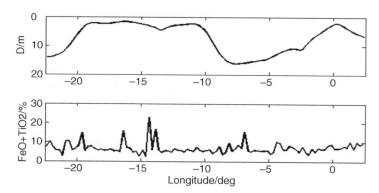

Figure 10.18 Parameters of the lunar surface. Reprinted from *Radio Science*, from AGU

$$\hat{z}(i_x, i_y) = \frac{\hat{t}_x \times \hat{t}_y}{\|\hat{t}_x \times \hat{t}_y\|}$$

$$\hat{t}_x = [2\Delta x \quad 0 \quad h(i_x + 1, i_y) - h(i_x - 1, i_y)]^{\mathrm{T}} \qquad (10.37)$$

$$\hat{t}_y = [0 \quad 2\Delta y \quad h(i_x, i_y + 1) - h(i_x, i_y - 1)]^{\mathrm{T}}$$

where $h(i_x, i_y)$ is the DEM at (i_x, i_y), and Δx, Δy are the resolution of the DEM.

Figure 10.19a–c show the image of co-pol *hh*, *vv* and cross-pol *hv* pulse echoes, respectively, where each pixel is of the resolution of about 4.1 km. The profile of the regolith layer

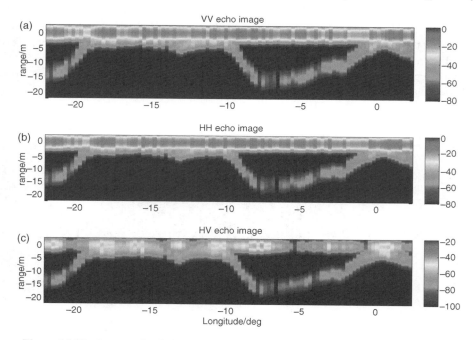

Figure 10.19 Images of polarimetric pulse echoes. Reprinted from *Radio Science*, from AGU

depth and the stone scatter layer can be identified. More information can be retrieved from fully polarimetric pulse echoes.

The time-domain Mueller matrix solution is derived from a vector radiative transfer formulation for a three-layer model of the lunar regolith media. The Mueller matrix solution contains seven scattering mechanisms of the stratified media: surface scattering from the top and bottom rough interfaces, volumetric scattering from random stone scatterers (spatially oriented oblate spheroids), and their multi-interactions by neglecting multiple resonant interactions. The interfaces can be tilted with respect to the principal coordinates.

As the radar pulse at 1.4 GHz is penetrating through the top random surface, attenuation through the low lossy medium 1, and polarimetric scattering through the scattering medium 2 and rough interfaces are formulated in our Mueller matrix solution. Model parameters are set according to the study of the lunar regolith structure.

The temporal characteristics and structure of the polarimetric echo profile are analyzed. The contributions of different scattering mechanisms, that is, surface, volumetric scattering and interactions, are analyzed, and their dependence on model parameters such as layer thickness and content of $FeO + TiO_2$, are also discussed. Simulation of pulse echoes displays an image of the underlying subsurface. This reveals information about the depth and other properties of the lunar regolith layer, and demonstrates a potential new way to explore the Moon's surface by spaceborne remote sensing in the future.

10.5 Monitoring Debris and Landslides

As an example of applications, areas with the potential occurrence of debris flows or landslides are modeled as layered land media embedded by random scatterers (stones or water) with randomly rough interfaces. Distinct features in radar range profiles can be directly attributed to changes in underground water content and/or water distribution. This proposed VHF radar seems promising as an early-warning system for geological hazards. The differences between radar images on a normal day and a "warning" day, for example, after a severe storm, could be used for predictions.

Triggered by intense rainfall or glacial melt, the land layers of unconsolidated soil (rock fragments, soil, mud, etc.) might suddenly lose their gravitational stability and flow along the slope to cause rapid landslides or massive debris flows. For example, during only two months, July and August 2010, more than 10 serious debris flows and landslides occurred in northwestern China, the most devastating one being in Zhou-Qu, Gangshu Province (http://news.cntv.cn/special/zhouqu), which claimed thousands of lives and caused a great loss of property. How to monitor and provide early warning of potential debris flows and landslides is critical in terms of taking precautions against natural calamities and reducing the damage caused by such hazards. Currently, conventional geological surveying is the major tool used to monitor highly dangerous locations where debris flows and landslides are likely to happen. Those traditional observations at discrete sites and discrete times (http://epaper.bjnews.com.cn/2010-08/20/) are quite restrictive in terms of timeliness and accuracy. Even the good forecast of a rainstorm, which is usually the trigger of a debris flow or landslide, cannot really determine whether a localized mudslide will happen or not.

Debris flows and landslides are mostly attributable to water saturation of unconsolidated soil, which loses its gravitational stability and ultimately results in rapid flow. The essential issue for the early warning of such disasters is to closely monitor the water change beneath the ground surface in localized stratified land media.

Microwave radar remote sensing has reached high spatial resolution of meters and below for fully polarimetric and all-weather all-time observations. The strong penetration capability of low-frequency radar enables the waves to propagate through lossy media. Some discussions about employing modern remote sensing technologies to monitor debris flows and landslides have been presented (Metternicht, Hurni, and Gogu, 2005; Fabbri *et al.*, 2003; Pradhan, Singh, and Buchroithner, 2006). However, almost all of these studies are restricted to terrain surface observation without water information underneath, and they are more focused on change detection of the terrain surface *after* the occurrence of a debris flow, landslide, or surface sedimentation. As the land soil moisture increases, the penetration depth at L band is quickly reduced to a few centimeters. Thus, to sense the change of water content below soil and make a plausible prediction of hazard occurrence, a much lower frequency will need to be employed, such as very high-frequency (VHF, ∼100 MHz) radar technology.

10.5.1 Modeling Debris Flows and Landslides

We propose to use VHF radar to probe the multi-layered land media for early warning of geological hazards. VHF radar waves can penetrate some tens of meters through the land media. As the water content in the land media increases, its dielectric constant increases accordingly. Thus, the returned radar echoes from multiple interactions of surface and volumetric scatterings would be able to indicate the characteristic structures of the low-loss media. In this section, the geological structure of the land is modeled as one layer of lossy dielectric media with rough top and bottom interfaces, embedded in between with randomly oriented scatterers such as stones or concentrated water bodies (Figure 10.20). Solving the temporal radiative transfer equations, the Mueller matrix solution taking into account

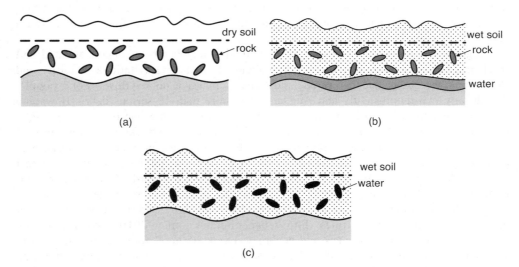

(a)   (b)

(c)

Figure 10.20 Model configurations for different geophysical conditions: (a) normal case; (b) warning case 1 (soil is porous and does not contain water); (c) warning case 2 (soil contains water). © [2011] IEEE. Reprinted, with permission, from *IEEE Geoscience and Remote Sensing Letters*

scattering from rough interfaces, discrete scatterers (stones, water body), and scatterer–surface interactions is obtained and numerically simulated, as presented in Section 10.4. It shows different radar range echoes as the water content below the surface changes significantly. The difference of the radar echoes between the normal day and a warning day, for example, after a rainstorm, can be used to provide early warning for the occurrence of a debris flow or landslide.

A landslide ranging from moderate to severe in grade may have a large landslide body that is over 20 m in depth and 200 m^2 in area. Underground water concentration and background water content are the two key factors in our model. In this paper, three situations are mimicked using the same model with different configurations. The model configurations for the three situations are shown in Figure 10.20: (a) normal case without severe precipitation; (b) warning case 1, underground water accumulated on bottom; and (c) warning case 2, water contained by soil layer.

In the normal case (Figure 10.20a), the ground layer above the bedrock is considered as dry soil mixed with randomly distributed stones. Some parameter selection guidelines are given in Table 10.1.

In warning case 1 (Figure 10.20b), under severe precipitation, if the soil layer has very good water penetrability, for example, sandstone, water sinks quickly, accumulates at the bottom, and finally forms an underground water strip (Figure 10.20b). As precipitation persists, this water strip might become thicker, while the background water content of the whole

Table 10.1 Model parameters

	Frequency 100 MHz		
	Normal case	Warning case 1	Warning case 2
Water strip depth	0 m	0.5 m	0 m
Layer 2 soil background permittivity ε_1	$3.5 + i0.05$	$(1 - r)(3.5 + i0.05) + r(84 + i0.6)$ where background water content $r = 0.1$	$(1 - r)(3.5 + i0.05) + r(84 + i0.6)$ where background water content $r = 0.1$
Scatterer permittivity ε_s	stone $6 + i0.1$	stone $6 + i0.1$	discrete water bodies $(1 - r)(3.5 + i0.05) + r(84 + i0.6)$, $r = 0.5$
Scatterer size (diameter)	10 cm × 10 cm × 5 cm	10 cm × 10 cm × 5 cm	20 cm × 20 cm × 10 cm
Scatterer orientation		Uniformly in all directions	
Scatterer fractional volume	0.1	0.1	0.1
Layer 3 permittivity ε_3	bedrock $6 + i0.1$	water strip $84 + i0.6$	bedrock $6 + i0.1$
Interface between layers 1 and 2	Gauss spectrum with root-mean-square height 4 cm and correlation length 20 cm		
Interface between layers 2 and 3	Gauss spectrum with root-mean-square height 2 cm and correlation length 40 cm		

layer keeps increasing as well. Strong wave power will be bounced back from the top of the water strip, and thus the structures below the water strip are in fact invisible to the radar. Based on this assumption, the part above the water strip can be modeled using the same model as developed in Section 10.2 and the parameters are selected as shown in Table 10.1.

In warning case 2 (Figure 10.20c), if the soil layer can well contain the water, for example, clay, the total precipitated water distributes across the whole layer. A reasonable assumption would be that water is contained by the soil non-uniformly and thus concentrates into discrete dense water bodies across the whole media. Neglecting the low-contrast stone scatterers, the high-contrast water bodies become the major scatterer embedded in the soil layer. As shown in Figure 10.20c, the problem is again modeled using the same model with different parameters as given in Table 10.1.

10.5.2 Echo Simulation of Nadir Looking Radar

The targeted scenario for radar monitoring of geological disasters is depicted in Figure 10.21. Mounted aboard an air platform, the radar emits frequency-modulated pulses toward the ground surface. Range resolution (toward the ground) can be achieved by pulse compression, while high resolution in the along-track dimension can be realized using synthetic aperture radar (SAR) processing. In the third dimension, that is, the cross-track dimension, a sparse antenna array might be employed to focus a very sharp beam. Another option is to perform additional SAR imaging in cross-track by multiple parallel pass flying.

To take advantage of the high penetration of the VHF band, we have to compromise with limitations on the engineering realization of the radar. First and foremost, the bandwidth is

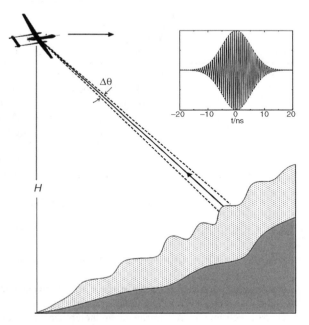

Figure 10.21 Nadir looking radar monitoring of debris flow and landslide area. © [2011] IEEE. Reprinted, with permission, from *IEEE Geoscience and Remote Sensing Letters*

limited by the carrier frequency. In this section, bandwidth is selected as 100 MHz and carrier frequency is also 100 MHz. Theoretically, this would achieve 1.5 m range resolution.

In addition, the antenna aperture is significantly limited by the physical size of the platform. A sharp beam is difficult to achieve at low frequency. Considering a 40 m wingspan of a nominal early-warning aircraft and a conformal sparse array, the theoretical limit of the beam width would be 0.1 rad at 100 MHz frequency. As geological disasters are often triggered by severe storms, the aircraft should be flying over the clouds to avoid weather hazards. The nominal altitude of cumulonimbus and/or stratocumulus is around 2000 m. Based on this consideration, the achievable footprint of the antenna beam is approximately 200 m. To achieve better resolution, either a lower platform or 3D SAR processing by multiple flying should be utilized.

The objective of this study is mostly emphasized on the theoretical basis of VHF radar early warning of geological disasters and demonstration of its feasibility. Radar system design and platform configuration can be examined in the future. In this section, we simulate the range profiles of the scattered radar wave from geological structures under the ground. The echoes from a 2D cross-section of the stratified media are presented.

According to pulse compression, the point spread function (PSF) has a sinc-like shape. Side-lobes of sinc PSF are often further suppressed by tapering in the frequency domain. For simplicity, a Gaussian-shaped narrow pulse is directly used in the simulation. The half-power pulse width is selected as 5 ns, which is equivalent to a 1.5 m range resolution.

As a proof-of-concept demonstration, the first simulation presented is based on some typical parameters, for example: layer depth 10 m; top soil depth 0.5 m; nadir looking angle 40°; local slope 20° (local incident angle 20°). The background water content in warning cases (corresponding to $\varepsilon_1 = 11.6 + 0.11i$) is chosen as follows: water strip depth in warning case 1 is 0.5 m; and water content of discrete water body in warning case 2 is 50% (corresponding to $\varepsilon_s = 44 + 0.33i$), respectively. Note that the total water contents under warning case 1 and warning case 2 configurations are about the same. The simulated range profiles of VV, HH, and VH channels for the normal case, warning case 1, and warning case 2 are plotted in Figure 10.22, respectively. We can make the following observations from these results:

(a) Compared with the normal case in the left column of Figure 10.22, the echo peaks from the bottom interface in the two warning cases are delayed due to the increase in the background water content in the landslide body. This delay is monotonically dependent on the increase of the background water content (increase of the media dielectric constants), which can be used as an indicator of soil water saturation.

(b) In warning case 1, the water strip significantly enhances the bottom peak, that is, it is ∼10 dB higher than for the normal case. However, the depth of the water strip cannot be simply retrieved from the peak position because it also depends on the background water content.

(c) In warning case 2, there is stronger scattering from the discrete water bodies present in both the middle (direct scattering from water bodies) and the end (multiple scattering via water bodies and the bottom interface), where the middle segment is enhanced by 15 dB and the end peak is enhanced by 10 dB as compared to the normal case.

(d) The direct return from the top surface is also enhanced by 5 dB more than the normal case due to the increase of background water content. This echo might be utilized to

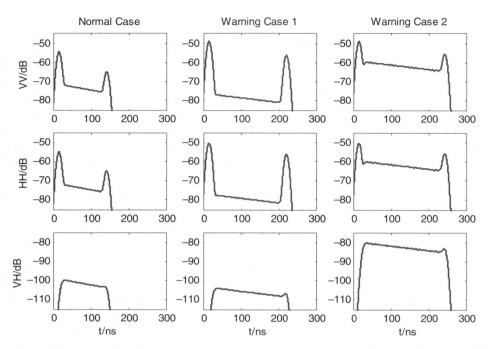

Figure 10.22 Simulated radar range profiles of geological structures with typical parameter settings. © [2011] IEEE. Reprinted, with permission, from *IEEE Geoscience and Remote Sensing Letters*

estimate the background water content, and thus would helpful to evaluate the water strip depth of warning case 1.

(e) Returns between the two peaks in warning case 2 are closely related to the discrete water bodies, and thus can be used to monitor the distribution of water across the landslide body.

(f) The cross-polarization (VH) channel is more sensitive to randomly oriented non-spherical scatterers (stones or water), given that these spheroids have strong polarity.

A cross-section of a geological layered structure as depicted in Figure 10.23 is proposed for 2D simulation of range echoes, which is formed by the range profiles collected at each horizontal pixel (1 m). The local slope of each pixel is calculated from the average gradients of the 20 nearby pixels. The layer 1 depth is then estimated from the vertical depth via trigonometric calculation. Both multiplicative (speckle) and additive (thermal) white noise are added to the final image. Finally, a moving average is taken along the horizontal dimension with a window size of 5 m. Note that the achievable radar image in reality might be much more blurred due to poor resolution in the horizontal dimension.

Figures 10.24–10.26 show the 2D images of radar range echoes in the three cases. Compared to the normal case, the most distinct signature of warning case 1 is the dramatically enhanced echoes from the top surface and the bottom interface. In warning case 2, the prominent characteristic is the enhanced returned strip corresponding to the water bodies. These features might be applicable to infer important information about the underground

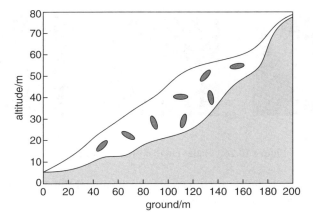

Figure 10.23   Cross-section of ground structure used in simulation. © [2011] IEEE. Reprinted, with permission, from *IEEE Geoscience and Remote Sensing Letters*

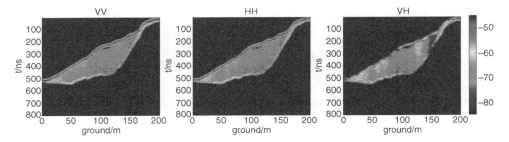

Figure 10.24   Radar range profiles of the normal case. Note that the VH channel is purposely enhanced by 30 dB to be correctly displayed in the same color scale as in VV and HH. © [2011] IEEE. Reprinted, with permission, from *IEEE Geoscience and Remote Sensing Letters*

water distributions and therefore to help identify and provide early warning of geological disasters such as landslides and debris flows.

This study aims to examine the theoretical basis and feasibility of VHF radar early warning of geological disasters. Radar system design and platform configuration, which can be either airborne, tower based or even satellite-borne, will be investigated in the future. Once

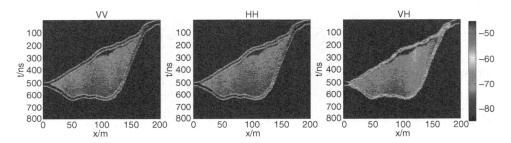

Figure 10.25   Radar range profiles of warning case 1

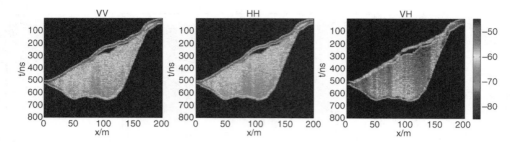

Figure 10.26   Radar range profiles of warning case 2

an experimental system is developed, fine tuning of the model and all the relevant parameters will be necessary to better match real-life situations.

Appendix 10A: Some Mathematics Needed in Derivation of Temporal Mueller Matrix

According to Taylor's formula (Abramowitz and Stegun, 1972, p. 14)

$$f(x+h) = f(x) + \sum_{m=1}^{\infty} \frac{1}{m!} f^{(m)}(x) h^m \tag{10A.1}$$

The error function $\mathrm{erf}(x+h)$ is expanded as

$$\mathrm{erf}(x+h) = \mathrm{erf}(x) + \sum_{m=1}^{\infty} \frac{1}{m!} \mathrm{erf}^{(m)}(x) h^m \tag{10A.2}$$

where $\mathrm{erf}^{(m)}(\cdot)$ denotes the mth-order derivative of the error function and can be written as (Abramowitz and Stegun, 1972, p. 298)

$$\mathrm{erf}^{(m)}(x) = (-1)^{m-1} \frac{2}{\sqrt{\pi}} H_{m-1}(x) \exp(-x^2) \tag{10A.3}$$

where the Hermite polynomials $H_{m-1}(x)$ can be written in terms of (Abramowitz and Stegun, 1972, p. 775)

$$H_{m-1}(x) = \sum_{\ell=0}^{[m/2]} (-1)^{\ell} \frac{(m-1)!}{\ell!(m-1-2\ell)!} (2x)^{m-1-2\ell} \tag{10A.4}$$

From Equations 10A.2–10A.4, it can be obtained that

$$\mathrm{erf}(x+h) = \mathrm{erf}(x) + \frac{2}{\sqrt{\pi}} \exp(-x^2) \sum_{m=1}^{\infty} \frac{(-1)^{m-1}}{m!} h^m \sum_{\ell=0}^{[m/2]} (-1)^{\ell} \frac{(m-1)!}{\ell!(m-1-2\ell)!} (2x)^{m-1-2\ell}$$

$$\tag{10A.5}$$

and then

$$\exp(x^2) \cdot [\text{erf}(x + h_1) - \text{erf}(x + h_2)]$$

$$= \frac{2}{\sqrt{\pi}} \sum_{m=1}^{\infty} \frac{(-1)^{m-1}}{m!} (h_1^m - h_2^m) \sum_{\ell=0}^{[m/2]} (-1)^\ell \frac{(m-1)!}{\ell!(m-1-2\ell)!} (2x)^{m-1-2\ell} \qquad (10\text{A}.6)$$

Taking $x = \sqrt{2}\sigma B_{n2}/|R_2|$, $h_1 = -|R_2|d_{n-1}/(\sqrt{2}\sigma)$, and $h_2 = -|R_2|d_n/(\sqrt{2}\sigma)$ and substituting them into Equation 10A.6, it is obtained that

$$\frac{1}{2|R_2|} \exp\left[\frac{2\sigma^2 B_{n2}^2}{R_2^2} - C_2\right] \times \left[\text{erf}\left(\frac{\sqrt{2}\sigma B_{n2}}{|R_2|} - \frac{|R_2|d_{n-1}}{\sqrt{2}\sigma}\right) - \text{erf}\left(\frac{\sqrt{2}\sigma B_{n2}}{|R_2|} - \frac{|R_2|d_n}{\sqrt{2}\sigma}\right)\right]$$

$$= \frac{1}{\sqrt{2\pi}\sigma} \exp(-C_2) \sum_{m=1}^{\infty} \frac{(2B_{n2})^{m-1}}{m!} [(d_n)^m - (d_{n-1})^m] \sum_{\ell=0}^{[m/2]} (-1)^\ell \frac{(m-1)!}{\ell!(m-1-2\ell)!} \left(\frac{2\sqrt{2}\sigma B_{n2}}{|R_2|}\right)^{-2\ell}$$

$$(10\text{A}.7)$$

Because

$$\lim_{|R_2|\to 0, \ell \geq 1} \left(\frac{2\sqrt{2}\sigma B_{n2}}{|R_2|}\right)^{-2\ell} = 0 \qquad (10\text{A}.8)$$

and keeping the $\ell = 0$ term of Equation 10A.7, this yields

$$\lim_{|R_2|\to 0} \frac{1}{2|R_2|} \exp\left[\frac{2\sigma^2 B_{n2}^2}{R_2^2} - C_2\right] \times \left[\text{erf}\left(\frac{\sqrt{2}\sigma B_{n2}}{|R_2|} - \frac{|R_2|d_{n-1}}{\sqrt{2}\sigma}\right) - \text{erf}\left(\frac{\sqrt{2}\sigma B_{n2}}{|R_2|} - \frac{|R_2|d_n}{\sqrt{2}\sigma}\right)\right]$$

$$= \lim_{|R_2|\to 0} \frac{1}{\sqrt{2\pi}\sigma} \exp(-C_2) \sum_{m=1}^{\infty} \frac{(2B_{n2})^{m-1}}{m!} [(d_n)^m - (d_{n-1})^m]$$

$$= \frac{1}{\sqrt{2\pi}\sigma} \exp\left[-\frac{(t - 2d_{N+1}\sec\theta_0/c_0)^2}{2\sigma^2}\right]$$

$$\times \sum_{m=1}^{\infty} \frac{[-\xi_{nk}(\theta,\varphi)\sec\theta + \xi_{ni}(\theta_0,\varphi_0)\sec\theta_0]^{m-1}}{m!} \cdot [(d_n)^m - (d_{n-1})^m]$$

$$(10\text{A}.9)$$

Using $e^x = \sum_{m=0}^{\infty} \frac{x^m}{m!}$, $\frac{1-e^x}{x} = -\sum_{m=1}^{\infty} \frac{x^{m-1}}{m!}$, we obtain

$$\lim_{|R_2|\to 0} \frac{1}{2|R_2|} \exp\left[\frac{2\sigma^2 B_{n2}^2}{R_2^2} - C_2\right] \times \left[\text{erf}\left(\frac{\sqrt{2}\sigma B_{n2}}{|R_2|} - \frac{|R_2|d_{n-1}}{\sqrt{2}\sigma}\right) - \text{erf}\left(\frac{\sqrt{2}\sigma B_{n2}}{|R_2|} - \frac{|R_2|d_n}{\sqrt{2}\sigma}\right)\right]$$

$$= \frac{1}{\sqrt{2\pi}\sigma} \exp\left[-\frac{(t - 2d_{N+1}\sec\theta_0/c_0)^2}{2\sigma^2}\right] \{\exp\{-d_{n-1}[\xi_{nk}(\theta,\varphi)\sec\theta - \xi_{ni}(\theta_0,\varphi_0)\sec\theta_0]\}$$

$$- \frac{\exp\{-d_n[\xi_{nk}(\theta,\varphi)\sec\theta - \xi_{ni}(\theta_0,\varphi_0)\sec\theta_0]\}}{[\xi_{nk}(\theta,\varphi)\sec\theta - \xi_{ni}(\theta_0,\varphi_0)\sec\theta_0]}\}$$

$$(10\text{A}.10)$$

Similarly, it can be obtained that

$$\lim_{|R_2|\to 0} \frac{1}{2|R_2|} \exp\left[\frac{2\sigma^2 B_{n3}^2}{R_2^2} - C_3\right] \times \left[\text{erf}\left(\frac{\sqrt{2}\sigma B_{n3}}{|R_2|} - \frac{|R_2|d_{n-1}}{\sqrt{2}\sigma}\right) - \text{erf}\left(\frac{\sqrt{2}\sigma B_{n3}}{|R_2|} - \frac{|R_2|d_n}{\sqrt{2}\sigma}\right)\right]$$

$$= \frac{1}{\sqrt{2\pi}\sigma}\exp\left[-\frac{(t-2d_{N+1}\sec\theta/c_0)^2}{2\sigma^2}\right] \times \{\exp\{-d_{n-1}[\xi_{ni}(\pi-\theta_0,\varphi_0)\sec\theta_0 - \xi_{nk}(\pi-\theta,\varphi)\sec\theta]\}$$

$$- \frac{\exp\{-d_n[\xi_{ni}(\pi-\theta_0,\varphi_0)\sec\theta_0 - \xi_{nk}(\pi-\theta,\varphi)\sec\theta]\}}{\xi_{ni}(\pi-\theta_0,\varphi_0)\sec\theta_0 - \xi_{nk}(\pi-\theta,\varphi)\sec\theta}\Bigg\}$$

$$(10A.11)$$

$$\lim_{R_1\to 0} \frac{1}{2R_1} \exp\left[\frac{2\sigma^2 B_{n1}^2}{R_1^2} - C_1\right] \times \left[\text{erf}\left(\frac{\sqrt{2}\sigma B_{n1}}{R_1} - \frac{R_1 d_{n-1}}{\sqrt{2}\sigma}\right) - \text{erf}\left(\frac{\sqrt{2}\sigma B_{n1}}{R_1} - \frac{R_1 d_n}{\sqrt{2}\sigma}\right)\right]$$

$$= \frac{1}{\sqrt{2\pi}\sigma}\exp\left(-\frac{t^2}{2\sigma^2}\right) \times \{\exp\{-d_{n-1}[\xi_{nk}(\theta,\varphi)\sec\theta + \xi_{ni}(\pi-\theta_0,\varphi_0)\sec\theta_0]\}$$

$$- \frac{\exp\{-d_n[\xi_{nk}(\theta,\varphi)\sec\theta + \xi_{ni}(\pi-\theta_0,\varphi_0)\sec\theta_0]\}}{\xi_{nk}(\theta,\varphi)\sec\theta + \xi_{ni}(\pi-\theta_0,\varphi_0)\sec\theta_0}\Bigg\}$$

$$(10A.12)$$

$$\lim_{R_1\to 0} \frac{1}{2R_1} \exp\left[\frac{2\sigma^2 B_{n4}^2}{R_1^2} - C_4\right] \times \left[\text{erf}\left(\frac{\sqrt{2}\sigma B_{n4}}{R_1} - \frac{R_1 d_{n-1}}{\sqrt{2}\sigma}\right) - \text{erf}\left(\frac{\sqrt{2}\sigma B_{n4}}{R_1} - \frac{R_1 d_n}{\sqrt{2}\sigma}\right)\right]$$

$$= \frac{1}{\sqrt{2\pi}\sigma}\exp\left(-\frac{t^2}{2\sigma^2}\right) \times \{\exp\{d_n[\xi_{nk}(\pi-\theta,\varphi)\sec\theta + \xi_{ni}(\theta_0,\varphi_0)\sec\theta_0]\}$$

$$- \frac{\exp\{d_{n-1}[\xi_{nk}(\pi-\theta,\varphi)\sec\theta + \xi_{ni}(\theta_0,\varphi_0)\sec\theta_0]\}}{\xi_{nk}(\pi-\theta,\varphi)\sec\theta + \xi_{ni}(\theta_0,\varphi_0)\sec\theta_0}\Bigg\}$$

$$(10A.13)$$

Substituting the above Equations 10A.10–10A.13 into Equations 10.19a–10.19d, it can be seen that if we let $t \to 0$, $c_0 \to \infty$ (i.e., $R_1, R_2 \to 0$), that is, a plane pulse becomes a continuous plane wave, and $N = 1$, $d_{N+1} = d_N$, that is, inhomogeneous random media degenerate to the homogeneous case, the temporal Mueller matrix of Equation 10.18 can return to the conventional Mueller matrix in Tsang and Kong (1990, p. 126).

References

Abramowitz, M. and Stegun, I.A. (1972) *Handbook of Mathematical Functions*, Dover, New York.

Boerner, W.-M., Mott, H., and Luneburg, E. (1997) Polarimetry in remote sensing: basic and applied concepts. In *Proceedings of the 1997 IEEE International Geoscience and Remote Sensing Symposium*, Singapore, Vol. **3**, IEEE, Piscataway, NJ, pp. 1401–1403.

Bondarenko, N.V. and Shkuratov, Yu.G. (1998) A map of regolith layer thickness for the visible lunar hemisphere from radar and optical data. *Solar System Research*, **32**(4), 301–309.

Chang, M. and Jin, Y.Q. (2002) The temporal Mueller matrix solution for polarimetric scattering from a layer of random non-spherical particles with a pulse incidence. *Journal of Quantitative Spectroscopy and Radiative Transfer*, **74**, 339–353.

Fa, W. and Jin, Y.Q. (2006) Simulation of multi-channel brightness temperature of lunar surface and inversion of lunar regolith layer thickness. *Progress in Natural Science*, **16**(1), 86–94.

Fabbri, A., Chung, C.J., Cendrero, A., and Remondo, J. (2003) Is prediction of future landslides possible with a GIS? *Natural Hazards*, **30**, 487–499.

Fung, A.K. (1994) *Microwave Scattering and Emission Models and Their Applications*, Artech House, Norwood, MA, pp. 112–116.

Heiken, G.H., Vaniman, D.T., and French, B.M. (1991) *Lunar Source-Book: A User's Guide to the Moon*, Cambridge University Press, Cambridge, UK.

Hong, S.T. and Ishimaru, A. (1977) Plane wave propagation through random media. *IEEE Transactions on Antennas and Propagation*, **25**(6), 822–830.

Ito, S. (1980) On the theory of pulse wave propagation in media of discrete random scatterers. *Radio Science*, **15**(5), 893.

Jin, Y.Q. (1994) *Electromagnetic Scattering Modeling for Quantitative Remote Sensing*, World Scientific, Singapore, pp. 13–64.

Jin, Y.Q. (2005) *Theory and Approach of Information Retrievals from Electromagnetic Scattering and Remote Sensing*, Springer, Berlin.

Jin, Y.Q. and Chen, F. (2003) Scattering simulation for inhomogeneous layered canopy and random targets beneath canopies by using the Mueller matrix solution of the pulse radiative transfer. *Radio Science*, **38**(6), 1107–1116.

Jin, Y.Q. and Xu, F. (2011) Monitoring and early warning the debris flow and landslides using HF radar echoes from multi-layered terrain media. *IEEE Geoscience and Remote Sensing Letters*, **8**(3), 575–579.

Jin, Y.Q., Chen, F., and Chang, M. (2004) Retrievals of underlying surface roughness and moisture for stratified vegetation canopy using polarized pulse echoes in the specular direction. *IEEE Transactions on Geoscience and Remote Sensing*, **42**(2), 426–433.

Liu, C.H. and Yeh, K.C. (1978) Pulse propagation in random media. *IEEE Transactions on Antennas and Propagation*, **26**, 561–566.

Metternicht, G., Hurni, L., and Gogu, R. (2005) Remote sensing of landslides: an analysis of the potential contribution to geo-spatial systems for hazard assessment in mountainous environments. *Remote Sensing of Environment*, **98**, 284–303.

Nozette, S., Lichtenberg, C.L., Spudis, P. *et al.* (1996) The Clementine bistatic radar experiment. *Science*, **274**(29), 1495–1498.

Oguchi, T. and Ito, S. (1990) Multiple scattering effects on the transmission and reflection of millimeter pulse waves in rain. *Radio Science*, **25**(3), 205–216.

Pradhan, B., Singh, R.P., and Buchroithner, M.F. (2006) Estimation of stress and its use in evaluation of landslide prone regions using remote sensing data. *Advances in Space Research*, **37**, 698–709.

Shkuratov, Yu.G. and Bondarenko, N.V. (2001) Regolith layer thickness mapping of the Moon by radar and optical data. *Icarus*, **149**, 329–338.

Stacyn, J.S., Campbell, D.B., and Ford, P.G. (1997) Arecibo radar mapping of the lunar poles: a search for ice deposits. *Science*, **276**(6), 1527–1530.

Sun, G. and Ranson, K.J. (2000) Modeling lidar returns from forest canopies. *IEEE Transactions on Geoscience and Remote Sensing*, **38**(6), 2617–2626.

Thompson, T.W. (1987) High-resolution lunar radar map at 70-cm wavelength. *Earth, Moon, and Planets*, **37**, 59–70.

Tsang, L. and Kong, J.A. (1990) *Polarimetric Remote Sensing*, Elsevier, New York.

Tsang, L., Kong, J.A., and Ding, K.H. (2000) *Scattering of Electromagnetic Waves*, Vol. **1**, *Theory and Applications*. Wiley Interscience, New York.

Whitman, G.M., Schwering, F., Triolo, A., and Cho, N.Y. (1996) A transport theory of pulse propagation in a strongly forward scattering random medium. *IEEE Transactions on Antennas and Propagation*, **44**(1), 118–124.

Xu, F. and Jin, Y.Q. (2006) Multi-parameters inversion of a layer of vegetation canopy over rough surface from the system response function based on the Mueller matrix solution of pulse echoes. *IEEE Transactions on Geoscience and Remote Sensing*, **44**(6), 1–13.

Zebker, H.A. and van Zyl, J.J. (1991) Image radar polarimetry: a review. *Proceedings of the IEEE*, **79**(11), 1583–1605.

11

Fast Computation of Composite Scattering from an Electrically Large Target over a Randomly Rough Surface

Calculation of the radar cross-section (RCS) and scattering analysis for electrically large targets located in the natural environment have been under study for many years. Complexities due to large target size and multiple interactions between the isolated target and surrounding random media, for example, a rough surface, make the model simulation much more difficult. Instead of a purely numerical computation, high-frequency approximations have been well applied with good accuracy and convincing physical insight, which may be listed as geometrical optics (GO), physical optics (PO), geometrical theory of diffraction (GTD), physical theory of diffraction (PTD), ray tracing, and so on (Deschamps, 1972; Keller, 1962; Ufimtsev, 1962; Knott, Shaeffer, and Tuley, 1993).

To carry out an RCS calculation under high-frequency approximation, one approach is to decompose the target into a group of components with primitive geometries and calculate scattering from all components separately. Youssef (1989) introduced the RECOTA (return from complex target) code for RCS calculation, where the target is first meshed to facets, and PO/PTD scattering/diffraction of facets/edges are taken into account, as well as GO–PO double scattering between every pair of directly facing facets. The software XPatch (Hazlett *et al.*, 1999) is a commercialized mature tool for RCS and imaging signature calculation, where multiple scatterings from concave structures are also taken into account, using the shooting and bouncing rays (SBR) method (Ling, Chou, and Lee, 1989). Moreover, GRECO (graphical electromagnetic computing) is a real-time tool for computation of single scattering of the target (Rius, Ferrando, and Jofre, 1993), which takes advantage of the hardware acceleration technique of 3D graphics visualization. More recently, some efforts have been made on hybrid approaches of numerical and high-frequency methods (Tzoulis and Eibert, 2005).

Polarimetric Scattering and SAR Information Retrieval, First Edition. Ya-Qiu Jin and Feng Xu.
© 2013 John Wiley & Sons Singapore Pte. Ltd. Published 2013 by John Wiley & Sons Singapore Pte. Ltd.

Regarding the coexistence of the target and rough surface, it becomes of great interest to understand and simulate their composite scattering and mutual interactions, for example, a ship over the sea surface, or a vehicle on the land surface (Dehmollaian and Sarabandi, 2006; Burkholder, Pino, and Obelleiro, 2001; Li and Jin, 2002; Ye and Jin, 2007). As a more complicated model, the RCS calculation of a 3D target over a randomly rough surface remains for further study.

In this chapter, the bidirectional analytic ray tracing (BART) method is developed to rapidly calculate the composite scattering of a 3D electrically large target over a randomly rough surface (Xu and Jin, 2009). The basic idea is to launch ray tubes from both the source and the observation, trace among the facets, and then record the illumination areas of rays shooting on facets and edges. For each pair of forward and backward rays illuminating the same facet/edge, a scattering path from source to observation is constructed by linking the forward and backward rays, and the corresponding scattering contributions are added up to give the total scattering. Compared with the conventional unidirectional tracing, more scattering mechanisms can be taken into account in bidirectional tracing.

In addition, analytic tracing of the polygon ray tubes is implemented to precisely calculate their illumination, reflection, and shadowing based on geometric computation, such as polygon union, intersection, and difference. Since all tracing and shadowing are analytically treated, the restriction of electrically small facets is released, and the facet meshing is uniquely determined by the target geometry and becomes independent of its electrical size.

Rough facets are similarly introduced to model rough surfaces. Instead of reflection and diffuse scattering of flat facets, coherent scattering and incoherent scattering from rough facets are considered.

The principle of BART is introduced in Section 11.1. Then, some numerical examples of BART at electrically small scale are presented in comparison with conventional exact numerical approaches in Section 11.2. Finally, a numerical simulation, angularly composite scattering from a 3D electrically large target, for example, a ship, over a randomly rough surface is presented and discussed.

11.1 Bidirectional Analytic Ray Tracing

11.1.1 Bidirectional Tracing

A complex target is described in terms of facets and edges. Its RCS is contributed by direct (single) and multiple scattering from all facets and edges. The single scattering can be calculated straightforwardly using PO/PTD. However, PO/PTD cannot be applied directly to multiple scattering since multiple angular integrals would be involved and significantly reduce the efficiency of the high-frequency method.

To take account of high-order scattering, ray tracing with multiple GO reflections plus the last hop of PO scattering is usually employed (as, for example, in Ling, Chou, and Lee, 1989; Knott, 1977), which, in this chapter, is recognized as unidirectional tracing. A bidirectional ray tracing mode is proposed here with the aim of improving the accuracy and completeness of high-order scattering calculations.

As shown in Figure 11.1, a ray is launched from the source (Tx) along the incident direction and is kept for tracing as it shoots onto facets and reflects. Meanwhile, another "ray" is traced along the converse direction of scattering from the observation

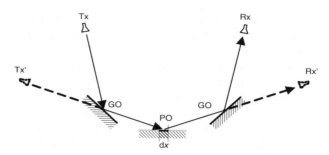

Figure 11.1 Bidirectional ray tracing. © [2009] IEEE. Reprinted, with permission, from *IEEE Transactions on Antennas and Propagation*

(Rx). When these two forward and backward rays meet at the same facet, it will compose a scattering path from the source to the observation after linking the forward and backward tracing paths by a PO scattering junction at this facet. If the forward ray takes n reflections, and the backward ray takes m reflections, then the ray path is experiencing $(m + n + 1)$th-order scattering.

As the facet is regarded as an integral of infinitesimal differential element dx, all GO reflections along the path can be treated equivalently by using the image method. This means that scattering from each differential element can be calculated equivalently as the wave is incident directly from its image source (Tx′) and scattered to the image observation (Rx′). Consequently, scattering calculation can be performed under the far-field approximation of PO scattering, taking into account the influences of the GO reflections applied to the incident and scattered fields.

The forward and backward rays illuminate a common area of the facet, and all the differential elements of this common area contribute in the same way to the $(m + n + 1)$th-order scattering. By carrying out the integral over this area, the nGO + PO + mGO scattering field vector is obtained as

$$\mathbf{E}_s = e^{ikr_0} \int e^{i(\mathbf{k}_i - \mathbf{k}_s)\cdot\mathbf{r}(x)}\, dx \prod_{b=1}^{m} \overline{\mathbf{R}}_b \cdot \overline{\mathbf{S}}_{\mathrm{PO}}^0(x) \cdot \prod_{f=1}^{n} \overline{\mathbf{R}}_f \cdot \mathbf{E}_i = e^{ikr_0} \prod_{b=1}^{m} \overline{\mathbf{R}}_b \cdot \overline{\mathbf{S}}_{\mathrm{PO}} \cdot \prod_{f=1}^{n} \overline{\mathbf{R}}_f \cdot \mathbf{E}_i$$

$$(11.1)$$

where $\overline{\mathbf{R}}$ denotes the GO reflection matrix, dx denotes a differential element, $\mathbf{r}(x)$ denotes its position with respect to the reference point $\mathbf{r}_0$, $\mathbf{k}_i, \mathbf{k}_s$ are the incident and scattered wavevectors, respectively, r_0 is the propagation distance of the reference point $\mathbf{r}_0$, $\overline{\mathbf{S}}_{\mathrm{PO}}^0(x)$ is the PO scattering matrix of one single differential element, and $\overline{\mathbf{S}}_{\mathrm{PO}}$ is the PO scattering matrix after integration.

Supposing a pair of forward and backward rays meet on an edge line, a $(n\mathrm{GO} + \mathrm{PTD} + m\mathrm{GO})$ scattering can be calculated in the same way according to diffraction theories of finite length edges, for example, the PTD, the method of equivalent current (MEC), and so on (Knott, Shaeffer, and Tuley, 1993). By integrating infinitesimal segments of equivalent edge currents, the diffracted field vector can be expressed identically as Equation 11.1, but with $\overline{\mathbf{S}}_{\mathrm{PO}}$ replaced by $\overline{\mathbf{S}}_{\mathrm{PTD}}$, if PTD is employed.

As rays are launched and traced simultaneously in both the forward and backward directions, all facets and edges that have been shot by the rays are recorded. As rays shoot at facets every time, new rays are generated by reflection and then traced, iteratively. The same process is repeated until the maximum tracing order as given is exceeded. Obviously, N orders of tracing will take into account $(2N + 1)$th-order multiple scattering.

Because the positions of the source and observation are always assumed in far-field regions, the rays can be seen as parallel lines or tubes. Suppose all facets are flat polygons; then the ray tubes generated during tracing can be described as polygonal cylinders. The area of the surface integral in Equation 11.1 is defined as the common area illuminated by the forward and backward polygonal ray tubes projecting on the polygon facet. Clearly, the geometry of this illumination area should be one or several polygons. Thus, $\overline{S}_{PO}$ can be derived directly using the analytical expression of PO scattering from a polygon surface (Gordon, 1975). Note that the derivations and expressions of PO/GO scattering matrices and the PTD diffraction coefficients are available in the literature (see, for example, Knott, Shaeffer, and Tuley, 1993; Gordon, 1975; Jin, 1994), and so will not be discussed any further in this chapter.

Likewise, it can be easily concluded that the common illumination area for two rays projecting on an edge line will be one or several line segments. Consequently, the analytical expression of $\overline{S}_{PTD}$ can be obtained directly, where the explicit forms of the PTD diffraction coefficients are available in Knott, Shaeffer, and Tuley (1993). The diffraction coefficients of other theories, for example, ILDC and FEC (Knott, 1985a,b; Michaeli, 1986), are essentially the same but with more complicated forms. Note that only diffractions of perfectly conducting facets/edges are considered in this chapter. As for targets with dielectric facets, more advanced diffraction theories for dielectric edges are available in a number of works in the literature (see, for example, Pelosi, Manara, and Nepa, 1998; Syed and Volakis, 1996).

It should be noted that the bidirectional tracing mode might cause the error of "double counting" of facet scattering. When a forward ray arrives at the facet, GO reflection occurs and tracing is continued. At last, as all scattering contributions are summed up, PO scattering of the facet is taken into account, which means the facet scattering is counted twice in different forms. On most occasions, there is no overlap between these two types of GO and PO scattering. Only when they have identical scattering angles, as shown in Figure 11.2, might the same scattering contribution be double counted.

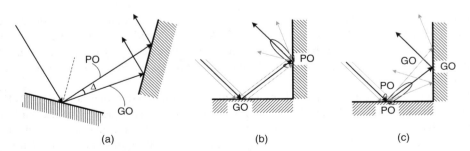

(a) (b) (c)

Figure 11.2 Double counting caused by bidirectional tracing. © [2009] IEEE. Reprinted, with permission, from *IEEE Transactions on Antennas and Propagation*

In fact, when the incident plane wave shoots at the facet, the scattered wave spreads out spherically over all directions under the PO approximation. But, under the GO approximation, the scattered wave propagates as a plane wave along the specular direction. So a facet located in the path of a GO reflection can receive the scattered powers of both GO and PO forms. Then, the scattering contribution will be double counted. As shown in Figure 11.2b and c, the middle rays have identical scattering paths. Generally speaking, PO scattering occurring near the specular direction is more likely to overlap with GO reflection.

In this section, the concept of "probability of occurrence" of GO/PO scattering is introduced to solve this conflict. As a forward ray is incident on a facet, it is regarded that both GO reflection and PO scattering could happen with the same probability. The probability is determined by the difference Δ between their scattering angles as

$$p = \min\left(\frac{1}{2} + \frac{\Delta}{\pi}, 1\right) \tag{11.2}$$

The final scattering contribution is then weighted and summed up according to the probabilities. So, in Equation 11.1, the PO scattering matrix is replaced by $p\overline{S}_{PO}$, and the forward reflection matrix is replaced by $p_f\overline{R}_f$, where p_f denotes the probability of occurrence of the fth-order forward reflection.

Following Equation 11.2, when PO and GO overlap in the specular direction, they are regarded as the same scattering, so the probabilities of occurrence are both 1/2 and the total probability of occurrence is 1. As PO scattering deviates away from the specular direction, they stand for different scattering mechanisms.

Nevertheless, one more accurate and deterministic way to cope with the conflict of PO/GO in bidirectional ray tracing would be to allocate the scattering power based on information about the exact interception solid angle of each neighboring facet. This would result in heavy computations due to the fact that the occurrence of PO/GO conflict closely depends on the specific geometry and facetization of the problem.

Since the high-frequency method is seen as an approximate approach, it might not be worth pursuing high precision with its extra cost of computational resources. Therefore, we propose this probabilistic method to reduce the PO/GO conflict. Although it might raise some non-deterministic errors, in real cases, in fact, the conflict of PO diffuse scattering and GO reflection only occurs in very rare situations and observing directions. The resulting error is not noticeable and is very likely to submerge into the background error of high-frequency methods.

With the prevention of possible overlapping of GO/PO scattering, both of the two types of scattering can be taken into account. It is somewhat better than conventional unidirectional tracing, in which PO is applied only to the last hop of scattering. Theoretically, under N orders of tracing, only N types of scattering are considered in unidirectional tracing, while bidirectional tracing takes into account N^2 types, including all N types of unidirectional tracing.

The accuracy of a method is determined by how it can take into account the most important or major scattering contributions. It closely depends on the target geometry and its situation in the environment. Since bidirectional ray tracing can include further extra scattering contributions, in general, it will give a higher accuracy and take into account the most important scattering contributions.

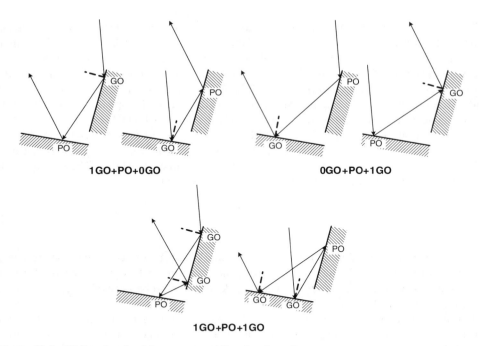

1GO+PO+0GO   0GO+PO+1GO

1GO+PO+1GO

Figure 11.3 Bidirectional tracing versus unidirectional tracing. © [2009] IEEE. Reprinted, with permission, from *IEEE Transactions on Antennas and Propagation*

For the case shown in Figure 11.3, four double scattering terms 1GO + PO + 0GO and 0GO + PO + 1GO, and two triple scattering terms 1GO + PO + 1GO, are all calculated in second-order bidirectional tracing, while only the two double scattering terms 1GO + PO + 0 GO are taken into account in third-order unidirectional tracing.

In fact, bidirectional tracing includes all scattering contributions composed by one diffuse scattering (PO/PTD) and several reflections (GO). Comparing with all-PO multiple scattering (Gressier and Balanis, 1987), the ignored scattering terms, which have experienced diffuse scattering more than twice, are actually very small and negligible.

11.1.2 Analytic Tracing

A critical issue of ray tracing is the calculation of shadowing and reflection. Conventionally, either a dense ray grid is launched or the surface is discretized into small elements; the rays are considered as lines without taking account of the space between them. Scattering contributed by a ray path can be counted by multiplying by an equivalent area of cross-section. In SBR, scattering is calculated by interpolating the equivalent fields among the ray grids. The accuracy is most likely influenced by the electrical fineness (relative to the wavelength) of the ray grid or the surface mesh, and the complexity is directly related to the target electrical size. A recent method (Weinmann, 2006) is to randomly send sparse rays and take the average for reducing the extra complexity. However, its accuracy can be affected by systematic errors (Weinmann, 2006).

An analytic tracing technique is employed in this section. Ray tubes are described as polygonal cylinders. Shadowing is accurately determined by geometric calculation of

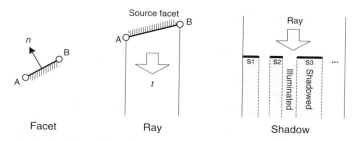

Figure 11.4 Definitions of the facet, ray, and shadow. © [2009] IEEE. Reprinted, with permission, from *IEEE Transactions on Antennas and Propagation*

polygons. Usually, an electrically large object can be modeled by large patches without losing precision if accurate analytic ray tracing is employed, so as to significantly reduce the complexity. To the best of our knowledge, only the Xpatch software uses the triangular operations to analytically calculate shadowing, but multiple scattering is still calculated by using the SBR technique.

As a concise example, we introduce the idea of analytic tracing first in the two-dimensional situation and then extend it to three dimensions. First of all, mathematical representations of the facet, ray, and shadow are defined in Figure 11.4. A facet is described by an ordered pair of points A–B, where its normal direction is on the left of A → B. A ray is defined by its direction $\hat{t}$ and its source facet A–B from which the ray is launched or reflected, that is, $R = [\hat{t}, A-B]$. The zeroth-order ray, for example, the initial incident ray, is simply seen as from afar. A shadow cast by the ray after being partially intercepted by different facets can be represented by a cluster of segments (intervals) $S = [s_1, s_2, \ldots]$, where each segment indicates a shadowed region and each gap indicates an illumination region.

As shown in Figure 11.5a, ray R (a ray tube) is traced in its local coordinates (w, t), where the axis $\hat{t}$ is the illuminating direction. Clearly, the coordinate t indicates the distance to the source facet along the illumination direction. Two positions with the same coordinate w have shadowing relationship, and the one with smaller t blocks the other.

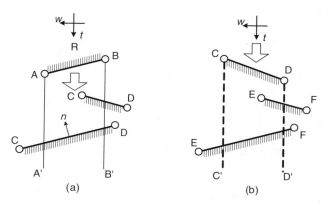

Figure 11.5 Ray tracing and shadowing calculation. © [2009] IEEE. Reprinted, with permission, from *IEEE Transactions on Antennas and Propagation*

The necessary condition for a facet C–D to be illuminated by the ray R is *Criterion I*:

- It faces the source facet of the ray, that is, $\hat{\imath} \cdot \hat{n} < 0$.
- At least one of its end points, C or D, is located within the illumination zone A′ABB′, or CD itself intersects ray AA′ or ray BB′.

According to *Criterion I*, those facets with no illumination are excluded for tracing at the first step.

Among the remaining facets, the shadowing relationship is determined. As shown in Figure 11.5b, a facet C–D produces a shadowing zone C′CDD′. Similarly, the necessary condition for another facet E–F to be blocked by C–D is *Criterion II*:

- At least one of its end points, E or F, is located within the shadowing zone C′CDD′, or EF itself intersects ray CC′ or ray DD′.

According to *Criterion II*, the remaining facets are sorted in a sequence where the facets behind never block the front facets.

The following procedure is performed to determine the illumination area cast by the ray R on each facet in the sequence:

1. Initialize the shadow as $S_R = [s_A, s'_{AB}, s_B]$, where s_A, s_B stand for the regions outside of the left of A and the right of B, respectively, and s'_{AB} is the shadow of its father ray R′ casting on A–B.
2. Visit each facet in the sequence. For the currently visited facet C–D, cut out the corresponding part S_{CD} between C and D from S_R, which is the shadow the ray R casts on the facet C–D, and the illumination area is the inverse of the shadow.
3. Overlap the shadow segment projected by the facet C–D onto S_R, and reorganize S_R. Then, return to step 2 until the end of the sequence.

Note that if a ray R′ illuminates a facet and reflects off to a higher-order ray R, they are referred to as the father ray and the child ray, respectively. Since the zeroth-order ray has no father ray, the initial shadow of the step 1 should be set to empty. Moreover, calculations of shadows can be realized straightforwardly by logical operations of intervals.

Each ray is traced by executing this procedure once. Information about the ray and its reflection is recorded when determining the shadow S_{CD} cast by the ray R on the facet C–D. A new higher-order ray is then generated from the reflection. All higher-order rays generated in this round will be traced in the next round. The entire process is repeated for the next order of tracing until the maximum number of orders is exceeded or no reflection occurs and no higher-order ray is generated during the entire round.

This routine can be realized recursively and is generally applicable to ray tracing problems with flat facets and straight edges. Note that an edge point is regarded as a facet with two end points overlapped and then can be handled in the same way. The difference is that the edges do not reflect a new ray and the first condition of *Criterion I* should be modified to "At least one of the two facets composing the edge faces the source facet of the ray".

After tracing in the forward and backward directions are finished, we have already recorded the shadows cast by forward rays, that is, S_f^+ ($f = 1, 2, \ldots, F$), and the shadows cast by backward rays, that is, S_b^- ($b = 1, 2, \ldots, B$). The common illumination area $I_{f,b}$ of the fth forward ray and the bth backward ray can be obtained by first overlapping their shadows S_f^+ and S_b^- and then calculating the inverse one. Following Equation 11.1, the corresponding scattering field vector is obtained by integrating over this common illumination area. Summing up all the scattering terms from all facets and edges, this yields the total scattering field, that is

$$\mathbf{E}_{st} = \sum_{\text{facets \& edges}} \sum_{f=1}^{F} \sum_{b=1}^{B} \mathbf{E}_s(\mathbf{I}_{fb}) \tag{11.3}$$

To extend the 2D solution to 3D, we only need to:

(i) define the facets as polygons, the edges as segments, the rays as polygonal cylinders, and the shadows as polygon sets;
(ii) modify *Criterion I* and *Criterion II*, correspondingly;
(iii) realize the shadowing calculation based on geometric operation of polygons.

In fact, as an arbitrary ray tube of a polygonal cylinder shoots on a polygon facet, it is easy to prove that the reflected ray tube is also a polygonal cylinder. Moreover, as it might propagate through and be blocked by several polygon facets, the projected shadow can be described by a large illuminated polygon, which contains a number of polygonal shadow regions. Thus, it straightforwardly extends to 3D shadowing calculation.

In fact, the basic calculations involved here are just the logical operations of polygons, that is, the union, intersection, and difference of two polygons, which can be implemented by using an algorithm in computational geometry (Zhou, 2005). For example, as shown in Figure 11.6, when a triangular ray tube shoots on a triangular facet, the illumination area is equal to the intersection of the two triangles, and so continues as the reflected ray tube. The passed region after blocking is a set of polygons that equals the difference of the two triangles.

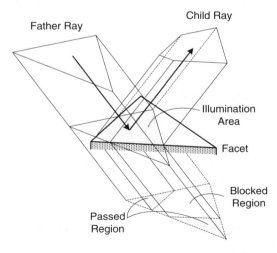

Figure 11.6 Three-dimensional analytic tracing and polygon calculations. © [2009] IEEE. Reprinted, with permission, from *IEEE Transactions on Antennas and Propagation*

11.1.3 Rough Facets

Under high-frequency approximation, a straightforward way is to mesh the rough surface with very small flat facets. This is actually equivalent to the conventional Kirchhoff approximation. However, fine meshing of small facets will cause extra complexity of computation, especially in electrically large problems.

In this chapter, an analytical rough surface model such as the integral equation method (IEM) (Fung, 1994) is employed. Based on the surface geometry and its conjunction with the target, the rough surface is first meshed into electrically large patches, that is, rough facets (see Figure 11.10). The rough facets are usually much larger than the correlation length of the rough surface. So we can ignore the inter-correlations among different rough facets and consider each of them as a complete randomly rough surface. Thus, analytic methods like IEM can be used to calculate scattering of all rough facets, independently.

Note that the whole exposed rough surface might be treated as one large patch, if it has no topographic fluctuation. However, in that case, the irregularity of the shape of the rough surface might cause an extra complexity and inconvenience in computation.

During ray tracing, rough facets are processed in the same way as flat facets except for their scattering calculation. Rough surface scattering includes two parts: coherent and incoherent (Fung, 1994). For a large rough surface, the coherent scattering is concentrated in the specular direction, and incoherent scattering diffusely spreads over all directions. They are somewhat similar to the GO and PO scattering of flat surfaces, respectively.

Coherent scattering of a rough facet is approximated as GO reflection modified with a factor determined by the roughness, that is

$$\overline{\mathbf{R}}_{\text{rough}} = e^{-4k_n^2\sigma^2} \cdot \overline{\mathbf{R}} \tag{11.4}$$

where k_n denotes the component of the wavevector in the normal direction of the facet and σ is the standard variance of the rough surface height.

Incoherent scattering is the second-order moment of random scattering fields. For a Gaussian rough surface, suppose the random scattering field obeys the Gaussian distribution

$$P(\mathbf{S}_{\text{rough}}) = \frac{1}{\pi^4 |\overline{\mathbf{C}}|} \exp\left(-\mathbf{S}_{\text{rough}}^+ \cdot \overline{\mathbf{C}}^{-1} \cdot \mathbf{S}_{\text{rough}}\right) \tag{11.5}$$

where $\mathbf{S}_{\text{rough}} = [S_{11}, S_{12}, S_{21}, S_{22}]_{\text{rough}}^{\text{T}}$ is the vectorized scattering matrix of the rough facet and its covariance matrix $\overline{\mathbf{C}}$ is equivalent to the Mueller matrix. Then, it can be recovered from the second-order statistical moment (Xu and Jin, 2006), that is

$$\mathbf{S}_{\text{rough}} = \overline{\mathbf{C}}^{1/2} \cdot \mathbf{s}^0, \quad s_j^0 = \xi_j + i\zeta_j, \quad j = 1, \dots, 4 \tag{11.6}$$

whereas ξ_j, ζ_j are independent Gaussian random numbers with zero mean and unit variance. The covariance matrix $\overline{\mathbf{C}}$ can be transformed from the Mueller matrix (Cloude and Pottier, 1996) and its lower triangular square root $\overline{\mathbf{C}}^{1/2}$ can be obtained by Cholesky decomposition (Novak, Burl, and Michael, 1990). The Mueller matrix of the IEM rough surface model has been given in Fung (1994).

Therefore, rough facets can be handled in the same way after replacing the GO reflection matrix by the coherent scattering matrix $\overline{\mathbf{R}}_{\text{rough}}$ in Equation 11.4, and replacing the PO scattering matrix $\overline{\mathbf{S}}_{\text{PO}}$ by the randomly generated scattering matrix $\overline{\mathbf{S}}_{\text{rough}}$ in Equation 11.6. Note that both coherent and incoherent scattering coexist in rough surface scattering, and thus no "double counting" issue will arise for rough facets.

The diffuse scattering of rough facets is stochastically generated based on the IEM-calculated covariance matrix. Theoretically, it is necessary to perform multiple realizations to obtain the averaged RCS of total scattering power. However, in practical operation, scattering contributions involved or not involved with the randomly generated scattering of rough facets are calculated, respectively. Those scattering contributions independent of random scattering are deterministic and only need calculating once. The scattering from the rough surface and interactive terms between the target and rough surface are calculated several times and then averaged to obtain the RCS. During each realization, the random numbers of Equation 11.6 for each rough facet are actually fixed in order to preserve the coherency among different scattering terms involved with the same rough facet.

11.2 Numerical Results

The instructions of the BART calculation are as follows:

1. Construct the geometric models of the target and rough surface using computer-aided design tools and mesh to facets, as large as possible to reduce the complexity of ray tracing. Basically, the flat part should be kept without further meshing, while the curved part should be meshed based on the compromise between accuracy and efficiency requirements.
2. Send the zeroth-order rays in both the forward and backward directions for the BART routine and record the tracing illumination of all facets and edges.
3. Calculate the scattering contributions from all facets and edges by linking every pair of forward and backward rays. Random scattering from random samples of rough facets is calculated and averaged.

To validate our method, the monostatic RCS of a perpendicular dihedral with two identical square plates ($5.6088\lambda \times 5.6088\lambda$, where λ denotes the wavelength) is calculated at 9.4 GHz. The results of BART, the method of moments (MoM), and measurement are shown in Figure 11.7. It can be seen that the BART result well matches MoM and measurement, except

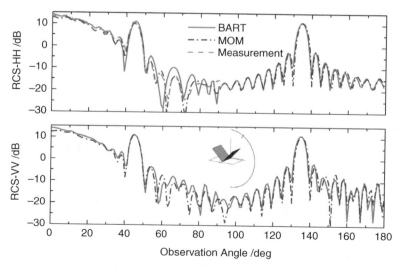

Figure 11.7 Comparison of BART, MoM, and measurement for RCS of a dihedral. © [2009] IEEE. Reprinted, with permission, from *IEEE Transactions on Antennas and Propagation*

for the region 45°–90° where the multiple diffraction contribution becomes significant, that is, diffracted over the first plate, scattered from the second plate, and finally diffracted back.

Another example is a 2λ cube over an $8\lambda \times 8\lambda$ square plate, for which Figure 11.8 also shows the good agreement of monostatic RCS of BART and MoM.

As a comparison with other methods dealing with the composite model of a target and a rough surface, such as GFBM/SAA in the 2D case (Li and Jin, 2002), a model of a ship-like target over the sea surface is considered in Figure 11.9a. The sea surface is regarded as a Gaussian rough surface with roughness $k\sigma = 0.63$ and correlation $k\ell = 7.54$. The IEM

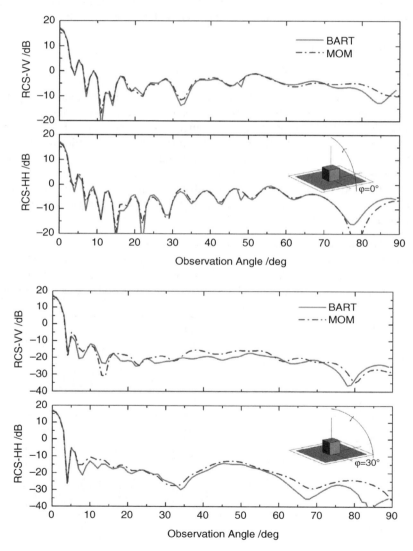

Figure 11.8 Comparison of BART and MoM for RCS of a cube on plane: (a) $\varphi = 0°$; (b) $\varphi = 30°$. © [2009] IEEE. Reprinted, with permission, from *IEEE Transactions on Antennas and Propagation*

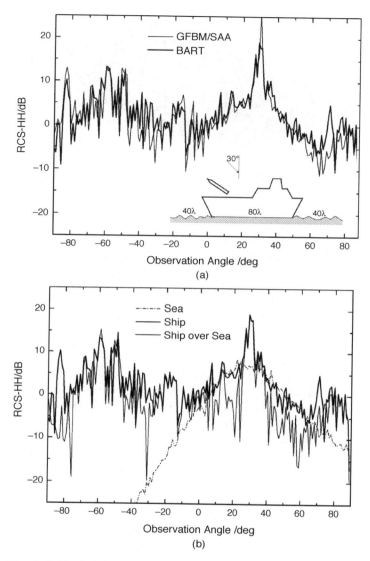

Figure 11.9 Bistatic RCS of a 2D ship over sea surface: comparison of BART versus GFBM/SAA. © [2009] IEEE. Reprinted, with permission, from *IEEE Transactions on Antennas and Propagation*

model is employed for rough facets in the BART calculation. The bottom length of the ship is 80λ and the sea surfaces on its two sides are both 40λ. Under the incidence of $30°$, the normalized bistatic RCS is calculated using BART and GFBM/SAA. The results are given in Figure 11.9a, which shows the good performance of BART.

To see the difference between composite scattering and independent scattering, the results of three cases, namely, rough sea without ship, only ship without sea, and ship over the sea, are shown in Figure 11.9b. It can be seen that interactions between the target and rough surface are prominent, especially around low grazing angles.

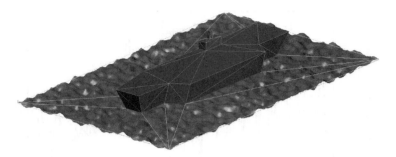

Figure 11.10 A 3D model of a ship over a rough sea surface. © [2009] IEEE. Reprinted, with permission, from *IEEE Transactions on Antennas and Propagation*

An alternative way to verify the correctness of the result is to examine energy conservation. In the case of Figure 11.9a, the normalized scattered energies of HH and VV polarization integrated over all directions are 0.9926 and 1.0126, respectively.

It is especially noted that BART is much faster than MoM and GFBM/SAA. In the case of Figure 11.9a, GFBM/SAA takes 41 minutes to calculate one realization using an ordinary personal computer with a 3 GHz processor, while BART needs only 2 seconds. Moreover, as mentioned above, how to mesh the facets is uniquely determined by the target geometry, not by the electrical size. If the target size becomes large, that would not increase computation complexity as long as the target geometry is kept invariant.

We now turn to the 3D case, as shown in Figure 11.10, where an example of the facet mesh is also shown. The maximum length and width of the ship are 200 m and 50 m, respectively. The deck is 20 m above the sea surface. The underlying surface is taken to be as large as 300 m × 150 m to ensure that potential interactions between the target and the rough surface will not be overlooked even at grazing angles as low as 70°. In order to evaluate the performance of our method, we first consider the scale-reduced model, for example, proportionally shrinking by 100 times, that is, the 200 m length becomes 2 m, at L band, 1.26 GHz.

In Figure 11.11, we compare the angular monostatic RCS of a 2 m ship target without the underlying surface calculated using BART and MLFMM (multi-layer fast multipole method), respectively. In Figure 11.12 is shown the monostatic RCS of a 2 m ship target with an underlying flat surface. The results of BART in Figures 11.11 and 11.12 show good accuracy as proved by MLFMM. Note that much stronger scattering is observed at its two sides due to large cross-section at the side views. Besides, the RCS with underlying flat surface is enhanced significantly in most directions. It is noticed that the asymmetric RCS of Figure 11.12c is caused by the asymmetric side structure of the ship.

Note that MoM is not used in these cases because too much CPU time is needed. In the case of Figure 11.12, the time for computing 180 observation angles using an ordinary personal computer (3 GHz processor) only needs 9 seconds for BART, but takes 3.6 hours for MLFMM, and more than 2 days for MoM. The memory required is about 3.0 MByte for BART, 58.4 MByte for MLFMM, and 2614.56 MByte for MoM.

The monostatic RCS of a ship only, a ship with underlying flat surface, and a ship with rough surface ($k\sigma = 0.63, kL = 7.54$) are presented in Figure 11.13. It can be seen

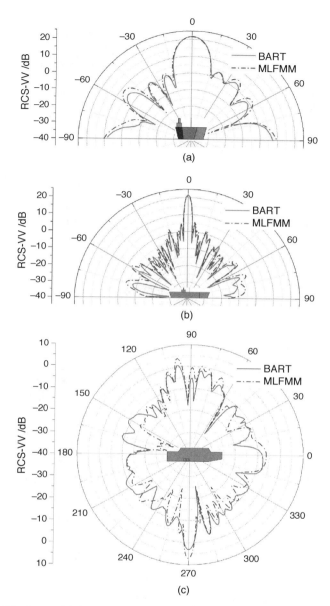

Figure 11.11 RCS of a model 2 m ship target alone, BART versus MLFMM: (a) incidence on the side; (b) incidence on the end; (c) azimuthal incidence at 50°. © [2009] IEEE. Reprinted, with permission, from *IEEE Transactions on Antennas and Propagation*

that the underlying surface, either flat or rough, enhances the RCS, especially in the directions perpendicular to the side face of the ship, for example, 0°, 45°, 90°, 135°, 180°, 270°. As the underlying surface becomes rough, the angular fluctuation of the RCS becomes smoother.

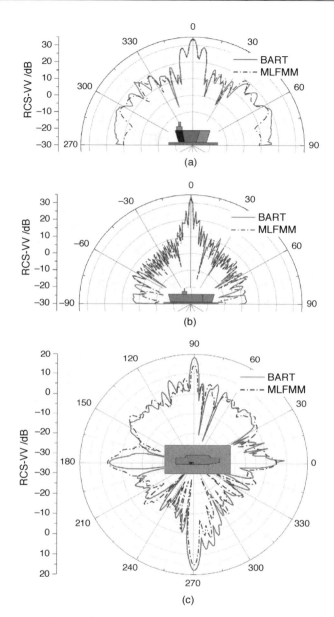

Figure 11.12 RCS of a model 2 m ship target on a plane, BART versus MLFMM: (a) incidence on the side; (b) incidence on the end; (c) azimuthal incidence at 50°. © [2009] IEEE. Reprinted, with permission, from *IEEE Transactions on Antennas and Propagation*

We now return to the original size, that is, a 200 m ship over 300 m sea surface at L band. Now only BART can work for the RCS computation. Note that hereafter all RCSs are normalized to the total illumination area.

In Figure 11.14, comparison between the RCSs of a ship only with the size 200 m and 2 m show the difference, especially the peaks at 0°, 90°, 180°, 270°. After investigating all the

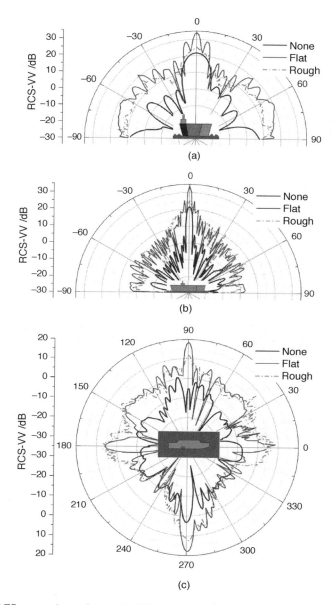

Figure 11.13 RCS comparison of target in different situations, i.e., no surface, flat surface, and rough surface: (a) incidence on the side; (b) incidence on the end; (c) azimuthal incidence at 50°. © [2009] IEEE. Reprinted, with permission, from *IEEE Transactions on Antennas and Propagation*

scattering terms, it is found that the 90° dihedral structures composed by the tower walls and the deck cause these peaks. Absolutely perpendicular structures are usually avoided in real design. Hence, we modify the inclination angle of the tower walls to 95°. Its RCS (green line) with smoother peaks is also shown in Figure 11.14.

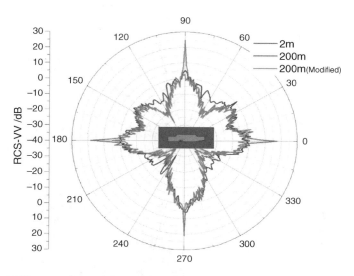

Figure 11.14 RCS comparison of 2 m model, 200 m model, and modified 200 m model. © [2009] IEEE. Reprinted, with permission, from *IEEE Transactions on Antennas and Propagation*

Figure 11.15 illustrates the different scattering terms, which have been taken into account in BART. Note that the contribution from the target itself might include the first, second, and third orders, the contribution from the sea surface could include only the first order, while their interactions might include the second, third, fourth, and fifth orders.

In Figure 11.16, we show different parts of the RCS in the case of the modified 200 m model. It can be seen that the angular fluctuation of the ship scattering is smoothed in the total RCS due to interactions of the ship and the rough surface. The interactions are prominent particularly in those directions facing straight toward the side faces of the ship.

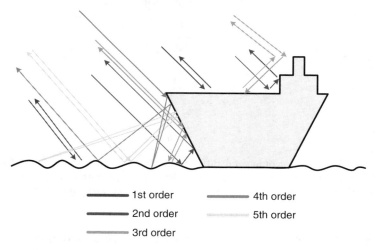

Figure 11.15 Different orders of scattering. © [2009] IEEE. Reprinted, with permission, from *IEEE Transactions on Antennas and Propagation*

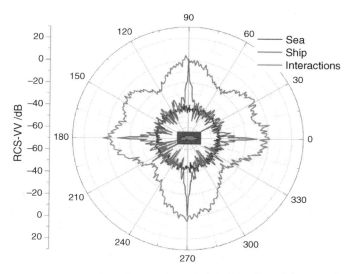

Figure 11.16 Contributions from sea, ship, and their interactions in the case of the modified 200 m model: azimuthal variation under incidence at 50°. © [2009] IEEE. Reprinted, with permission, from *IEEE Transactions on Antennas and Propagation*

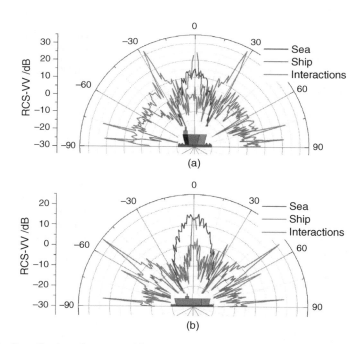

Figure 11.17 Contributions from sea, ship, and their interactions in the case of the modified 200 m model: (a) incidence on the side; (b) incidence on the end. © [2009] IEEE. Reprinted, with permission, from *IEEE Transactions on Antennas and Propagation*

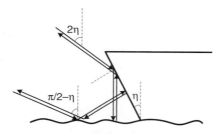

Figure 11.18 Causes of third-order scattering peaks. © [2009] IEEE. Reprinted, with permission, from *IEEE Transactions on Antennas and Propagation*

Figure 11.17 shows the angular distribution of the RCS in the side and end planes. The interactions play a dominant role at most angles except for small angles around 20°–30°, where rough surface scattering is more significant. Note the peaks at ±85° are directly reflected from the tower walls after modification.

In addition, the interaction peaks at some angles, for example, ±28° and ±76°, happen to be due to the third order of interactive scattering. As explained in Figure 11.18, if the side face of the ship has an inclination angle of η, there exist two special monostatic observation angles 2η and $\pi/2 - \eta$, under which the third-order scattering paths are composed by two specular reflections and one 0° reflection.

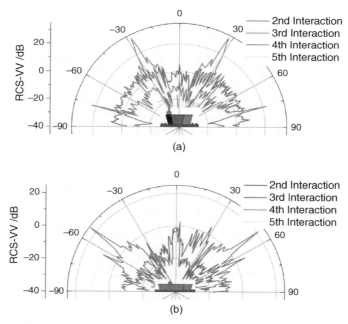

Figure 11.19 Different orders of interactions between ship target and sea surface: (a) incidence on the side; (b) incidence on the end. © [2009] IEEE. Reprinted, with permission, from *IEEE Transactions on Antennas and Propagation*

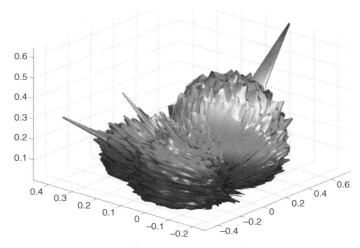

Figure 11.20   Angular pattern of bistatic scattering RCS

In Figure 11.19, the different orders of interactions corresponding to Figure 11.16 are shown separately. Except for these peaks caused by the third-order interaction, the main contributions are from the second-order interaction. Note that higher-order interactions around 0° are contributed by the concave faces on the left side of the ship. The contributions from the fourth and fifth orders of scattering are negligible.

Multiple scattering plays a prominent role here, not only due to the large structure of the complex target but also due to interactions between the target and the rough surface. As a conclusion, a high-frequency method that is capable of taking into account multiple scattering up to the third order is required for RCS prediction of an electrically large complex target or a target over a rough surface environment.

An example of a normalized bistatic RCS pattern is shown in Figure 11.20, where the incidence angle is taken as (50°, 0°). It can be seen that there is significant scattering at specular and backward directions. Figure 11.21a–d show the backscattering from a ship over a rough sea surface, where the individual parts show total backscattering, and contributions from the ship, the ocean surface, and interactions of ship and surface, respectively.

In this chapter, the bidirectional analytic ray tracing (BART) method has been developed to calculate the composite RCS of a 3D electrically large ship-like target over a randomly rough surface. Meshing the target and surface with facets and edges, bidirectional ray tracing is performed along both the forward and backward directions. As a pair of forward and backward rays meets on the same facet or edge, a diffuse scattering path of this facet or edge is constructed. Scattering from rough surface facets is calculated using an IEM model. Analytical ray tracing is realized to precisely describe the ray tubes and calculate the shadowing, which makes complexity independent of its electrical size. Therefore, rapid computation is implemented for RCS calculations of a complex target with rough surface interactions.

The BART is validated by comparing RCS calculations with other numerical methods and measurements in the case of low-frequency observations. It is demonstrated that

BART can achieve good accuracy with a very low requirement of computational resources. BART is then applied to RCS calculation and analysis of a 3D electrically large ship-like target over a rough sea surface. It is found that the RCS contributed by interactions between the target and the rough surface is very significant, and the

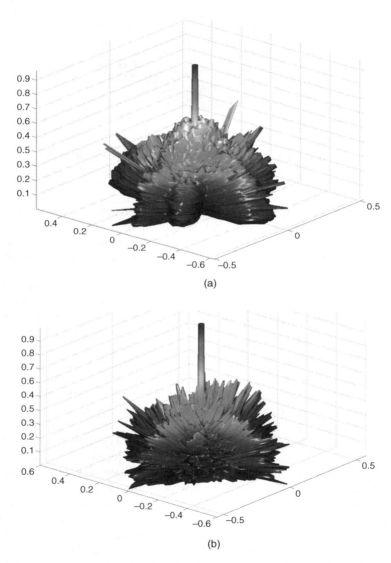

Figure 11.21 Backscattering contributed by different mechanisms: (a) total backscattering; (b) scattering from a ship; (c) scattering from underlying rough surface; (d) scattering due to interactions of ship and rough surface

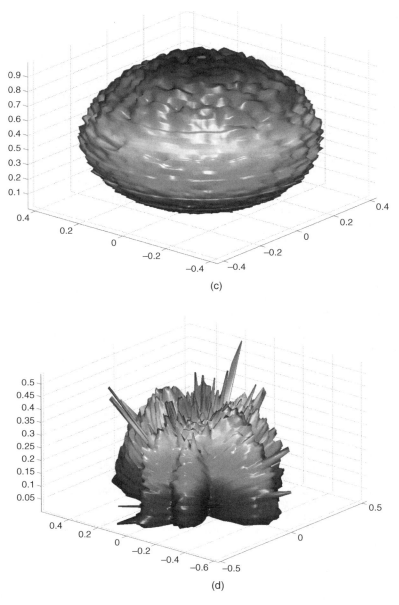

(c)

(d)

Figure 11.21 (*Continued*)

influence of the rough surface on the composite RCS is also prominent. A high-frequency method that is capable of taking into account multiple scattering up to the third order is required for RCS prediction of an electrically large complex target or a target over a rough surface environment.

References

Burkholder, R.J., Pino, M.R., and Obelleiro, F. (2001) A Monte Carlo study of the rough-sea-surface influence on the radar scattering from two-dimensional ships. *IEEE Antennas and Propagation Magazine*, **43**(2), 25–33.

Cloude, S. and Pottier, E. (1996) A review of target decomposition theorems in radar polarimetry. *IEEE Transactions on Geoscience and Remote Sensing*, **34**(2), 498–518.

Dehmollaian, M. and Sarabandi, K. (2006) Electromagnetic scattering from foliage camouflaged complex targets. *IEEE Transactions on Geoscience and Remote Sensing*, **44**(10), 2698–2709.

Deschamps, G.A. (1972) Ray techniques in electromagnetics. *Proceedings of the IEEE*, **60**(9), 1022–1035.

Fung, A.K. (1994) *Microwave Scattering and Emission Models and Their Applications*, Artech House, Norwood, MA.

Gordon, W.B. (1975) Far field approximation of the Kirchhoff–Helmholtz representation of scattered fields. *IEEE Transactions on Antennas and Propagation*, **AP-23**(7), 864–876.

Gressier, T. and Balanis, C.A. (1987) Backscatter analysis of dihedral corner reflectors using physical optics and the physical theory of diffraction. *IEEE Transactions on Antennas and Propagation*, **AP-35**(10), 1137–1147.

Hazlett, M.A., Andersh, D.J., Lee, S.W. *et al.* (1999) XPATCH: a high-frequency electromagnetic scattering prediction code using shooting and bouncing rays. *Proceedings of SPIE*, **2469**, 266–275.

Jin, Y.Q. (1994) *Electromagnetic Scattering Modeling for Quantitative Remote Sensing*, World Scientific, Singapore.

Keller, J.B. (1962) Geometrical theory of diffraction. *Journal of the Optical Society of America.*, **52**, 116–119.

Knott, E.F. (1977) RCS reduction of dihedral corners. *IEEE Transactions on Antennas and Propagation*, **AP-25**(3), 406–409.

Knott, E. (1985a) The relationship between Mitzner's ILDC and Michaeli's equivalent currents. *IEEE Transactions on Antennas and Propagation*, **33**(1), 112–114.

Knott, E.F. (1985b) A progression of high-frequency RCS prediction techniques. *Proceedings of the IEEE*, **73**, 252–264.

Knott, E.F., Shaeffer, J.F., and Tuley, M.T. (1993) *Radar Cross Section*, 2nd edn, Artech House, London.

Li, Z. and Jin, Y.Q. (2002) Bistatic scattering and transmitting through a fractal rough surface with high permittivity using the physics-based two-grid method in conjunction with the forward–backward method and spectrum acceleration algorithm. *IEEE Transactions on Antennas and Propagation*, **50**(9), 1323–1327.

Ling, H., Chou, R.C., and Lee, S.W. (1989) Shooting and bouncing rays: calculating the RCS of an arbitrarily shaped cavity. *IEEE Transactions on Antennas and Propagation*, **AP-37**(2), 194–205.

Michaeli, A. (1986) Elimination of infinities in equivalent edge currents, part I: Fringe current components. *IEEE Transactions on Antennas and Propagation*, **AP-34**(7), 912–918.

Novak, L., Burl, M., and Michael, C. (1990) Optimal speckle reduction in polarimetric SAR imagery. *IEEE Transactions on Aerospace Electronic Systems*, **26**(2), 293–305.

Pelosi, G., Manara, G., and Nepa, P. (1998) Electromagnetic scattering by a wedge with anisotropic impedance faces. *IEEE Antennas and Propagation Magazine*, **40**(6), 29–34.

Rius, J.M., Ferrando, M., and Jofre, L. (1993) GRECO: graphical electromagnetic computing for RCS prediction in real time. *IEEE Antennas and Propagation Magazine*, **35**(2), 7–17.

Syed, H.H. and Volakis, J.L. (1996) PTD analysis of impedance structures. *IEEE Transactions on Antennas and Propagation*, **44**(7), 983–988.

Tzoulis, A. and Eibert, T.F. (2005) A hybrid FEBI-MLFMM-UTD method for numerical solutions of electromagnetic problems including arbitrarily shaped and electrically large objects. *IEEE Transactions on Antennas and Propagation*, **53**, 3358–3366.

Ufimtsev, P.Ya. (1962) *Metod krayevykh Voln v Fizicheskoy Teorii Difraktsii*, Sovetskoye Radio, Moscow, pp. 1–243 (in Russian) [transl. Foreign Technology Division, USAF (1971) *Method of Edge Waves in the Physical Theory of Diffraction*, NTIS, Springfield, VA].

Weinmann, F. (2006) Ray tracing with PO/PTD for RCS modeling of large complex objects. *IEEE Transactions on Antennas and Propagation*, **54**(6), 1797–1806.

Xu, F. and Jin, Y.Q. (2006) Imaging simulation of polarimetric SAR for a comprehensive terrain scene using the mapping and projection algorithm. *IEEE Transactions on Geoscience and Remote Sensing*, **44**(11), 3219–3234.

Xu, F. and Jin, Y.Q. (2009) Bidirectional analytic ray tracing for fast computation of composite scattering from an electric-large target over randomly rough surface. *IEEE Transactions on Antennas and Propagation*, **57**(5), 1495–1505.

Ye, H.X. and Jin, Y.Q. (2007) A hybrid analytic–numerical algorithm of scattering from an object above rough surface. *IEEE Transactions on Geoscience and Remote Sensing*, **45**(5), 1174–1180.

Youssef, N.N. (1989) Radar cross section of complex targets. *Proceedings of the IEEE*, **77**(5), 722–733.

Zhou, P.D. (2005) *Computational Geometry–Algorithm Design and Analysis*, Tsinghua University Press, Beijing.

12

Reconstruction of a 3D Complex Target using Downward-Looking Step-Frequency Radar

Target detection and reconstruction have been of great interest in both civilian and defense applications (Firoozabadi et al., 2007; Wang, Zhang, and Amin, 2006; Yoon and Amin, 2008; Chang et al., 2009; Chan, Kuga, and Ishimaru, 1999; Kidera, Sakamoto, and Sato, 2008, 2009; Du et al., 2010; Mahafza and Sajjadi, 1996; Jun et al., 2010; Matthias and Markus, 2010; Lopez-Sanchez and Fortuny-Guasch, 2000; Fortuny-Guasch and Lopez-Sanchez, 2001; Xu and Narayanan, 2001; Budillon, Evangelista, and Schirinzi, 2011; Qi et al., 2008; Fornaro, Reale, and Serafino, 2009), for example, for reconnaissance and surveillance of objects in the natural environment, such as a vehicle under foliage, or a ship on a rough sea surface, and for rescue missions for searching for survivors after an earthquake or avalanche, and so on.

Some techniques of two-dimensional (2D) imaging for target reconstruction have been well reported (e.g., Wang, Zhang, and Amin, 2006; Yoon and Amin, 2008; Chang et al., 2009). The 2D radar imaging was acquired by projecting a 3D target onto a certain 2D plane, where some geometric distortions (e.g., foreshortening, sheltering, layover, etc.), clutter blurring, or target position shift can be found.

The 3D imaging synthetic aperture radar (SAR) system has been proposed (Chan, Kuga, and Ishimaru, 1999; Kidera, Sakamoto, and Sato, 2008; Du et al., 2010; Mahafza and Sajjadi, 1996; Jun et al., 2010; Matthias and Markus, 2010; Lopez-Sanchez and Fortuny-Guasch, 2000; Fortuny-Guasch and Lopez-Sanchez, 2001; Xu and Narayanan, 2001; Budillon, Evangelista, and Schirinzi, 2011; Solimene et al., 2010). The interferometric synthetic aperture radar (In-SAR) technique has been applied to retrieve the height of the target in terms of strong scattering points (Xu and Narayanan, 2001). However, it is difficult to treat 3D imaging of complex-shaped electrically large targets. Extending the conventional SAR technique to synthesize 1D linear aperture with pulse compression for 2D imaging, one can move the radar platform in a 2D virtual aperture to realize 3D SAR imaging. For example, the radar platform can move in a circular trajectory to perform 3D SAR imaging (Chan, Kuga, and Ishimaru, 1999). Alternatively, the platform can be mounted with a linear antenna

Polarimetric Scattering and SAR Information Retrieval, First Edition. Ya-Qiu Jin and Feng Xu.
© 2013 John Wiley & Sons Singapore Pte. Ltd. Published 2013 by John Wiley & Sons Singapore Pte. Ltd.

array and move toward the cross direction to synthesize a 2D aperture for 3D imaging (Mahafza and Sajjadi, 1996; Jun *et al.*, 2010). A novel tomographic imaging technique (Budillon, Evangelista, and Schirinzi, 2011) based on compressive sampling theory (Candès and Wakin, 2008) was proposed to enhance the elevation resolution using a small number of measurements with a given limited aperture extent. The ARTINO (airborne radar for three-dimensional imaging and nadir observation) system was designed to implement the experiment of 3D imaging of radar system (Matthias and Markus, 2010), whose data have not been made available to the public.

As mentioned above, either 2D or 3D imaging is only focused on the image formation of simple objects, for example, idealized point, spherical or cubical targets, and so on. Also, the assumption of constant reflectivity of the scattering objects usually adopted in the imaging algorithm is certainly oversimplified.

In this chapter, we present a numerical train of simulated scattering–imaging–reconstruction for a complex-shaped electrically large object above a dielectric rough surface. An efficient approach to the calculation of the polarized backscattering fields of a complex target using our bidirectional analytic ray tracing (BART) code (Xu and Jin, 2009) is developed in Chapter 11, and synthesized to yield volumetric geometry features in simulating a 3D target image. Instead of the conventional approach to treat the clutter from a rough surface as Gaussian noise, it models and calculates the composite polarized scattering from both the volumetric target and the randomly rough surface, and our imaging algorithm is then developed to detect or reconstruct the 3D target from the rough surface background. In BART calculations, more scattering mechanisms are taken into account, the implementation of analytic tracing of polygon ray tubes significantly enhances the computational efficiency, and the composite scattering of an electrically large target over a randomly rough surface can be implemented.

The step-frequency radar (SF radar) technique (Taylor, 1995) is employed to 3D SAR imaging using the downward-looking spotlight mode to alleviate the sheltering and layover effect. Finally, some simulation examples of 3D imaging and reconstruction of perfect electric conductor (PEC) targets, that is, a square frustum, a tank-like target, and a tank-like target over a rough surface, as well as discussions on the accuracy and computational efficiency are presented.

12.1 Principle of 3D Reconstruction

Nowadays, step-frequency radar (SF radar) has often been adopted in localizing man-made targets because of its high resolution of wide-band systems, as well as the merits of the narrow-band system.

12.1.1 Principle of Imaging based on Point Scattering

As a sequence of narrow pulses of the SF radar is transmitted, it is modulated by increasing the discrete frequency of successive pulses linearly over the operating bandwidth. The range and cross-profiles are, respectively, obtained by pulse compression and by the radar performing in a 2D aperture. They are then used for 3D target reconstruction.

Suppose we employ a group of K narrow pulses with constant frequency increment, Δf, the kth pulse of which is written as

$$S_t = A_1 \exp[-j2\pi(f_0 + k\Delta f)t] \tag{12.1}$$

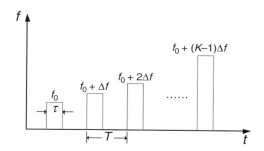

Figure 12.1 The waveform of a sequence of modulated signals including K pulses. The duration of the instantaneous pulse is τ, and T is the interval between successive pulses. © [2011] IEEE. Reprinted, with permission, from *IEEE Transactions on Geoscience and Remote Sensing*

Figure 12.1 shows the features of the step-frequency waveform (Taylor, 1995). Since the pulse duration τ is small, the receiver bandwidth is also small, which may simultaneously reduce the bandwidth of the noise. In addition, the receiver may reject the multiple scattering from clutter outside corresponding narrow bandpass. Thus, it yields high signal-to-noise ratio. The SF radar with fine resolution is feasible to reconstruct the profile of a complex target even with low radar cross-section (RCS) in environmental clutters.

Following the model of the strong scattering centers (Mensa, 1991), a volumetric target is usually divided numerically into finite point units. Let the radar send an SF signal from location A, as shown in Figure 12.2. The radar echo from a strong scattering point at $B(x, y, z)$ is expressed as

$$S_r = A_1\delta \exp[-j2\pi(f_0 + k\Delta f)(t - 2R/c)] \tag{12.2}$$

where δ is the reflection coefficient of the scattering point, f is the incident electromagnetic (EM) frequency, $t_D = 2R/c$ is the round-trip delay, and R is the distance between A and B, that is

$$
\begin{aligned}
R &= \sqrt{(R_0 \sin\theta \cos\varphi - x)^2 + (R_0 \sin\theta \sin\varphi - y)^2 + (R_0 \cos\theta - z)^2} \\
&= \sqrt{R_0^2 - 2R_0(x \sin\theta \cos\varphi + y \sin\theta \sin\varphi + z \cos\theta) + (x^2 + y^2 + z^2)}
\end{aligned}
\tag{12.3}
$$

where R_0 is the distance between A and the origin O, and φ and θ denote the azimuth and elevation angles, respectively.

The kth sampling output at the detector can be modeled as the product of the reference signals, which are the same as the transmitting signal, and the received signal passing through the lowpass filter, expressed as

$$V(\theta, \varphi, f) = \delta \exp(-j4\pi Rf/c) \tag{12.4}$$

where the amplitude of the output signal is simply ignored, $f = f_0 + k\Delta f$, and K sampling data correspond to K pulses in a sequence.

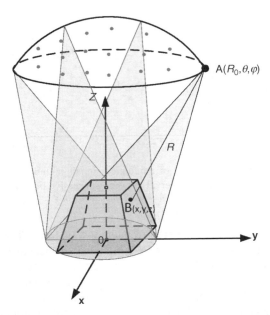

Figure 12.2 Geometry of 3D spotlight-mode SAR imaging for an object in a Cartesian coordinate system. A is along the trajectory of radar motion; B is an arbitrary point unit of the object. © [2011] IEEE. Reprinted, with permission, from *IEEE Transactions on Geoscience and Remote Sensing*

The total scattering electric field, $G(\theta, \varphi, f)$, received by the radar receiver at location A is contributed by the integral of all the point units as (Mensa, 1991)

$$G(\theta, \varphi, f) = \iiint g(x, y, z)\exp(-j4\pi Rf/c)\, dx\, dy\, dz \qquad (12.5)$$

The function $g(x, y, z)$ is the complex reflectivity, which describes the 3D distribution of the EM scattering capability of the target. The function $G(\theta, \varphi, f)$ is the complex backscattering field, with dependences on frequency, azimuth, and elevation angles. It can be seen that $g(x, y, z)$ is the continuous spectral Fourier transform of $G(\theta, \varphi, f)$. The goal of the SAR system is to produce samples of the spatial spectrum, denoted as $G(\theta_m, \varphi_n, f_k)$, which are obtained in terms of amplitude and phase in the SF radar system. From this discrete and finite spectral domain information, the reconstructed image of the complex reflectivity function can be acquired in discrete form, denoted as $g(x_i, y_j, z_k)$.

Suppose the radar transmitter and receiver move in a 2D finite circular arc aperture as shown in Figure 12.2, where the red dots in a circle are the discrete sampling locations. The incident EM wave upon the imaging area yields an $M \times N$ scattering matrix, where M and N are the sampling numbers along the azimuth and elevation directions, respectively.

The radar echo at each location A along the circular arc trajectories (indicated by the red dots in Figure 12.2), $G(\theta_m, \varphi_n, f_k)$, is numerically calculated by the BART approach (Xu and Jin, 2009). Meanwhile, the commercial simulation software FEKO is also employed for validation and comparison.

12.1.2   Imaging Algorithm using 3D Fast Fourier Transform

Using the Taylor series expansion of Equation 12.3 and the far-field approximation, $R_0 \gg x, y, z$, it yields

$$R \approx R_0 - (x \sin \theta \cos \varphi + y \sin \theta \sin \varphi + z \cos \theta)$$

Thus, the scattering field received at each location is written as

$$G(\theta, \varphi, f) = \iiint g(x, y, z) \exp\left[j4\pi \frac{f}{c} (x \sin \theta \cos \varphi + y \sin \theta \sin \varphi + z \cos \theta) \right] dx\, dy\, dz \quad (12.6)$$

where the term $\exp(-j4\pi R_0/\lambda)$ is simply omitted. Substituting

$$X = 2f \sin \theta \cos \varphi/c, \quad Y = 2f \sin \theta \sin \varphi/c, \quad Z = 2f \cos \theta/c \quad (12.7)$$

into Equation 12.6 yields

$$G(X, Y, Z) = \iiint g(x, y, z) \exp[j2\pi(xX + yY + zZ)]\, dx\, dy\, dz \quad (12.8)$$

It can be seen that $G(X, Y, Z)$ of Equation 12.8 is the 3D inverse Fourier transform of $g(x, y, z)$. Thus, it gives

$$g(x, y, z) = \iiint G(X, Y, Z) \exp[-j2\pi(xX + yY + zZ)]\, dX\, dY\, dZ \quad (12.9)$$

We extend the fast Fourier transform (FFT) of conventional 2D image reconstruction, that is, 2D polar format algorithm (PFA), to the 3D imaging case. This approach assumes that the unknown spatial spectrum outside the measuring region is zero. The fan region in Figure 12.3

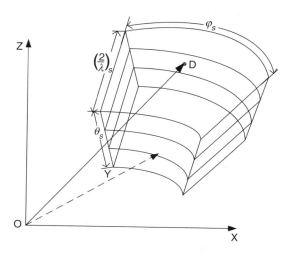

Figure 12.3   Geometry for 3D collected data in X, Y, Z coordinates. The data D are evenly distributed along the clusters of concentric arcs with evenly discretized θ_m, φ_n, f_k within the extension of θ_s, φ_s, f_s.
© [2011] IEEE. Reprinted, with permission, from *IEEE Transactions on Geoscience and Remote Sensing*

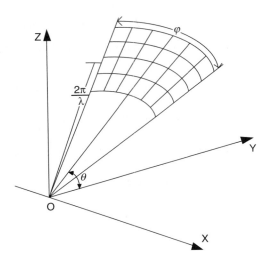

Figure 12.4 Geometry for collected data (located in the grids) on a plane with fixed θ value. © [2011] IEEE. Reprinted, with permission, from *IEEE Transactions on Geoscience and Remote Sensing*

illustrates the collected data in *X, Y, Z* coordinates, and, as an example, Figure 12.4 gives one set of the collected data on a plane with fixed θ value.

A rectangular cuboid region, as shown in Figure 12.5, is inscribed to enclose the fan region. Before implementing the 3D FFT, interpolation is first employed to make the resampling data uniformly distributed in Cartesian raster from the evenly sampled data in polar

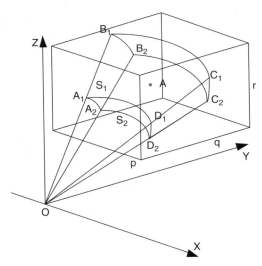

Figure 12.5 Geometry for the resampled data: the uniform grids of the rectangular region for the Fourier transform (the grid details are not shown). Resampled data A (X_A, Y_A, Z_A) may be located between the adjacent sectors S_1 and S_2. © [2011] IEEE. Reprinted, with permission, from *IEEE Transactions on Geoscience and Remote Sensing*

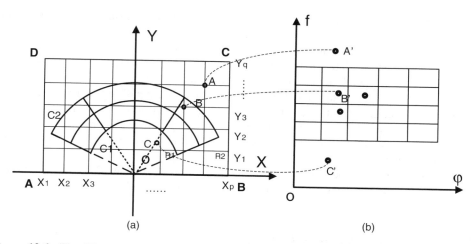

Figure 12.6 The 2D sample mapping from the plane X–Y to the plane f–φ: (a) uniformly sampling data on a plane; (b) the plane for implementing 2D interpolation. © [2011] IEEE. Reprinted, with permission, from *IEEE Transactions on Geoscience and Remote Sensing*

raster, as shown in Figure 12.6a. The interpolation is carried out in spatial spectral domain, as explained in Figure 12.6b. The steps are as follows:

1. The bandwidth of the incident wave is divided into K step frequencies uniformly distributed within $(f_1 - f_2)$. The backscattering fields, $G(\theta_m, \varphi_n, f_k)$, for different incident angles (θ_m, φ_n) and each frequency f_k are measured.
2. Determine the resampling region and density. To fully use the measuring data and uniform sample data for FFT, a rectangular cuboid is inscribed close to the measuring data, as shown in Figure 12.5. The widths of the region to be mapped are W_p, W_q, W_r. The resampling density is similar to the measuring data. The region is equally divided in the X, Y, and Z directions, and is denoted by (p, q, r), and the numbers are P, Q, and R. It describes the grid coordinates as

$$X_p = X_1 + (p - 1)\Delta X, \quad Y_q = Y_1 + (q - 1)\Delta Y, \quad Z_r = Z_1 + (r - 1)\Delta Z$$

where X_1, Y_1, Z_1 are the minimum coordinates in the X, Y, Z axes, respectively, and determined by the range of f, θ, φ, and $\Delta X = \Delta Y = \Delta Z \equiv 2\Delta f/c$.
3. Some relations are especially given as

$$f = c\sqrt{X^2 + Y^2 + Z^2}\big/2$$
$$\varphi = \tan^{-1}(Y/X), \quad \theta = \tan^{-1}(\sqrt{X^2 + Y^2}\big/Z) \tag{12.10}$$

Mapping (X_p, Y_q, Z_r) into $(\theta_m, \varphi_n, f_k,)$ according to Equation 12.10 and using tricubic interpolation in the 3D spectral domain (Candès and Wakin, 2008), one can acquire the resampled data in the $f - \theta - \varphi$ domain. A 2D case can be found in Dai and Jin (2009).

In Figure 12.6, the points A and C are outside of the measuring data region, and we set $G(X_p, Y_q) = 0$. The electric field at B' mapped from B might be unknown as not recorded. It is feasible to obtain it by using interpolation on the $f - \varphi$ plane from the measured data for the real and imaginary parts, respectively. To keep the balance between computational efficiency and accuracy, 2D bicubic interpolation is employed. One can obtain the 3D resampled data by extending this approach to the 3D case. Because of data truncation due to the imposition of a rectangular window on the sampling data, the side-lobe effect might be raised. To reduce these side-lobes, the resampled data can be weighted by a well-designed window before interpolation.

4. Apply the 3D FFT to the uniformly resampled data on the rectangular cuboid region to yield the 3D reconstructed image, which is displayed as the magnitude of the resulting 3D array.

12.1.3 Resolution and Sampling Criteria

The resolution of the 3D image reconstruction acquired by SF radar is dependent on the frequency bandwidth and the dimension of the synthesis aperture. The range resolution, Δz, is written as

$$\Delta z = \Delta R = c/2B \tag{12.11}$$

where B is the frequency bandwidth. The cross-resolutions in the x and y directions are, respectively, written as

$$\Delta x = \lambda_{\min}/2\theta_s, \quad \Delta y = \lambda_{\min}/2\varphi_s \tag{12.12}$$

where $\lambda_{\min}$ is the minimum working wavelength, and θ_s, φ_s are the dimensions of the synthetic aperture.

The reconstructed region of interest is usually confined within a rectangular cuboid with the dimension of $D_x \times D_y \times D_z$. According to the Nyquist sampling criterion in a spotlight-mode SAR system (Lopez-Sanchez and Fortuny-Guasch, 2000; Fortuny-Guasch and Lopez-Sanchez, 2001), the sampling intervals are represented as

$$\Delta f \le \frac{c}{2\sqrt{D_x^2 + D_y^2 + D_z^2}}, \quad \Delta\theta \le \frac{\lambda_{\min}}{2\sqrt{D_x^2 + D_y^2}} \frac{1}{\sin\theta_s}, \quad \Delta\varphi \le \frac{\lambda_{\min}}{2\sqrt{D_x^2 + D_y^2 + D_z^2}} \tag{12.13}$$

12.2 Scattering Simulation and 3D Reconstruction

Three examples, (case I) a PEC square frustum (as shown in Figure 12.2), (case II) a PEC tank-like target, and (case III) a tank-like target over a rough surface (see Figure 12.13) are presented. Numerical simulation of the scattering fields $G(\theta, \varphi, f)$ is performed using the BART solver, and also compared with the FEKO results. The calculations were performed on Dell PowerEdge computers.

12.2.1 Model of Square Frustum (Case I)

The BART and FEKO are both used to calculate the scattering fields from a PEC square frustum, as shown in Figure 12.2, and, to verify the flexibility, 3D imaging is presented. The target geometry is described as follows: the lengths of the top and bottom sides are 0.4 m and 1.0 m, respectively; the height is 1.0 m.

The horizontally polarized (H) plane wave is incident upon the region of interest, as shown in Figure 12.2. The SF signals between 1.0 and 3.0 GHz with 10 MHz interval are transmitted from the radar moving in a 2D circular aperture (i.e., $\theta = -45°$ to $45°$, $\varphi = -45°$ to $45°$) every $0.5°$ in both directions; the total sampling number is 181×181. The co-polarized (HH) and cross-polarized (HV) backscattering electric fields are obtained. The setups satisfy the requirements $\Delta f \leq 30$ MHz, $\Delta\theta \leq 0.67°$, $\Delta\varphi \leq 0.55°$ and $D_x = D_y = D_z = 3$ m from Equation 12.13.

In the BART simulation, every quadrilateral facet is divided into two triangular patches, and the number of total patches is only 12; whereas in the FEKO calculation, the target surfaces are meshed into 47 974 triangles, as the sides of triangles are acquired to be shorter than about one-sixth of the wavelength to obtain higher accuracy.

The comparisons of BART and FEKO in Figures 12.7 and 12.8 show that both the RCS and the phase of the backscattering fields using the two approaches are matched well in most regions. A few deviations in some small regions might need further code study.

The reconstructions of the square frustum are shown in Figures 12.9–12.12. Those pixels whose scattering coefficients are greater than the fixed threshold (here we choose one-tenth

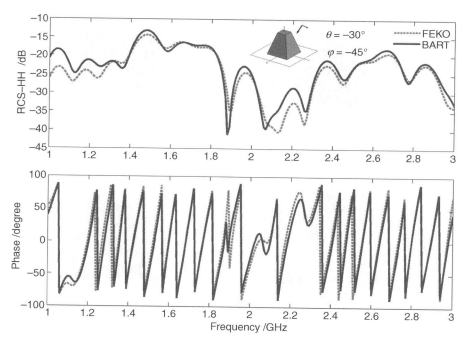

Figure 12.7 Comparison of BART and FEKO-PO for RCS and phase of a PEC square frustum. The incident angle is: $\theta = -30°, \varphi = -45°$. © [2011] IEEE. Reprinted, with permission, from *IEEE Transactions on Geoscience and Remote Sensing*

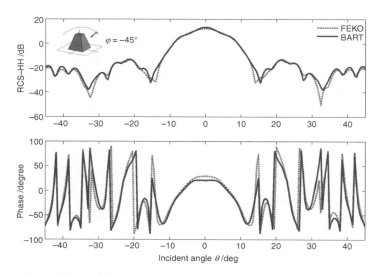

Figure 12.8 Comparison of BART and FEKO-PO for RCS and phase of a square frustum. The frequency of the wave is 2.0 GHz, and the incident angle is: $\varphi = -45°$, $-45° \leq \theta \leq 45°$. © [2011] IEEE. Reprinted, with permission, from *IEEE Transactions on Geoscience and Remote Sensing*

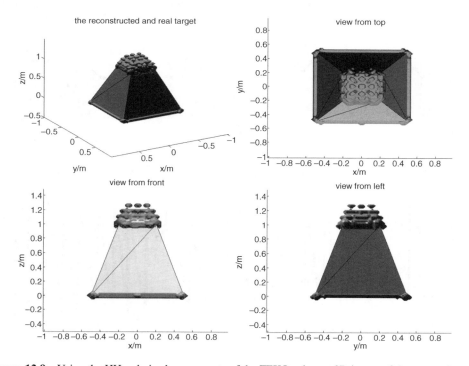

Figure 12.9 Using the HH-polarized components of the FEKO solver, a 3D image of the square frustum is reconstructed from the views from the front, top, and left (drawn in fixed contour level, −40 dB). © [2011] IEEE. Reprinted, with permission, from *IEEE Transactions on Geoscience and Remote Sensing*

of the maximal value) are selected to reconstruct the profile of the complex target shape. The reconstructed results of both co-polarized and cross-polarized components display well the profiles of the square frustum, except for those two facets without EM wave illumination.

Figures 12.9 (HH-polarized) and 12.10 (HV-polarized) are produced by the FEKO solver, and Figures 12.11 (HH-polarized) and 12.12 (HV-polarized) are produced by BART, respectively. Both BART and FEKO can reconstruct well the top surface, whose reflection is intense, and the four bottom sides, whose diffractions are also strong.

Figure 12.10 acquired from HV-polarized components performed by FEKO displays the tilted surfaces well, while the eight corners are shown in Figure 12.10 from the data implemented by BART. The FEKO software takes into account so-called fringe waves (Jakobus and Landstorfer, 1995), which occur on flat surfaces. The HV fringe wave is much less than the HH component of the scattering field, but comparable with the cross-polarized component. As the fringe wave in the FEKO solver is accounted for, the surfaces illuminated by EM wave are clearly represented. However, we can only observe the strong reflective corners

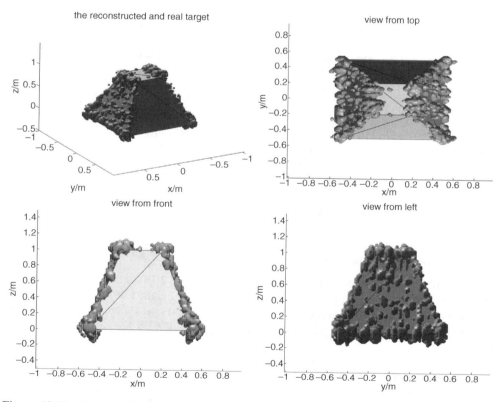

Figure 12.10 Using the HV-polarized components of the FEKO solver, a 3D image of the square frustum is reconstructed from the views from the front, top, and left (drawn in a fixed contour level, −60 dB). © [2011] IEEE. Reprinted, with permission, from *IEEE Transactions on Geoscience and Remote Sensing*

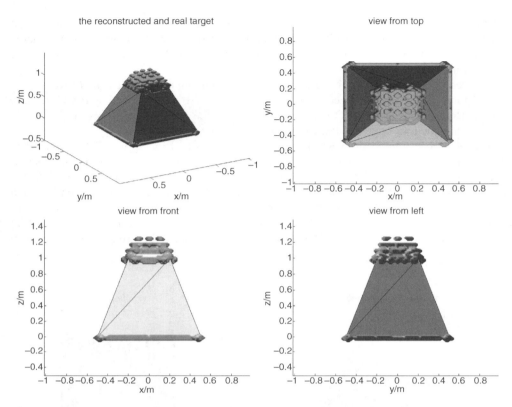

Figure 12.11   Using the HH-polarized components of the BART solver, a 3D image of the square frustum is reconstructed from the views from the front, top, and left (drawn in fixed contour level, −40 dB). © [2011] IEEE. Reprinted, with permission, from *IEEE Transactions on Geoscience and Remote Sensing*

in Figure 12.11 using the BART solver without consideration of the fringe wave, which demands further work.

12.2.2   Model of Tank-Like Target (Case II)

The 3D profile reconstruction of a tank-like target (Figure 12.13) is also investigated. The dimensions of the object are: length 3.3 m, width 1.9 m, and height 1.46 m. The model of the tank-like target for BART is depicted by 278 triangular patches, as shown in Figure 12.13. In the FEKO-PO calculation, the geometric surfaces of the tank must be meshed into 280 702 triangles.

The configuration of the radar in case II is the same as in case I. The reconstructed tank profile is presented in Figures 12.14–12.16

It can be seen that the acquired images depict well the main features of the tank target. The results in Figures 12.14 and 12.16 derived from co-polarized HH components, respectively, using FEKO and BART solvers, show the strong reflection from the turret, main gun, top hull, and track, and less reflection from the front glacis plate. Since a perfect electric

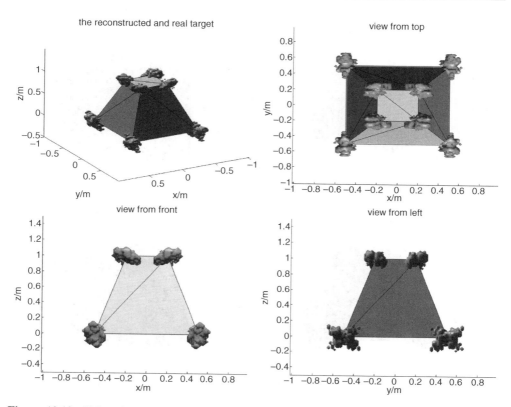

Figure 12.12 Using the HV-polarized components of the BART solver, a 3D image of the square frustum is reconstructed from the views from the front, top, and left (drawn in a fixed contour level, −60 dB). © [2011] IEEE. Reprinted, with permission, from *IEEE Transactions on Geoscience and Remote Sensing*

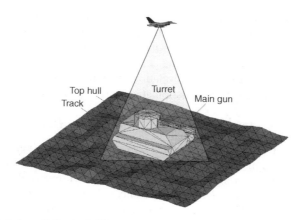

Figure 12.13 Geometry of the tank-like target over a randomly rough surface. © [2011] IEEE. Reprinted, with permission, from *IEEE Transactions on Geoscience and Remote Sensing*

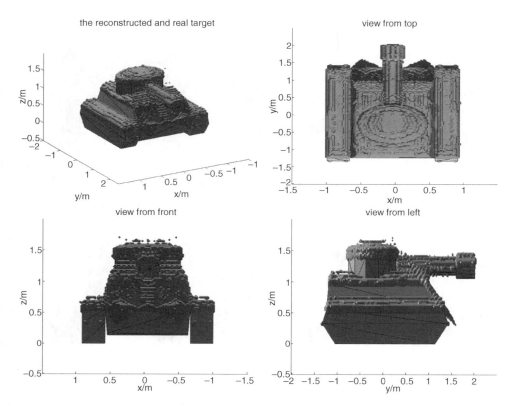

the reconstructed and real target

view from top

view from front

view from left

Figure 12.14 Using the HH-polarized component of the FEKO solver, a tank-like 3D target is reconstructed from the views from the front, left, and top sides (drawn in a fixed contour level, −40 dB). © [2011] IEEE. Reprinted, with permission, from *IEEE Transactions on Geoscience and Remote Sensing*

conductor is impenetrable, the interior and underlying structures, for example, the bottom of the main tank body and tracks, cannot be seen. Because those sharp structures of the target might cause significant cross-polarized echoes, Figure 12.15 using the HV component describes and reconstructs the whole tank profile well. Similar results can be obtained when the VV and VH components are employed.

By synthesizing the comprehensive information from these two different polarizations, the main characteristics of the target can be automatically reconstructed.

12.2.3 Tank-Like Model over Rough Surface (Case III)

The tank-like model in case II is now placed over a randomly rough surface, as shown in Figure 12.13. A rough surface with Gaussian correlation function of root-mean-square height 0.2 m and correlation length 0.5 m is generated (Pak, Tsang, and Chan, 1995; Toporkov, 1998). This rough surface truncated at −4 m and 4 m in both x and y directions is divided into 1800 dielectric triangles. The dielectric constant of the surface medium is assumed as $6 + i0.1$. It is especially noted that the scattering problem for a composite model of a PEC target over a dielectric rough surface can be well solved by

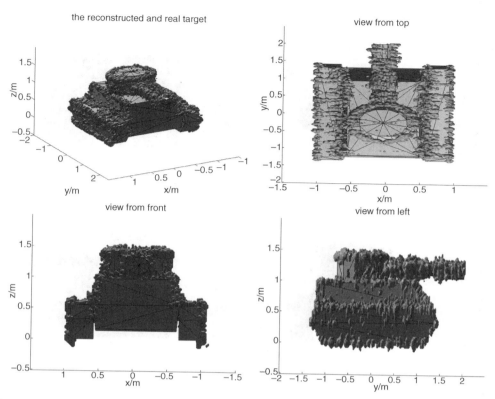

Figure 12.15 Using the HV-polarized component of the FEKO solver, a tank-like 3D target is reconstructed from the views from the front, left, and top sides (drawn in a fixed contour level, −60 dB). © [2011] IEEE. Reprinted, with permission, from *IEEE Transactions on Geoscience and Remote Sensing*

using the BART solver (Candès and Wakin, 2008), while it would be very difficult if FEKO were adopted.

The configuration of the radar in this case is the same as case I. Figure 12.17 is the reconstruction from the composite scattering fields of the target over a rough surface. One can see that the reconstructed profile has been blurred by the strong clutters from the underlying rough surface. However, the target can be clearly identified above the rough surface, while the target can be seriously blurred in a conventional 2D SAR image due to the layover effect and multiple scattering interactions between the volumetric tank target and the underlying rough surface. This demonstrates well the advantages of the 3D SAR reconstruction technique.

The four parts of Figure 12.18 present the horizontal cross-section of the target in the X–Y plane at different altitudes. For example, the reflections of the rough surface shown in Figure 12.18a are comparable with the reflections from the target. The interactions between the tank tracks, the front glacis plate (enclosed by the red rectangles in Figure 12.18b and c), and the rough surface are significantly apparent. All of these effects might reduce the image quality, especially on the edges, and can be revealed well using the 3D imaging technique.

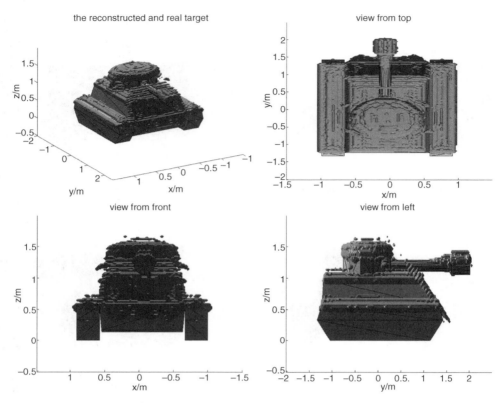

Figure 12.16 Using the HH-polarized component of the BART solver, a tank-like 3D target is reconstructed from the views from the front, left, and top sides (drawn in a fixed contour level, −40 dB). © [2011] IEEE. Reprinted, with permission, from *IEEE Transactions on Geoscience and Remote Sensing*

The computational efficiency of BART is much better than that of FEKO in these examples. The comparison has been implemented for three cases: a square frustum, a tank-like target, and a tank-like target over a flat PEC plane. Table 12.1 lists the computational times for the FEKO and BART simulations for one incident angle with 401 frequencies, including solving and memorizing time. In the table, the first number is the running time (seconds) and the second one indicates the number of target facets or meshes in the calculation corresponding to BART or FEKO, respectively. The dimension of the facets in BART is usually taken so that the extent of each facet is fine enough to describe the object's geometrical feature, while

Table 12.1 Comparison of the computational efficiency of the BART and FEKO solvers, for the three cases considered. The first number is the running time (seconds) and the second one indicates the number of target facets or meshes in the calculation

	Square frustum	Tank-like target	Tank-like target over PEC plane
BART	0.1/12	7.9/278	10.1/302
FEKO	33/47 974	300/280 702	926/748 200

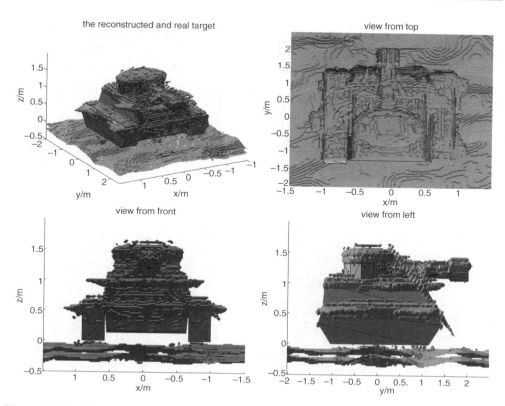

Figure 12.17 Using the HH-polarized component of the BART solver, a 3D tank-like target over a rough surface is reconstructed from the views from the front, left, and top sides (drawn in a fixed contour level, −40 dB). © [2011] IEEE. Reprinted, with permission, from *IEEE Transactions on Geoscience and Remote Sensing*

the triangle sizes meshed in FEKO have to be fine enough to satisfy the accuracy demand, that is, usually less than one-sixth of the wavelength. As the target scale increases, the computational advantages of the BART solver become more obvious. Especially, the underlying dielectric rough surface is involved in modeling.

Numerical simulations of polarized scattering, 3D imaging, and reconstruction of a 3D electrically large complex target shape are developed. Using the step-frequency radar with synthesizing 2D circular arc aperture and downward-looking spotlight mode, co-polarized and cross-polarized EM scattering fields produce 3D target imaging. The BART solver is mainly applied to the calculation of polarized scattering, and is validated and compared with FEKO in some simple cases. A 3D tricubic interpolation in spatial spectral domain is implemented to resample the uniform data. The FFT makes the algorithm more suited for real-time requirements. Reconstructions of two targets, for example, a square frustum and a tank-like target over rough surface, have been presented.

The examples demonstrate that, using BART, numerical computation of composite scattering of a PEC target over a dielectric rough surface is effectively carried out for 3D imaging,

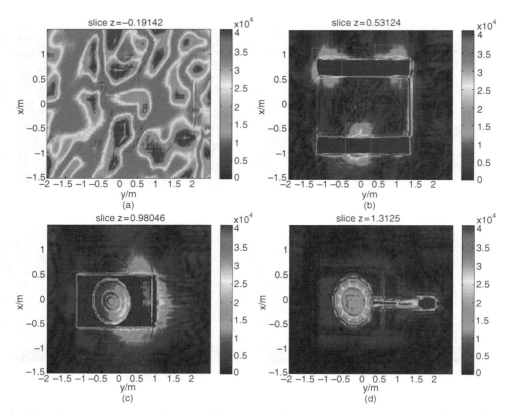

Figure 12.18 Cross-sections in the *X–Y* plane of a tank-like target over a rough surface at altitudes (a) $z = -0.2$ m, (b) $z = 0.53$ m, (c) $z = 1.0$ m, and (d) $z = 1.31$ m. © [2011] IEEE. Reprinted, with permission, from *IEEE Transactions on Geoscience and Remote Sensing*

and the reconstruction technique presents well the profile of the volumetric target from background clutter.

This work presents a train of numerical simulation of scattering–imaging–reconstruction. However, it certainly needs further work, for example, to design a more feasible and practicable way to perform the platform flight (Jun *et al.*, 2010). The sparse linear antenna array technique (Du *et al.*, 2010; Mahafza and Sajjadi, 1996; Budillon, Evangelista, and Schirinzi, 2011) might be applicable to the issue of reducing the amount of observation data. How to apply full polarization information to synthesize more information for reconstructing the target profile, including the fringe wave into BART, and so on, remain for further study.

References

Budillon, A., Evangelista, A., and Schirinzi, G. (2011) Three-dimensional SAR focusing from multipass signals using compressive sampling. *IEEE Transactions on Geoscience and Remote Sensing*, **49**(1), 488–499.

Candès, E. and Wakin, M.B. (2008) An introduction to compressive sampling. *IEEE Signal Processing Magazine*, **25**(2), 21–30.

Chan, T.K., Kuga, Y., and Ishimaru, A. (1999) Experimental studies on circular SAR imaging in clutter using angular correlation function technique. *IEEE Transactions on Geoscience and Remote Sensing*, **37**(5), 2192–2197.

Chang, P.C., Burkholder, R.J., Volakis, J.L. *et al.* (2009) High-frequency EM characterization of through-wall building imaging. *IEEE Transactions on Geoscience and Remote Sensing*, **47**(5), 1375–1387.

Dai, J. and Jin, Y.Q. (2009) Scattering and image simulation for reconstruction of 3D PEC objects concealed in a closed dielectric box. *Progress in Electromagnetics Research M*, **9**, 41–52.

Du, L., Wang, Y., Hong, W. *et al.* (2010) A three-dimensional range migration algorithm for downward-looking 3D-SAR with single-transmitting and multiple-receiving linear array antennas. *EURASIP Journal on Advances in Signal Processing*, **2010**, 957916.

Firoozabadi, R., Miller, E.L., Rappaport, C.M., and Morgenthaler, A.W. (2007) Subsurface sensing of buried objects under a randomly rough surface using scattered electromagnetic field data. *IEEE Transactions on Geoscience and Remote Sensing*, **45**(1), 104–117.

Fornaro, G., Reale, D., and Serafino, F. (2009) Four-dimensional SAR imaging for height estimation and monitoring of single and double scatterers. *IEEE Transactions on Geoscience and Remote Sensing*, **47**(1), 224–237.

Fortuny-Guasch, J. and Lopez-Sanchez, J.M. (2001) Extension of the 3-D range migration algorithm to cylindrical and spherical scanning geometries. *IEEE Transactions on Antennas and Propagation*, **49**(10), 434–1444.

Jakobus, U. and Landstorfer, F.M. (1995) Improved physical optics approximation for flat polygonal scatterers. Proceedings of 11th International Zurich Symposium on Electromagnetic Compatibility, March.

Jun, S., Zhang, X.L., Yang, J.Y., and Chen, W. (2010) APC trajectory design for "one-active" linear-array three-dimensional imaging SAR. *IEEE Transactions on Geoscience and Remote Sensing*, **48**(3), 1470–1486.

Kidera, S., Sakamoto, T., and Sato, T. (2008) High-resolution and real-time UWB radar imaging algorithm with direct waveform compensations. *IEEE Transactions on Geoscience and Remote Sensing*, **46**(11), 3503–3513.

Kidera, S., Sakamoto, T., and Sato, T. (2009) High-resolution 3-D imaging algorithm with an envelope of modified spheres for UWB through-the-wall radars. *IEEE Transactions on Antennas and Propagation*, **57**(11), 3520–3529.

Lopez-Sanchez, J.M. and Fortuny-Guasch, J. (2000) 3D radar imaging using range migration techniques. *IEEE Transactions on Antennas and Propagation*, **48**(5), 728–737.

Mahafza, B.R. and Sajjadi, M. (1996) Three-dimensional SAR imaging using linear array in transverse motion. *IEEE Transactions on Aerospace Electronic Systems*, **32**(1), 499–510.

Matthias, W. and Markus, G. (2010) Initial ARTINO radar experiments. Proceedings of the 8th European Conference on Synthetic Aperture Radar, Aachen, Germany.

Mensa, D.L. (1991) *High Resolution Radar Cross-Section Imaging*, Artech House, Norwood, MA.

Pak, K., Tsang, L., and Chan, C.H. (1995) Backscattering enhancement of electromagnetic waves from two-dimensional perfectly conducting random rough surface based on Monte-Carlo simulations. *Journal of the Optical Society of America A*, **12**(11), 2491–2499.

Qi, W., Xing, M., Lu, G., and Bao, Z. (2008) High-resolution three-dimensional radar imaging for rapidly spinning targets. *IEEE Transactions on Geoscience and Remote Sensing*, **46**(1), 22–30.

Solimene, R., Buonanno, A., Soldovieri, F., and Pierri, R. (2010) Physical optics imaging of 3-D PEC objects: vector and multipolarized approaches. *IEEE Transactions on Geoscience and Remote Sensing*, **48**(4), 1799–1808.

Taylor, J.D. (1995) *Introduction to Ultra-Wideband Radar Systems*, CRC Press, Boca Raton, FL.

Toporkov, J.V. (1998) *Study of Electromagnetic Scattering from Randomly Rough Ocean-Like Surface Using Integral-Equation-Based Numerical Technique*, Virginia Polytechnic Institute and State University, Blacksburg, VA.

Wang, G., Zhang, Y., and Amin, M. (2006) New approach for target locations in the presence of wall ambiguities. *IEEE Transactions on Aerospace Electronic Systems*, **42**(1), 301–315.

Xu, F. and Jin, Y.Q. (2009) Bidirectional analytic ray tracing for fast computation of composite scattering from electric-large target over a randomly rough surface. *IEEE Transactions on Antennas and Propagation*, **57**(5), 1495–1505.

Xu, X.J. and Narayanan, R.M. (2001) Three-dimensional interferometric ISAR imaging for target scattering diagnosis and modeling. *IEEE Transactions on Image Processing*, **10**(7), 1094–1102.

Yoon, Y.S. and Amin, M.G. (2008) High-resolution through-the-wall radar imaging using beamspace MUSIC. *IEEE Transactions on Antennas and Propagation*, **56**(6), 1763–1774.

Index

Polarimetric Scattering and SAR Information Retrieval, First Edition. Ya-Qiu Jin and Feng Xu.
© 2013 John Wiley & Sons Singapore Pte. Ltd. Published 2013 by John Wiley & Sons Singapore Pte. Ltd.